Heat Treatment, Structure and Properties of Nonferrous Alloys

Heat Treatment, Structure and Properties of Nonferrous Alloys

CHARLIE R. BROOKS
Professor of Metallurgical Engineering
The University of Tennessee

AMERICAN SOCIETY FOR METALS
Metals Park, Ohio 44073

First printing, October 1982
Second printing, July 1984
Third printing, October 1986
Fourth printing, July 1990
Fifth printing, February 1995

Library of Congress Catalog Card Number: 82-72484
ISBN: 0-87170-138-3
SAN: 204-7586

PRINTED IN THE UNITED STATES OF AMERICA

Preface

I believe that engineers involved extensively in the selection and use of metallic materials should have some understanding of the reasons for the recommended heat treatments that are employed. This understanding requires a knowledge of the principles and of the effects of phase equilibria and the kinetics of the phase transformations on the microstructure, and hence properties, of the particular alloys under consideration. This book, *Heat Treatment, Structure and Properties of Nonferrous Alloys,* is designed to present these principles for several classes of nonferrous alloys.

The book is written with the only prerequisite to its use an introductory course in metallurgy or materials science. Thus, it is designed to be used in courses for students in metallurgy and materials science, as well as in service courses for students in related areas such as mechanical engineering. The first three chapters review three areas which are of prime importance in understanding the principles of the heat treatment of nonferrous alloys: plastic deformation and annealing, solidification, and precipitation hardening. With these chapters as a convenient review source, this book should be useful to practicing engineers who wish to learn about the heat treatment of nonferrous alloys yet have not covered this introductory material recently.

I have made no serious attempt to convert data to the SI system; some conversion factors are given in Appendix 2. The magnification of the microstructures is given by a marker whose real length is shown in micrometers (μm), units of 10^{-4} cm. The method of imaging the microstructures dictates to some extent their appearance, so I have indicated those which have been obtained by using an optical microscope by OM, a scanning electron microscope by SEM, and a transmission electron microscope by TEM.

I want to express my appreciation to my colleague Professor E. E. Stansbury for allowing me to use some of his ideas in this book. Appreciation is also expressed to my colleague Professor W. T. Becker for reading part of the manuscript and making suggestions. Mr. Chris North, Mr. Mark Koss, Mr. Jim Andrus and Mr. Ashok Choudhury assisted in preparing some of the metallographic specimens. The support of the Chemical, Metallurgical and Polymer Engineering Department of the University of Tennessee in writing this book is appreciated. A special thanks goes to Mr. Timothy Gall of the American Society for Metals and Mrs. Nancy Tenney of Carnes Publication Services for their editorial assistance.

CHARLIE R. BROOKS

Contents

3 PRECIPITATION HARDENING *95*

4 ALUMINUM-BASE ALLOYS *115*

5 NICKEL-BASE ALLOYS *139*

6 COBALT-BASE ALLOYS *229*

7 MAGNESIUM-BASE ALLOYS *253*

8 COPPER-BASE ALLOYS *275*

9 TITANIUM-BASE ALLOYS *329*

CHAPTER 1

Plastic Deformation and Annealing

In this chapter some of the principles and observations of plastic deformation and annealing are briefly reviewed. Emphasis is on microstructural observations and, qualitatively, on the atomic mechanisms involved.

1-1. General Stress-Strain Behavior

In this review of the plastic deformation of metals, some discussions in terms of simple tensile behavior are presented. Other configurations are quite important, especially in practical behavior, but in order to illustrate the principles it is sufficient to limit attention to simple tensile behavior.

Some of the mechanical property terms that can be derived from a tension test are described in this section. This is given in the framework of tension tests for typical polycrystalline metallic materials. We are concerned here with the response of a material to increasing load, and it is usual to perform the test in such a manner that the strain rate is kept approximately constant; the load is monitored as a function of the elongation of the specimen. If the load is divided by the original cross-sectional area (a constant) normal to the load axis, the engineering stress is obtained. The engineering strain is defined by the change in length divided by the original length, $(l - l_o)/l_o$. Figure 1-1 shows a typical stress-strain curve. The definitions of yield strength, tensile (ultimate) strength, elastic modulus and elongation at fracture are shown in the figure.

It should be noted that on the engineering stress–engineering strain curve the stress is initially proportional to the load. Beyond a certain load, called the elastic limit, plastic strain occurs. Under proper test conditions (to

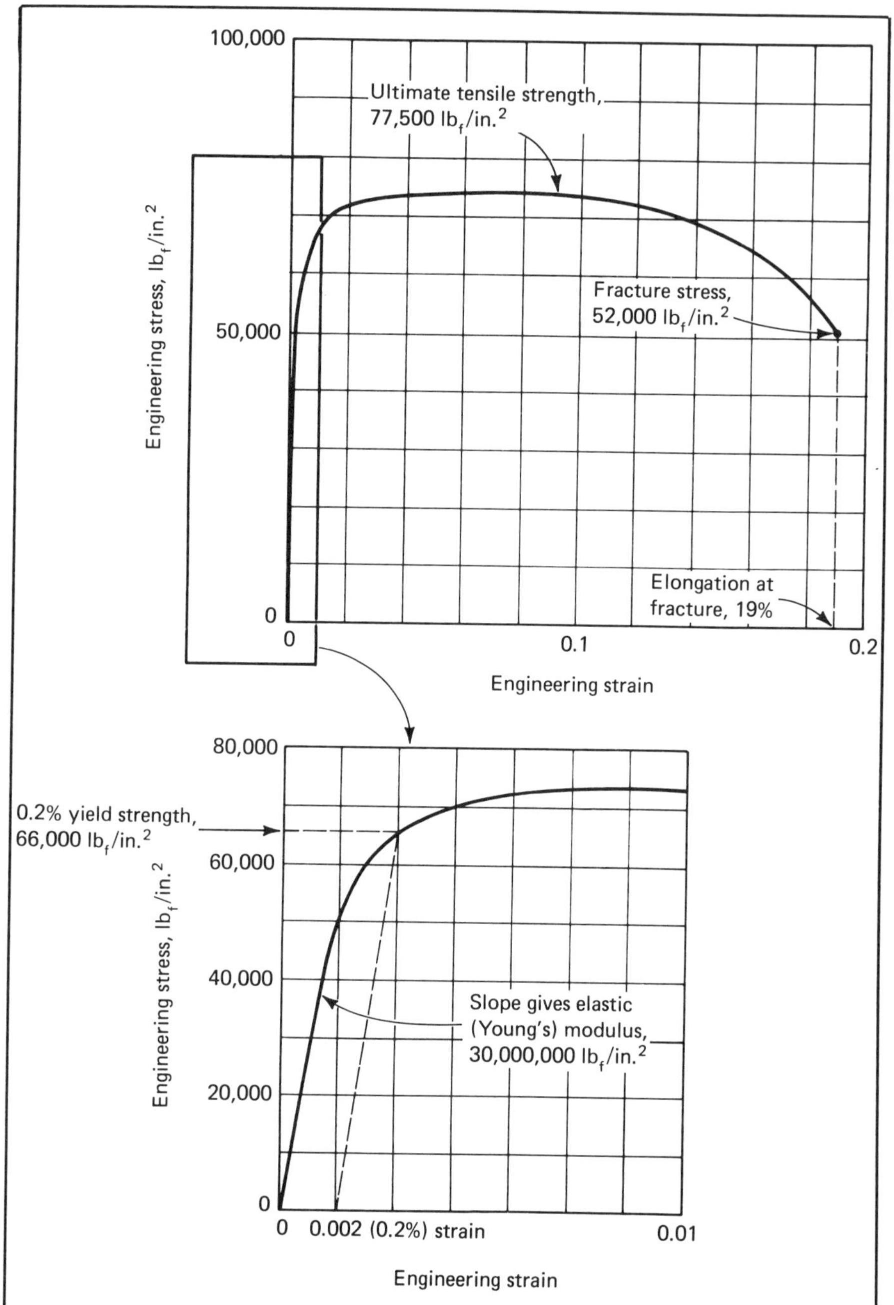

Fig. 1-1. Engineering stress–engineering strain curve of 0.2% C plain carbon, cold worked, steel, showing definition of mechanical property terms.

be discussed), the material gets stronger as it is plastically deformed, as evidenced by the positive slope to the stress-strain curve in the plastic region. However, the plastic strain reaches a level where the localized stress (due to such factors as slightly smaller specimen diameter) is sufficiently high that further plastic deformation is restricted to this one region. Thus, from this point until fracture, the material in this region continues to reduce in area (referred to as necking), and the load required to maintain a constant strain

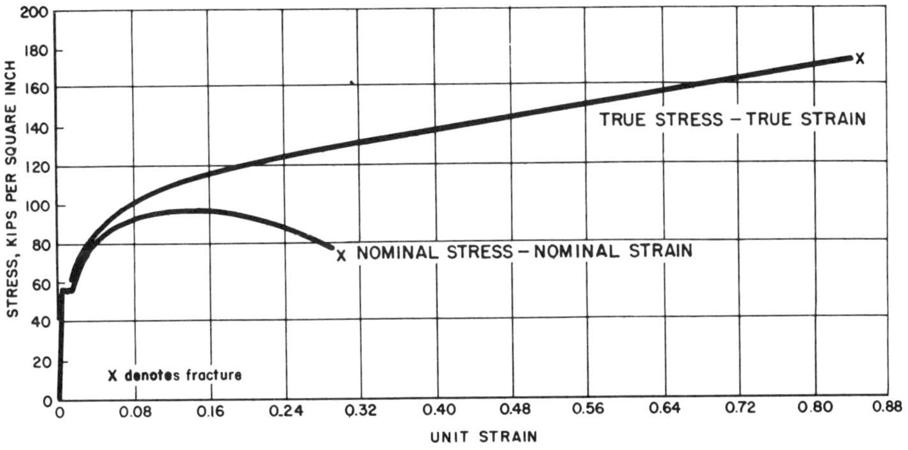

Fig. 1-2. Comparison of the true stress–true strain curve with the engineering (nominal) stress–engineering (nominal) strain diagram. (From The Making, Shaping and Treating of Steel, ed. H. E. McGannon, U. S. Steel Corporation, Pittsburgh, 1971)

rate decreases. The engineering stress is proportional to the load, since the original cross-sectional area is a constant; thus the engineering stress also decreases from this point on, and the engineering stress–engineering strain curve passes through a maximum, which defines the tensile (ultimate) tensile strength.

It is possible to monitor the cross-sectional area of the sample during the test, so that the load can be divided by the actual minimum area in order to obtain the true stress. This is usually plotted against the true strain, which is given by $ln\ (l/l_o)$. Figure 1-2 compares the two types of curves. Note that the true stress–true strain curve contains no maximum.

A property of importance is the stress at which plastic deformation (yielding) begins. Since this is difficult to determine experimentally, for most materials it is taken, by definition, to be the stress at 0.2% plastic strain. This is obtained by extending a straight line from a strain of 0.2% with a slope equal to the elastic modulus until it intersects the stress-strain curve. (Note that 0.2% strain means that the original gage length is now 0.2% longer.)

Some materials, notably certain steels under certain heat treated conditions, have a stress-strain curve which shows a well-defined yield point (Fig. 1-3). For such cases, the 0.2% yield strength is not used to define yielding.

1-2. Single Crystal Behavior

Now consider a single crystal of a hexagonal close-packed (HCP) metal, such as zinc or magnesium, in the shape of a cylinder. Upon tensile loading, plastic deformation occurs by regions slipping past one another on parallel planes (slip plane) and in a parallel direction (slip direction). This generates

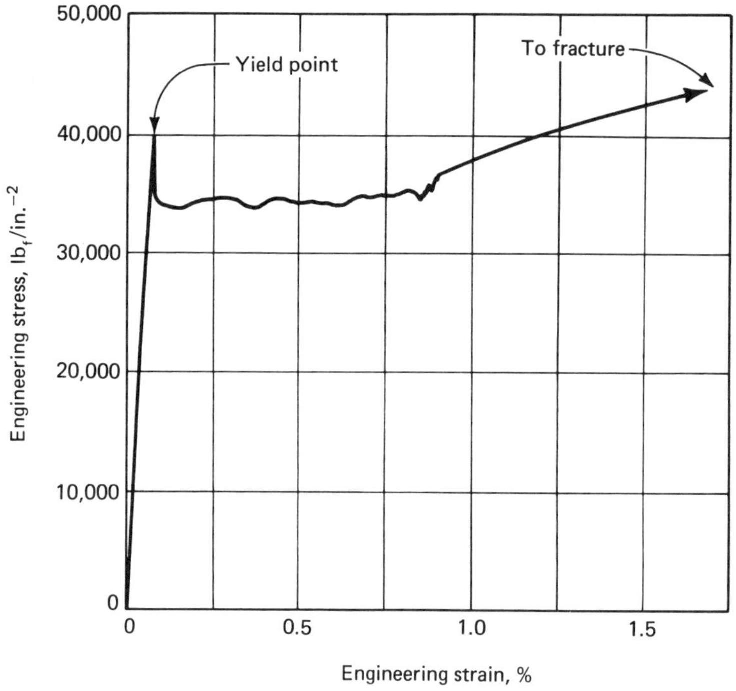

Fig. 1-3. The stress-strain curve for an annealed 0.18% C steel, showing yield-point behavior.

a series of elliptical offsets on the surface of the crystal (Fig. 1-4). We will define the orientation of the slip process relative to the cylinder by the angles shown in Fig. 1-5. For example, Fig. 1-6 shows the normal stress-strain curve for a magnesium crystal for which $\lambda = 58°$ and $\phi = 31°$. Note that the yield strength (see lower arrow) is about 108 gm/mm^2. Another magnesium crystal gave the other stress-strain curve shown in Fig. 1-6. This crystal also exhibited elliptical parallel offsets, for which $\lambda = 9°$ and $\phi = 81°$, and the yield strength (see upper arrow) is about 310 gm/mm^2.

Now, the observation that the plastic deformation in these single crystals occurs by the parallel offset leads to the viewpoint that the deformation occurs when the resolved shear stress on some crystallographic plane attains a critical value. That is, as the tensile load is increased, the resolved shear stress on any plane, in any direction, increases. When the resolved shear stress on a certain crystallographic plane and in a certain direction in that plane reaches a *critical* value, slip occurs. It is expected that this *critical resolved shear stress* (CRSS) will be the same independent of the orientation of the crystal relative to the cylinder. This can be tested by calculating the critical resolved shear stress from the two yield strength values from Fig. 1-6. The values are 49 and 48 gm/mm^2, identical within experimental uncertainty. In the example of magnesium given above, the fact that the same critical resolved shear stress was calculated for the two different crystals suggests that crystallographically the slip plane and direction in the two cases must be the same. It is found that for the HCP metals this is the (001) plane (basal plane) and one of the $\langle 100 \rangle$ or the $\langle 110 \rangle$ direction (Fig. 1-7a).

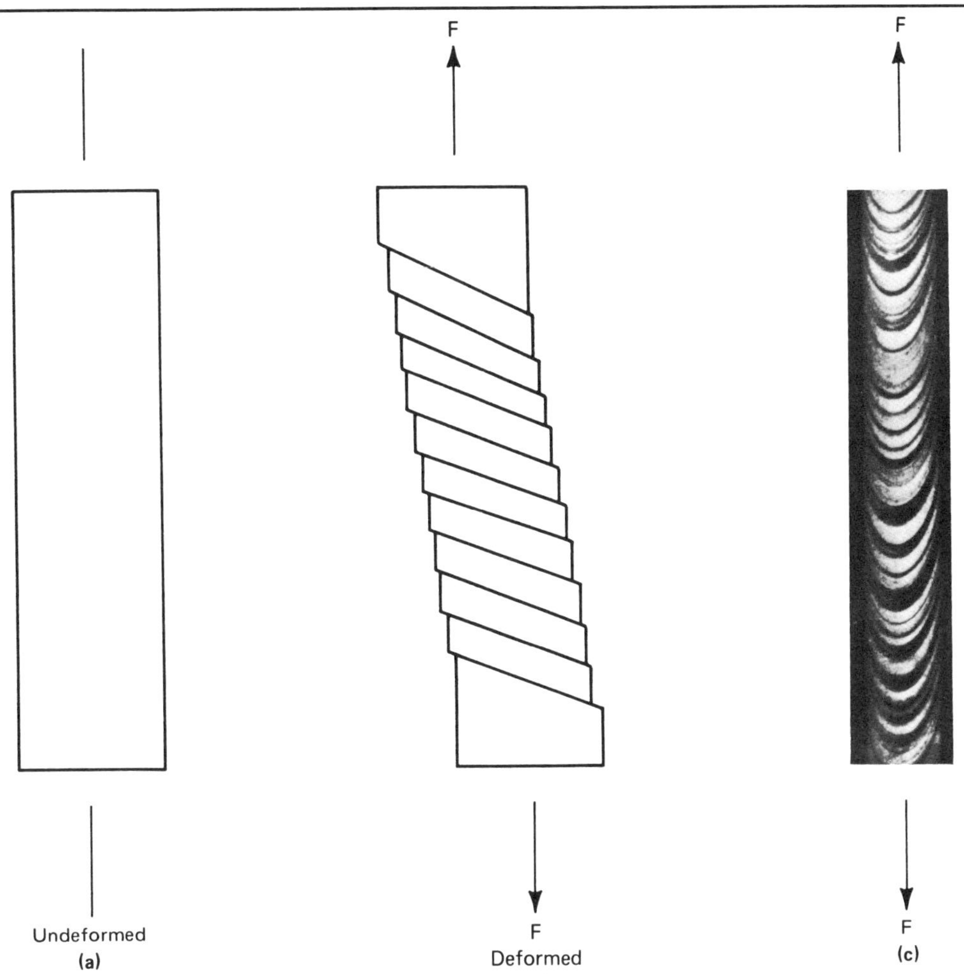

Fig. 1-4. Schematic illustration of plastic deformation of a single crystal (a) undeformed; (b) deformed showing slip. (c) Photograph of a deformed single crystal of zinc, showing elliptical offsets due to slip. (Adapted from W. Boas and E. Schmid, Z. Physik, Vol 61, p 767, 1930).

Since this is the close-packed plane and direction for this structure, it then would be predicted correctly that for face-centered cubic (FCC) structures the slip plane and direction would be $\{111\}$ and $\langle 110 \rangle$ (Fig. 1-7b). Now, in the HCP structure, there is only one slip plane and three directions, or three *slip systems*. However, in the FCC structure, there are four $\{111\}$ planes and three $\langle 110 \rangle$ directions in each plane, or 12 slip systems (Fig. 1-7d). This difference makes the tension test behavior of FCC metals more complicated than that of HCP metals. This is partly because in a tension test of this type the bottom of the crystal must stay aligned with the top. Thus, when slip begins, the crystal must rotate (Fig. 1-8), and this allows the resolved shear stress on another slip system to increase to the critical value and activate slip on this new system. Thus, these FCC crystals will usually show multiple slip; i.e., slip on more than one slip system. In the body-centered cubic (BCC) structure (Fig. 1-7c) generally the slip plane $\{110\}$ and direction $\langle 111 \rangle$ are the closest packed, giving also 12 slip systems (Fig. 1-7e).

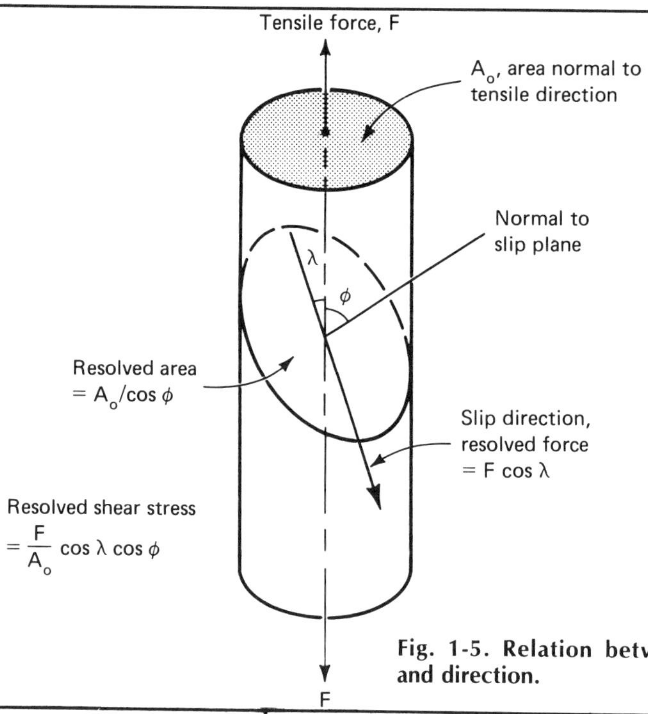

Fig. 1-5. Relation between cylindrical specimen and slip plane and direction.

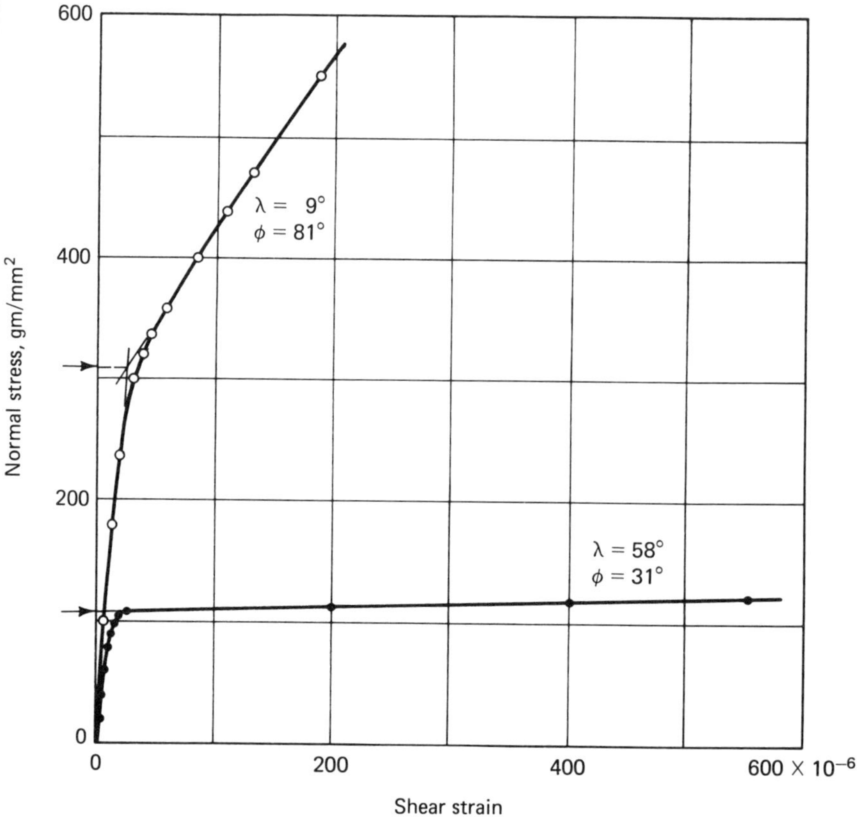

Fig. 1-6. Stress-strain curve for cylindrical single crystals of magnesium. Arrows indicate yield strength. (Adapted from E. C. Burke and W. R. Hibbard, Trans. AIME, Vol 194, p 295, 1952)

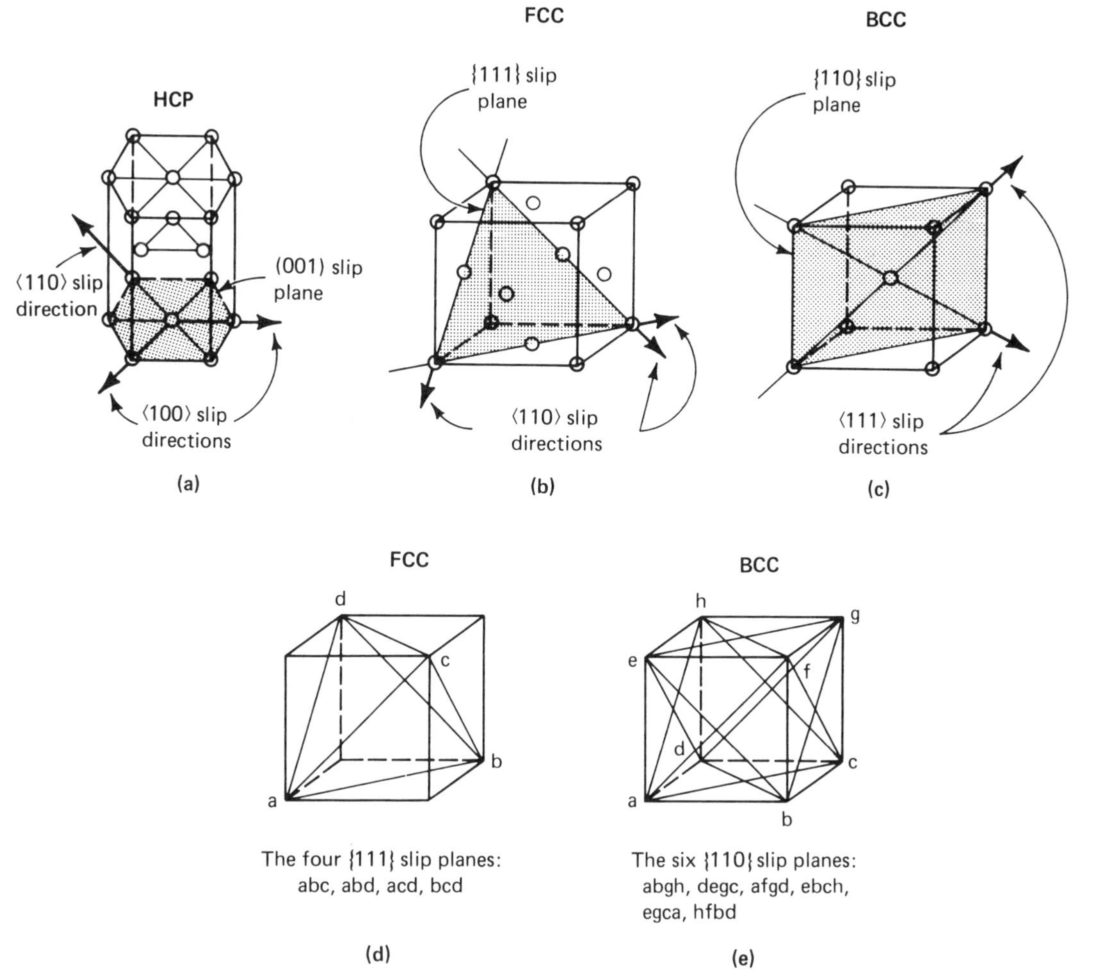

Fig. 1-7. Illustration of slip plane, slip direction and slip systems in HCP, FCC and BCC structures.

The close-packed direction as the slip direction can be predicted if the atoms are considered to be rigid spheres. The shear stress moves the row of atoms into a trough; the closer together the atoms are, the shallower this trough and the easier it is to move the row back out of the trough. (In the FCC and HCP structures, the path is not straight, but zigzag, as the atoms move across a saddle between the other atoms.)

1-3. Dislocations

If slip involves the movement of large segments of atoms as a group past each other, the shear strength can be estimated if the shear modulus is known. It was recognized early that the calculated strength was three or four orders of magnitude greater than the observed strength. It was eventually recognized that the low strength could be accounted for by the presence of

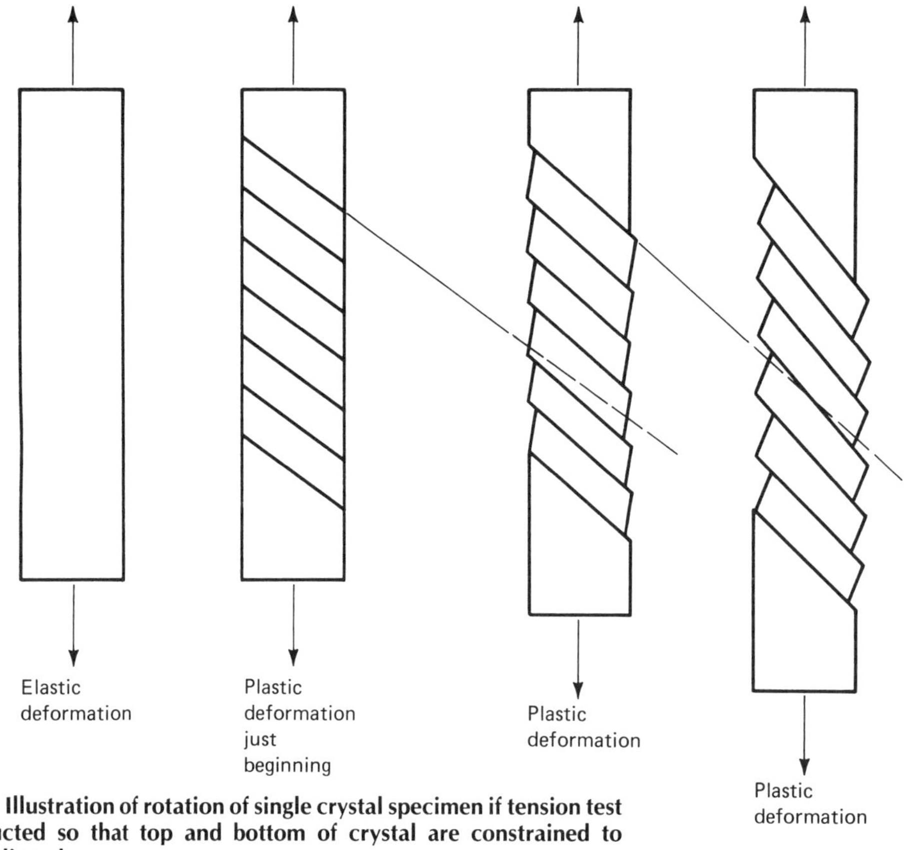

Fig. 1-8. Illustration of rotation of single crystal specimen if tension test is conducted so that top and bottom of crystal are constrained to remain aligned.

a defect in the lattice that makes the crystal less resistant to plastic flow. We will illustrate this defect by centering attention on the *edge dislocation*. Such a defect can be thought of as an extra half-row of atoms inserted in the crystal (Fig. 1-9). The core of the dislocation is elastically strained and thus has higher energy than regions far removed from it, where the crystal is essentially perfect. The dislocation is characterized by the Burgers vector, which is obtained by enclosing the dislocation along an atom-by-atom path. If the region enclosed were perfect, upon closure the starting position would be reached. However, the presence of the dislocation requires an additional step for closure. The direction and magnitude of the required step gives the Burgers vector (Fig. 1-9).

When a crystal containing a dislocation is mechanically loaded, the shear stress on the slip plane at the dislocation will reach a value so that the atoms are moved, and the bonding shifts one Burgers vector (Fig. 1-10a and b). This process will continue until the dislocation exits at the free surface (Fig. 1-10c). (Of course, there must be many dislocations moving to the surface to give visible offsets, such as those seen in Fig. 1-4c.) Considerably less energy is required to produce slip than to move the segments as a group past each other; thus such a defect present in a crystal could account for the observed low strength.

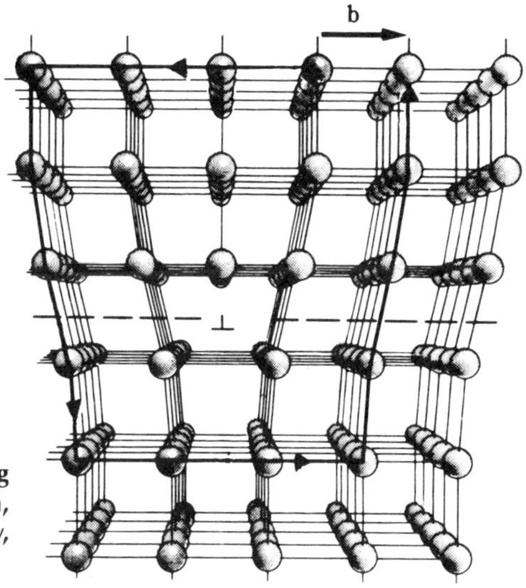

Fig. 1-9. Atomic model of a simple edge dislocation, showing the Burgers vector b. (Adapted from A. G. Guy and J. J. Hren, Elements of Physical Metallurgy, ©1971, Addison-Wesley, Reading, Mass., p 78. Reprinted with permission.)

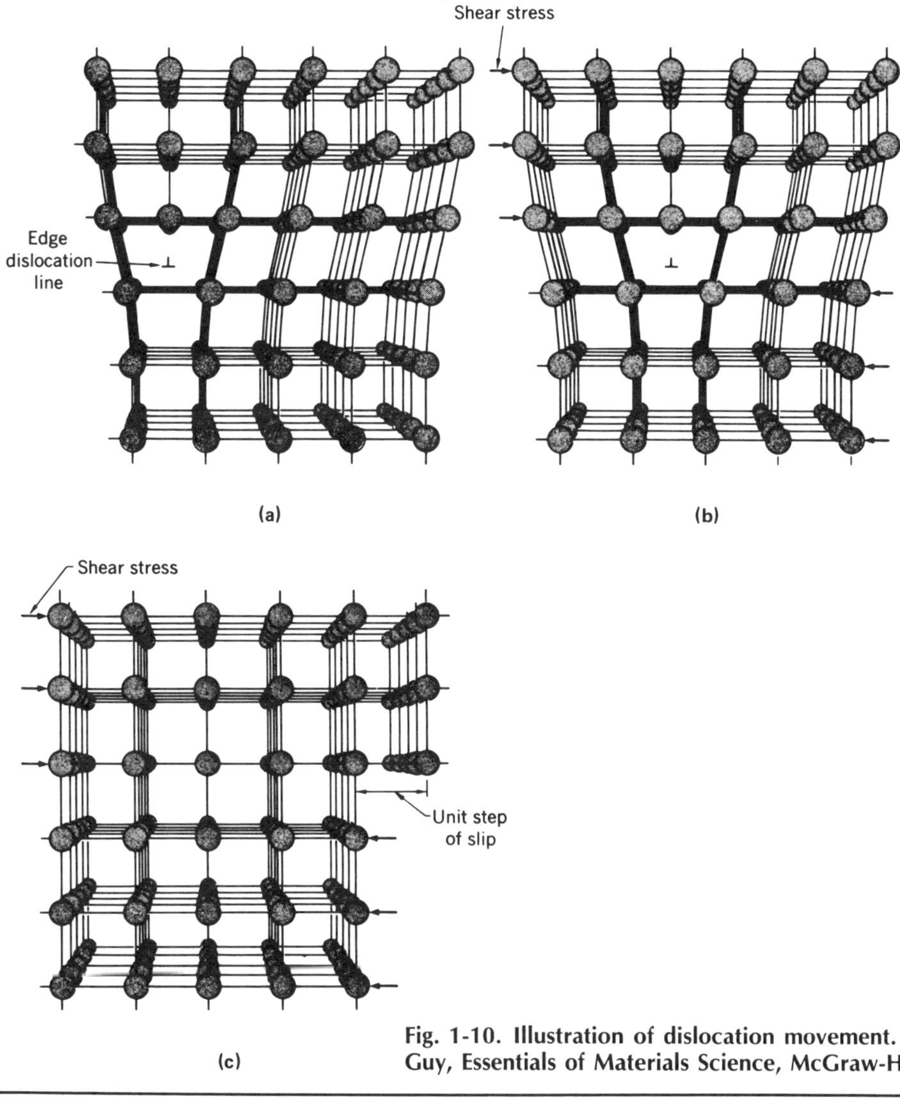

Fig. 1-10. Illustration of dislocation movement. (Adapted from A. G. Guy, Essentials of Materials Science, McGraw-Hill, New York, 1976)

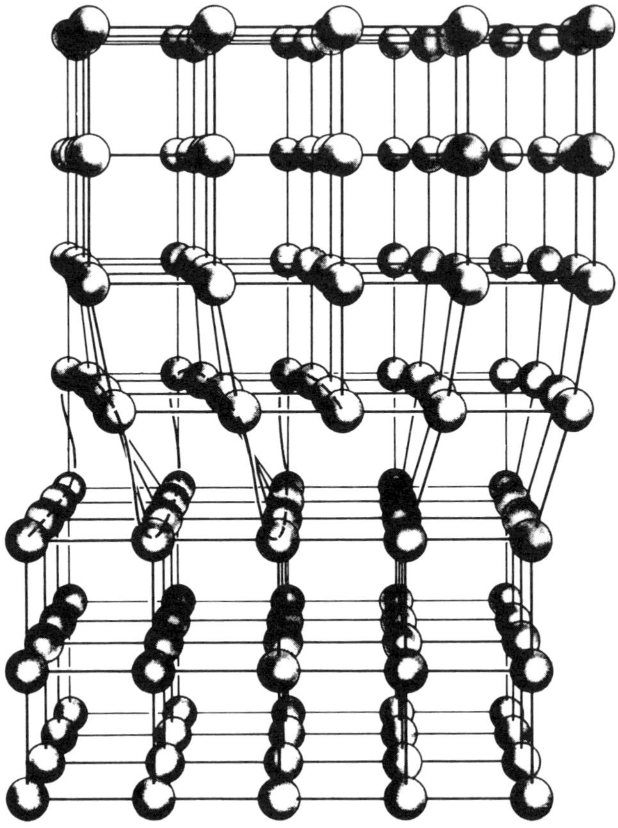

Fig. 1-11. Atomic model of a screw dislocation. (Adapted from p 84 of same source as Fig. 1-9)

The elastic energy stored in the core of the dislocation is proportional to the square of the Burgers vector. Thus, the dislocation which has the smallest Burgers vector will increase the energy of the crystal less, and this predicts that the close-packed direction will be the slip direction, as was found in the previous section.

Another type of dislocation is the screw dislocation, illustrated in Fig. 1-11. Most dislocations are a combination of screw and edge, as shown in Fig. 1-12. The movement of such a dislocation to produce slip is illustrated in Fig. 1-13.

An annealed crystal will have a dislocation density of approximately 10^7 cm of dislocation core length per cubic centimetre of material (or 10^7 dislocation/cm^2). Even if all of these dislocations move to the surface, this cannot account for the amount of plastic deformation observed. Although the additional deformation could be accounted for by the introduction of new dislocations from the free surface, the force required to do this is too high. Thus, there must be an internal mechanism for generating dislocations. The mechanism most easily visualized is called the Frank-Read source, and is illustrated in Fig. 1-14. Here a dislocation line DD′ is prevented from moving (i.e., it is pinned) by some means. For example, the lines BD and

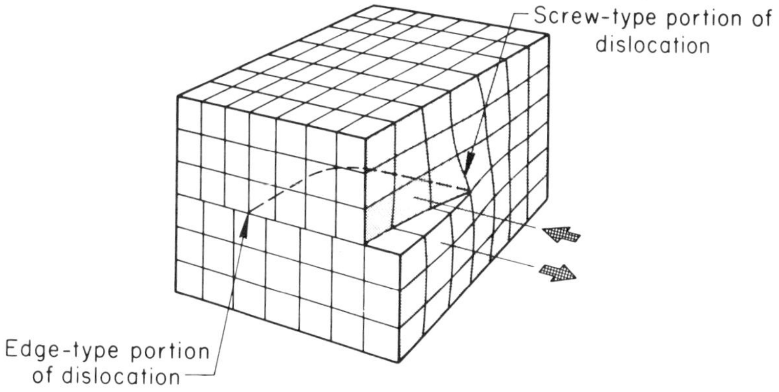

Fig. 1-12. Illustration of a mixed edge and screw dislocation to form a dislocation loop. (From Metals Handbook, 8th Ed., Vol 8, American Society for Metals, Metals Park, Ohio, 1973)

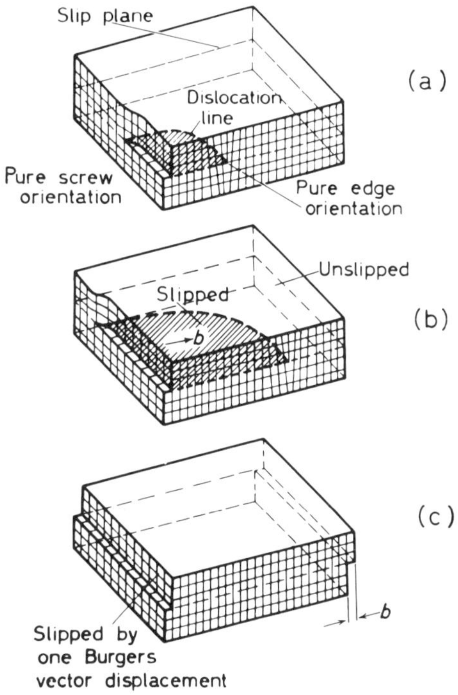

Fig. 1-13. Process of slip by movement of a dislocation loop. (From R.E. Smallman, Modern Physical Metallurgy, Butterworths, London, 1963)

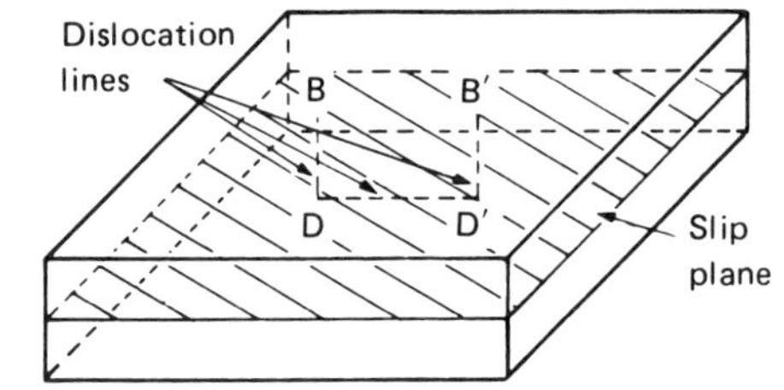

(a) Initial condition, with the dislocation line D—D′ leaving the slip plane at B and at B′.

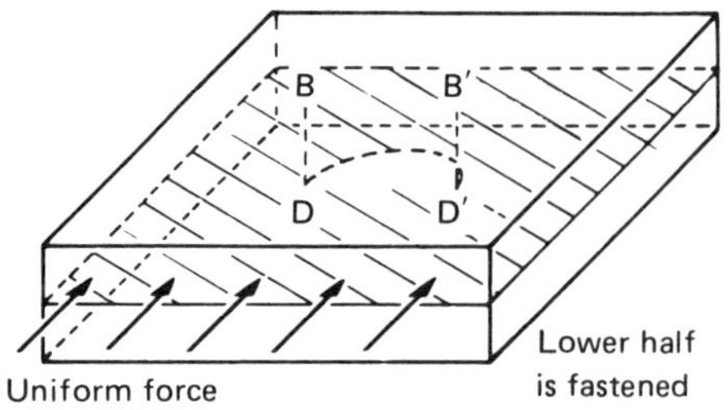

(b) A uniform force tending to shear the top half of the crystal causes D—D′ to begin sweeping over the slip plane by bowing outward.

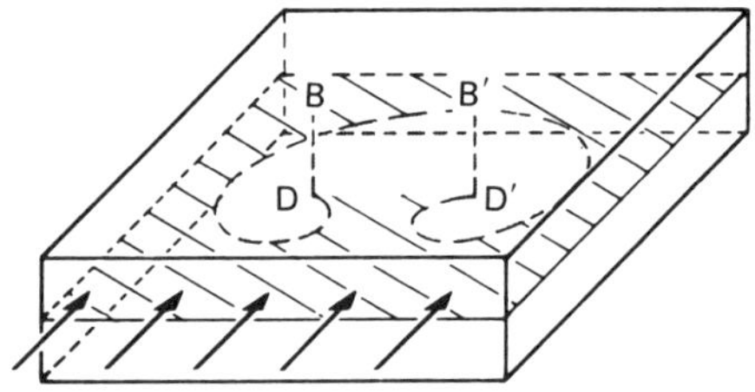

(c) As the dislocation line continues to expand it spirals about D and D′.

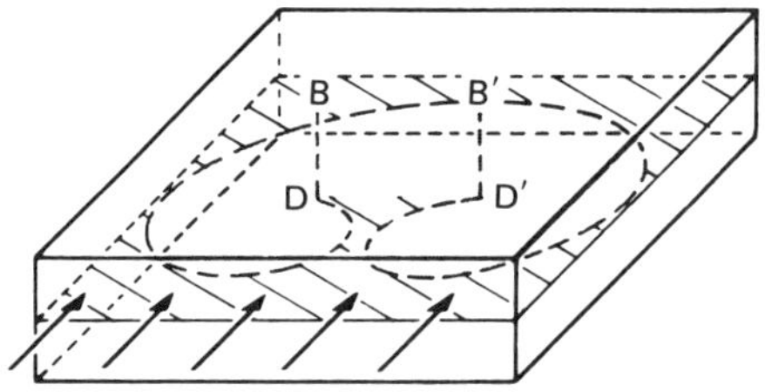

(d) Further expansion of the dislocation line brings the two spiral sections together.

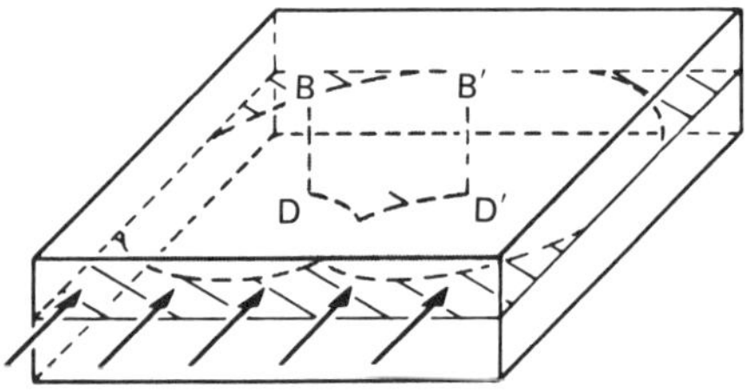

(e) As the two sections join they produce an outer loop that continues to expand and an inner line similar to D—D′ in (a).

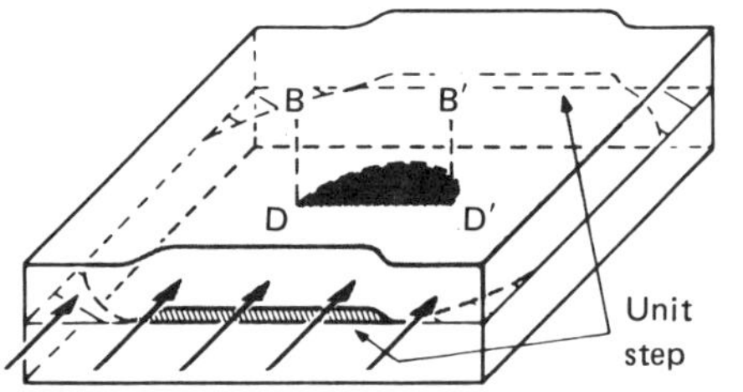

(f) The outer loop produces a unit step as it leaves the surface of the crystal; D—D′ begins the cycle again.

Fig. 1-14. The Frank-Read mechanism for the production of dislocation loops. (Adapted from p 99 of same source as Fig. 1-9)

B′D′ could be a grain boundary, or a hard particle could be present at D and at D′. As the shear stress increases, the dislocation bows out, and then bows back on itself (c). When the segment touches (e), the alignment is perfect, although translation has occurred in the direction of slip. When the moving edge of the dislocation loop meets the front and back free surface, displacement occurs. However, a dislocation is now regenerated at DD′, and the process can repeat itself. Such a mechanism has been observed in the electron microscope and is well established. Figure 1-15 shows a source which has generated three loops, none of which has yet reached the free surface.

There is no question that dislocations play the major role in controlling mechanical properties, and thus it is important to have at least a qualitative understanding of dislocation behavior. In this section we have presented a simple description of these defects, but this is sufficient to serve as a basis for further discussion.

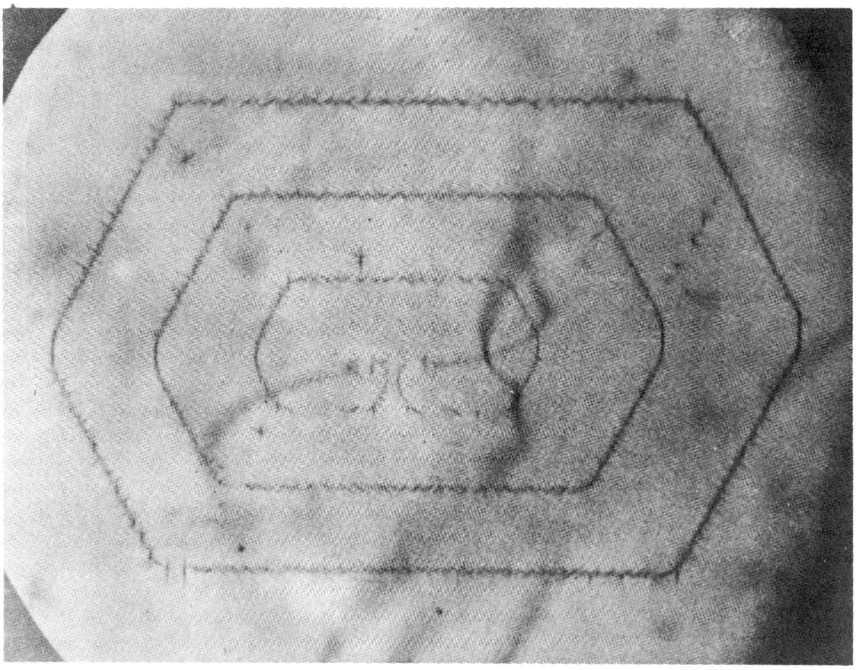

Fig. 1-15. Dislocation loops from a Frank-Read source in deformed silicon. This is a low-magnification photograph, and the dislocations were made visible by precipitation of copper particles along the dislocations. (From W. C. Dash, in Dislocations and Mechanical Properties of Crystals, ed. J. C. Fisher, W. G. Johnston, R. Thomson and T. Vreeland, Wiley, New York, 1957)

Most observations of dislocations are made using transmission electron microscopy. The core of the dislocations shows up because it is elastically strained, and locally has a different lattice parameter from that of the bulk of the material. Thus, under proper conditions, this region will diffract electrons which deviate from the optical axis and are stopped by an aperture. Hence, in the image of the crystal this diffracting region appears dark. In a sense, the dislocation casts a "shadow" of its core. The imaging process is illustrated schematically in Fig. 1-16. Two dislocations whose cores lie along aa′ are shown in a section of a specimen. If a thin slice is cut from the specimen, so that it contains the two dislocation cores (Fig. 1-16b) running from the bottom surface to the top surface of the slice (bb′), and then this slice is thinned so that it is transparent to the electrons, the dislocation cores will image as shadows if the thin film is oriented so that the cores diffract. This gives an image of the dislocation, such as that illustrated in Fig. 16(c), an electron micrograph of an austenitic stainless steel containing several dislocations. Those which are lined up all lie on the same slip plane. Each of these short lines runs from the bottom surface of the thin film to the top surface.

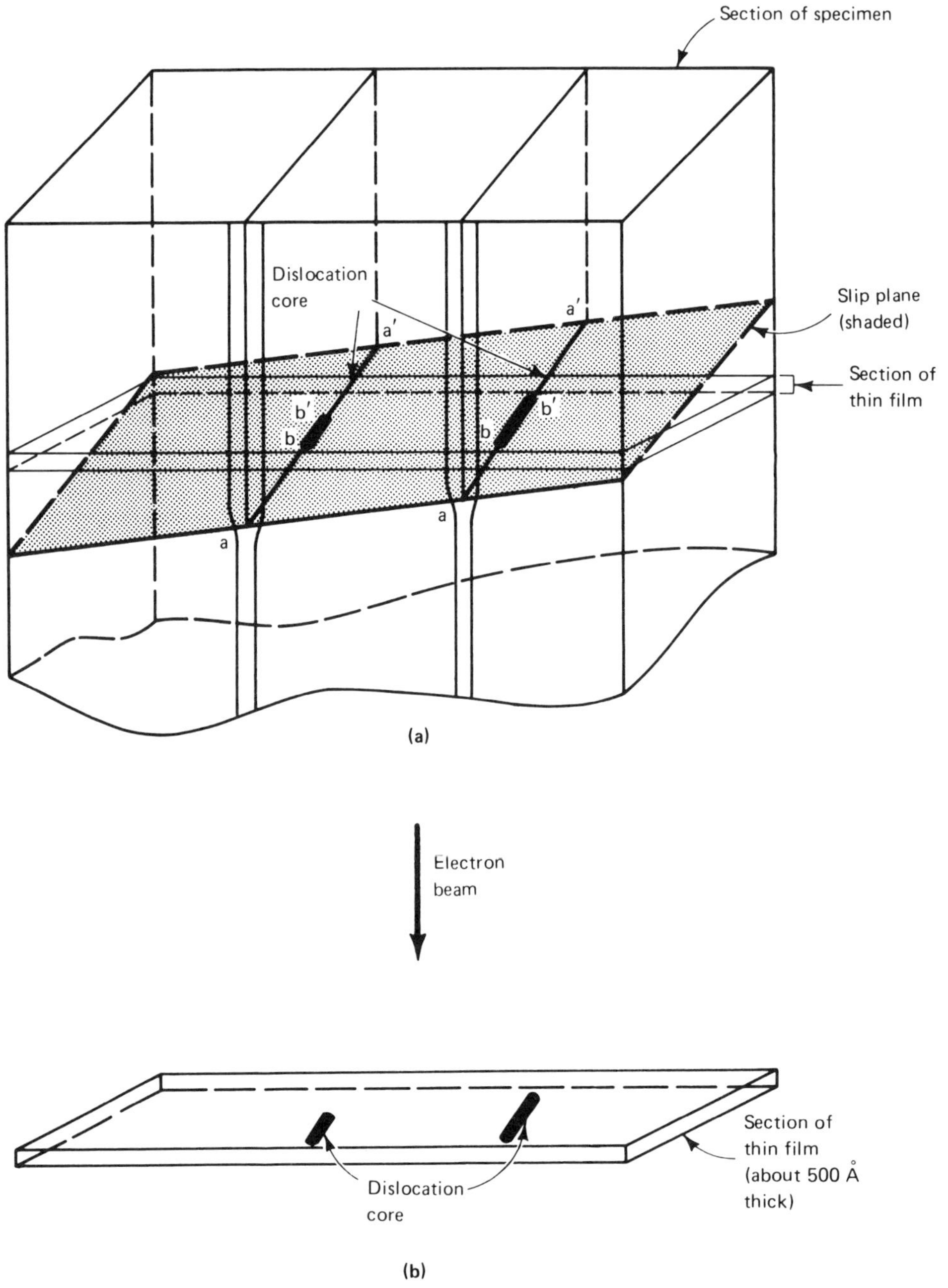

Fig. 1-16. Illustration of the imaging of dislocations in transmission electron microscopy. (a) Two dislocations in a section of a specimen. A thin section is cut out of the specimen such that the dislocations pass from the bottom of the section to the top (b). (c) Electron micrograph of dislocations in an austenitic stainless steel. (From G. Dupouy and F. Perrier, J. de Microscopie, Vol 1, p 167, 1962).

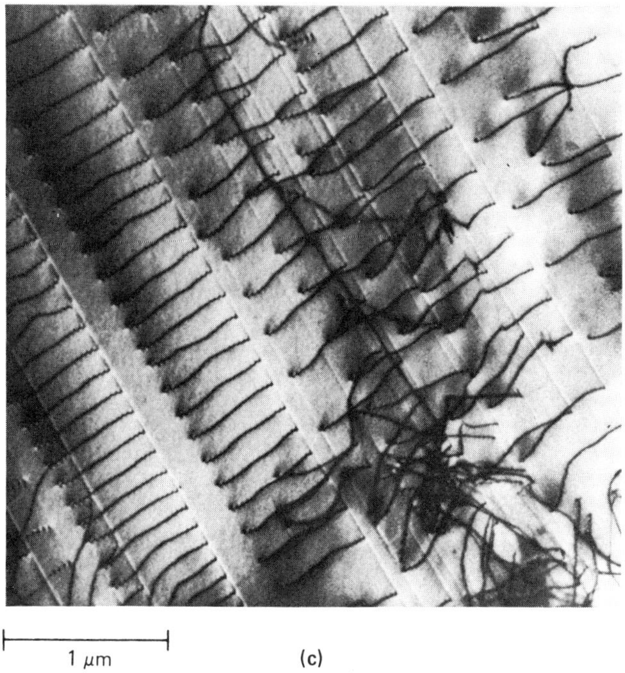

Fig. 1-16 (continued).

1-4. Strengthening Mechanisms

We are now in a position to discuss methods of strengthening. Anything done to the crystal to impede dislocation movement should cause strengthening, and we can discuss four general methods of importance with such a mechanism in mind: work (strain) hardening, grain size effect, particle effect and solid solution effect.

Work (Strain) Hardening

If the orientation of a crystal, relative to the external forces, is such that more than one slip system is activated, then dislocation intersections will occur. This generates a complex stress state at their junction, and their motion is impeded. In a polycrystalline sample, some crystals statistically will be oriented for multiple slip. In addition, crystal rotation will occur (see Fig. 1-8), which will bring other slip systems into operation. Thus, as deformation proceeds, more and more dislocation interaction occurs, and hence an increasing stress is required to continue dislocation movement. This is the cause of strain (work) hardening (see Fig. 1-1). Figure 1-17 shows the dislocation structure of iron as a function of the amount of cold work. Note the increased dislocation tangling as the amount of deformation increases, and the correlation between increasing dislocation density and yield strength.

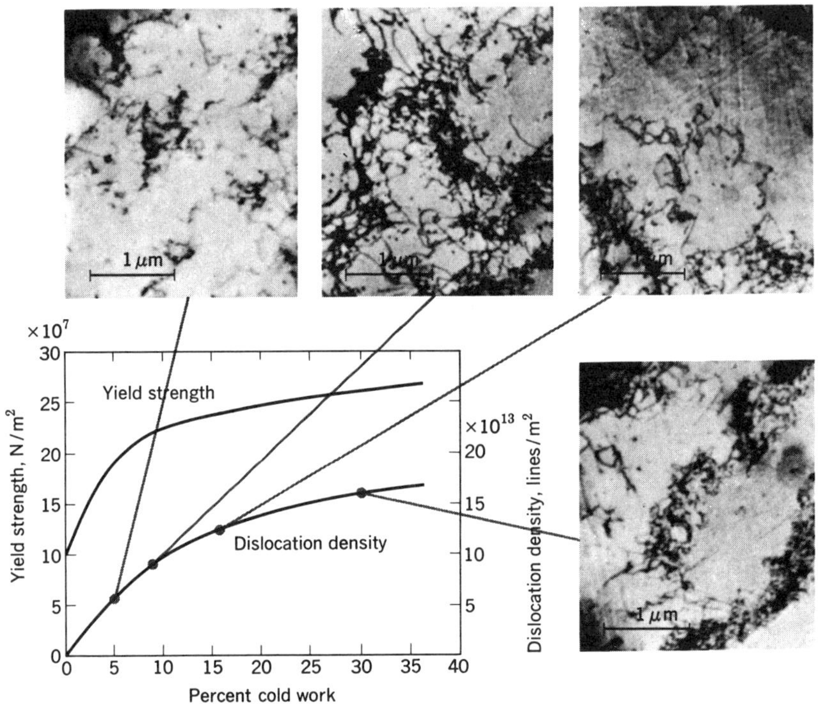

Fig. 1-17. Relation between dislocation structure, dislocation density and yield strength as a function of amount of cold work for pure iron. (From A. S. Keh, in Direct Observations of Imperfections in Crystals, ed. J. B. Newkirk and J. H. Wernick, Interscience, New York, 1962, as given by A. G. Guy, Introduction to Materials Science, McGraw-Hill, New York, 1972)

Grain Size Effect

When a dislocation moving on the slip plane encounters a grain boundary, movement is retarded because of the orientation change. To move another dislocation on the same plane toward the boundary requires an increase in stress because two dislocations of like sign repel each other. Thus, to continue to generate dislocations from a source requires an increasing stress, and thus work hardening occurs. The dislocation pile-up is illustrated in Fig. 1-18. The farther the source is from the boundary, the more dislocations can be generated before the back stress on the dislocation balances the external applied shear stress. Since locations of interaction of dislocations at the boundary can act as dislocation sources, then (if these are the only sources) the maximum length for slip is the grain diameter. Thus it would be expected that the strength would be inversely proportional to the grain size. Usually, it is found to vary with the $(-1/2)$ power.

Particle Effect

If hard particles are present in a softer matrix they hinder dislocation motion. If the stress becomes sufficiently high the dislocation may cut through the particle, or the dislocation may move out of the slip plane onto another and climb around the particle. However, the more likely process is for the dislocation to bow around the particle, as illustrated in Fig. 1-19. Note that adjacent particles are acting similar to Frank-Read sources (Fig. 1-14). Since the core of the dislocation is elastically strained, the bowing increases the length of this strained core and hence the energy of the crystal. Thus additional stress must be applied to force the dislocation to move through the barrier, and this gives strengthening. If the particles are rigid, this type of hardening is referred to as dispersion hardening. It turns out that the more closely spaced the particles are the more difficult it is to force the dislocation to bow. Thus the strength is inversely proportional to the spacing.

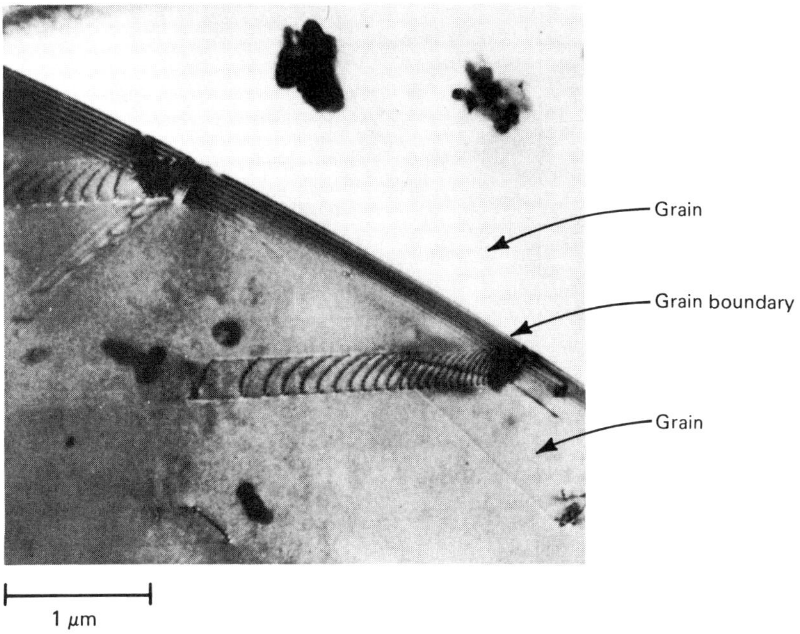

Fig. 1-18. Transmission electron micrograph showing dislocation pile-up at a grain boundary in austenitic stainless steel. (Adapted from same source as Fig. 1-16c)

Solid Solution Effect

Most pure metals can be strengthened by the addition of solute atoms, either substitutionally or interstitially. There are several reasons for this, but the most important and obvious one is that the strain energy of the crystal will be reduced if the solute atom is located along the core of the dislocation. Thus, it requires more stress to move the dislocation away from the dislocation than through the pure matrix.

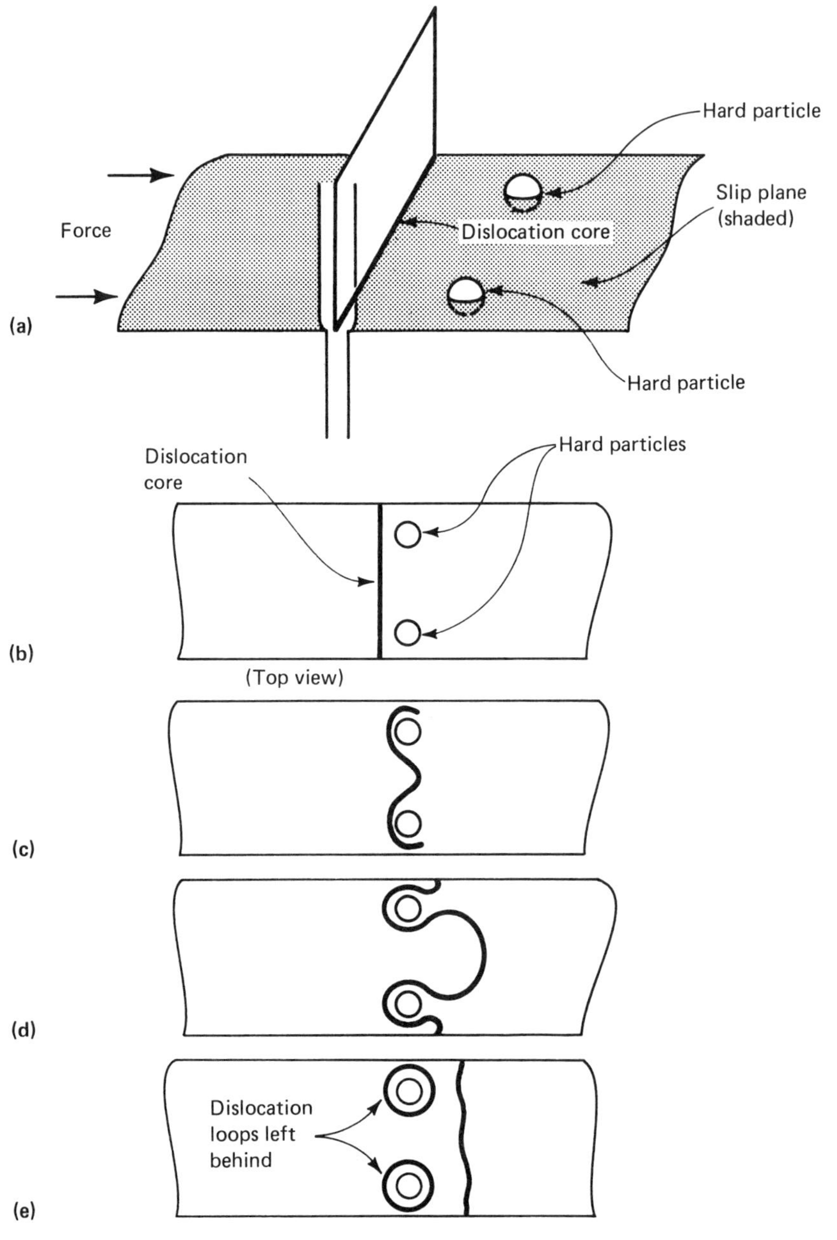

Fig. 1-19. Illustration of movement of a dislocation through two hard particles.

1-5. Microstructural Effects

The effects of mechanical and thermal treatment of metals and alloys are usually followed by observation of the microstructure, especially at a level of resolution using the optical microscope. Recall that this involves carefully polishing the specimen, then etching the surface to develop a surface topology which is related to the underlying structure. For example, the grain boundaries will be etched more rapidly than the inside of a grain, creating a groove from which light is reflected outside the optical system so that the boundary appears as a dark line.

Now, the stress field around the dislocation should make it susceptible to etching where it cuts the surface. The region of the dislocation core is elastically strained, and hence the etch should preferentially remove (dissolve) atoms faster in this location, developing a pit. When such an etched surface is examined by optical microscopy, the light striking the pit will be reflected outside the optical system, so that this location will appear dark. Figure 1-20 shows etch pits where dislocations on the same slip plane have

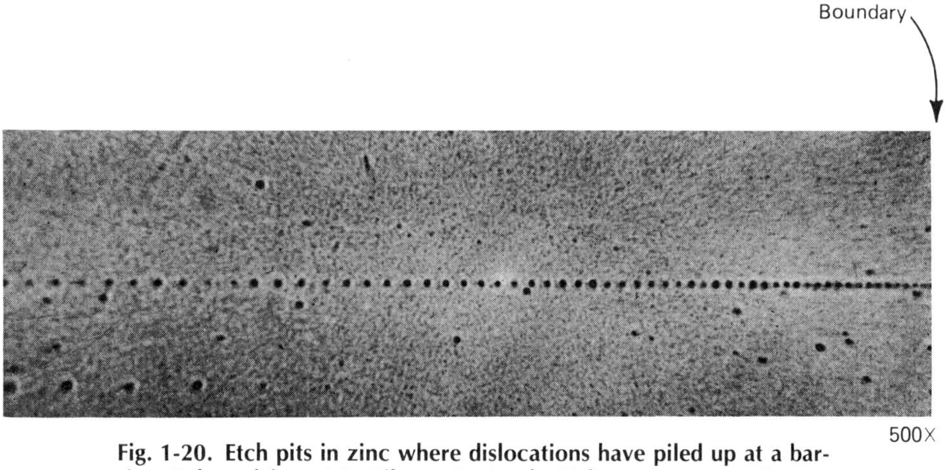

Fig. 1-20. Etch pits in zinc where dislocations have piled up at a barrier. (Adapted from J. J. Gilman, J. Metals, Vol 8, p 1000, 1956)

piled up at an obstacle. However, when the amount of plastic deformation exceeds about 10%, the dislocation structure becomes so complex that where it cuts the surface, grooved regions called *deformation bands* are more likely to develop than simple pits. Figure 1-21 illustrates a typical microstructure of a cold worked metal. The grains are elongated in the direction of rolling, and the annealing twins (see Section 1-13) are bent. In addition, deformation bands are prominent.

The development of the elongated grains and deformation bands with the amount of cold work is illustrated for a Cu-5Zn alloy in Fig. 1-22. Note that the grains become progressively elongated, and the annealing twins (see Section 1-13) more bent and fragmented. The deformation bands are most apparent for intermediate amounts of deformation; when the deformation is high, each grain has considerable deformation, and hence sufficiently high dislocation density, to cause numerous, very closely spaced deformation bands, so that they can not be so easily differentiated. At 80% cold work, the structure shows only a general flow pattern. The hardness data for this alloy, as well as for pure copper and a Cu-20Zn alloy, are shown in Fig. 1-23. Here the work hardening for each material is seen, as well as the pronounced solid solution hardening.

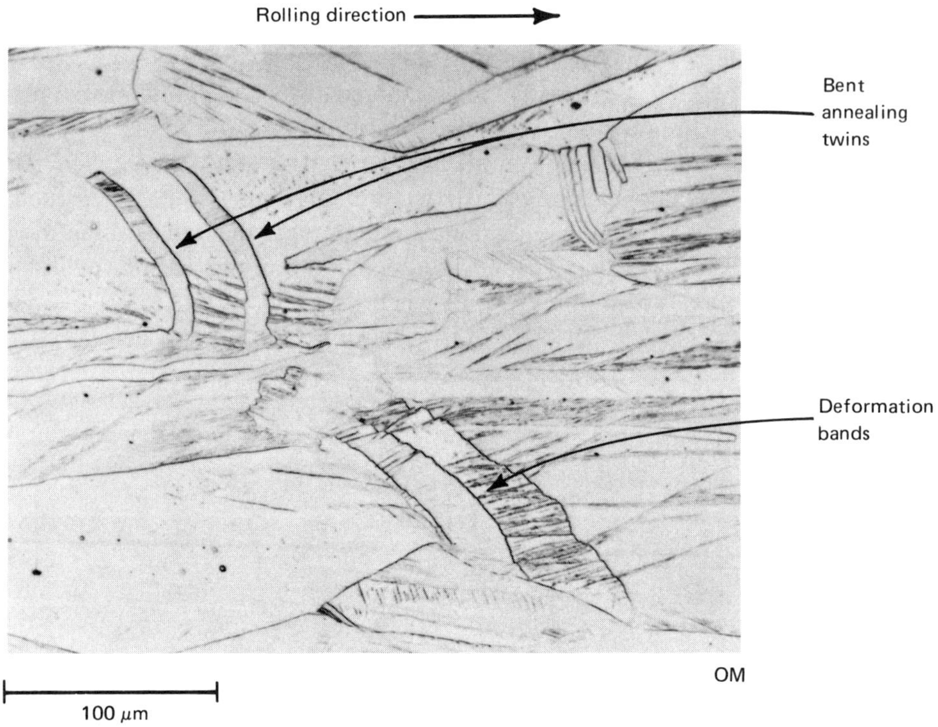

Fig. 1-21. Microstructure of a Cu-5Zn alloy cold rolled at 25 °C to a 40% reduction in thickness, showing deformation bands and bent annealing twins revealed by etching the polished surface.

1-6. Deformation Textures

When plastic deformation is carried out in such a way that material flow is prevented in some direction (constraint), and this is usually the case in practical situations, the crystals overall must undergo a lattice rotation. This was illustrated in Fig. 1-8 for the case of tensile deformation of a single crystal where the ends of the specimen were constrained to remain aligned. Thus, upon plastic deformation of a polycrystalline sample of originally random orientation, the crystals take up a non-random orientation relative to the deformation geometry (e.g., relative to the rolling plane and rolling direction). It is important to realize that even with large amounts of deformation, and the presence of a high dislocation density and very elongated grains, the material is still highly crystalline. Thus the properties of the deformed material become anisotropic.

The non-random structure is referred to as having preferred orientation, and the material is said to have developed a deformation texture. The most common way to indicate the degree of preferred orientation is by the use of pole figures. By use of x-ray diffraction it is possible to determine the relative intensity from crystals with given poles (direction) pointing in any direction relative to the deformation geometry. For example, if the material is a rolled plate, the [100] pole figure gives the concentration of [100] poles in every direction from the plate, relative to the rolling direction. The data are plotted as a stereographic projection, similar to plotting on polar coordinates.

Figure 1-24 shows the [100] pole figure for copper cold rolled 96% reduction in thickness. If the crystals were oriented randomly, the intensity distribution would be random. However, note that there are four symmetric positions which have a high concentration of [100] poles. The rolling texture affects the texture present after annealing (discussed in Section 1-12), and thus it is quite important to follow the deformation texture even if the material will be annealed.

1-7. Recovery, Recrystallization and Grain Growth

We have seen that plastic deformation introduces a variety of lattice defects, most noticeably a complex array of dislocations. The deformed structure is unstable thermodynamically, and thus given sufficient time will return to an equilibrium state of lower energy. This state consists of a single crystal with no dislocations but with the equilibrium number of vacancies present. In reality the deformed material will return to a polycrystalline state consisting of vacancies of approximately equilibrium concentration and a relatively low dislocation density (about 10^7 lines/cm^2), unless held extremely long at elevated temperatures. The atomic mechanisms that bring about this change vary depending upon the material, the deformation process and the heat treatment. In this section, the effect of these changes on some of the properties of metals and alloys is discussed, and then a discussion of mechanisms is presented.

We will consider a metal plastically deformed at a low enough temperature to retain the defects introduced during the deformation. (That is, the metal has been cold worked.) To bring about a significant change in this structure requires heating to elevated temperature, a process called *annealing*.* The temperature required to bring about these changes depends upon the material. For most metals and alloys, for a reasonable annealing time, it is above 25 °C but for some metals (e.g., lead) may be below this temperature.

To illustrate the annealing process, we consider an alloy of copper with 5% zinc (a single-phase solid solution) which has been rolled at 25 °C. As the reduction in thickness increases, the hardness increases, and the microstructure develops a complex array of dislocations. In the optical microscope, this manifests itself in a high density of deformation bands (Fig. 1-22), bent annealing twins and elongated grains.

If the deformed alloy is held at 400 °C, and the hardness periodically measured, the data give a curve such as that shown in Fig. 1-25. The

*Annealing is one of those metallurgical terms which have multiple meanings.

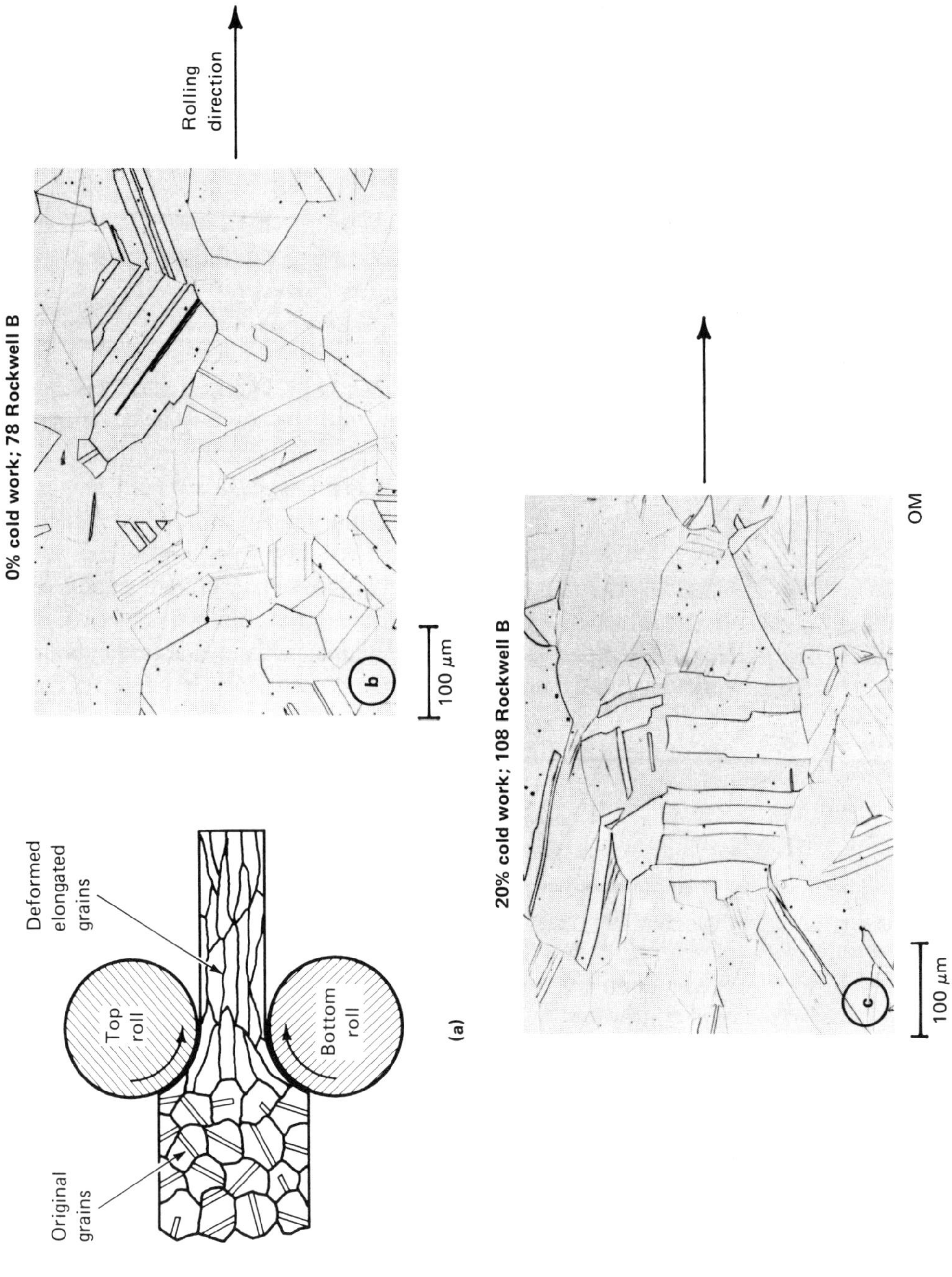
Original grains
Top roll
Bottom roll
Deformed elongated grains
(a)
0% cold work; 78 Rockwell B
b
100 μm
Rolling direction
20% cold work; 108 Rockwell B
c
100 μm
OM

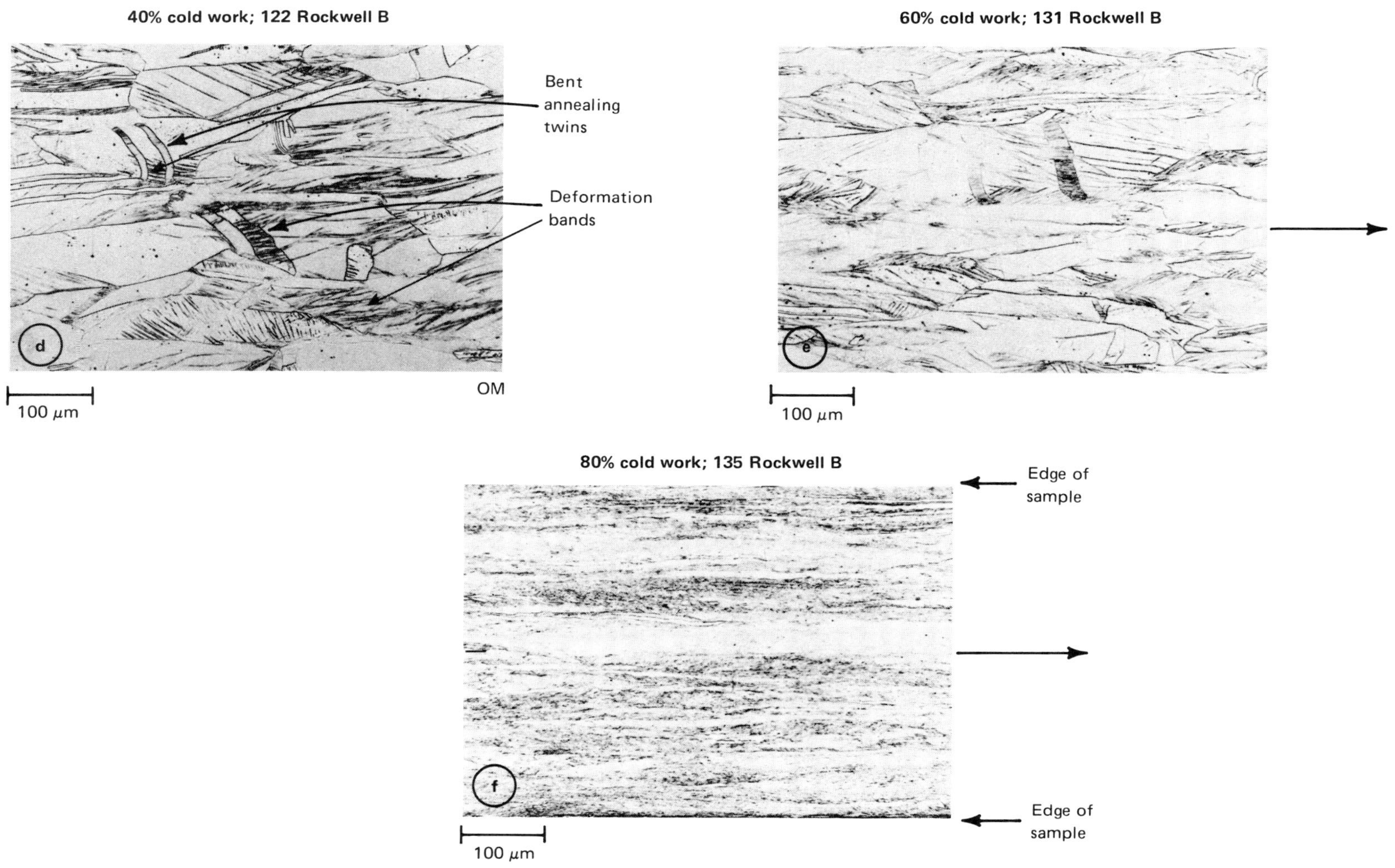

Fig. 1-22. Optical micrographs (all at same magnification) of a cold worked Cu-5Zn solid solution alloy, showing Rockwell B hardness at each stage. Percentages for cold work are based on reduction in thickness. Rolling direction is from left to right.

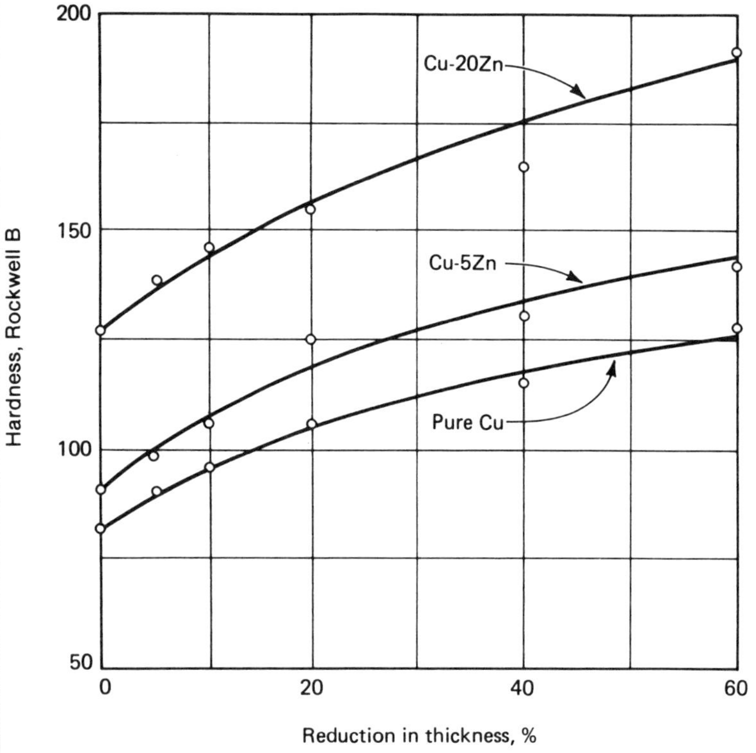

Fig. 1-23. Hardness as a function of reduction in thickness for pure Cu and two Cu-Zn solid solution alloys, all cold rolled at 20 °C.

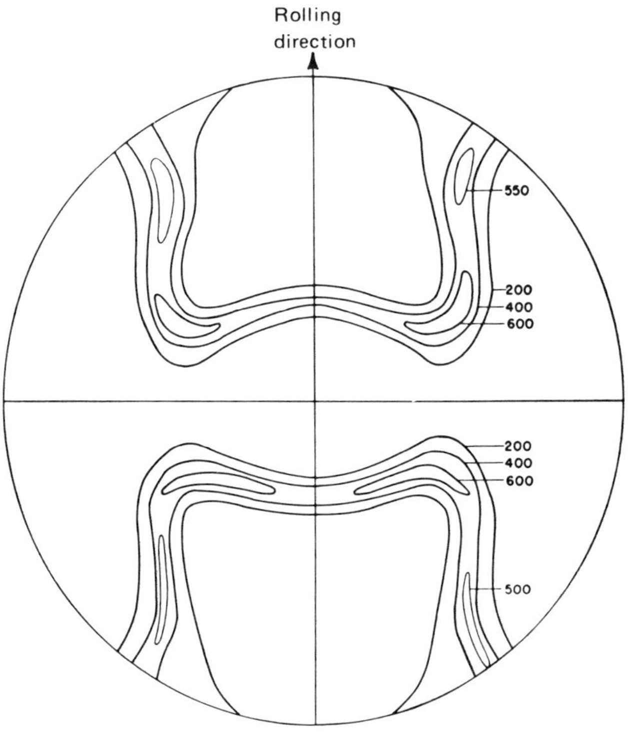

Fig. 1-24. [100] pole figure for copper cold rolled to a 96% reduction in thickness. The plane of the paper is parallel to the rolling plane, and the rolling direction is vertical. The numbers give the intensity of the x-rays diffracted by the (100) planes. (Adapted from H. Hu, P. R. Sperry and P. A. Beck, Trans. AIME, Vol 194, p 78, 1952)

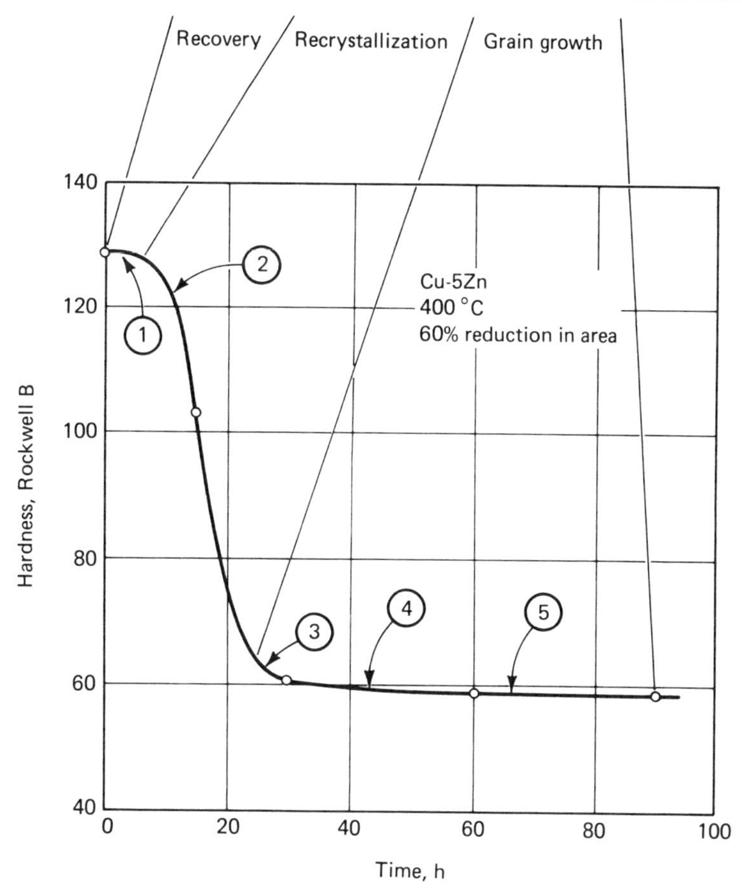

Fig. 1-25. Effect of annealing time (at 400 °C) on the hardness of a Cu-5Zn solid solution alloy cold worked 60%.

annealing curve is divided into three regions, labeled *recovery, recrystallization* and *grain growth.* In the recovery stage the hardness remains constant, or may increase slightly. This is followed by a drastic decrease in hardness during recrystallization. The hardness continues to decrease, but much more gradually, during grain growth.

Although hardness is not affected greatly during recovery, other properties (e.g., electrical resistivity) are, showing that significant structural changes occur during this stage, but they are changes to which the hardness is not very sensitive. The term *recovery* is used because recovery of some of the original properties occurs during this short time. During recrystallization, it is observed that relatively soft, strain-free (low dislocation density) crystals form (nucleate) and grow in the plastically deformed matrix (Fig. 1-26). (This is shown more clearly in Fig. 1-27, which shows small, isolated crystals having just formed in the deformed matrix.) Once the deformed matrix is consumed by these new crystals, further annealing only causes grain growth.

It is emphasized that more than one mechanism bringing about the structural changes may be operating simultaneously. For example, once a

128 Rockwell B

No recrystallization yet; still in recovery

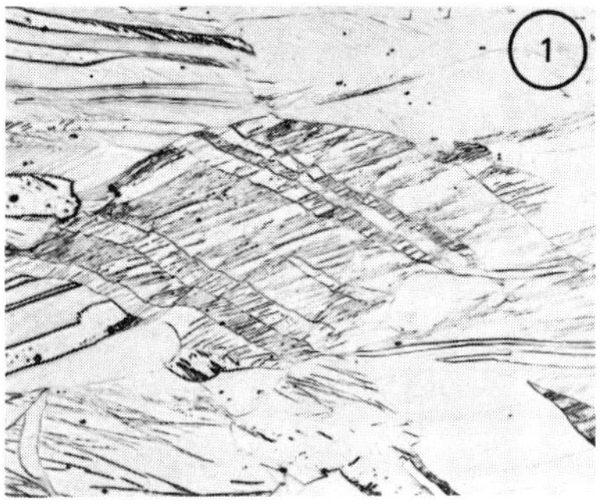

100 μm

127 Rockwell B
Recrystallization just beginning
(see Fig. 1-27)

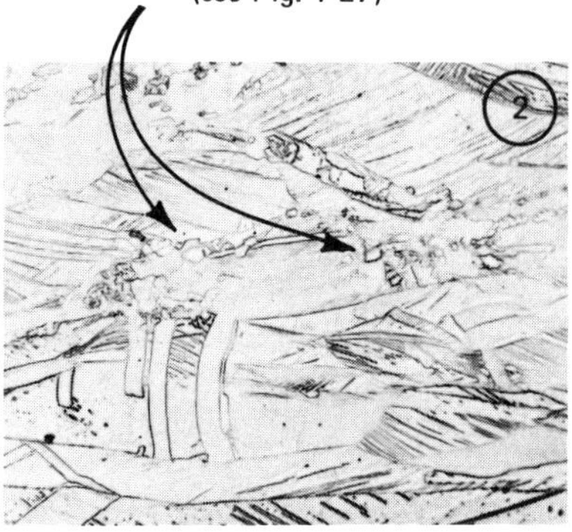

100 μm

OM

63 Rockwell B
Recrystallization essentially complete;
grain growth beginning

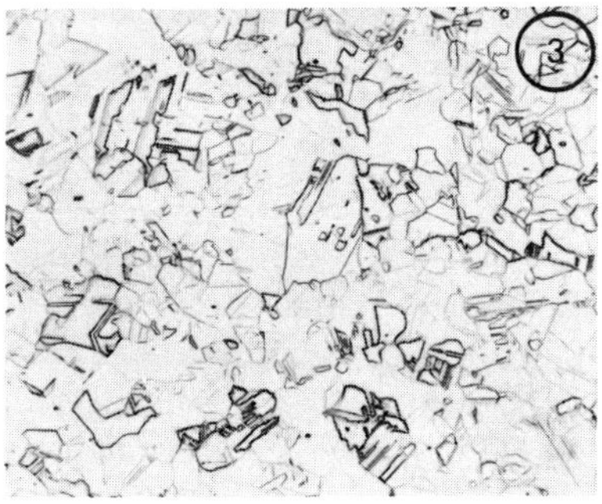

100 μm

OM

60 Rockwell B

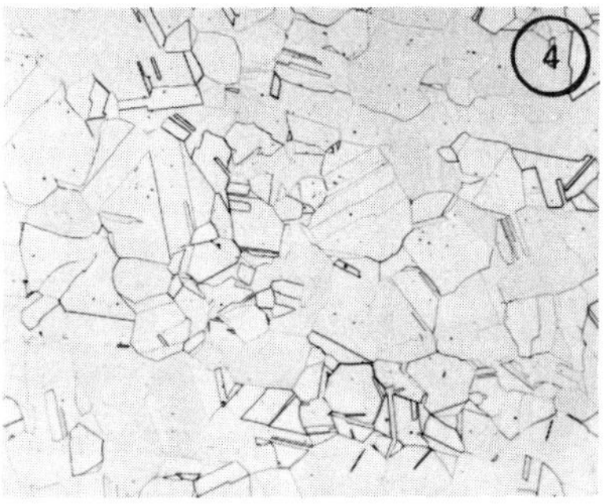

200 μm

OM

58 Rockwell B

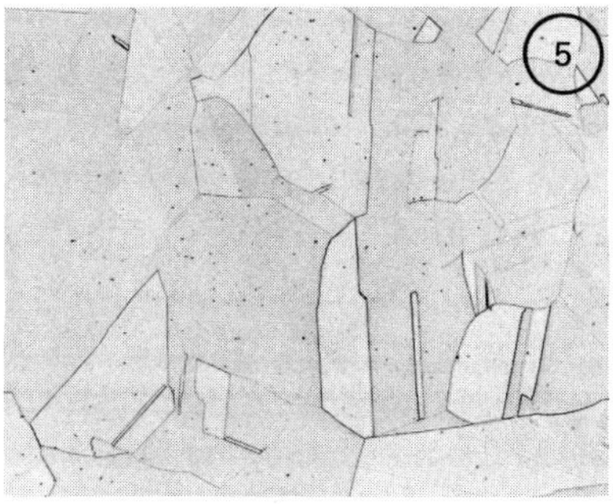

200 μ

OM

Fig. 1-26. Microstructure of a Cu-5Zn alloy, cold rolled to 60%, then annealed for different times at 400 °C. The numbers refer to locations on the curve in Fig. 1-25.

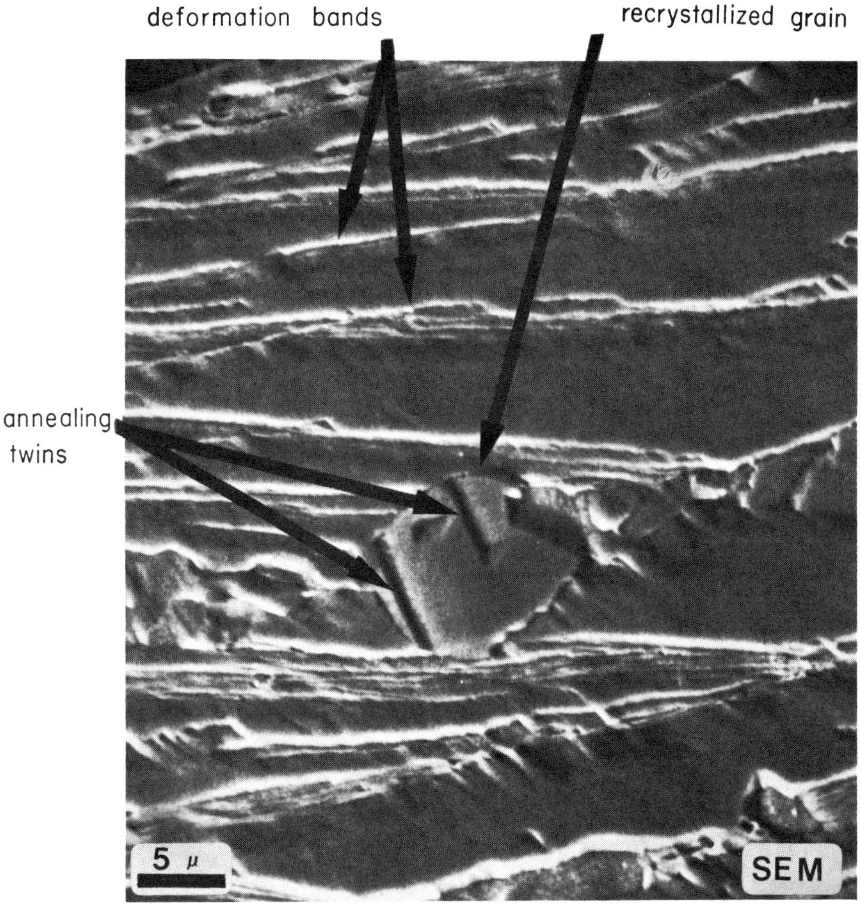

Cu-5% Zn alloy, cold worked by rolling at 20 °C to reduction in thickness of 60%; annealed 60 min. at 350°C.

Fig. 1-27. Microstructure of a Cu-5Zn alloy showing recrystallization just beginning. Note that annealing twins form in these small crystals.

small strain-free crystal forms (the beginning of recrystallization), it then grows into the deformed matrix, which at the same time is continuing to undergo recovery. The particular annealing process is a consequence of all of the mechanisms operating, and this is dependent upon the material, the processing history and the annealing procedure.

Annealing behavior is usually presented by plotting property against annealing temperature, for a fixed annealing time (Fig. 1-28). The same type of curve results, with the stages of recovery, recrystallization and grain growth present. Figure 1-29 shows the reduction in electrical resistivity in the recovery period, even though the hardness remains about constant. In this case, only after the resistivity has returned to its prior cold worked range does recrystallization begin.

This type of curve is used to determine the *recrystallization temperature,* a term that can be defined several ways. It can be defined as the

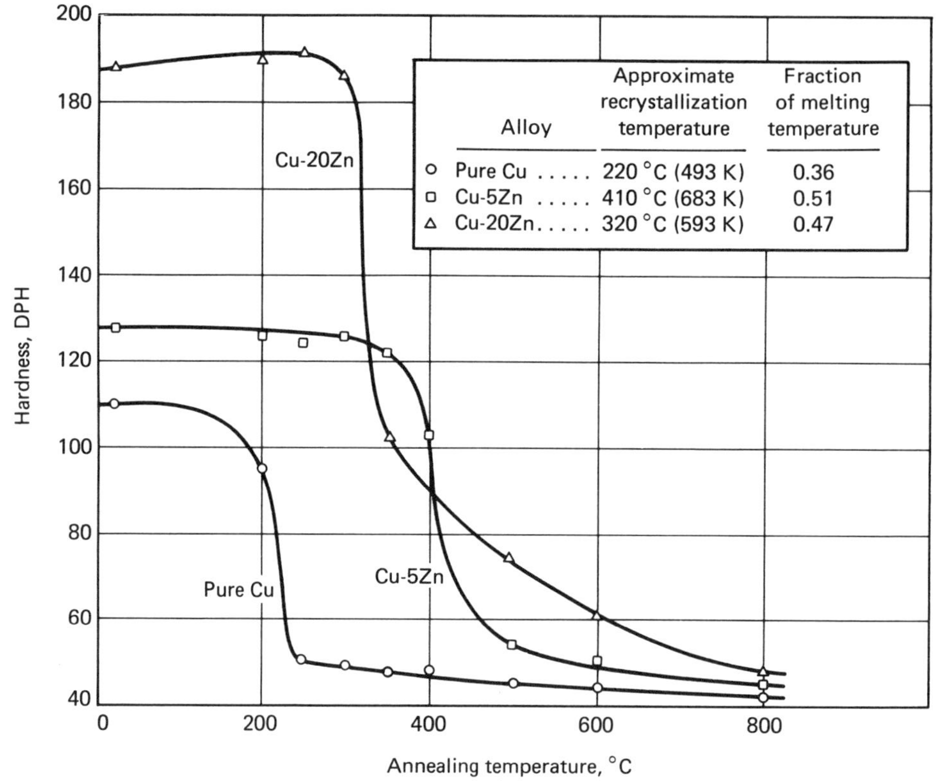

Fig. 1-28. Hardness as a function of annealing temperature for 15-min annealing time. The alloys were originally rolled at 25 °C to a 60% reduction in thickness. The recrystallization temperatures listed are based on the inflection point.

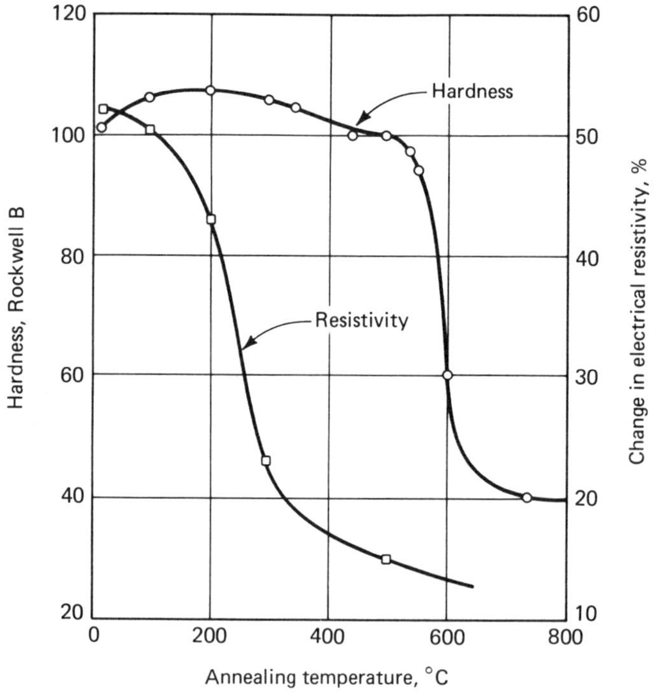

Fig. 1-29. Effect of annealing temperature on the hardness and electrical resistivity of nickel. The material had been worked at 25 °C almost to fracture. Annealing time 1 h. (Adapted from J. E. Wilson and L. Thomassen, Trans. ASM, Vol 22, p 769, 1934)

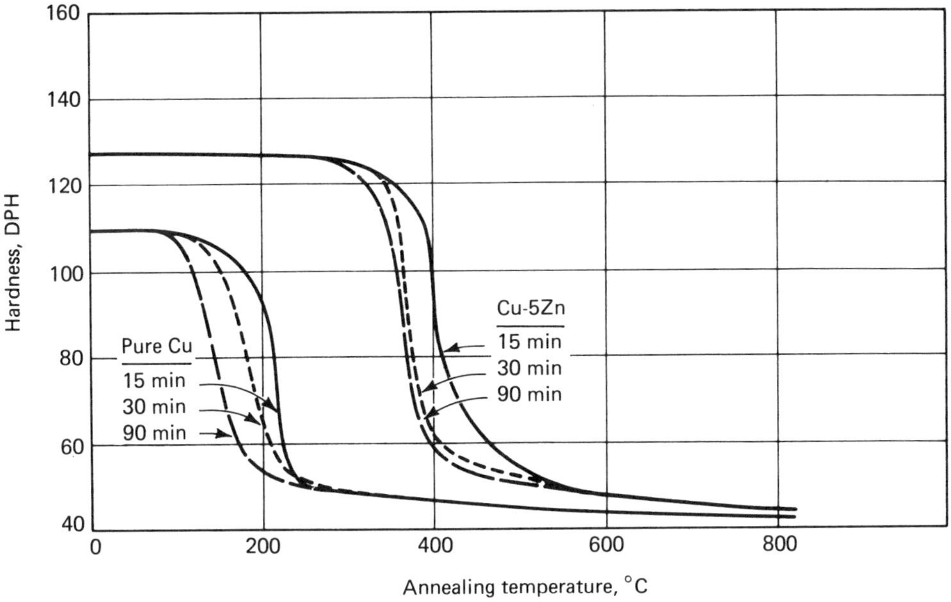

Fig. 1-30. Hardness as a function of annealing temperature for pure Cu and a Cu-5Zn alloy, using three different annealing times. Both materials were originally cold rolled at 25 °C to a 60% reduction in thickness.

approximate minimum temperature at which complete recrystallization of a cold worked metal occurs within a specified time.* It can be defined also as the temperature at which recrystallization just begins, as the temperature at which recrystallization is half completed, or as the inflection point of the hardness-annealing temperature curve. This last definition gives the approximate recrystallization temperatures listed in Fig. 1-28.

The recrystallization temperature depends, of course, upon several variables. The time at the annealing temperature (Fig. 1-30), the chemical composition of the alloy (Fig. 1-28), and the amount and type of plastic deformation (Fig. 1-31) all must be specified. Any variable that increases the energy of the cold worked material will make it more unstable relative to the annealed condition. This energy difference is the "driving force" for the return to the soft, strain-free condition, and usually the larger this energy difference is, the more rapidly the process occurs. Thus even such variables as the original grain size and the temperature of deformation affect the recrystallization temperature. Table 1-1 summarizes these effects.

It has been found that recrystallization will occur in a reasonable time (e.g., 1 h) when the temperature is between 0.3 and 0.6 of the *absolute* melting point. This is only an approximation, but the data in Fig. 1-28 agree with this rule.

---

*Metals Handbook, 8th Ed., Vol 1, p 30, American Society for Metals, Metals Park, Ohio, 1961.

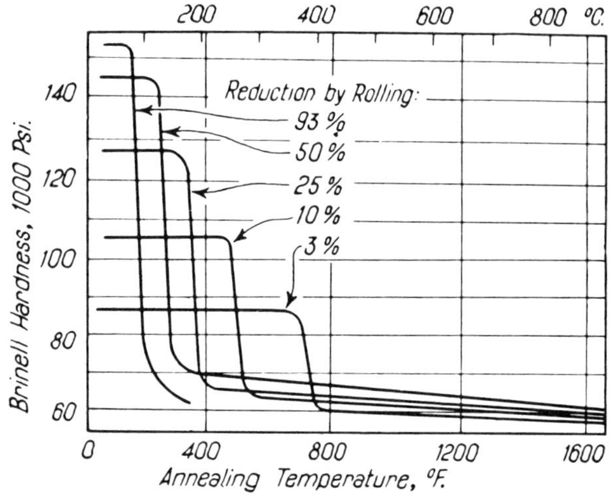

Fig. 1-31. Curves for copper showing the effect of the amount of the initial deformation of the annealing response. (From G. Sachs and K.R. Van Horn, Practical Metallurgy, p 143, American Society for Metals, Metals Park, Ohio, 1951)

Table 1-1. Variables Which Affect the Recrystallization Temperature

Variable	Effect on recrystallization temperature
Time at annealing temperature	Increasing time decreases temperature
Amount of plastic deformation by cold working	Increasing amount of cold work decreases temperature
Solute concentration in solid solution	Generally increases temperature; depends upon effect on melting temperature
Presence of second phases	Second-phase particles decrease temperature
Original grain size before cold working	Decreasing grain size decreases temperature
Rate of plastic deformation	Increasing rate decreases temperature
Temperature of cold working	Decreasing temperature of cold working decreases recrystallization temperature

1-8. An Example of the Use of Annealing Data

The use of annealing data is illustrated by an example. Suppose that a shipment of brass containing about 5% Zn (gilding metal) is to be used to manufacture a product. The initial thickness of the bar is 0.250 in. It is to be plastically deformed by rolling so that the final thickness is 0.062 in. In this condition the tensile strength must exceed 40,000 psi, the hardness must exceed 80 HRF, and the elongation at fracture must exceed 8%. The data in

Fig. 1-32 are available for this type of alloy. Cold rolling mills and annealing furnaces are the equipment available.

One possible procedure appears to be to cold roll the as-received bar to 0.062 in. However, the as-received condition was not specified; the material may already be cold rolled. The data in Fig. 1-32 do not exceed 70% reduction in thickness, probably because beyond this amount cracking begins to be a serious problem. Thus, it is best to limit the reduction to less than 70%. If the hardness of the bar is greater than around 40 HRF, then the material is certainly already cold rolled, and it would be wise to give the material an initial annealing to put it in the soft condition. Although the recrystallization curve in Fig. 1-32 is only for 50% reduction, it is clear that annealing for 1 h at about 800 °C will completely recrystallize the material even if the amount of cold work is small. Thus, once the material is softened, it is tempting to try to roll it to 0.062 in. However, this corresponds to a reduction in thickness of 75%, which is too high, and an intermediate step must be used.

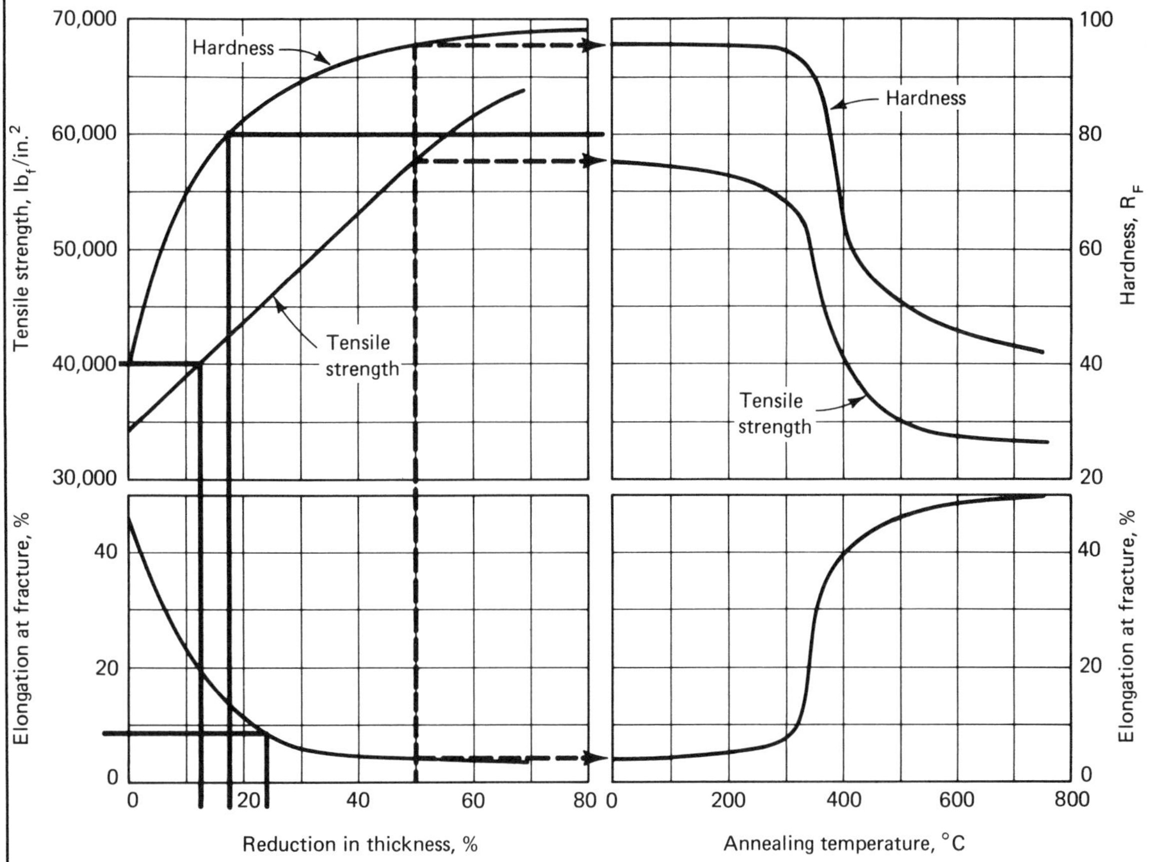

Fig. 1-32. Properties as a function of cold work and annealing typical for Cu-5Zn brass (gilding metal). Annealing data are for 50% reduction in thickness by rolling at 20 °C and 1-h annealing time. (Adapted from R. A. Wilkins and E. S. Bunn, Copper and Copper Base Alloys, McGraw-Hill, New York, 1943)

The intermediate step is chosen by examining how to obtain the desired properties in the 0.062 in. thick strip. Figure 1-32 is used to determine that the allowed final cold reduction must lie between about 18 and 24%. Let us choose 20%. Thus if the bar is processed so that the thickness is 0.078 in. and is in the annealed condition, then the *final* step will be to cold roll the brass to 0.062 in.

Then it appears that we want to take the as-received brass, anneal it, then cold roll it to 0.078 in., anneal it again, and then give it the final cold rolling to 0.062 in. However, we must ensure that going from 0.250 to 0.078 in. does not exceed 70% reduction. This is a 69% reduction. Since this is quite close to the maximum 70%, another intermediate step is suggested. It is arbitrarily chosen to be rolling to a thickness of 0.125 in.

Figure 1-33 shows the process chosen. It is best to look upon this as a preliminary decision, and run some material through the process to ensure that it is appropriate for this application. Instead of cold rolling and annealing, hot rolling could be used. The exact process used depends upon the equipment available and on economic factors.

1-9. Hot Working and Cold Working

We can now understand the difference between cold working and hot working. One way to distinguish them is to define cold working as plastic deformation such that strain (work) hardening occurs, and hot working as plastic deformation such that strain hardening does not occur. In hot working, strain hardening does not occur because the temperature of deformation (for the specific deformation process) is sufficiently high that as soon as plastic deformation commences, with dislocation generation, grain elon-

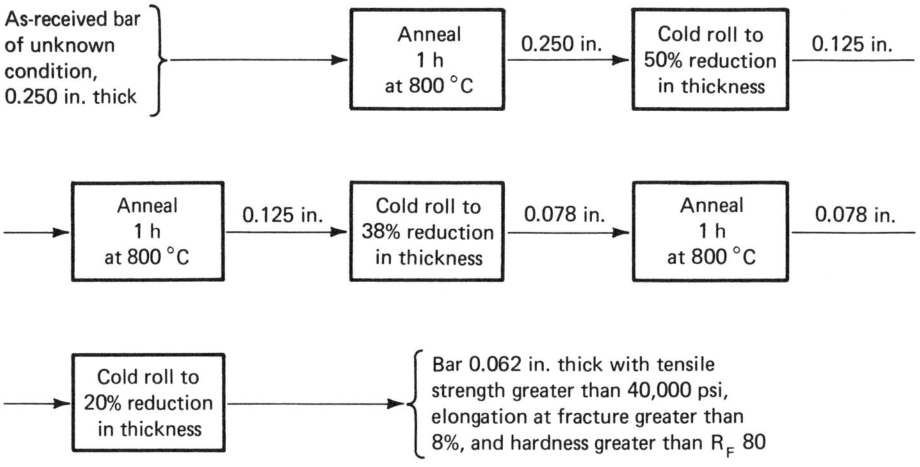

Fig. 1-33. Flow diagram of a process to reduce a Cu-5Zn brass from 0.250- to 0.062-in. thickness with specific properties.

gation, etc., recrystallization commences, and the material softens before it can appreciably harden. This is illustrated in Fig. 1-34.

Another way to distinguish hot and cold working is to define hot working as plastic deformation above the recrystallization temperature, and cold working as plastic deformation below the recrystallization temperature. Whether a process is hot or cold working depends upon the specific process, especially the amount of deformation (e.g., percentage reduction in thickness), the rate of deformation and the temperature.

1-10. Mechanisms of Annealing

Figure 1-35 is a schematic summary of the driving force and the mechanism associated with recovery, recrystallization and grain growth. The "driving force" is the thermodynamic energy difference between the plastically deformed state and the structure at each stage. In this section, the defect structure in the three stages of annealing is described, and the mechanism leading to this condition is considered. The approach taken is to give a qualitative description of the processes in order to give some insight into the mechanisms. Although these descriptions are, in some cases, simplified, they are designed to emphasize the importance of appreciating the atomic nature of the processes.

Recovery

First, we will examine some experiments which have considerably clarified the structural rearrangements that occur during recovery. Recall that in

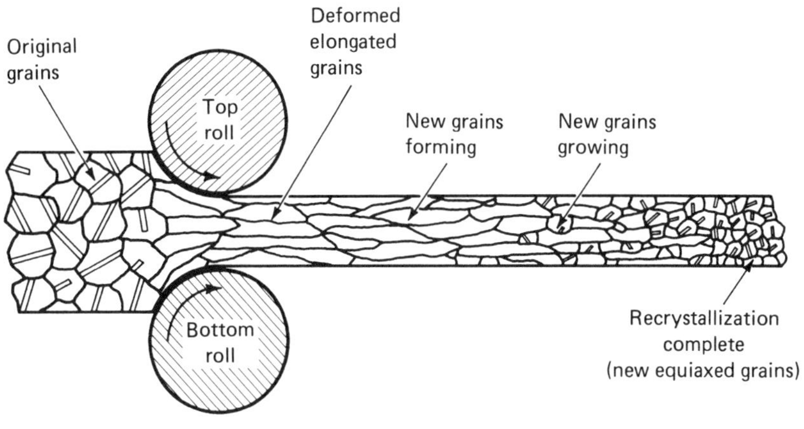

Fig. 1-34. Schematic illustration of the change in grain structure upon hot rolling. (From R. A. Grange, in Fundamentals of Deformation Processing, ed. E. A. Backofen, J. J. Burke, L. F. Coffin, N. T. Reed and V. Weiss, Syracuse Univ. Press, Syracuse, N. Y., 1964, as adapted from J. M. Camp and C. B. Francis, The Making, Shaping and Treating of Steel, 5th Ed., U. S. Steel Corp., Pittsburgh, 1940)

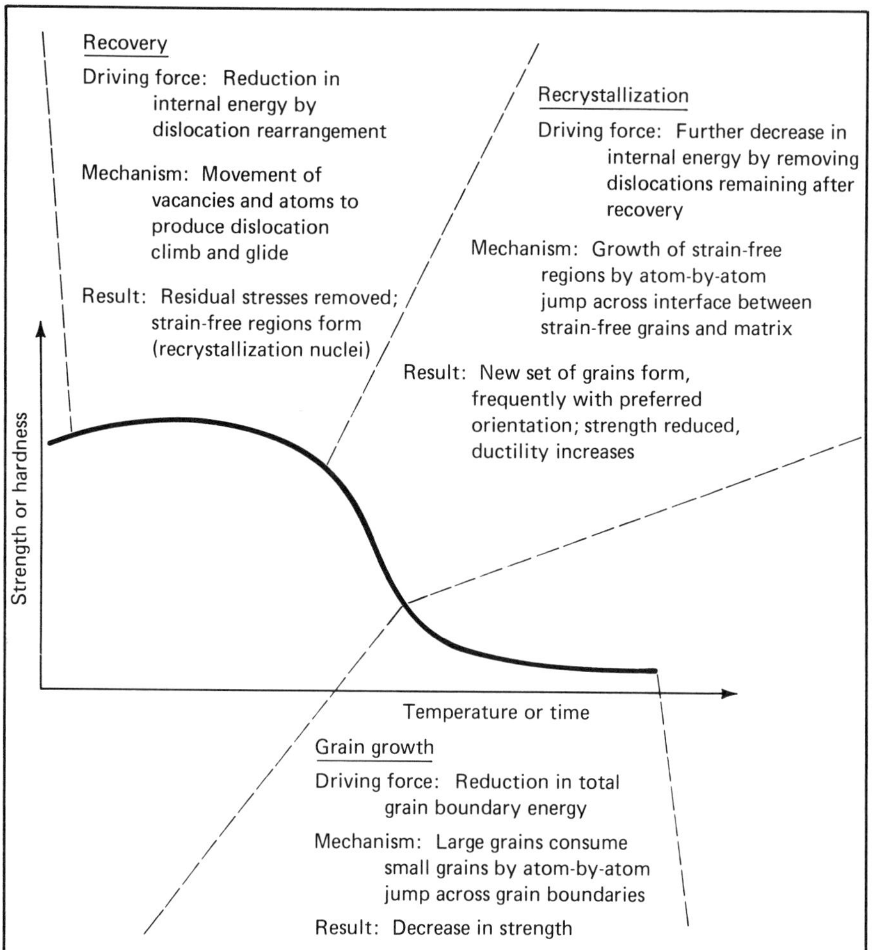

Fig. 1-35. Schematic summary of the driving force, mechanism and result for recovery, recrystallization and grain growth. (Adapted, courtesy of Prof. E. E. Stansbury)

crystals in which edge dislocations are arranged above one another, a misorientation exists between the two regions separated by the row of dislocations (Fig. 1-36a). (Such a row of dislocations is called a low-angle boundary, or a sub-boundary.) If the atomic spacing for the crystal and the angle of misorientation are known, the average separation distance between dislocation cores can be calculated. Now, if the surface on which the dislocations exit is placed in a suitable etching agent, etch pits will develop. Thus a row of etch pits is seen (Fig. 1-36b), and the model proposed for the dislocations can be substantiated by comparing the average pit separation distance to the calculated value. For such cases, the agreement is excellent. The point is that this is a method of revealing dislocation cores, especially if the dislocation density is not too high and the configuration is simple.

Now consider an iron – 3.25% silicon (solid solution alloy) single crystal, with the $(\bar{2}11)$ plane carefully polished and the slip plane $(01\bar{1})$ oriented as shown in Fig. 1-37(a). If the crystal is bent around an axis normal to the

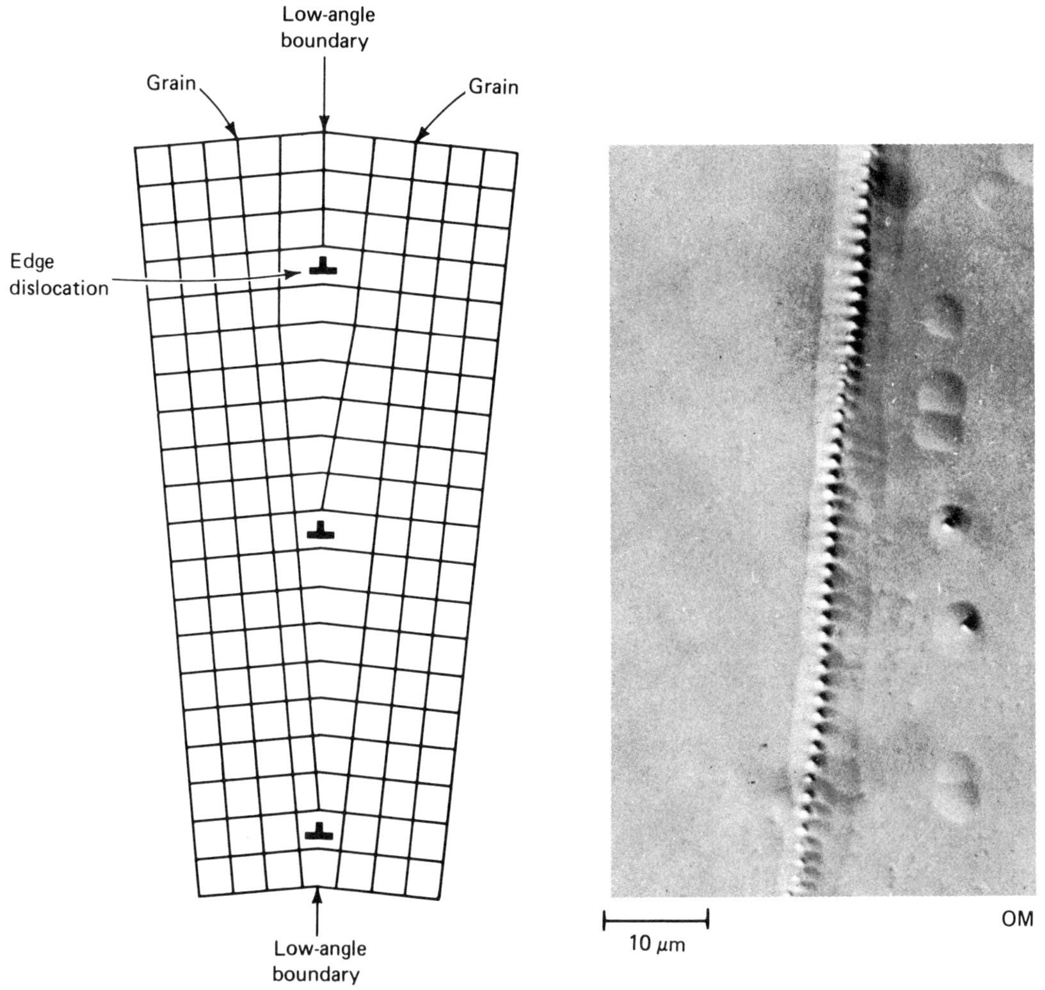

(a) (b)

Fig. 1-36. (a) Geometry of a row of edge dislocations, causing a misorientation between the two sections of the crystal. (b) Polished and etched surface of a germanium crystal revealing a sub-boundary by the row of etch pits associated with the dislocation cores; (from Metals Handbook, 8th Ed., Vol 8, American Society for Metals, Metals Park, Ohio, 1973)

plane of the paper, dislocations should be generated so that the cores lie along the $(01\bar{1})$ slip plane, as shown in Fig. 1-37(b), and should exist from the surface (where the shear stress is highest) to a position near the center at which the stress is not sufficient to propagate the dislocations. In the photomicrograph of the etched crystal shown in Fig. 1-37(c), the etch pits lie on (110) planes and do not extend into the center.

When the crystal is bent plastically, a shear stress gradient exists along the slip plane in the slip direction, being a maximum at the surface and vanishing at the neutral axis. The dislocations are present only in the region

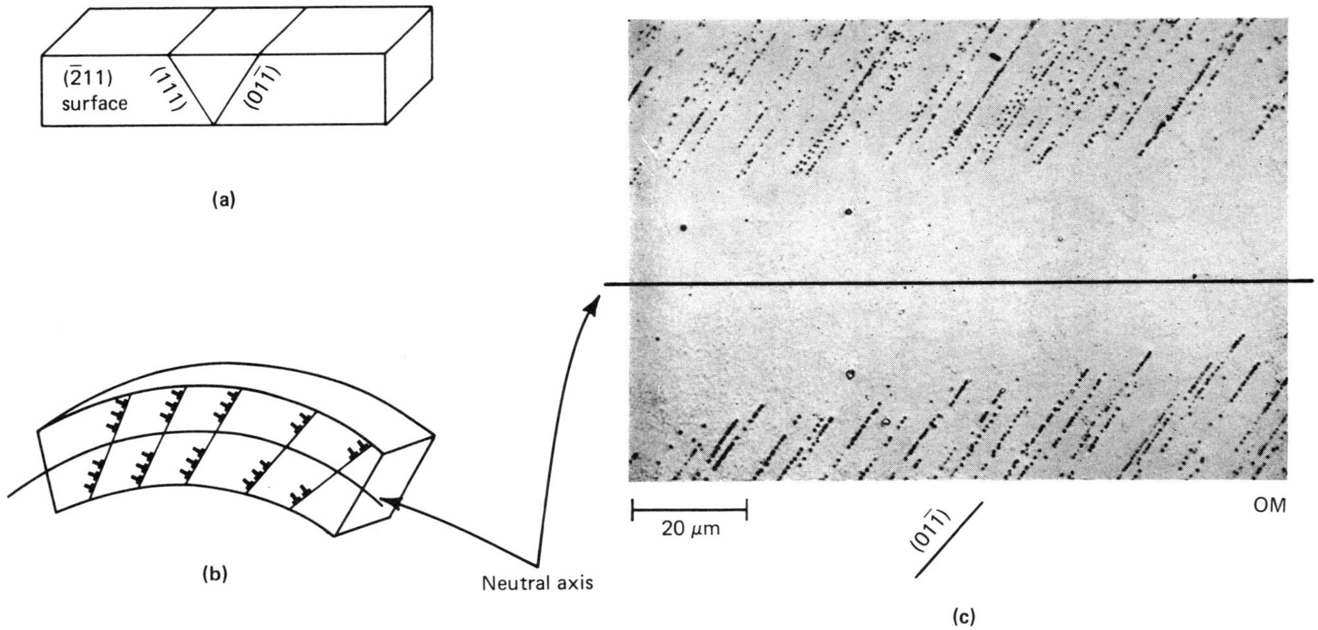

Fig. 1-37. (a) and (b) Geometry of Fe-3.25Si crystal bent plastically to induce dislocation movement on (01$\bar{1}$) plane. (c) Region near the neutral axis, where the dislocation cores are revealed by the etch pits. (Adapted from W. R. Hibbard, Jr., and C. G. Dunn, Acta Met., Vol 4, p 307, 1956)

where the critical resolved shear stress was exceeded. Upon removal of the external bending force, the dislocations remain in position, and their presence is the origin of the residual stress remaining in the plastically bent crystal.

To understand the behavior of the dislocations upon annealing, we must examine the interaction of the stress fields of dislocations as they come in the neighborhood of one another. The description given here is simplified, but it serves to illustrate the principles involved. Figure 1-38 shows two edge dislocations of the same sign. It is obvious that in the x-direction the lattice is in tension just below each dislocation, and in compression just above. Thus, if the two dislocations are on the same plane, the total elastic energy is reduced as they move apart. Heating to a sufficiently high temperature will allow this to occur by increasing lattice vibrations so that the repelling force of neighboring dislocations can overcome the local atom binding. This process, shown schematically in Fig. 1-39, is referred to as *thermal glide*. This mechanism is responsible for the change in the dislocation appearance upon annealing the bent iron-silicon crystals at 600 °C (Fig. 1-40a).

Now, as shown in Fig. 1-41, the energy of the crystal can be reduced also if the dislocations can arrange themselves one above the other, to form a low-angle boundary. Initially two dislocations exist on the same slip plane, at O and A. If, by a mechanism discussed below, the dislocation at A can *climb* to position B, then the energy of the combined stress fields can be lowered if the dislocation moves to C by thermal glide to align above the

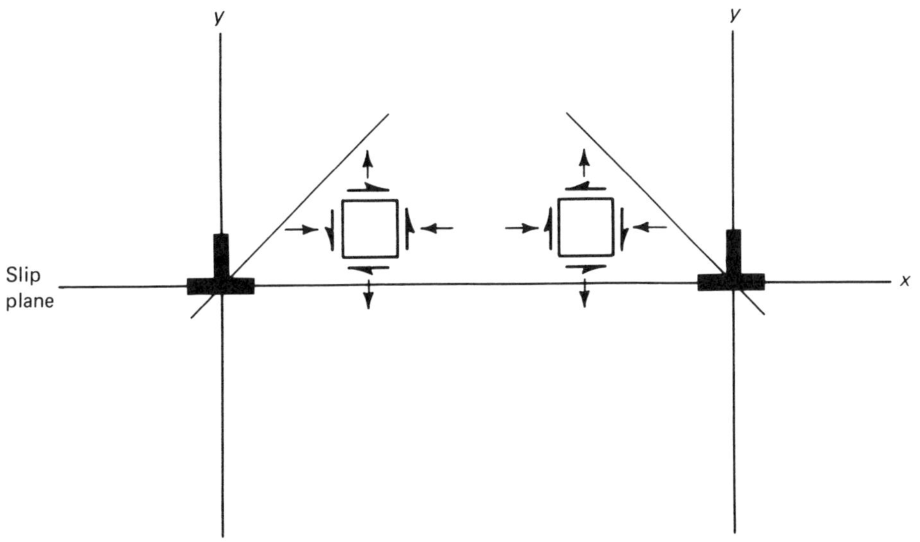

Fig. 1-38. Schematic illustration of the interaction of the stress fields of two dislocations of the same sign on the same slip plane. The stress field of the isolated dislocations is shown, where it is seen that as they approach each other the compressive stress in the x-direction increases, and thus they repel each other.

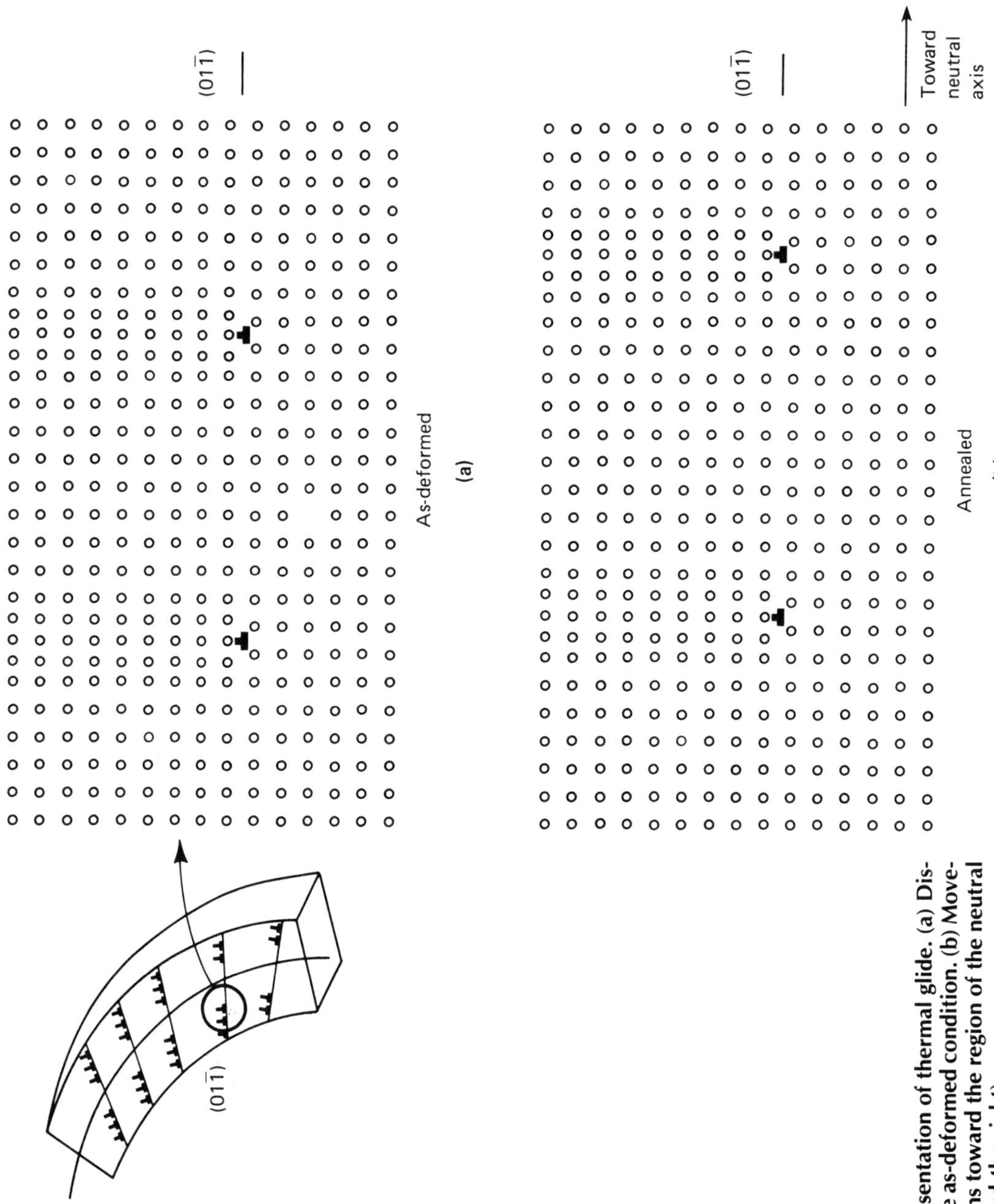

Fig. 1-39. Schematic representation of thermal glide. (a) Dislocation arrangement in the as-deformed condition. (b) Movement of the two dislocations toward the region of the neutral axis upon annealing (toward the right).

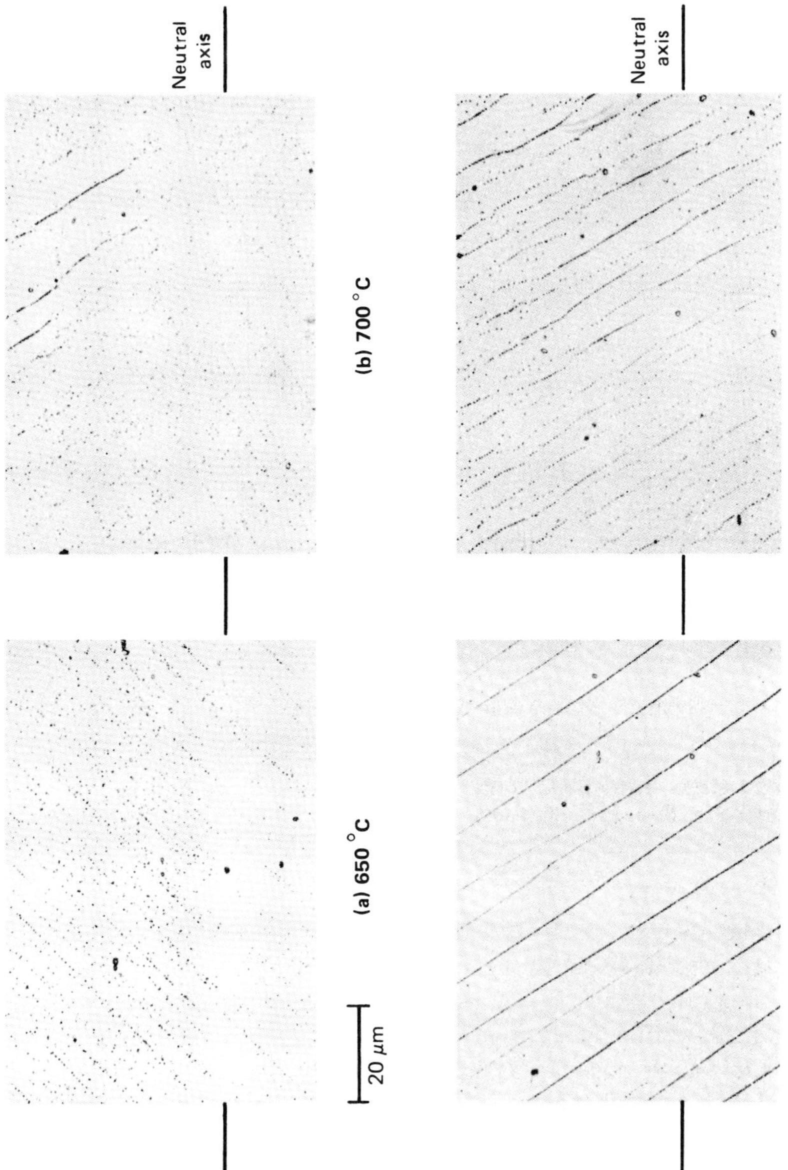

Fig. 1-40. Etched surface of bent Fe-Si crystals after annealing for 1 h at the temperatures indicated. Optical micrographs, at same magnification. (Adapted from same source, p 311 and 312, as Fig. 1-37)

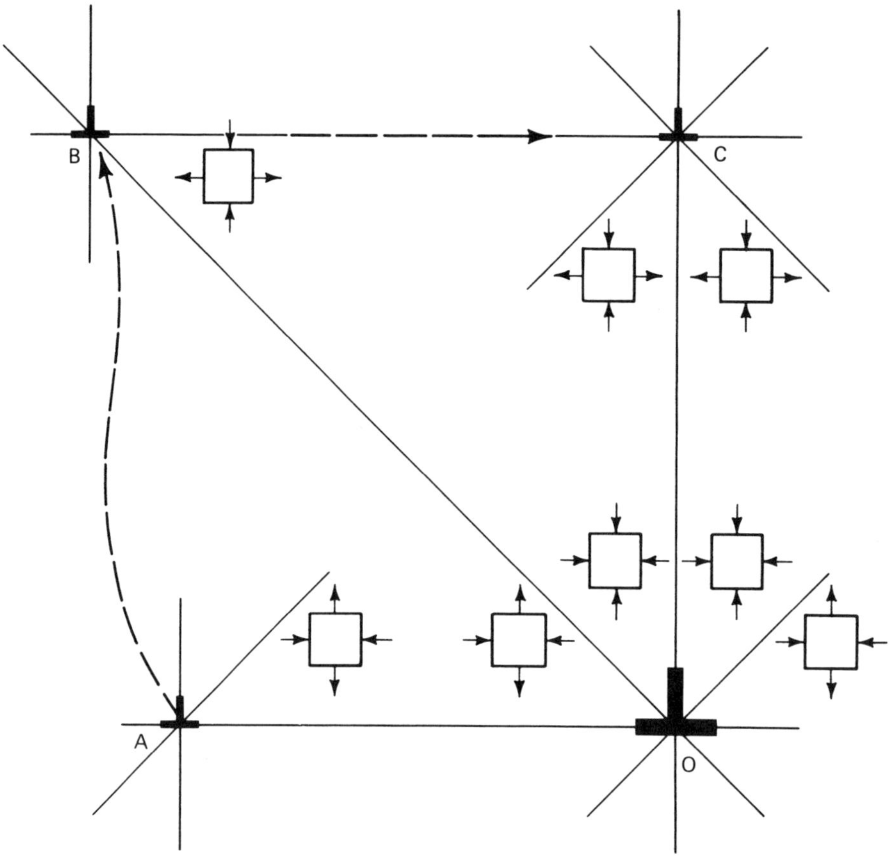

Fig. 1-41. Schematic illustration of the reduction in total energy associated with two dislocations of the same sign originally on the same slip plane, if one climbs and glides to locate above the other. In the localized stress distribution, only the normal stresses are shown. (Adapted after Prof. E. E. Stansbury)

other dislocation. It is seen that as the dislocation moves from B to C, the tensile component in the x-direction of the stress field tends to cancel the compressive component in the x-direction of the stationary dislocation at O, thus lowering the energy. If several dislocations are aligned in this manner, a low-angle boundary is formed. (See Fig. 1-36.) As shown in Fig. 1-40(b), (c), and (d), this is what occurs upon higher-temperature annealing of the bent iron-silicon single crystal.

The mechanism of climb involves the movement of the vacancies, some of which are present before deformation and others generated during deformation. Recall that a vacancy is not stationary. There is a finite probability, which increases rapidly with temperature, that a neighboring atom will vibrate so that it moves into the hole, and thus the vacancy moves. If a vacancy is near a dislocation, its movement is influenced by the stress field of the dislocation. For example, in Fig. 1-42(b) any of the four closest neighboring atoms could move into the hole, but the movement of "1" into the hole lowers the energy of the crystal by reducing the compressive stress

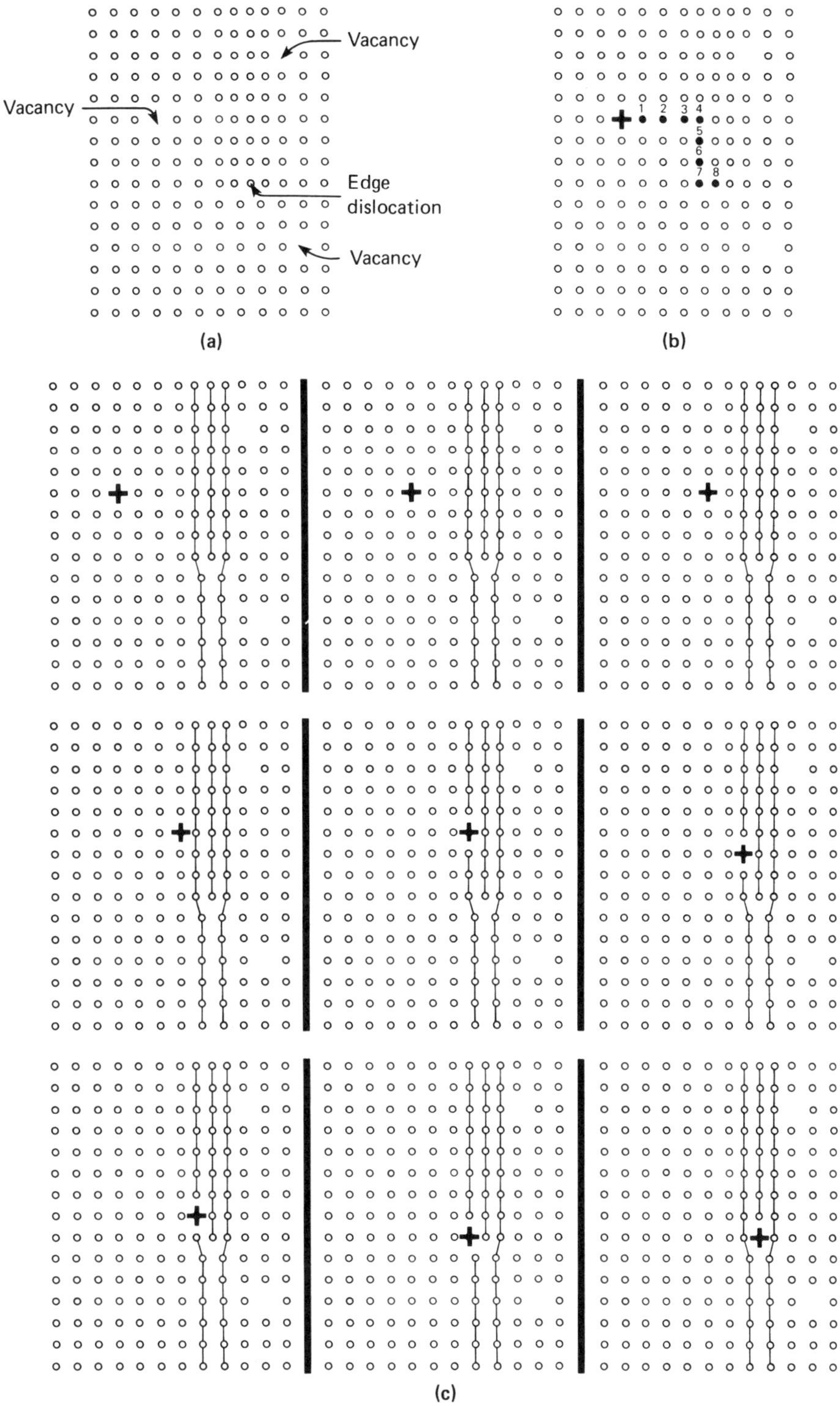

Fig. 1-42. Schematic illustration of movement of a vacancy to the core of a dislocation, causing the dislocation to climb by one atom position. (a) shows the location of the vacancy and the dislocation, and (b) shows the atoms that move into the vacancy (marked by +) as the vacancy moves to the dislocation core. The total sequence is shown in (c).

associated with the dislocation. Obviously, this reduction will continue as the vacancy moves toward the dislocation core, and the maximum reduction occurs when the vacancy takes the position which was occupied by the atom at "8." This movement is a rather statistical process, but the wandering of the vacancy is highly favored for the general path toward the dislocation core.

The vacancy allows the dislocation to move up one atom position in the plane of the picture in Fig. 1-41. For significant climb to occur, of course, there must be a considerable vacancy flux. These excess vacancies are generated during the plastic deformation, as discussed earlier.

In polycrystalline metals deformed under complex loading, the dislocation structure is a complicated network of dislocations. Thus, the change in this structure during recovery is more complex than that described above. However, the mechanisms of glide and climb operate, and the dislocations move into a network of sub-boundaries. This increases considerably the mean free path of the electrons, and the electrical resistivity decreases. However, these networks are sufficiently stable, and the subgrains sufficiently small, that dislocation movement and multiplication are difficult, and the strength and hardness remain constant during recovery or increase slightly (see Fig. 1-29). Figure 1-43 shows the dislocation structure of high-purity aluminum deformed at 20 °C, with dislocation tangles which are developing into cells. Also shown is the dislocation structure developed after annealing for 100 h at 120 °C. This is a typical appearance of a recovered structure in a cold worked metal.

Recrystallization

Although during recovery the energy of the material decreases, it is still relatively high because of the dislocation network of the sub-boundaries. This stored energy is the "driving force" for recrystallization (more specifically called primary recrystallization). The atomic mechanism whereby these new, strain-free crystals form has not been clearly established. Either a small group of atoms rearrange themselves into a strain-free crystal, or a small, dislocation-free region, existing among the dislocations, serves as the nucleus for a recrystallized grain. Whatever the nucleus, it must be clearly recognized that the growth of the crystal involves the slight movement of the atoms on the matrix side of the crystal-matrix interface. With this movement they attach themselves to the crystal and thus the crystal grows.

The beginning of recrystallization is illustrated schematically in Fig. 1-44. The nucleus is assumed to form in the vicinity of a grain boundary and two dislocations (Fig. 1-44a and b). A group of atoms rearrange themselves by a slight movement of each, forming a nucleus (Fig. 1-44c and d) free of dislocations and having a different orientation from that of either original grain. Then this nucleus grows by the slight movement of atoms along the matrix-grain interface (Fig. 1-44e and f).

Figure 1-27 shows a recrystallized grain in the early stages of formation in the deformed matrix. However, evidence of recovery is difficult to see in this micrograph, as only the polished and etched surface is shown. Figure

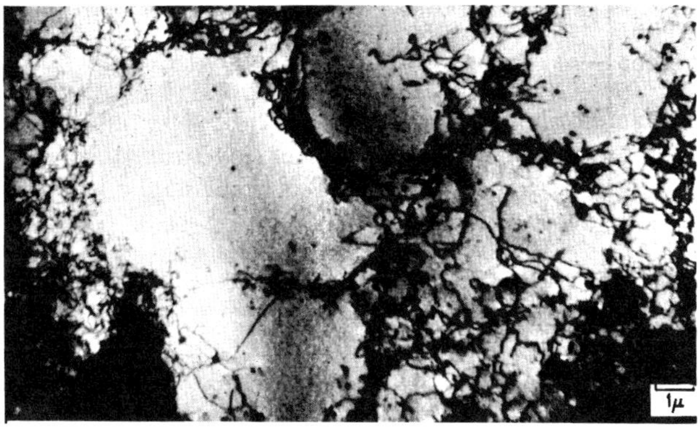

(a) Strained 10% in tension at 20 °C

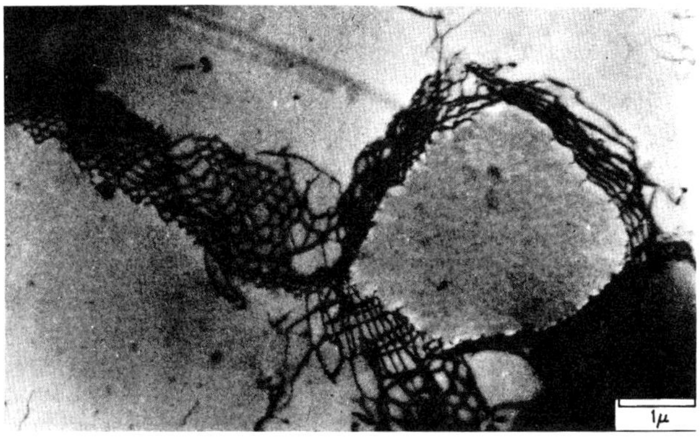

(b) Strained 10% in tension at 20 °C
then annealed for 100 h at 120 °C

Fig. 1-43. Transmission electron micrographs showing the dislocation structure of 99.995% Al after straining 10% in tension, then after annealing at 120 °C for 100 h. (Adapted from J. L. Lytton, K. H. Westmacott and L. C. Potter, Trans. AIME, Vol 233, p 1760 and 1762, 1965)

1-45(b) is a transmission electron micrograph in which a region containing several recrystallized grains, and the matrix into which these are growing, are shown. The matrix has already undergone recovery, as evidenced by the small, subgrain structure. Note that the dislocation density of the recovered region is considerably higher than that of the recrystallized grains.

It is during this early stage of growth that annealing twins form in FCC metals (hence the term annealing twins). The normal stacking sequence on the (111) planes is ABCABC, referring to the location of subsequent atoms over the holes in the layer below. When the third layer is placed over the first layer (ABABAB packing), the HCP structure is generated; if the third layer is placed in the other location, the (ABCABC) stacking of the FCC structure is generated. However, as the interface of the recrystallized grain moves, the probability is rather high that the ABCABC stacking sequence will be inter-

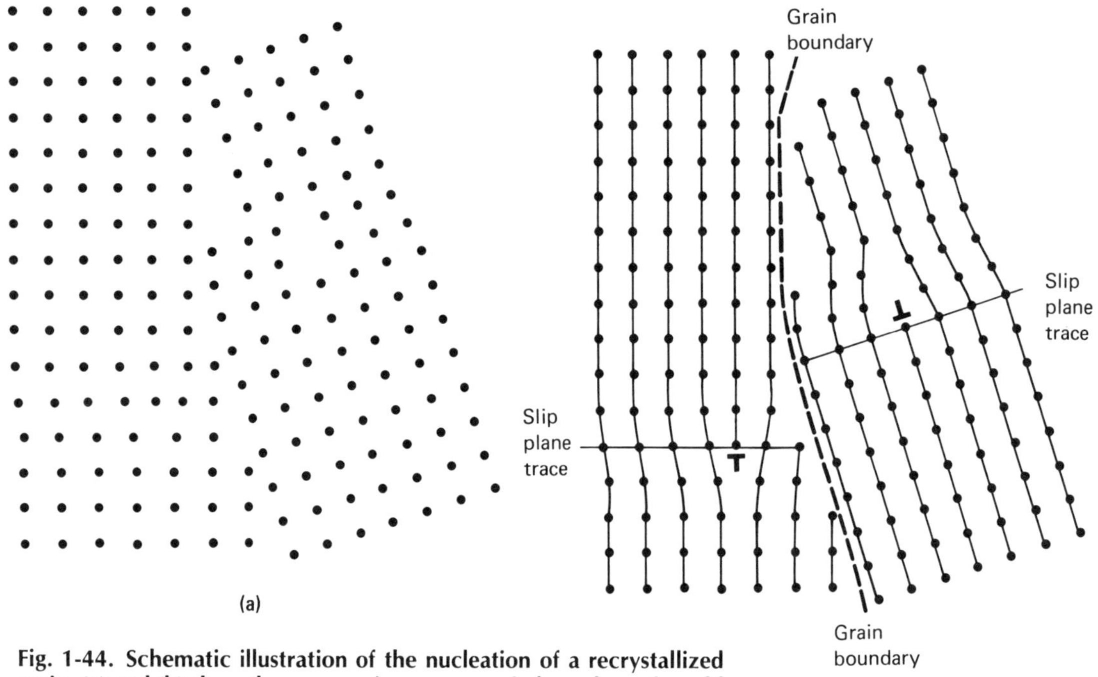

Fig. 1-44. Schematic illustration of the nucleation of a recrystallized grain. (a) and (b) show the same region near a grain boundary of a cold worked material, with the grain boundary and two dislocations marked in (b). In (c), the atoms marked by arrows move to new positions indicated by open circles. This gives the array outlined in (d), a recrystallized grain. In (e), further atom rearrangement at the matrix-grain interface is shown by the arrows, the atoms moving to the new positions indicated by the open circles. The new crystal size is outlined in (f).

rupted, to give ABCABABC, generating a twin orientation. Thus, it is common to see annealing twins form when the recrystallized grains are quite small (Fig. 1-27 and 1-45b). (Annealing twins are described in more detail in Section 1-13.)

The process of recrystallization involves the nucleation of the new crystals, and their growth. Thus, the rate of recrystallization is controlled by the rate of nucleation N and the rate of growth G. Assuming that the growth rate and the nucleation rate are constant, that nucleation is random and that the grains are spheres, and taking into account the problem when growing crystals impinge, the following expression for the fraction x of the matrix recrystallized with time t is obtained:

$$x = 1 - \exp\left(-\frac{\pi G^3 N t^4}{3}\right)$$

Figure 1-46 shows curves for the recrystallization data for aluminum which were used to determine a value for N and G to give the calculated curve shown. It is seen that the equation gives the general correct shape; the theory

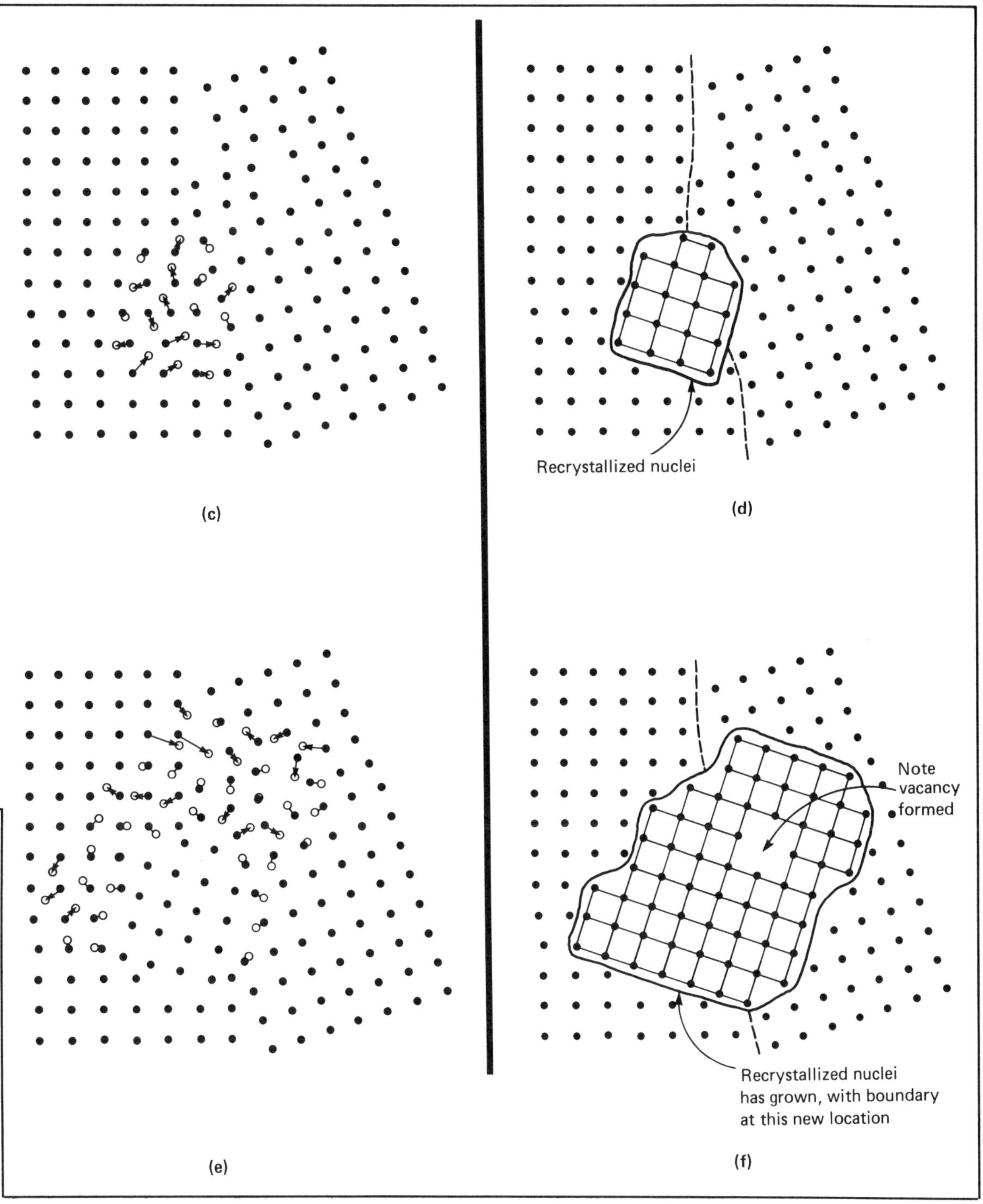

can be modified to give a closer fit to the data. Both N and G are temperature dependent, being related to temperature by the form $e^{-(Q/RT)}$, where Q is an activation energy for either growth or nucleation, R is the gas constant and T is the absolute temperature. This exponential behavior makes the recrystallization-time behavior very sensitive to the temperature.

Fig. 1-45. Transmission electron micrographs of the structure of a Ni-20Cr solid solution alloy. (a) Cold worked structure after about 90% reduction in area by swaging. (b) Annealed structure after 1500 h at 500 °C, showing recrystallized grains and recovered structure. (Courtesy of R. Gaylord)

Grain Growth

Upon completion of recrystallization the structure consists of a polycrystalline aggregate of equiaxed grains. The grain boundary regions constitute regions of atomic disorder, and hence have higher energy than the bulk of the crystals. Thus, the energy of the system will be reduced by grain growth, which removes some of the high-energy grain boundary area. This grain growth is illustrated in Fig. 1-26. (Indeed, it is quite important to recognize that grain growth will occur whenever a polycrystalline structure is at a sufficiently high temperature; its occurrence does not depend upon prior plastic deformation.)

The atomic mechanism of grain growth is that of the atoms in the boundary taking up the specific arrangement of the atoms in one crystal, thus moving the boundary away from this crystal and into the other. This is illustrated in Fig. 1-47. Here, the boundary moves toward its center of curvature because there is a greater net attraction on the atoms which are bounded by the grain toward the center of the curvature. Note that there is a tendency for a curved boundary to straighten. At the same time, surface energy considerations in terms of surface tension show that the included angle will tend toward 120°. Thus, if the grains have straight boundaries in

(b)

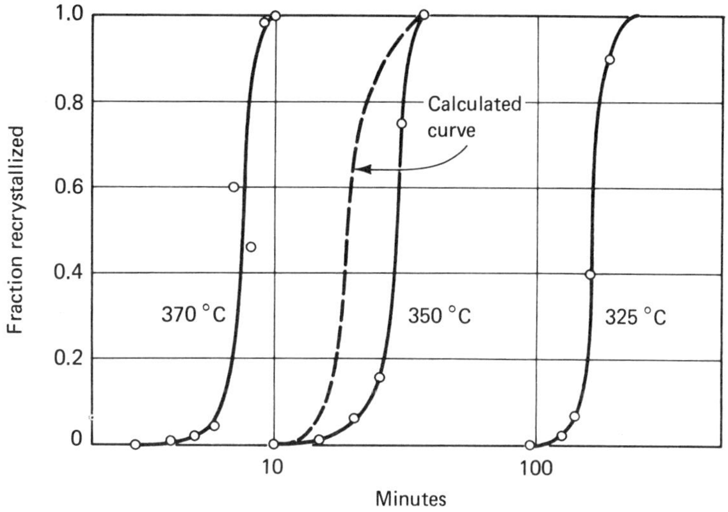

Fig. 1-46. Recrystallization curves for aluminum elongated 15%. The dashed curve is a calculated curve based on a nucleation and growth model. (Adapted from W. A. Anderson and R. F. Mehl, Trans. AIME, Vol 161, p 155, 1945)

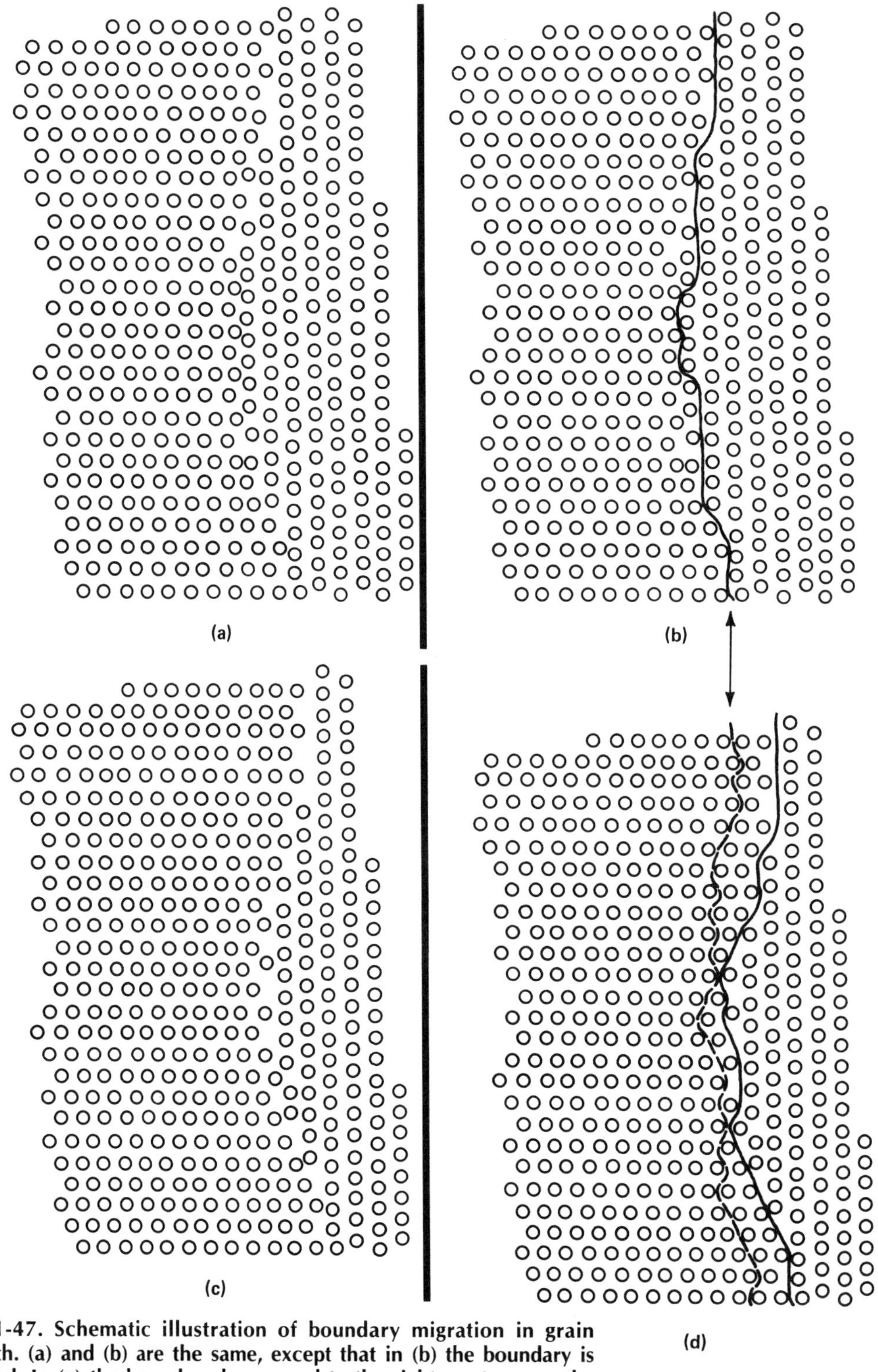

Fig. 1-47. Schematic illustration of boundary migration in grain growth. (a) and (b) are the same, except that in (b) the boundary is marked. In (c) the boundary has moved to the right as atoms on the right of the boundary have taken up the orientation of the atoms of the grain on the left of the boundary. Thus, the boundary moves toward the right, and the grain on the left increases in size. In (d), the original boundary is shown dashed.

the vicinity of the grain junctions, transfer across the boundary will occur as the boundary attempts to attain 120°. This causes the boundary to curve, leading to boundary movements toward the center of curvature. This changes the angle from 120°, and the process repeats. Thus, this process will continue until the boundary impinges another boundary. The direction of movement depends upon whether the initial angle is greater or less than 120°.

It has been shown from many detailed studies that grain growth is by its nature a complex statistical process. This ensures that the process will be much slower than when the grains are growing into a deformed matrix, and it is consistent with the fact that energy reduction upon grain growth is considerably smaller than that which occurs during recrystallization. Thus grain growth is a relatively slow process, becoming slower as the grain size increases.

1-11. Abnormal Grain Growth

Under certain conditions, the grain size after annealing can become very large, much larger than that normally encountered at the end of the grain growth region discussed previously. There are generally two causes of the abnormal grain size, one associated with a small, critical amount of strain before annealing, and the other with the continued annealing of a structure of fine uniform grains.

As an external load is applied (e.g., by rolling) to a polycrystalline metal, those crystals which are oriented so that the critical resolved shear stress is exceeded will begin to slip, with an increase in dislocation density. If the load is not too great, the critical resolved shear stress will not be exceeded in those crystals not favorably oriented, and these will have a low dislocation density. For this small amount of deformation the metal will consist mostly of grains with some plastic deformation, and a few grains with very little deformation. These few crystals are much more stable than the surrounding plastically deformed grains, and hence their boundaries will move into the deformed grains, giving rise to a very large annealed grain size. (Note that nucleation of strain-free grains is not involved.) This process is illustrated in Fig. 1-48. Here one large grain is expanding and consuming the slightly cold worked grains.

Formation of abnormal grains by this mechanism requires a critical deformation, as illustrated in Fig. 1-49 and 1-50. For example, for the aluminum annealed at 630 °C (Fig. 1-50b), a strain of about 6% will yield grains about 10 mm in diameter. For strains much larger than this, there is not sufficient energy difference between the grains for boundary migration to occur, and instead recrystallization occurs, leading to a small grain size. Note that for strains from about 20 to 55%, the grain size is about 0.3 mm. In this range, the grain size actually decreases as the strain increases; on the scale shown the effect is slight, which emphasizes just how large the abnormal grains are.

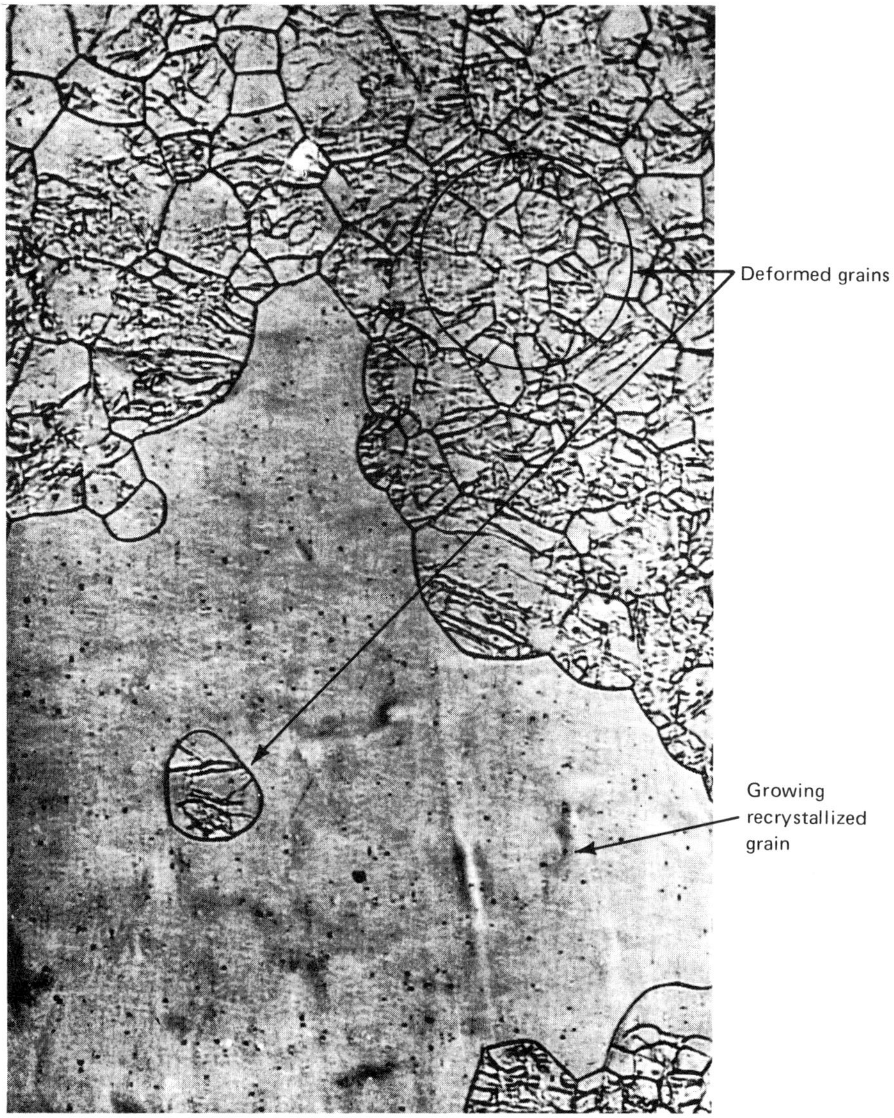

Fig. 1-48. Microstructure of an Al-Zn solid solution alloy, strained and annealed at 450 °C. Optical micrograph; 50×. (Adapted from P. Lacombe and A. Berghezan, Metaux et Corrosion, Vol 25, 1949)

Abnormal grain growth by the mechanism described above is caused by a critical-strain anneal and is sometimes called germinative grain growth.

Abnormal grain size can develop in already recrystallized structures under certain conditions. This is referred to as secondary recrystallization, whereas the development of the strain-free grain in the deformed matrix described in the previous sections is called primary recrystallization. Figure 1-50 illustrates grain growth due to secondary recrystallization. For example, if the aluminum is deformed 80% and then annealed at 630 °C, during the annealing time the deformed material will undergo recovery, then

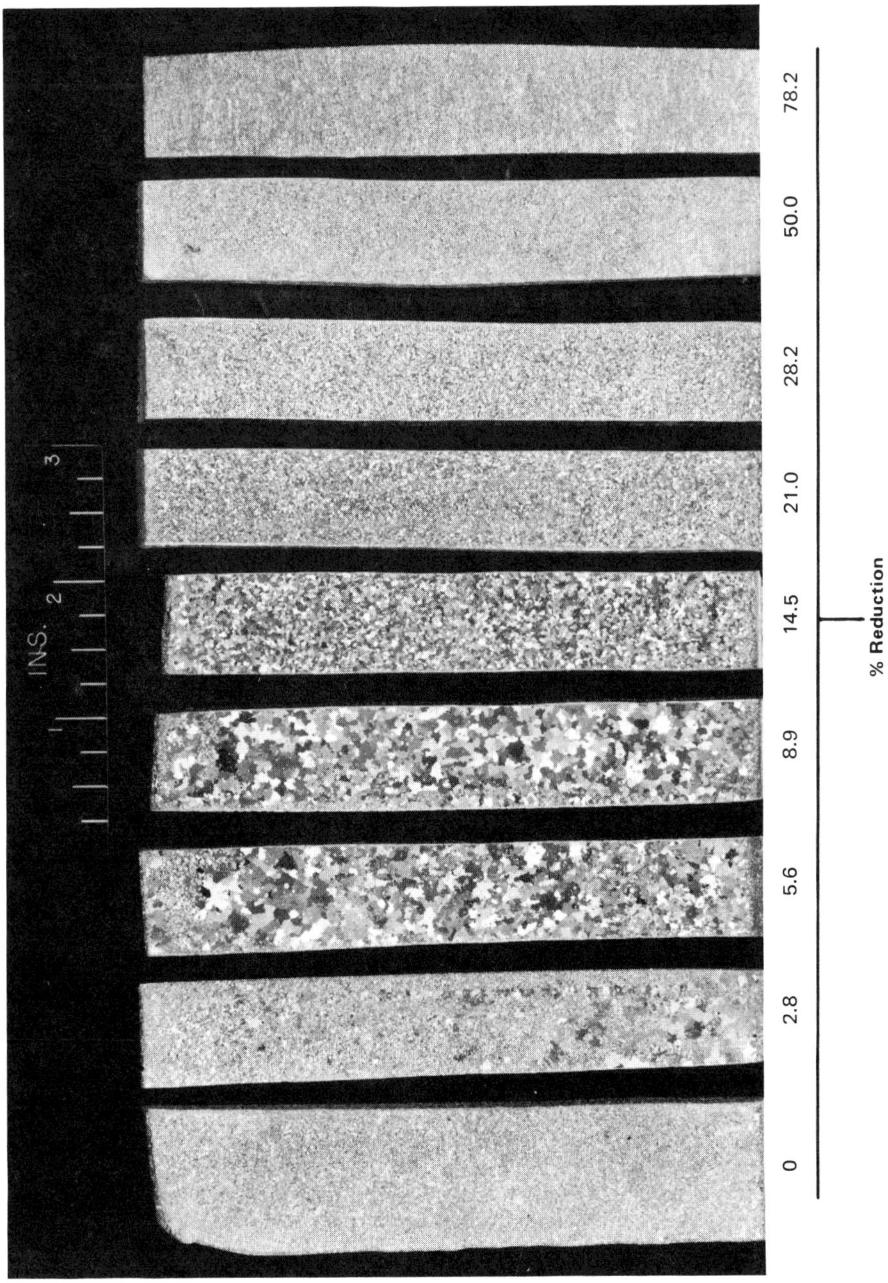

Fig. 1-49. Macrostructure of pure aluminum sheet showing the grain size after reduction in thickness for the amounts shown, then annealed. Abnormal grain growth is most prominent between 3 and 10% reduction. Above 50% reduction, the structure has undergone primary recrystallization to establish a small grain size. (Courtesy of Prof. E. E. Stansbury and Dr. L. K. Jetter, Oak Ridge National Laboratory)

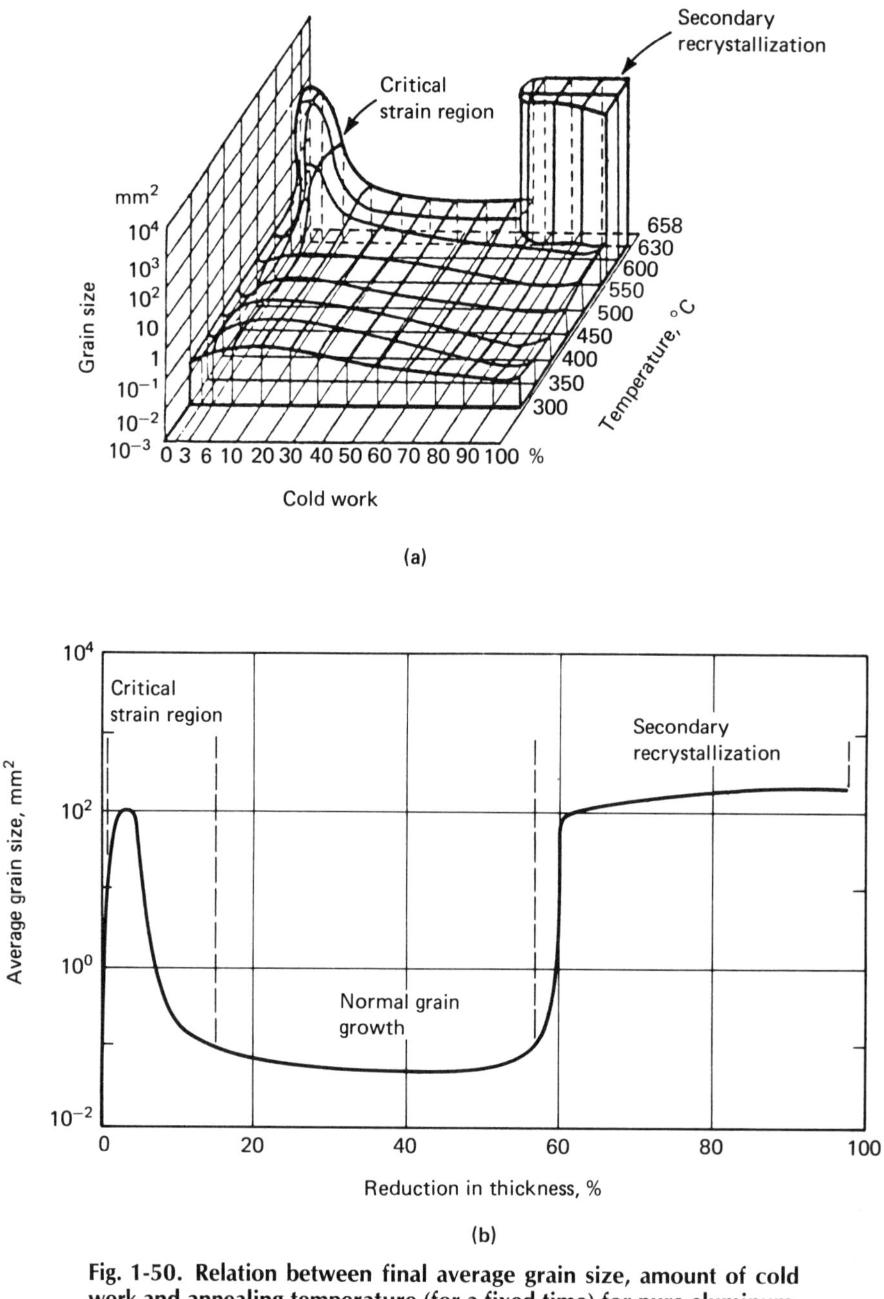

Fig. 1-50. Relation between final average grain size, amount of cold work and annealing temperature (for a fixed time) for pure aluminum. The curve in (b) is from (a) for an annealing temperature of 630 °C. (Adapted from M. Van Lancker, The Metallurgy of Aluminum Alloys, Wiley, New York, 1967, taken from Congre's de l'Al, p 258, 1954)

primary recrystallization to a uniform, fine grain size, then some normal grain growth to a grain size of about 0.3 mm, and then secondary recrystallization to develop abnormal grains about 10 mm in diameter.

Although the driving force for secondary recrystallization is the reduction in grain boundary area, it is not clear what favors those grains which grow at the expense of the others. It is known that a well-defined minimum temperature must be exceeded, and that abnormal grain formation is favored by the presence of grain growth inhibitors, such as insoluble particles. Those grains in the polycrystalline material which do grow are appreciably larger than the average. Also, there are observed orientation effects involved (see Section 1-12). All these factors suggest that the mechanism of secondary recrystallization is very subtle.

1-12. Annealing Textures

Just as preferred orientation develops upon cold working, it is also frequently present after annealing, but usually not the same texture. The causes of the texture development are still being examined. Here we will illustrate a texture and its implication.

If certain FCC metals and alloys are heavily cold rolled, then annealed at high temperature, a highly developed preferred orientation develops, referred to as "cube texture." The name refers to the fact that a high percentage of grains have their [001] direction parallel to the rolling direction and their (100) plane parallel to the rolling plane. Figure 1-51 shows a pole

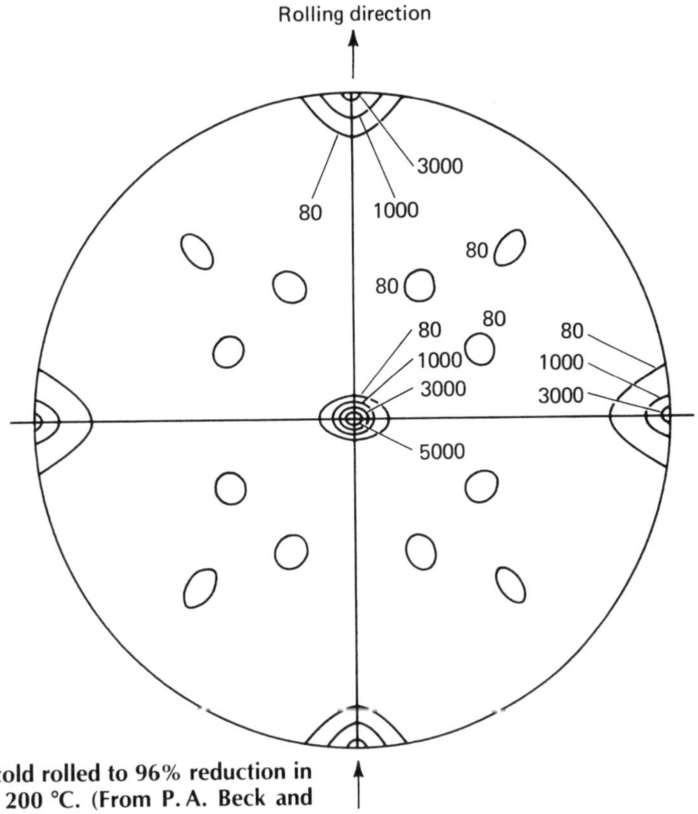

Fig. 1-51. (100) pole figure of copper cold rolled to 96% reduction in thickness, then annealed for 5 min at 200 °C. (From P. A. Beck and H. Hu, J. Metals, Vol 4, p 83, 1952)

figure of copper cold rolled to a 96% reduction in thickness and then annealed for 5 min at 200 °C. Figure 1-52 is a graph of the effect of the amount of cold work and of the annealing temperature on the degree of cube-texture development.

Such a strong texture clearly will make the material quite anisotropic. Figure 1-53 shows mechanical properties of sheet material as a function of the direction, relative to rolling direction, along which the tensile samples were cut. The properties are very sensitive to direction. The practical result of this dependency is seen in Fig. 1-54, which shows cups deep drawn from a sheet of copper-nickel solid solution alloy with a strong cube texture. The unevenness of the rim of some of the cups (referred to as *earing*) is the result of higher ductility in certain directions, and is of course undesirable if an even edge is required. Note that the material annealed at the lowest temperature, so that only recovery has occurred, shows earing; this is due to the original deformation texture. The ears do not occur at the same angle to the rolling direction as those which develop upon the high-temperature annealing, because the texture is different. An intermediate anneal (e.g., 525 °C) minimizes the development of a strong texture and consequently minimizes the earing.

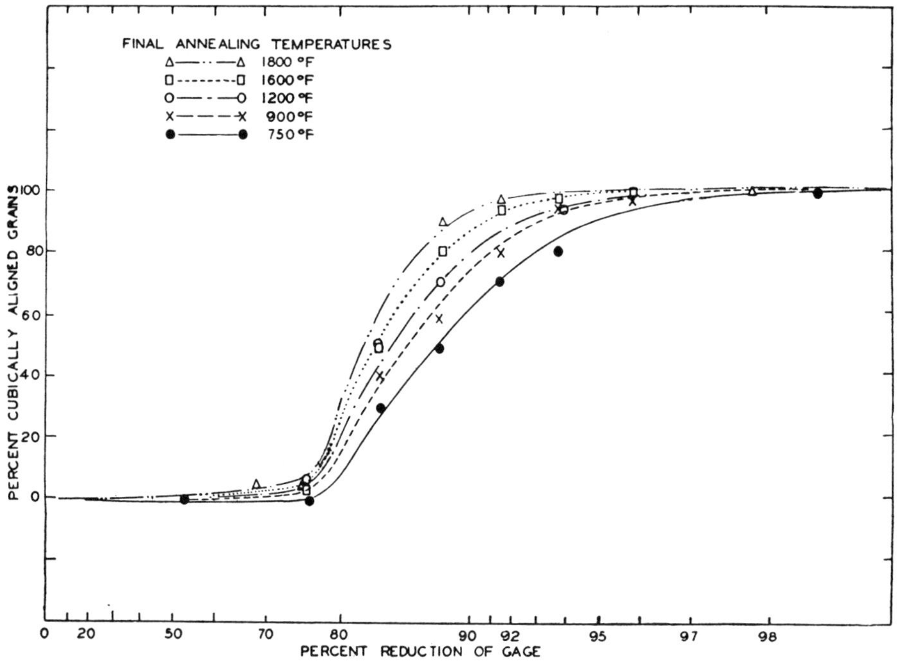

Fig. 1-52. Effect of amount of cold work and annealing temperature on the degree of cube-texture development in copper. (From W. M. Baldwin, Trans. AIME, Vol 166, p 599, 1946)

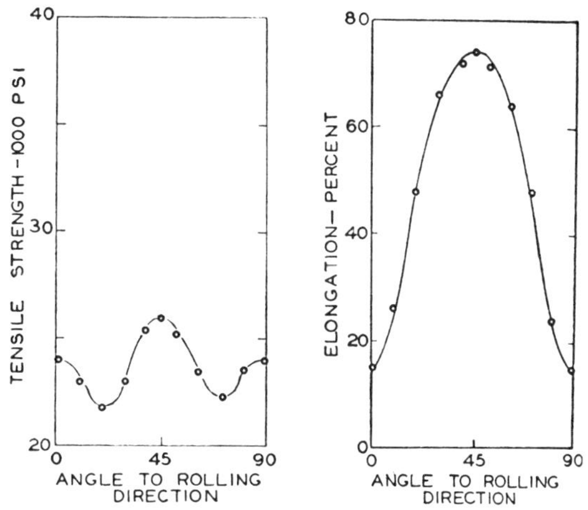

Fig. 1-53. Dependency of mechanical properties on the direction of testing of copper sheet with a strong cube texture. (From same source, p 606, as Fig. 1-52)

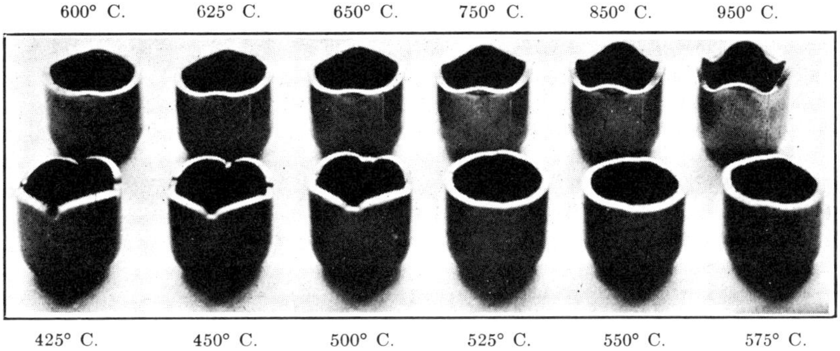

Fig. 1-54. Illustration of the development of "ears" during deep drawing of sheet with a strong cube texture. The alloy was cupronickel, a Ni-Cu solid solution alloy. (From W. H. Bassett and J. C. Bradley, Trans. AIME, Vol 104, p 181, 1933)

1-13. Annealing Twins

The annealing twins commonly seen in the microstructure have been referred to several times in this chapter. In this section they are defined in more detail.

We first consider the stacking of close-packed planes. In Fig. 1-55(a) are shown identical arrangements to obtain a close-packed plane. To build up a crystal in three dimensions, we need to add spheres or atoms onto this plane. For the second layer, we get the arrangements shown in Fig 1-55(b). When the third layer is added, however, we have two choices. If we choose the sites on the left in Fig. 1-55(c), the atoms will be over holes in the first layer; if we choose the other possible sites, shown on the right in Fig. 1-55(b), the atoms will be directly over those in the first layer. Clearly

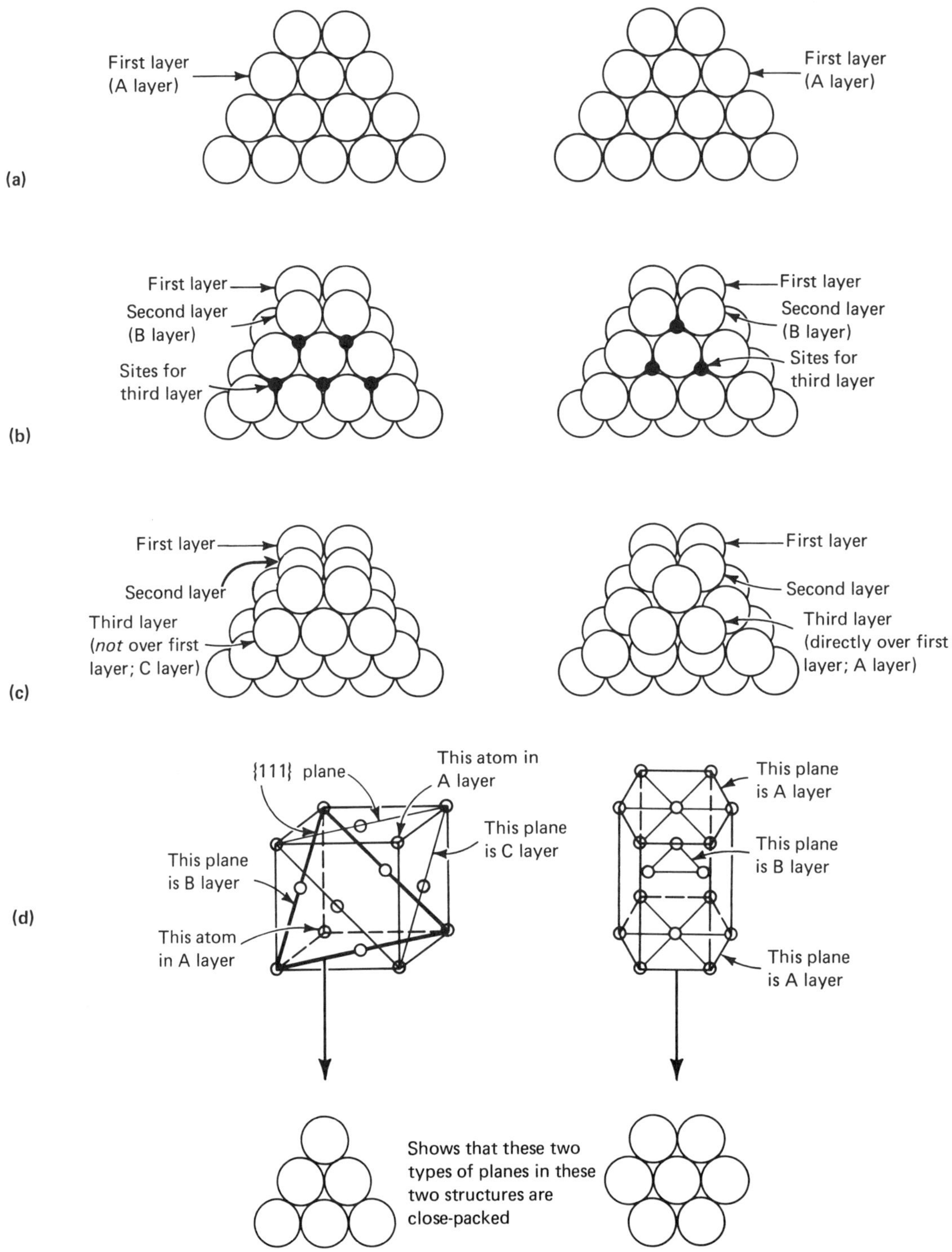

Fig. 1-55. Illustration of the generation of either FCC or HCP structure, depending upon the location of the third layer of close-packed atoms.

the two arrangements are not identical, although both are close-packed structures.

We designate the stacking sequence by denoting the first layer A, the second B, and the third, if the atoms are over holes in the first (A) layer, as C. Thus, in the arrangement on the left in Fig. 1-55(c), the sequence is ABCABC . . . and the sequence in the arrangement on the right is ABABAB. . . . The difference in stacking sequence generates two separate crystal structures, FCC and HCP (Fig. 1-54d). For the former, the close-packed planes are of the {111} type, and for the latter they are the basal plane.

If, in building up the FCC arrangement shown in Fig. 1-55, we had, upon adding a plane, located the atoms in the incorrect position and then repeated the correct sequence for successive layers, we would have generated the arrangement shown in Fig. 1-56. This is visualized by using a {110} type plane, with the atoms in it numbered 1, 2, 3, 4, 5 and 6. It is redrawn in the plane of the paper (Fig. 1-56b) so that one set of {111} planes is perpendicular to the plane. These planes have the sequence ABCA. However, if a stacking error is made, and then the correct sequence resumed, the arrangement of atoms outlined by 1′. 2′, 3′, 4′, 5′ and 6 is obtained. Note that the {111} planes now are mirror images of the left-hand part of the

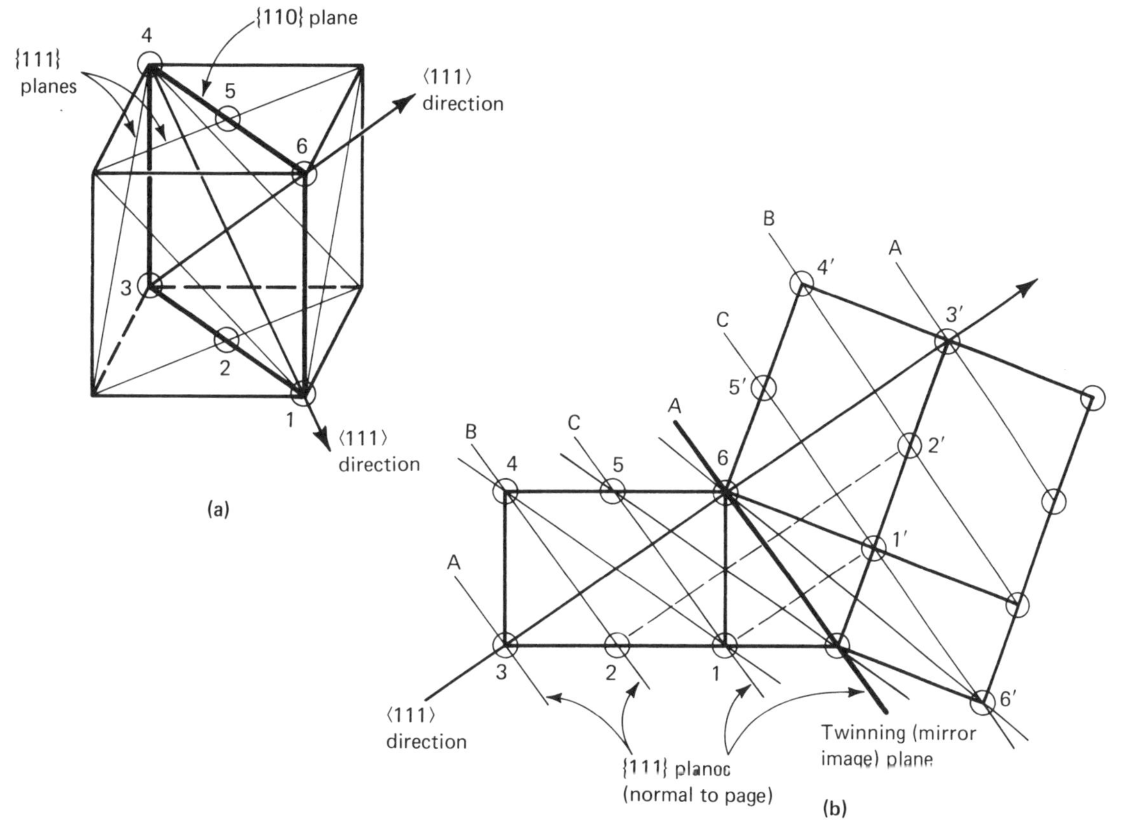

Fig. 1-56. Illustration of the formation of a twin by a stacking error.

arrangement. For example, atom 1 is the same perpendicular distance from the mirror image plane as 1′, 2 as 2′, etc. Thus, the left-hand part of the crystal is considered to be a twin of the right-hand part, and the mirror plane is called the twinning plane. The {111} plane stacking sequence is not ABCABCABC but ABCACBA. . . . In a sense, the crystal has a small HCP region between the FCC regions.

The presence of such a twin causes the left-hand part of the crystal to have a different orientation from that of the right-hand part, even though both share perfectly a common twinning plane. This is seen by noting the two ⟨111⟩ type directions in the {110} plane outlined by atoms 123456. The direction through the atoms 36 continues as a ⟨111⟩ direction in the twinned crystal (on the right, Fig. 1-56b) through atoms 63′. However, the other ⟨111⟩ direction, which passes through atoms 14 (see Fig. 1-56a), does not contain the same linear density in the twinned crystal (along 66′). Thus, the orientation of the two is indeed not identical.

Figure 1-57 demonstrates how twins are revealed in the microstructure. If a polycrystalline sample with grains containing twins (Fig. 1-57a) is cut for preparation for polishing, etching and metallographic observation, the plane of cut will pass through many twins (Fig. 1-57b and c). Thus, the probability is very high that the cut surface will reveal a different orientation on each side of a twin. Since the etching rate may depend upon the orientation of the surface in contact with the etchant (that is, the arrangement or planar density of the atoms on the surface in contact with the etchant), then the two parts of the twinning plane will be revealed as a line. This is shown in the microstructure in Fig. 1-57(e).

In the FCC structure, there are four types of {111} planes. Thus, we would expect to see occasionally up to four traces of twinning planes in a grain, although the large grain in Fig. 1-57(e) shows only traces of several parallel twinning planes. That these are traces of parallel planes is evident when a grain contrast etch is used (as in Fig. 1-57e), because the contrast will alternate as twin traces are crossed. Also note that some twins terminate inside grains.

What leads to the stacking error which generates the twins? In solidification, as atoms from the liquid attach themselves onto a {111} plane of a growing crystal, a stacking error can clearly occur when a new layer is commenced. However, the frequency of twins in as-cast materials is not high, but they are present. They are much more numerous after annealing of cold worked material. During the early stages of recrystallization, the twins become apparent and common when the recrystallized grains are in the quite early stages of formation. This was illustrated in Fig. 1-27 and 1-45. As the interface between the relatively dislocation-free recrystallized grain moves into the cold worked region (see Fig. 1-44), the stacking error can occur relatively easily, so that a twin may appear early in the recrystallization stage. Hence, these twins are commonly referred to as annealing twins.

1-14. Mechanical Twinning

The mechanism of plastic deformation described earlier in this chapter was confined to slip, as this is the most prominent process in metals and alloys. However, some crystals can relieve stresses not only by slip but by mechanical twinning, a mechanism which is important in some alloys, especially those that are non-cubic (e.g., HCP). In this section, a brief description of the crystallography and effects of twinning is given.

In a mechanical twin, a mirror-image arrangement of atoms is obtained just as in annealing twins, except the process is caused by mechanical loading. Such twinning can occur in FCC, HCP and BCC metals, but it is more common in HCP and BCC crystals. To describe mechanical twinning, we shall examine its occurrence in HCP metals. The description is simplified in that only a single twinning, slip and cleavage plane (although not necessarily the same crystallographic plane) is assumed. In some HCP crystals, this is not so, leading to a considerably more complicated process than described here.

In HCP metals the twinning plane is the $(10\bar{1}2)$, shown in Fig. 1-58 in an HCP cell. The atoms in the $(\bar{1}2\bar{1}0)$ plane are shown in Fig. 1-59, with the twinning plane normal to the paper. Now if a force is applied to the crystal such that there is a resolved shear component in the twin plane, then the atoms may move as shown in Fig. 1-59. If they do, they become mirror images across the $(10\bar{1}2)$ plane of some of the atoms, and hence a twin forms.

The changes in shape that accompany mechanical twinning can be large and hence significant. Figure 1-60 illustrates this by depicting how a twin can be formed by cutting two HCP crystals along the twinning plane (1-60a), removing one half of each (1-60b) and then fitting the remaining halves together (1-60c) to form a twinned crystal (1-60d). Note that if a crystal with the z-axis parallel to the force is twinned *completely,* orientation changes to that with the z-axis normal to the force. This also involves a change in length. For example, for zinc ($c/a = 1.865$), the calculated length decrease amounts to about 7%, which is rather low.

The effect of twinning on the stress-strain curve of a tension test is illustrated in Fig. 1-61. For the case shown in 1-61(a), the crystal was oriented so that slip occurred initially. However, as slip progressed, rotation occurred (see Fig. 1-8), and eventually the resolved shear stress on the twinning plane allowed twinning to occur, with a sudden decrease in the applied load and an increase in length. It is clear from the shape of the stress-strain curve that mechanical twinning occurs heterogeneously (microstructures will be discussed later). Indeed, this can be demonstrated quite graphically by bending a piece of tin (polycrystalline material will do) so that plastic deformation occurs. A grinding noise will be heard, as local regions

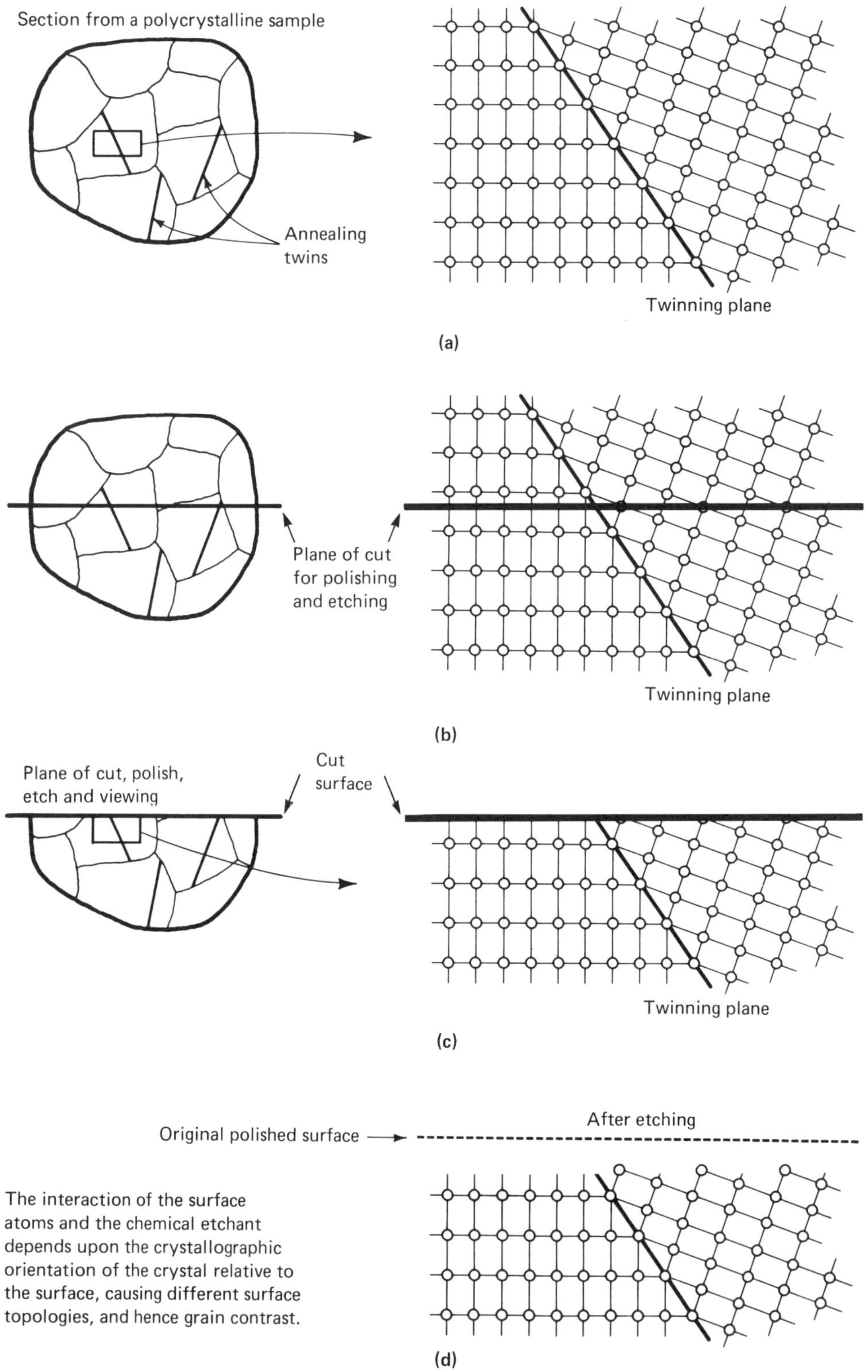

Fig. 1-57. Illustration of the reason that annealing twins are revealed upon etching a polished surface. The microstructure shown in (e) is that of a Cu-5Zn alloy in the annealed condition.

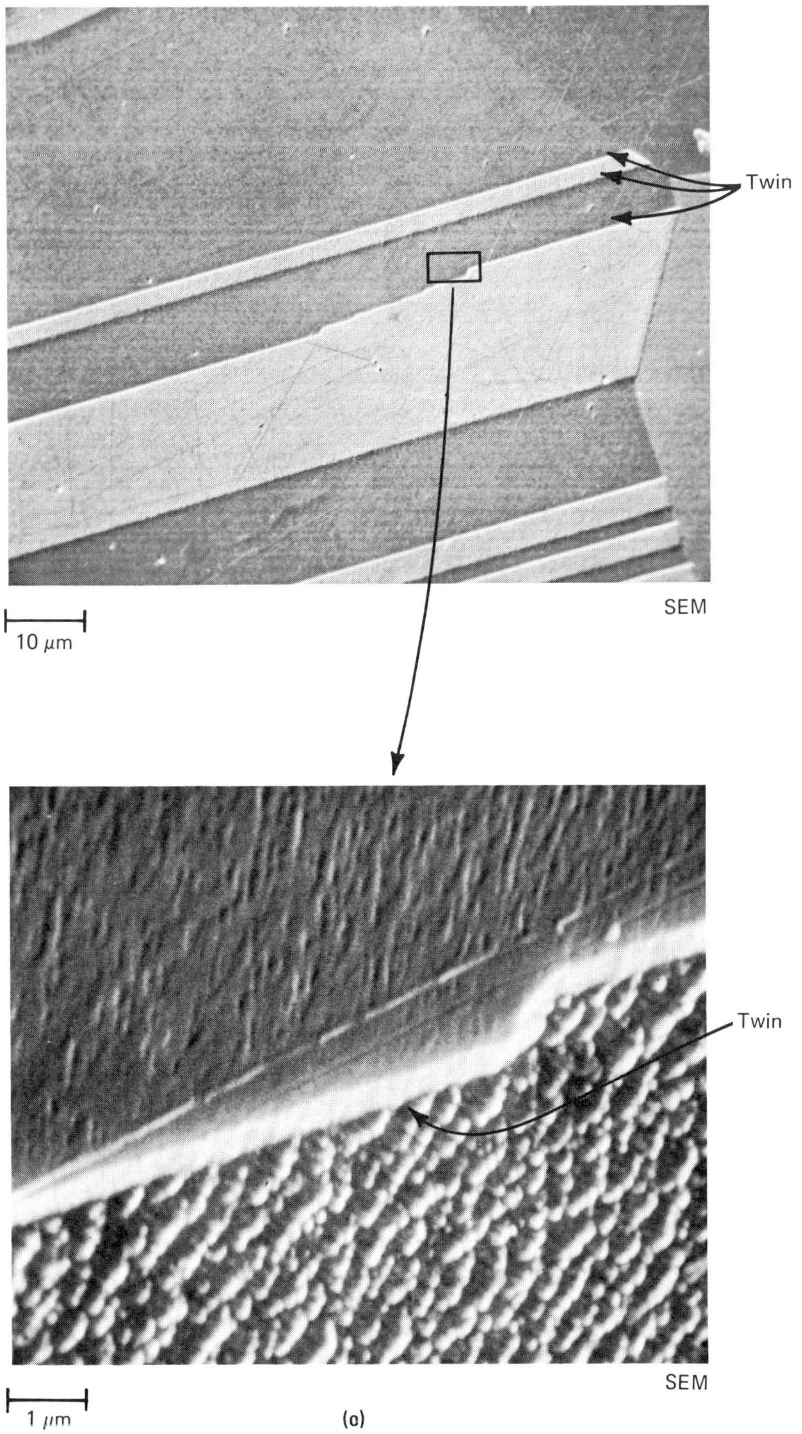

(a)

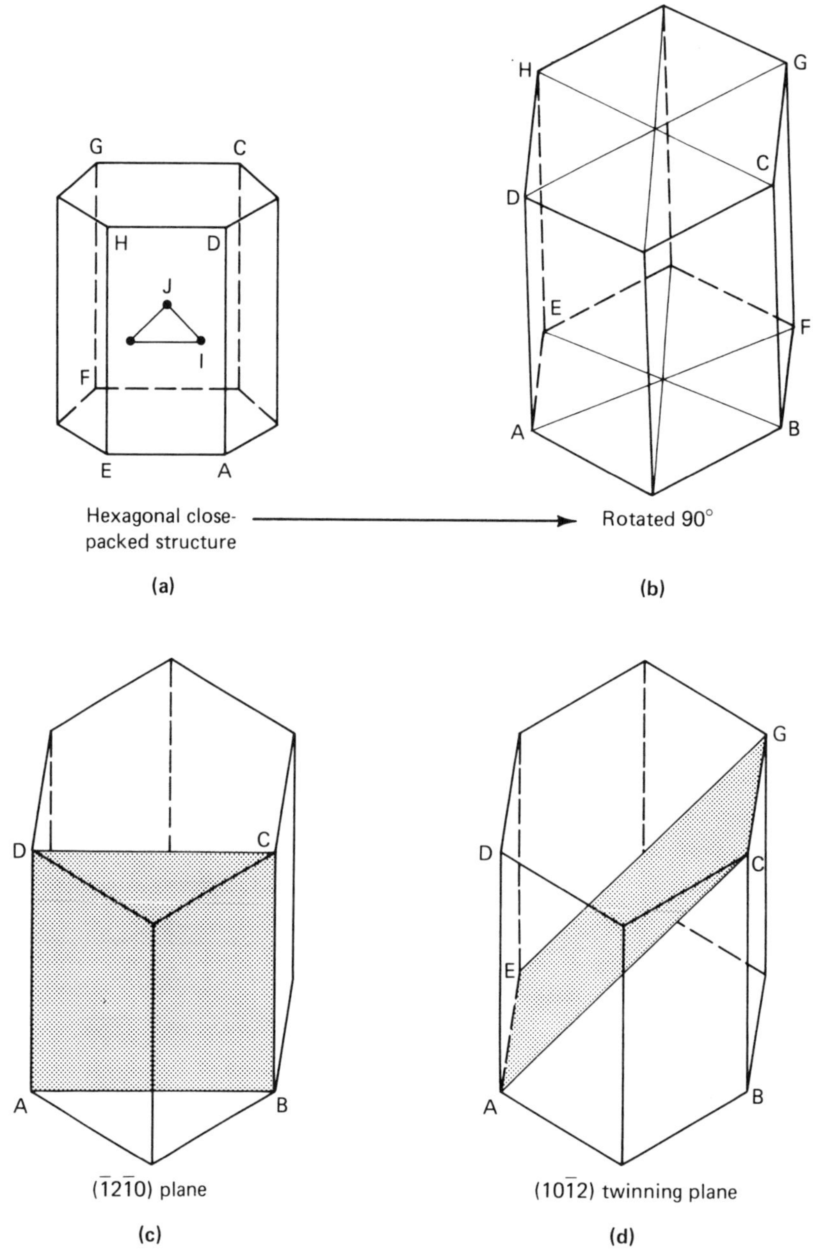

Fig. 1-58. Illustration of the $(10\bar{1}2)$ twinning plane in the HCP structure. The $(\bar{1}2\bar{1}0)$ plane is shown in detail in Fig. 1-59.

twin. The propagation of the atoms as the twinned region thickens occurs with the speed of sound in the material. Upon reaching the surface, a sound wave, referred to as tin crying, is generated.

The c/a ratio strongly influences the twinning behavior of HCP crystals. To illustrate this, Fig. 1-62 shows atom positions similar to those shown in Fig. 1-59 for two different c/a ratios. Figure 1-62(a) is for a c/a = 1.59,

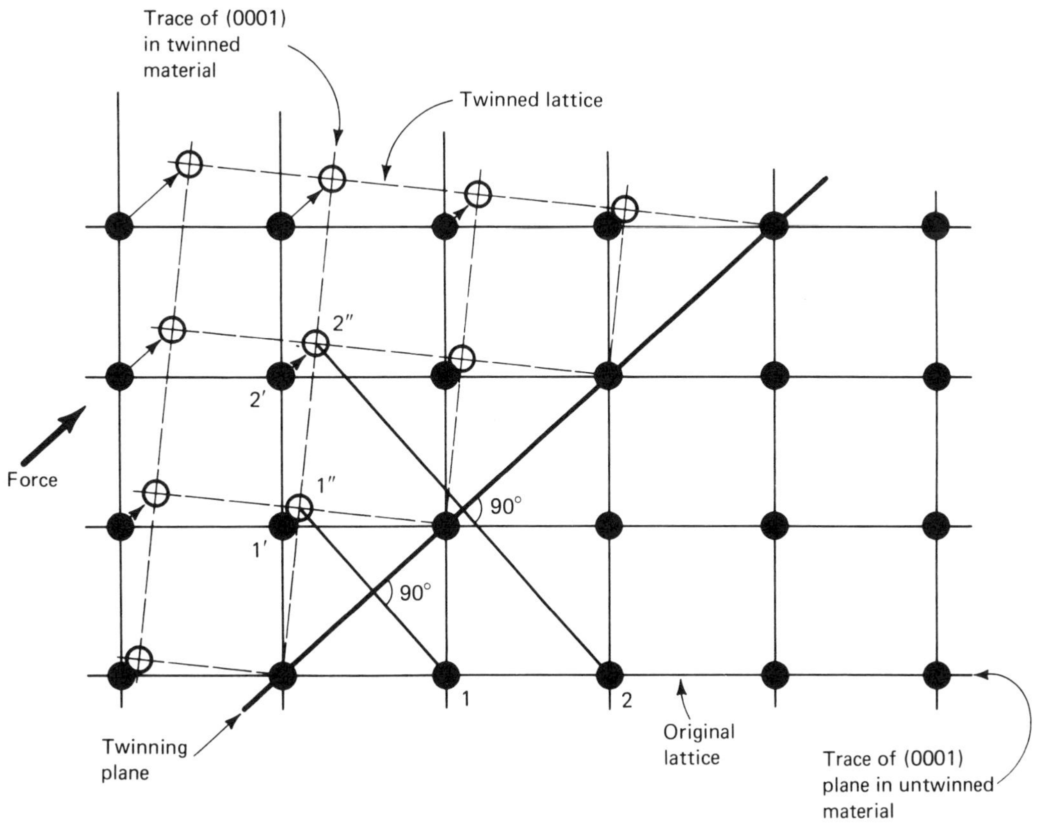

Fig. 1-59. Illustration of atom shifts during mechanical twinning in HCP crystals. As a result of shear force, atom 1′ moves to position 1″, 2′ to 2″, etc., forming a lattice (open circles) that is a mirror image of the original lattice (closed circles) below the twinning plane.

and 1-62(b) for c/a = 1.85. The case is considered with the force parallel to the basal plane, which gives a resolved shear force on the twinning plane to force atom D to position D′, at which location it is the mirror image of B, and hence a twin has formed. However, note that in 1-62(b) the resolved shear force is in the wrong direction for D to move to the mirror-image position. For this c/a ratio, the force would have to be reversed. Thus, in Fig. 1-62(b), twinning cannot occur. If c/a = 1.732, atom D would already be a mirror image of B. Thus, for c/a < 1.732, twinning can occur for the configuration shown in Fig. 1-62(a); for c/a > 1.732 it cannot.

Figure 1-63 shows the resolved shear force on the twinning plane for single crystals in tension and in compression along either the *z*-axis [force normal to the basal (slip) plane] or normal to the *z*-axis. This information can be used to deduce the effect of the c/a ratio on the possibility of twinning for HCP crystals when loaded in tension normal and parallel to the basal plane. If the crystal has the tensile force parallel to the basal plane, slip cannot occur. If c/a <1.732, twinning cannot occur, and when the force is sufficiently high, cleavage will occur, with almost no plastic deformation.

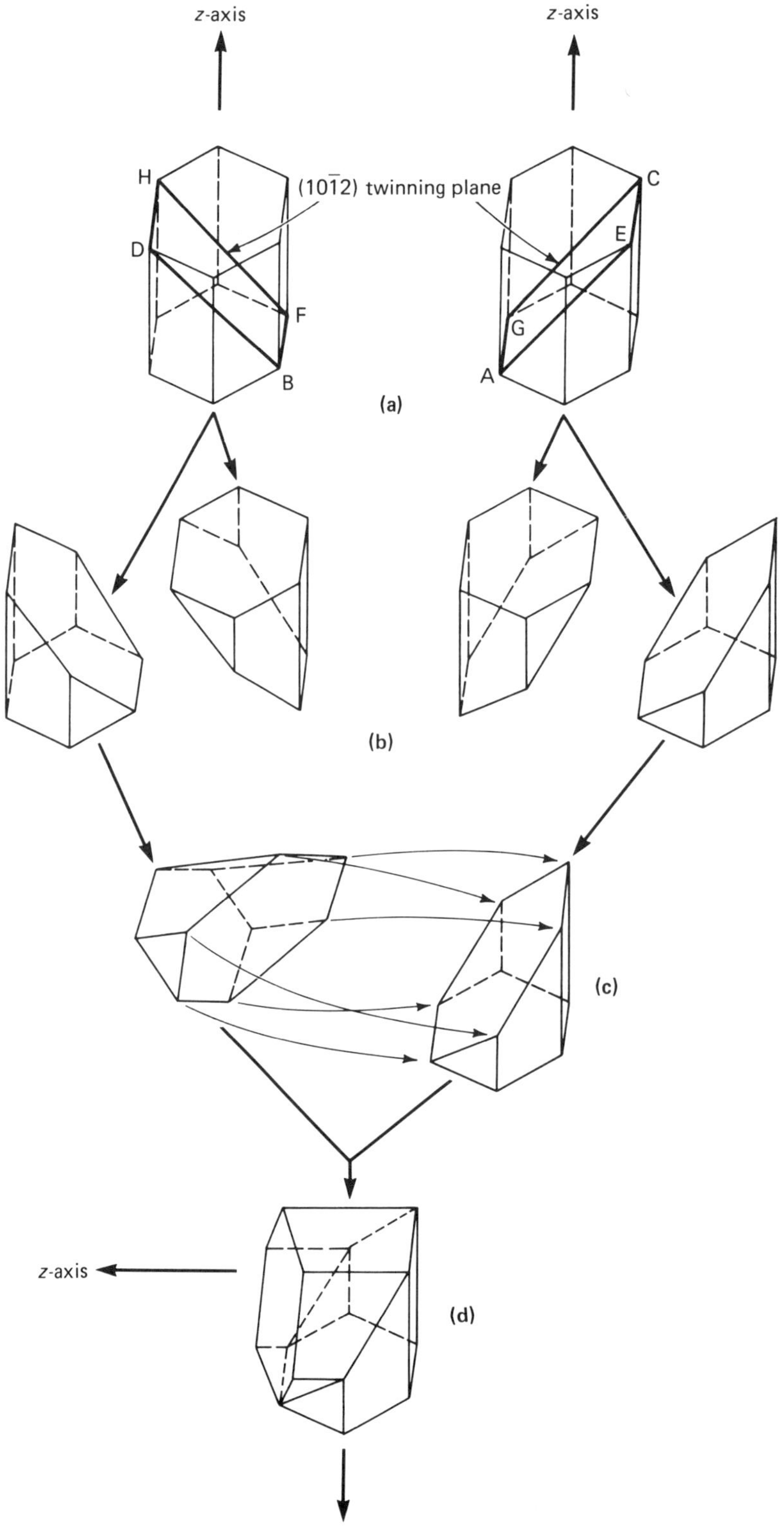

Fig. 1-60. Illustration of twin formation on (10$\bar{1}$2) plane in HCP crystals.

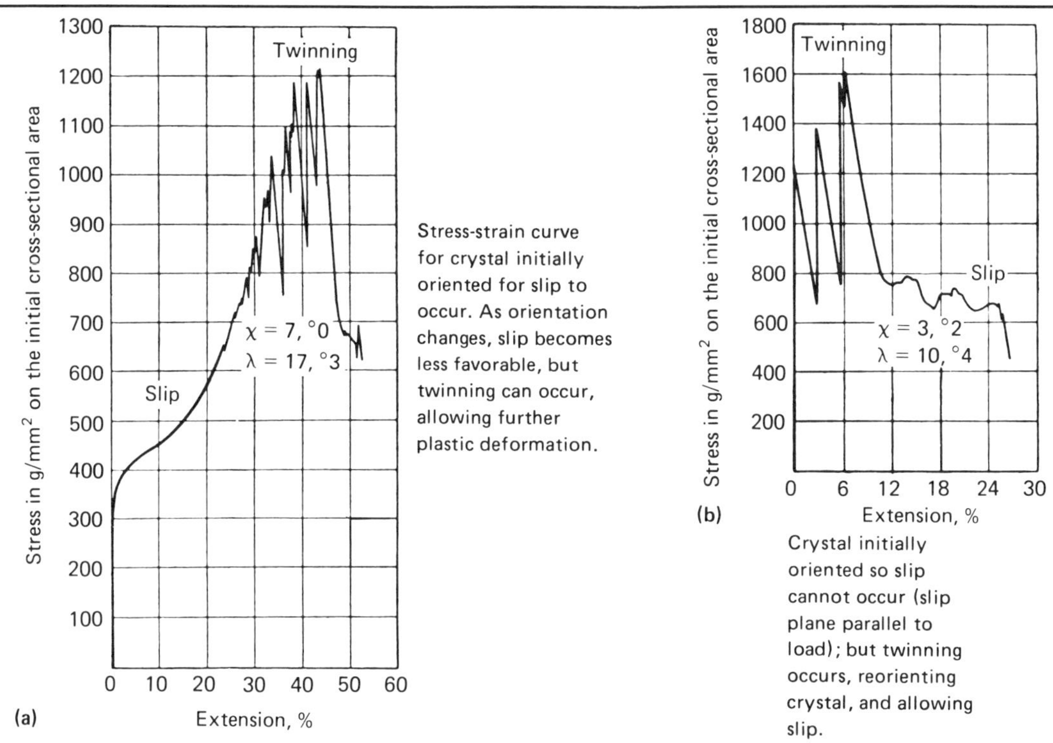

Fig. 1-61. Effect of crystal orientation on stress-strain behavior in a tension test. The material is cadmium with c/a = 1.886. (Adapted from E. Schmid and W. Boas, The Plasticity of Crystals, Chapman and Hall, London, 1950)

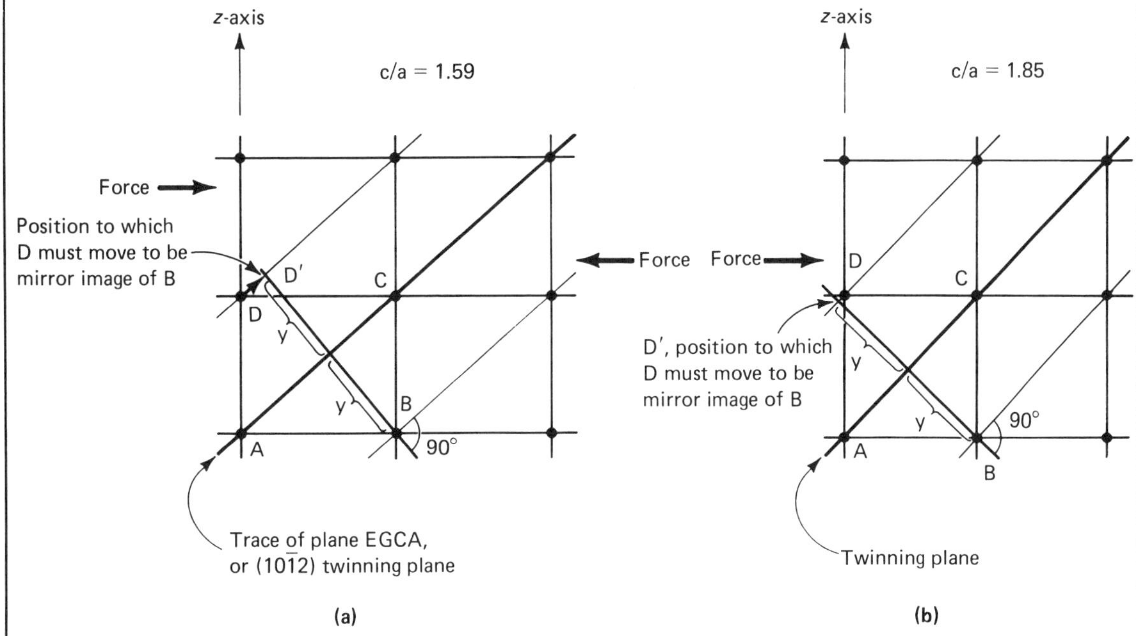

Fig. 1-62. Schematic illustration of the effect of the c/a ratio on the atom movements to give twinning on a (10$\bar{1}$2) plane in the HCP structure.

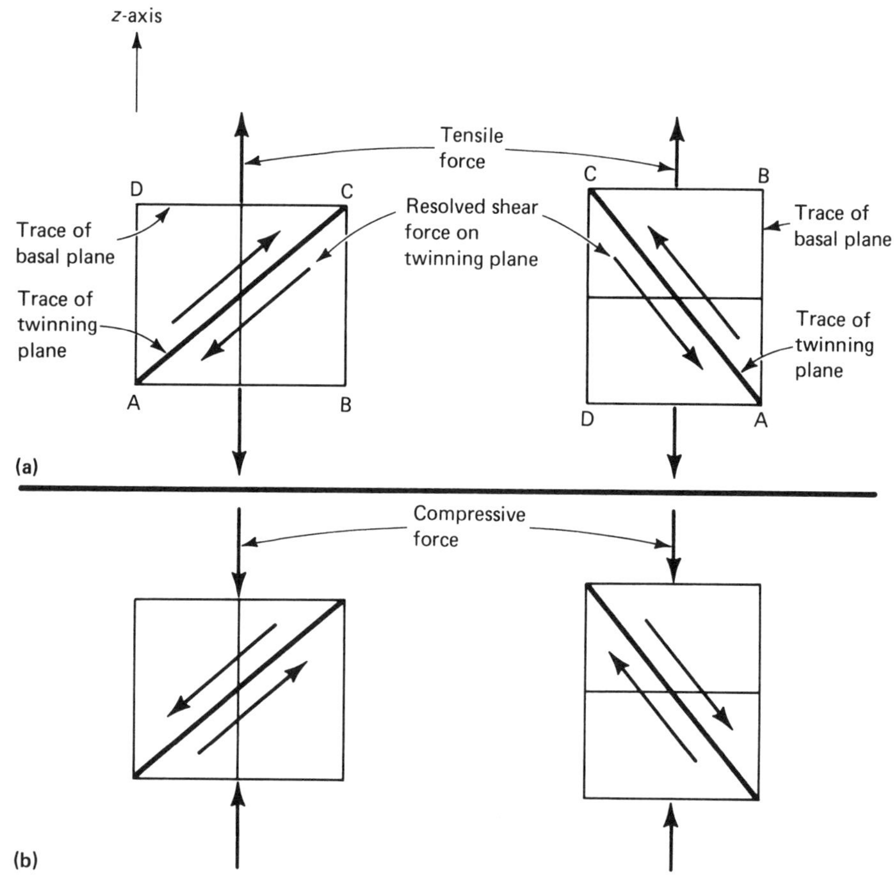

Fig. 1-63. Effect of crystal orientation and load direction on the direction of the resolved shear force on the twinning plane.

However, if c/a > 1.732, twinning can occur, and the twinned region which forms now has its basal plane inclined to the force, so that there is a resolved shear stress on the basal plane, and slip can occur. If the tensile force is normal to the basal plane, slip cannot occur as there is no resolved force on this plane. If c/a > 1.732, twinning cannot occur, and again brittle fracture occurs. If c/a < 1.732, twinning can occur, which allows slip to then occur. These possibilities are summarized in Fig. 1-64. (Remember, this description assumes that the basal plane is the only slip plane and the only cleavage plane.)

As described previously (see Fig. 1-8), during slip in such tensile loading, the slip plane tends to rotate parallel to the tensile force. Thus, if a crystal with c/a < 1.732 is oriented with the tensile force normal to the basal (slip) plane, twinning will occur, followed by slip in the twinned regions. However, these regions upon continued slip tend to rotate to a position with the force parallel to the slip plane, and, according to Fig. 1-64(b), neither twinning nor slip can occur, and brittle fracture then follows. Any plastic deformation must occur during the stage following

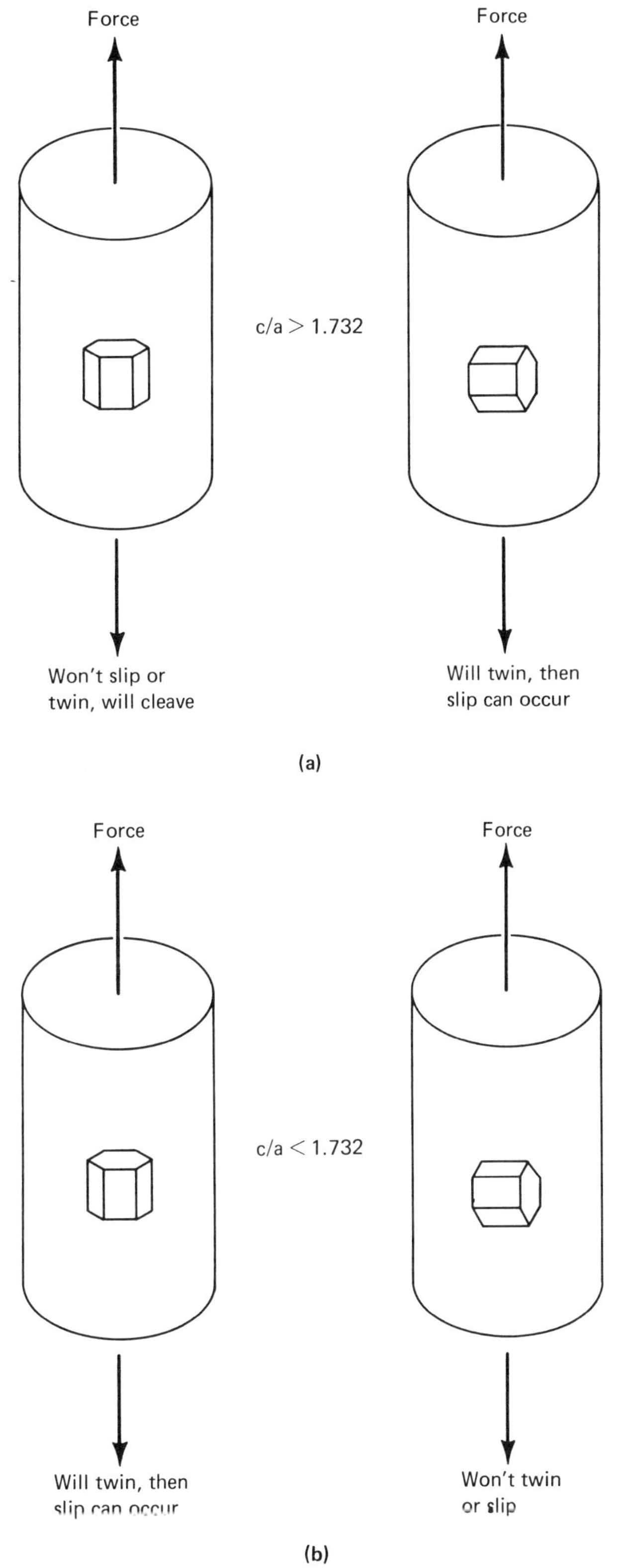

Fig. 1-64. Illustration of the effect of the orientation of HCP single crystals on whether slip or twinning can occur.

twinning. If, for this case of $c/a < 1{,}732$, the crystal is initially inclined to the tensile force, then there is a resolved shear stress for slip, and slip will initially occur. However, upon continued slip, the slip plane rotates parallel to the force, and again neither twinning nor further slip can occur, and brittle fracture occurs.

If $c/a > 1.732$, and the crystal is initially inclined to the tensile force, then slip can occur. However, in this case (Fig. 1-65a), when the basal (slip) plane rotates parallel to the tensile force, twinning can occur, and then

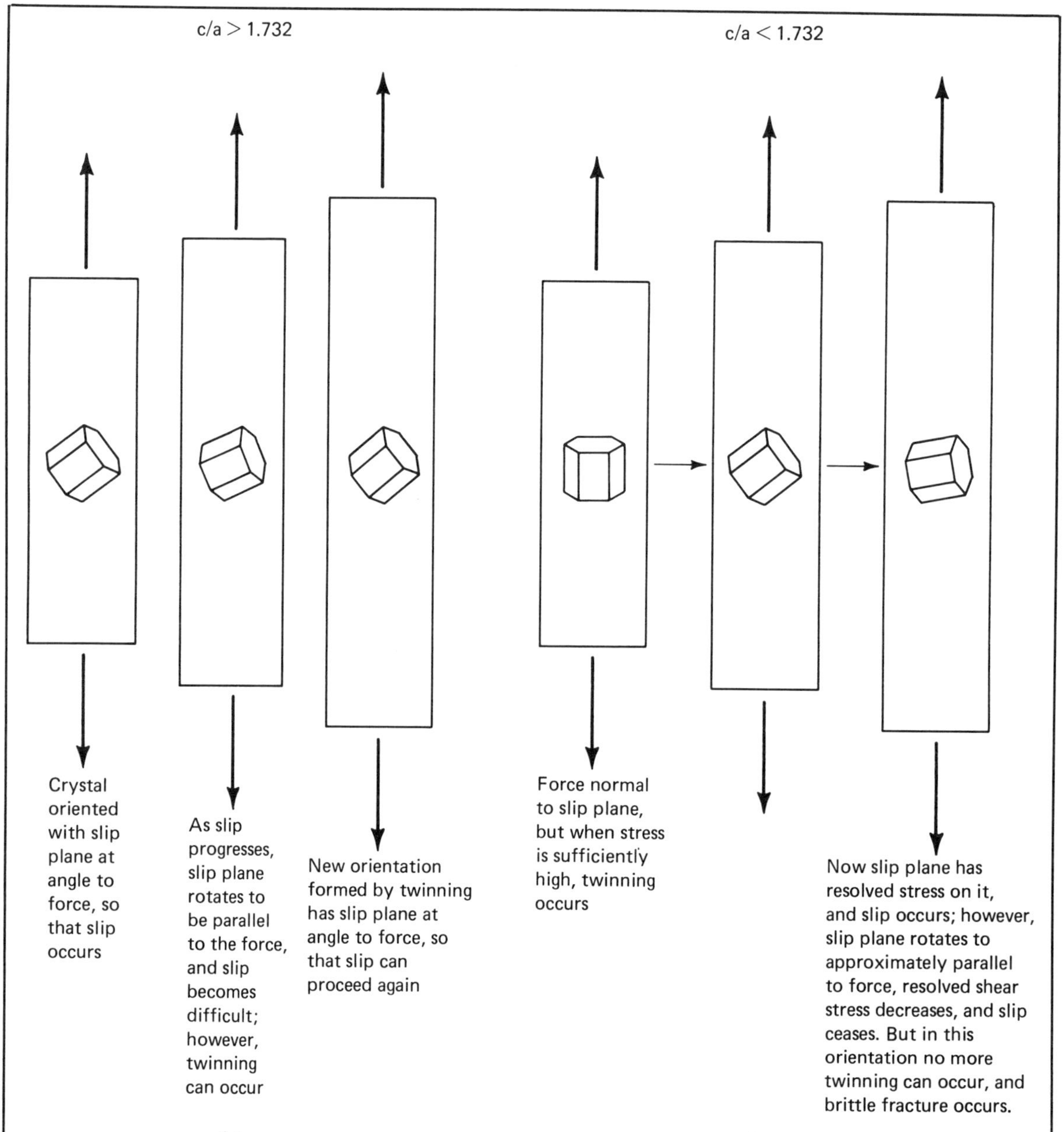

Fig. 1-65. Schematic illustration of the effect of c/a ratio on the slip and twinning behavior of single crystals in a tension test.

Table 1-2. Effect of c/a Ratio on the Tensile Ductility of Polycrystalline HCP Metals

Metal	c/a	Approximate yield strength, lb in.$^{-2}$	Approximate tensile ductility at 25 °C, %
Be	1.586	35,000	2
Zr	1.589	50,000	35
Ti	1.601	30,000	50
Mg	1.623	14,000	10
Zn	1.856	20,000	40
Cd	1.886	10,000	50

further slip. Thus, for $c/a > 1.732$, greater plastic deformation should be possible in tension than for $c/a < 1.732$. These processes are illustrated in Fig. 1-65.

If the HCP material is polycrystalline, and $c/a < 1.732$, then those crystals with their basal (slip) plane inclined to the tensile force will slip, but eventually they will rotate toward the tensile force, and further slip becomes difficult. More importantly, twinning is difficult with this orientation, and hence brittle fracture occurs, with a relatively small amount of plastic deformation at fracture. If $c/a > 1.732$, then those crystals inclined to the tensile force will slip, but when they rotate closely parallel to the tensile force, twinning can occur, allowing further slip, and hence the amount of plastic deformation at fracture should be relatively high in this case.

Table 1-2 compares approximate values for the tensile ductility of polycrystalline HCP metals. The data are consistent with the description in the previous paragraphs, although other factors are involved, such as the operation of other slip and twin systems.

Although HCP metals with low c/a ratios are relatively brittle in tension, they can be plastically deformed in compression, such as by rolling. This is because the slip plane tends to rotate parallel to the rolling plane, which puts a compressive force normal to the basal plane, allowing twinning to occur (Fig. 1-64b).

The microstructural appearance of twins is illustrated in Fig. 1-66. Because of the change in orientation, the twinned and untwinned regions etch differently, just as for annealing twins. Note that the twins are rather lenticular in shape and tend to terminate at other twins or at grain boundaries (not shown in Fig. 1-66). In BCC crystals, the twinned regions are quite narrow, and are referred to as Neumann bands.

1-15. Stacking Faults

Consider a simple edge dislocation in an FCC crystal. If the stacking below the plane containing the dislocation is examined, so that the "missing" row of atoms is revealed, the arrangement shown in Fig. 1-67 is seen. Plastic deformation will occur by the movement of such dislocations along the ⟨110⟩

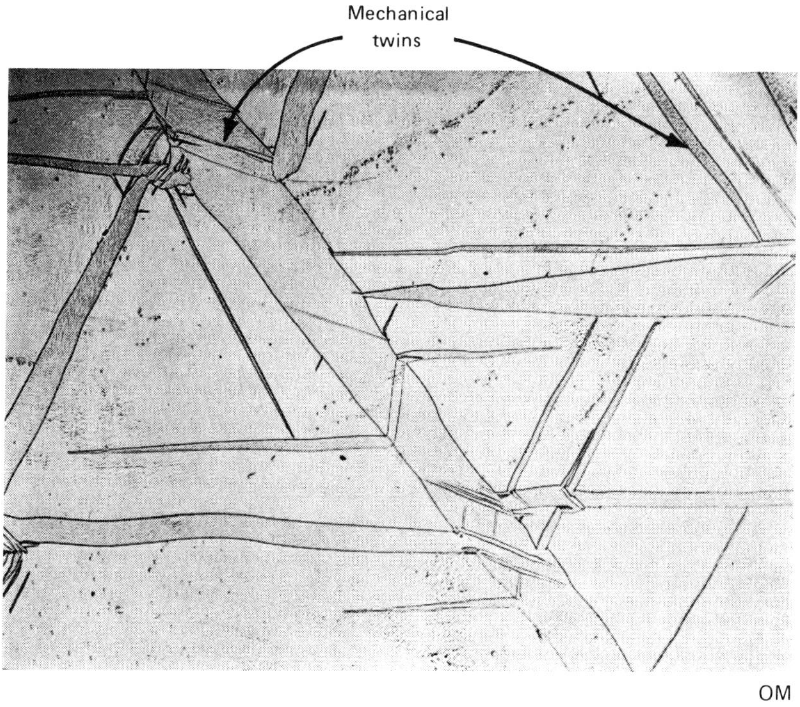

Fig. 1-66. Microstructure of plastically deformed magnesium, showing mechanical twins. (Adapted from C. S. Roberts, Magnesium and Its Alloys, Wiley, New York, 1960)

direction (in the plane of the paper). Thus the row of atoms numbered 1 through 4 will move to the new positions denoted by the arrows. However, examination of the arrangement shows that movement will require less energy if the row moves first to the position marked "c" in Fig. 1-67(b), then to the position "d."

In fact, the energy of the dislocation ("missing" row) in Fig. 1-67(b) is lowered if the atoms move to the configuration shown in Fig. 1-67(c). In a sense, on each side of the row 1-2-3-4 is a missing half-plane. Such a configuration is known as a partial dislocation. The energy of the configuration in Fig. 1-67(c) is lowered further if the atoms on the left of the row of atoms numbered 1-2-3-4 move to the configuration shown in Fig. 1-67(d). This configuration is called an extended dislocation.

It appears that the more extended the dislocation is, the lower the energy. However, note that because the atoms in the extended region have moved to the intermediate position, they are now in the incorrect position for proper FCC stacking on {111} planes. This is depicted in Fig. 1-68. This stacking error, or stacking fault, raises the energy of the crystal. Thus the partial dislocations will separate only until the energy reduction due to this separation equals the energy increase due to the formation of the stacking fault.

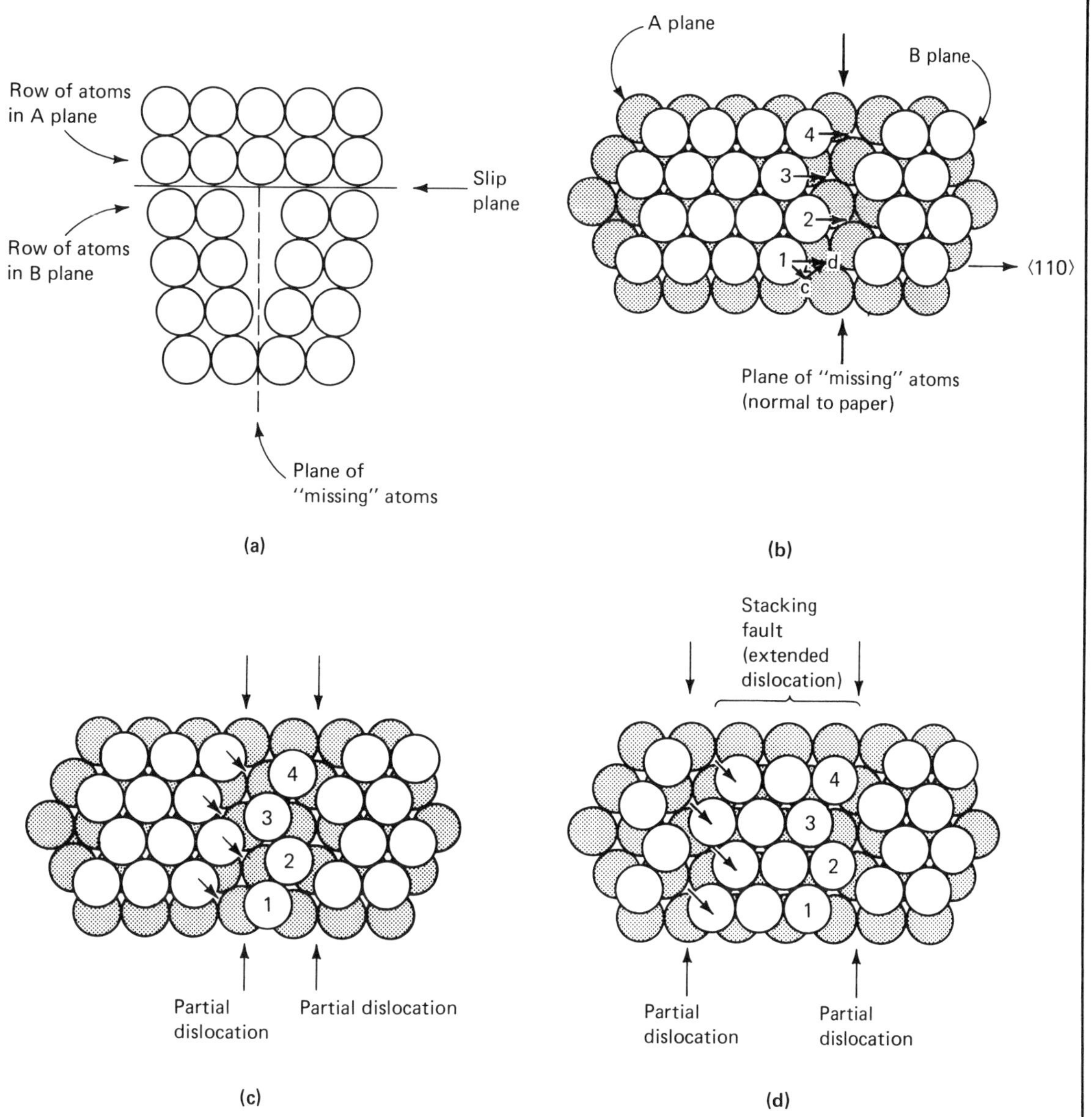

Fig. 1-67. Illustration of the formation of a stacking fault in a FCC crystal. Shown are two {111} planes, with a "missing" row representing the dislocation. The energy of the dislocation is lowered if the configuration shown in (d) is formed. This gives rise to an error in the stacking fault sequence in the stacking arrangement of the {111} planes. See Fig. 1-68. (Adapted from Physical Metallurgy Principles, 2d Ed., by R. E. Reed-Hill. Copyright © 1973, Van Nostrand Reinhold Company, New York. Used with the permission of Brooks/Cole Publishing Company)

Stacking faults can be revealed by transmission electron microscopy. An example showing several parallel stacking faults is given in Fig. 1-69. The stacking fault plane extends from one side of the thin foil to the other.

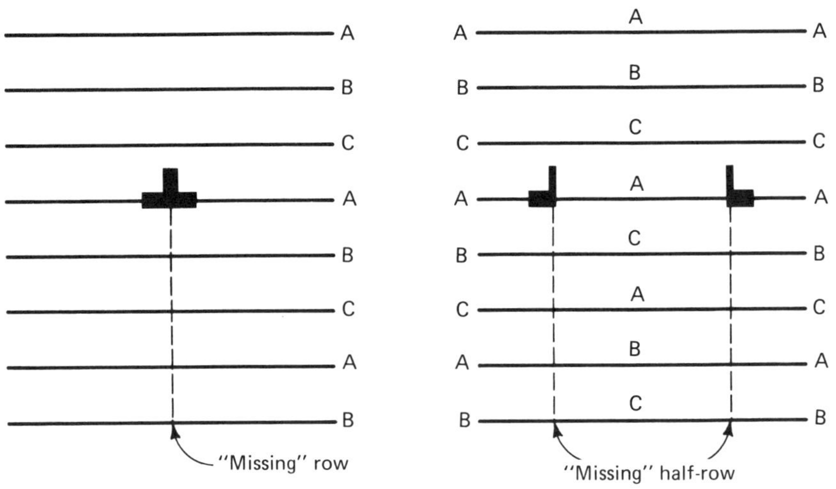

Fig. 1-68. Illustration of the {111} stacking sequence by which an error in the stacking fault sequence develops when the dislocation separates into two partial dislocations.

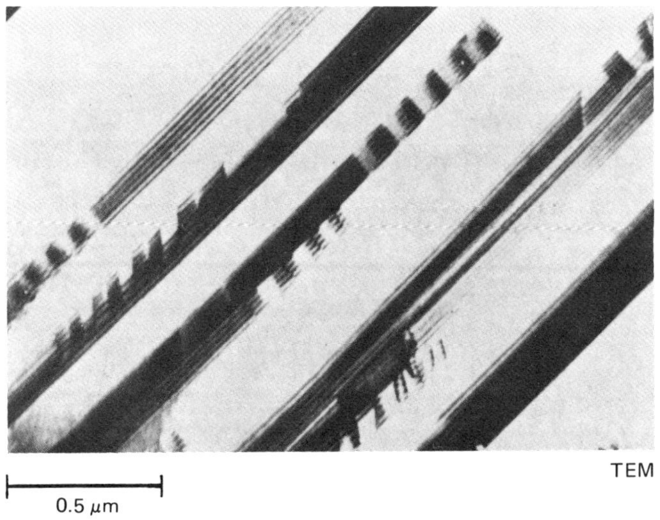

Fig. 1-69. Stacking fault bundles in a Cu-4Si alloy. If the faults are on closely neighboring planes and overlap, then an alternating contrast may be seen which reveals the stacking faults. (From P. R. Swann, in Electron Microscopy and Strength of Crystals, ed. G. Thomas and J. Washburn, Wiley, New York, 1963)

An important feature of stacking faults is their influence on strength. For the dislocation to shift to another slip plane, the two partial dislocations must be compressed into a single dislocation. This requires work, and the more extended the dislocation (the larger the stacking fault area), the more work is required to compress the partial dislocations into a single dislocation. Thus, the more extended the partial dislocations, the stronger the crystal.

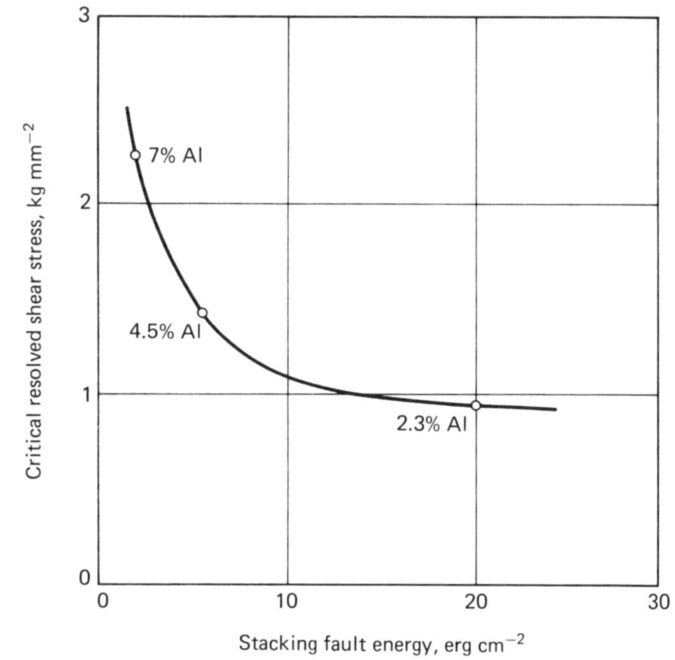

Fig. 1-70. Effect of stacking fault energy on the strength of Cu-rich Cu-Al solid solution alloys. Increasing the Al content increases the stacking fault energy and the strength. (Adapted from A. Howie, in Direct Observation of Imperfections in Crystals, ed. J. B. Newkirk and J. H. Wermick, Interscience, New York, 1962)

The extension of the partial dislocations is controlled by the stacking fault energy; the lower the stacking fault energy, the more extended the partials can be. Therefore, a strengthening mechanism in FCC crystals is to control the chemical composition to reduce the stacking fault energy (Fig. 1-70). In some of the following chapters, this effect will be considered.

References

Transformations in Metals, P. G. Shewmon, McGraw-Hill, New York (1969).

Physical Metallurgy Principles, 2d Ed., R. E. Reed-Hill, Van Nostrand Reinhold, New York (1973).

CHAPTER 2

Solidification

An understanding of solidification is important because this process controls the structure, and hence the properties, of castings and dictates heat treatment prior to working into wrought products. In this chapter the principles of the formation of crystals from liquid are briefly reviewed. A rather detailed description of the origin of coring in alloys is presented, as this will be referred to in the material in later chapters dealing with specific alloys.

2-1. The Formation of Crystals

The atoms in a liquid are in constant agitation, caused by the thermal energy they contain. When the temperature of the liquid is lowered, the thermal agitation decreases, the bonding among the atoms becomes stronger, and the material decreases in specific volume. Eventually, a sufficiently low temperature is reached where the material is more stable if the atoms are arranged in a repetitive order of the crystalline state than if they exist in the random arrangement of the liquid. In some location in the liquid, a small group of atoms takes on the crystalline arrangement, giving off heat which is the difference in energy of this group of atoms when in the crystalline form and in the liquid form. This is the beginning of crystallization. The formation of these crystal nuclei requires a finite undercooling, since some of the energy of the phase change is used in creating the surface. For the nucleus to be stable, it must grow to sufficient size for the energy liberated to compensate for the surface energy. Thus, until the nuclei attain this critical size, they may redissolve. The formation of these stable nuclei is governed by the net rate of attachment of atoms from the liquid onto the surface of the crystal nucleus. This is proportional to the expression $\exp(-Q/RT)$, where

Q is the activation energy for the atom attachment (a constant), R is the ideal gas constant and T is the absolute temperature.

Once the nuclei attain the critical size, their growth rate is governed by the rate at which atoms from the liquid are attached to the surface. This again is exponentially temperature dependent, but with a different activation energy than that for the nucleation rate. The processes of nucleation and of growth are illustrated schematically in Fig. 2-1.

The rate of attachment of atoms in the liquid onto the crystal surface varies with the orientation of the crystal (that is, with the type of crystallographic plane facing the liquid). Since atoms are attached more easily on some faces than on others, the crystal does not grow as a sphere but with a geometry in which facets are present. As the crystal grows, side branches form, usually repeating and continuing, and the crystal takes a form called a *dendrite*. This is illustrated in Fig. 2-2.

The process of solidification can be followed by determining a cooling curve, the temperature of the metal versus time. Figure 2-3 shows schematically two such curves. The metal is heated to a temperature above the melting point and then cooled. When a temperature is reached such that nucleation of crystals begins, the energy difference between those regions

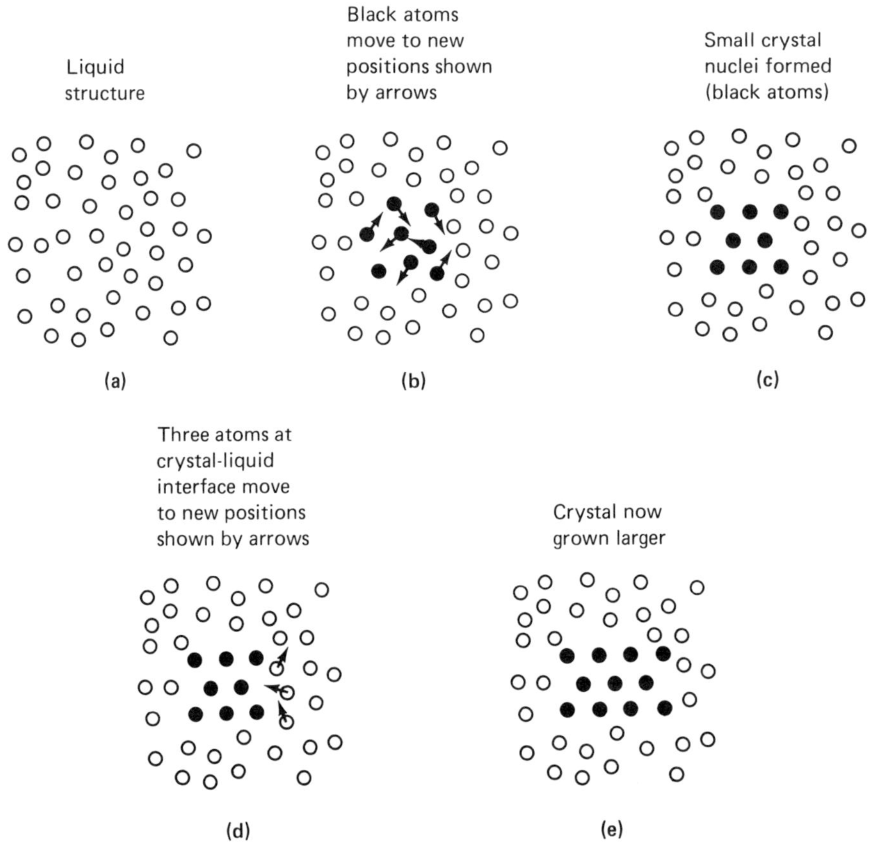

Fig. 2-1. Schematic representation of formation of a crystal nucleus and its growth in a liquid.

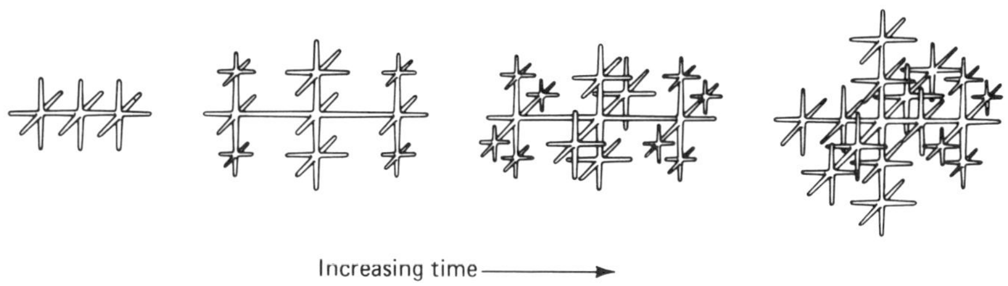

(a)

(b)

Fig. 2-2. (a) Schematic illustration of the growth of a dendritic crystal (adapted from P. S. Hurd, Metallic Materials, Holt, Rinehart and Winston, New York, 1968). (b) Photograph of a large, 9 in. long, dendrite which formed in a steel (from A. Sauveur, The Metallography and Heat Treatment of Iron and Steel, 4th Ed., The University Press, Cambridge, Eng., 1935).

which are now crystalline and were liquid is converted to heat. If the rate of cooling is sufficiently low, this compensates for the heat lost by the metal to the surroundings, and the temperature remains constant during the entire solidification process (Fig. 2-3a). (This is the method by which the melting point is determined.) Once solidification is complete, the temperature again decreases.

If significant undercooling occurs (as can be brought about by rapid cooling, for example), then the heat released will not be sufficient to completely compensate for the heat lost, and the cooling curve looks more like that in Fig. 2-3(b).

2-2. As-Cast Grain Size and Shape

The as-cast grain size is controlled by the rate of nucleation and rate of growth of the crystals, and their shape. Polycrystalline structures develop in castings when the nucleation rate is sufficiently high for a large number of nuclei to form in the liquid before the first nucleus can grow to consume the liquid to give a single crystal. The process of nucleation and growth

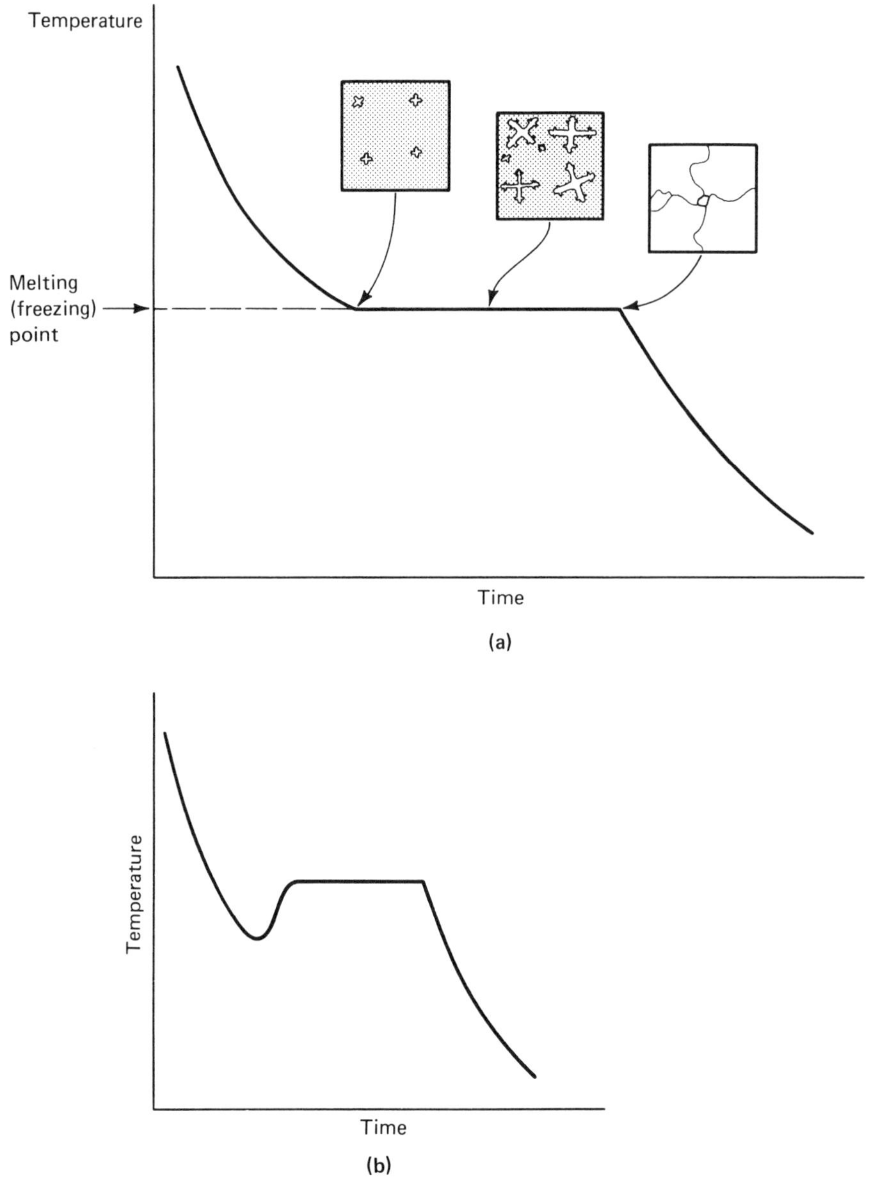

Fig. 2-3. Schematic cooling curves. (a) Curve in which negligible undercooling occurs. (b) Curve showing undercooling.

is illustrated in Fig. 2-4(a). Note that nucleation continues as long as liquid remains.

The effect of the nucleation rate (N) compared to the growth rate (G) can be best examined by considering the ratio N/G. When N/G is high a fine grain size is expected; a low N/G gives a larger grain size, as illustrated in Fig. 2-4(a) and (b).

The nucleation process depicted in Fig. 2-1 is homogeneous nucleation, since no foreign surface is involved. In casting, however, the mold walls, as

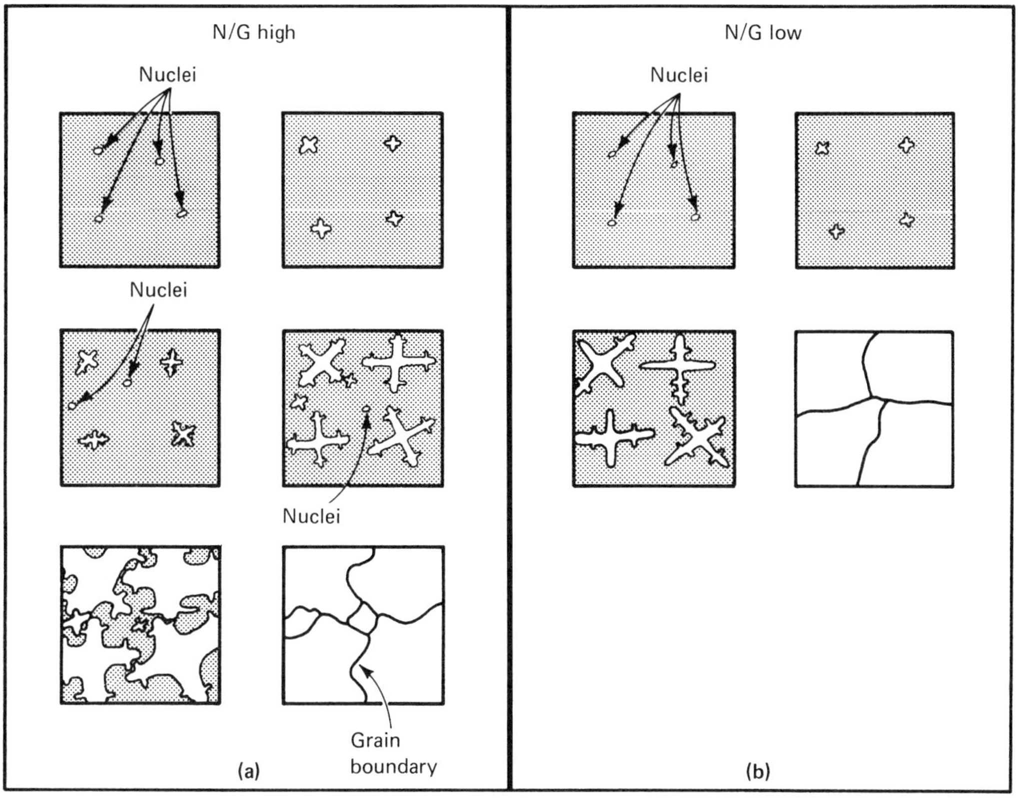

Fig. 2-4. Schematic illustration of the effect of the ratio of the nucleation rate (*N*) to the growth rate (*G*) on the final grain size. The lower *N/G* ratio in (b) gives a larger grain size.

well as insoluble solid particles (e.g., oxides) which are frequently present in the liquid, serve as nucleating surfaces. This is heterogeneous nucleation, and it usually has a higher rate than homogeneous nucleation. Thus inoculating the melt with very fine, crystalline particles called grain refiners to increase the rate of nucleation is a method of reducing the grain size, as illustrated in Fig. 2-5.

When the molten metal is poured into a cold mold, the metal at the surface cools rapidly, and heterogeneous nucleation occurs readily on the mold wall. The crystals grow sideways and outwards, but they quickly contact other crystals in the sidewise growth. The rate of growth into the liquid depends upon the crystal orientation, and only those which have the most rapid growth direction normal to the mold wall will survive the growth process. These will grow as long crystals, called columnar grains. The process is illustrated schematically in Fig. 2-6, and an example of columnar grains is shown in Fig. 2-7.

In some cases, the surface consists of columnar grains and the center of equiaxed grains. An example is shown in Fig. 2-8. The origin of this effect is not the nucleation of grains in the liquid before the columnar grains can

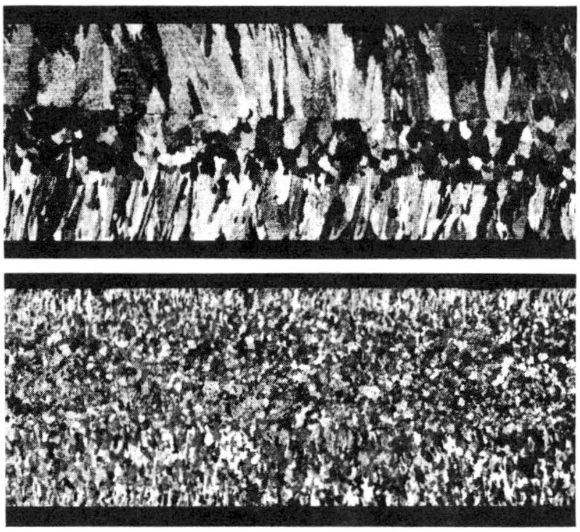

Fig. 2-5. Effect of addition of a grain refiner on the cast grain size and shape of commercially pure aluminum (1100 aluminum) cast by a two-belt method. Both pictures are of longitudinal sections, shown at actual size. The upper slab was cast without a grain refiner and the lower with a grain refiner addition (e.g., titanium or titanium and boron). (From Metals Handbook, 8th Ed., Vol 8, American Society for Metals, Metals Park, Ohio, 1973)

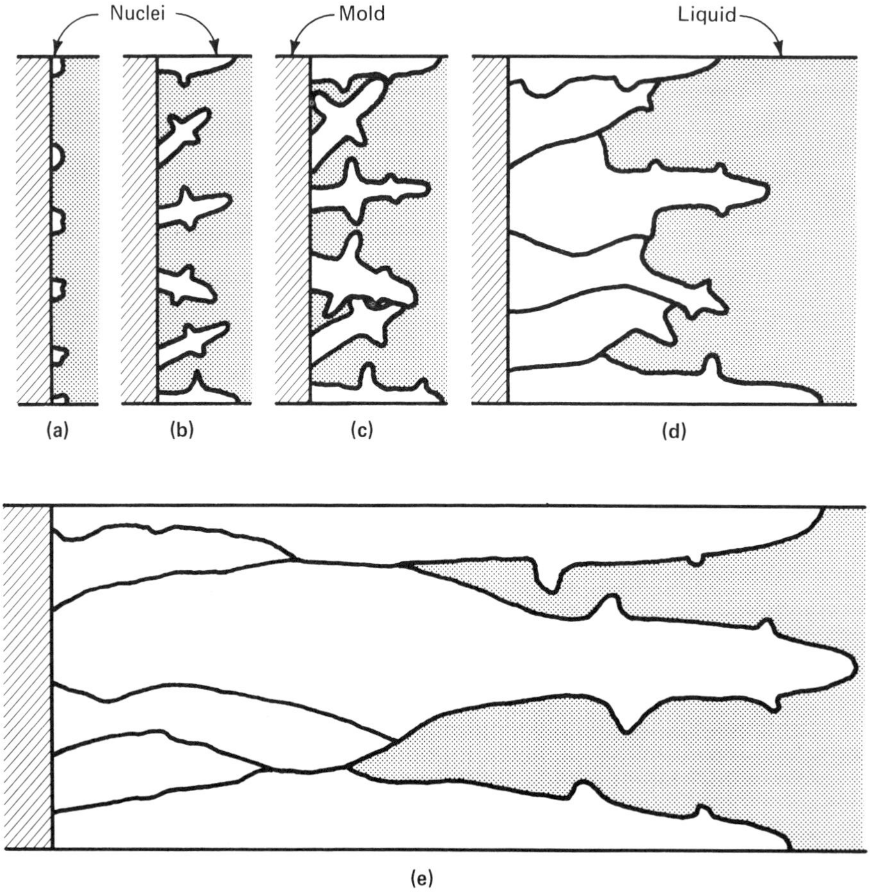

Fig. 2-6. Schematic illustration of the formation of columnar grains.

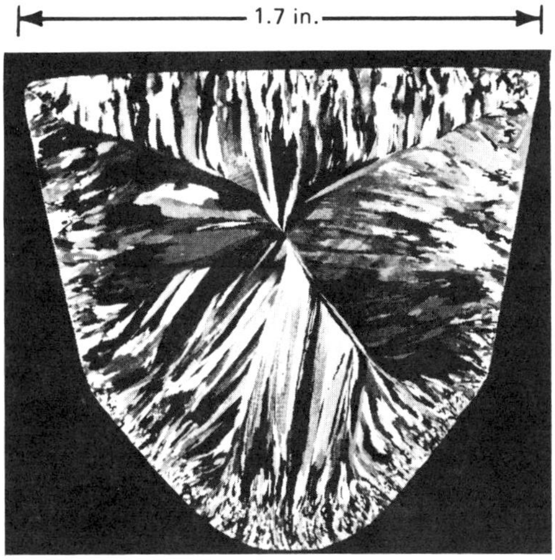

Fig. 2-7. Macrostructure of a transverse section through an ingot of commercially pure aluminum (1100 aluminum) cast by the wheel-and-belt method. Note that columnar grains have formed perpendicular to the mold wall, and also note the small grains at the surface which were outgrown by the other grains. (Adapted from same source as Fig. 2-5)

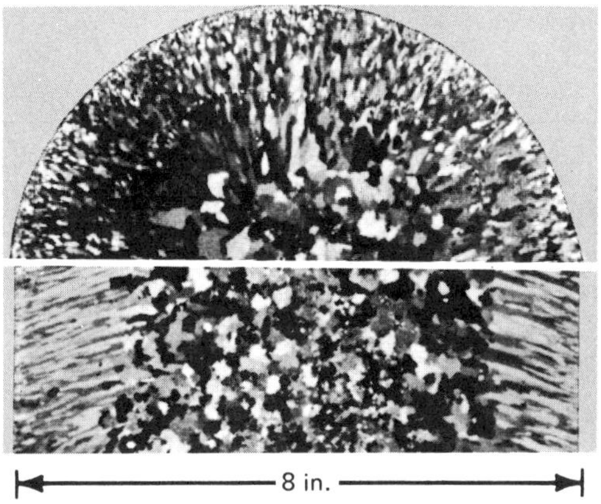

Fig. 2-8. Macrostructure of a static cast ingot of a free-cutting brass (alloy 360), showing columnar grains growing in from the surface and equiaxed grains in the center. The top picture is a transverse section, the bottom longitudinal. (Adapted from same source as Fig. 2-5)

completely consume the liquid. Instead, small pieces of the dendrite arms are separated from the dendrite and swept into the center. This is illustrated in Fig. 2-9. As the side branches of the dendrites form, the heat released at the interface is sufficient to raise the temperature locally, and a side arm melts off. This small crystal is swept by convection into the center of the casting to serve as the nucleus for the formation of the equiaxed grains.

Another factor that affects grain size is the cooling rate. Rapid cooling apparently gives a higher N/G ratio than slow cooling and leads to a finer grain size. Also, mechanical effects such as vibration may increase the nucleation rate, leading to finer grains, probably because the vibration has caused side arms of the dendrites to break off.

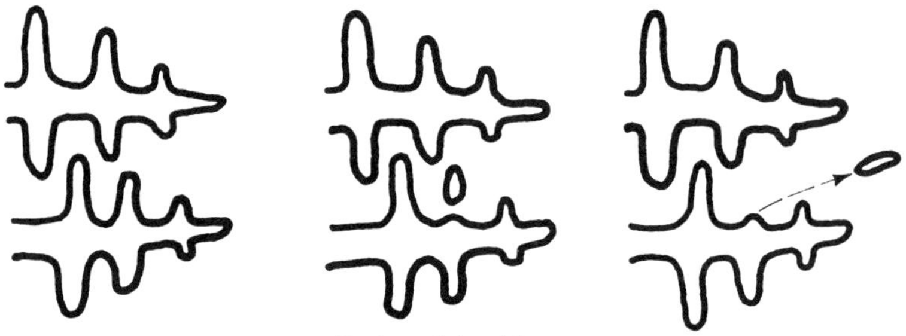

Fig. 2-9. Schematic illustration of the remelting of a side arm of a dendrite and movement of the side arm into the center of the casting to serve as the nucleus of an equiaxed grain.

2-3. Coring

Consider a binary system of components A and B whose phase diagram is shown in Fig. 2-10(a). We examine the behavior of an alloy whose composition is 30% B. Upon cooling from the liquid, the alloy passes through the two-phase region, then into the single solid phase. Thus, freezing occurs over a temperature range, not as it did with the pure metal in Fig. 2-3. Consider cooling sufficiently slowly so that equilibrium is attained. This means that the chemical composition of the liquid and solid at any temperature is that given by the phase diagram. Thus, upon cooling, when the temperature reaches T_0 (actually slightly below T_0 to initiate nucleation), the first crystals to form have a composition 10% B (Fig. 2-10a). Now, when freezing is complete at T_2, the composition of the solid obviously has to be that of the alloy, 30% B. Thus, as the temperature decreases from T_0 the solid composition must change continuously from 10% toward 30%. When the temperature reaches T_1 the solid composition is 20%. Note that the chemical composition of the liquid in equilibrium with the solid also increases in B content (from 30% to 57% to 70%). When the temperature is just above T_2, the crystals have a composition just less than 10% and are surrounded by liquid of composition 70%. When the temperature reaches exactly T_2, the solid must have a composition of exactly 30%. Thus, when this B-rich liquid layer freezes it must lose B atoms (and gain A atoms) to the crystal to bring its composition everywhere to 30%.

This process is illustrated schematically in Fig. 2-11. At just below T_0 the first crystals contain 10% B, and the liquid contains slightly greater than 30% B. (In Fig. 2-11 the density of the liquid is shown much too low.) By

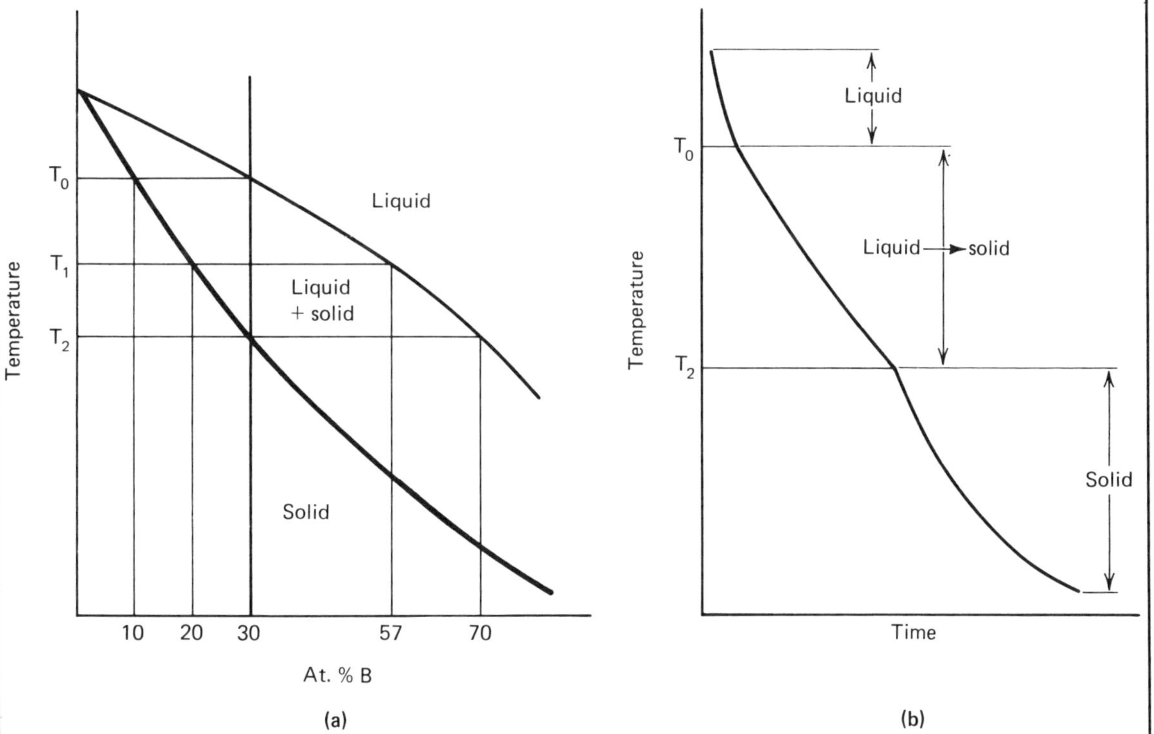

Fig. 2-10. (a) Phase diagram for system A-B in which the liquid freezes to give a single solid phase by passing through a liquid-solid region. (b) Cooling curve for a 30% B alloy.

the time the alloy has cooled to T_1 the two crystals have increased in size, and they now contain 20% B. Note that on an atomic scale the crystal contains uniformly 20% B. The region in the solid (dashed outline) which did contain 10% B at T_0 now contains 20% B, and thus during cooling from T_0 to T_1 some B atoms have diffused to, and some A atoms away from, this region. Upon cooling to a temperature just above T_2, only a small layer of liquid surrounds each solid crystal. The solid is now almost 30% B, and the liquid is 70% B. When the temperature is lowered to exactly T_2, the solid crystals will have uniformly 30% B. Although the process is continuous, one can visualize the atoms of this liquid layer joining the arrangement of the crystal, then the excess B atoms diffusing into the crystal to bring its composition to exactly 30% B. It is emphasized that this description is for equilibrium cooling, where at each infinitessimal change in temperature the alloy attains the equilibrium composition of the phases given by the phase diagram.

It is seen in Fig. 2-11 that when the temperature decreases from T_0 to T_1, part of the crystal which formed at T_0 must change its composition from 10 to 20% B. This means that B atoms must move into this part of the crystal, and A must move out. This diffusion process requires a finite time, and it is expected that the process shown in Fig. 2-11 will require a very low

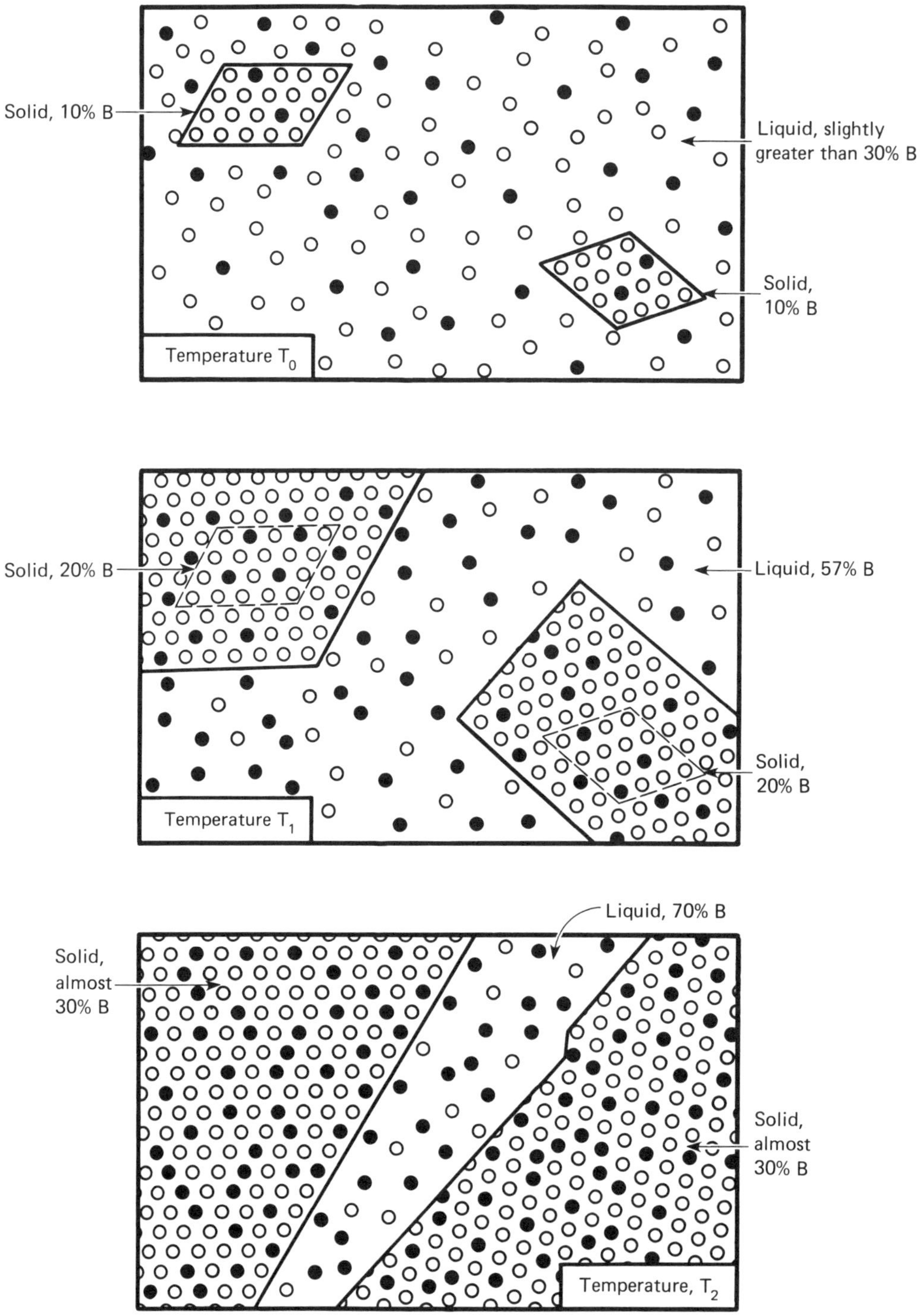

Fig. 2-11. Schematic illustration of the growth of two crystals in a 30% B alloy during equilibrium cooling. The chemical compositions of the solid and liquid are obtained from the phase diagram in Fig. 2-10(a). For simplicity, the crystals are not shown as dendrites.

cooling rate. If cooling is not slow, then little or no diffusion will occur, and a nonequilibrium structure will develop.

The atomic process during rapid cooling is illustrated schematically in Fig. 2-12. It is assumed here that at the interface between the liquid and the solid, each phase has the chemical composition given by the phase diagram, but no diffusion is allowed in the solid. In the liquid, it is assumed that the composition is uniform, a reasonable assumption because of convection. Upon rapid cooling, the first nucleus will appear at finite undercooling, but here it is assumed to be small. Thus, the first crystal forms at temperature T_0 and has a composition of 10% B. The liquid will contain slightly greater than 30% B. If the temperature drops rapidly to T_1, the crystal grows in size, but the surface region will have a composition of 20% B and the liquid 57% B. (See the phase diagram in Fig. 2-12f.) The inside of the crystal, however, still has the composition 10% B since the rapid cooling has allowed no solid-state diffusion. Upon cooling from T_1 to T_2, the crystal grows in size; the outside layer now has the composition 30% B. Note that upon slow cooling, this is the temperature at which freezing is completed. However, upon rapid cooling, at this temperature the *average* composition of the solid is somewhere between 10% B (at the center) and 30% B (at the surface), and thus freezing cannot be completed since the average is not yet 30% B. Upon cooling to T_3, the outside of the crystal has 40% B and the liquid 70% B. When this last liquid region freezes, it remains very high in B. If we allow a slight amount of solid-state diffusion, the composition of the last layer to form will be about 40% B. The average composition of the solid is now 30% B, and freezing is completed.

Note that the description in Fig. 2-12 shows the crystal growing as a dendrite, and when freezing is complete the two dendrite arms have across them a concentration gradient, as shown in Fig. 2-12(e). Thus, when crystals form upon rapid cooling for such an alloy system, not only the grain boundaries but also the interdendritic regions have a high solute content. When a sample of such an alloy is carefully polished and then etched, the surface is attacked unevenly, so that the surface topology acquires high and low regions caused by the regions of higher and lower (than the average) B content. The regions of the etched surface which are not parallel to the surface will reflect light from an optical microscope out of the optical system, and these regions will appear dark. This etching process is illustrated in Fig. 2-13 and 2-14. The microstructure appears to be two-phase, but it is a single-phase structure. The dendrite arms when viewed end-on appear to have "cores," and hence this type of structure is referred to as cored, and the process of the formation of the chemically segregated crystals is called coring. Figure 2-15 shows the microstructure of a Ni-30Cu alloy which has been nonequilibrium cooled. Note that the dendrite structure is revealed, and the cored structure is easily seen. It is emphasized that the exact appearance depends upon the etchant used and the alloy.

The segregation in such an alloy can be removed by heat treatment. Since the grains contain concentration gradients, diffusion will occur with the solute-rich and solute-poor regions approaching uniformly the average

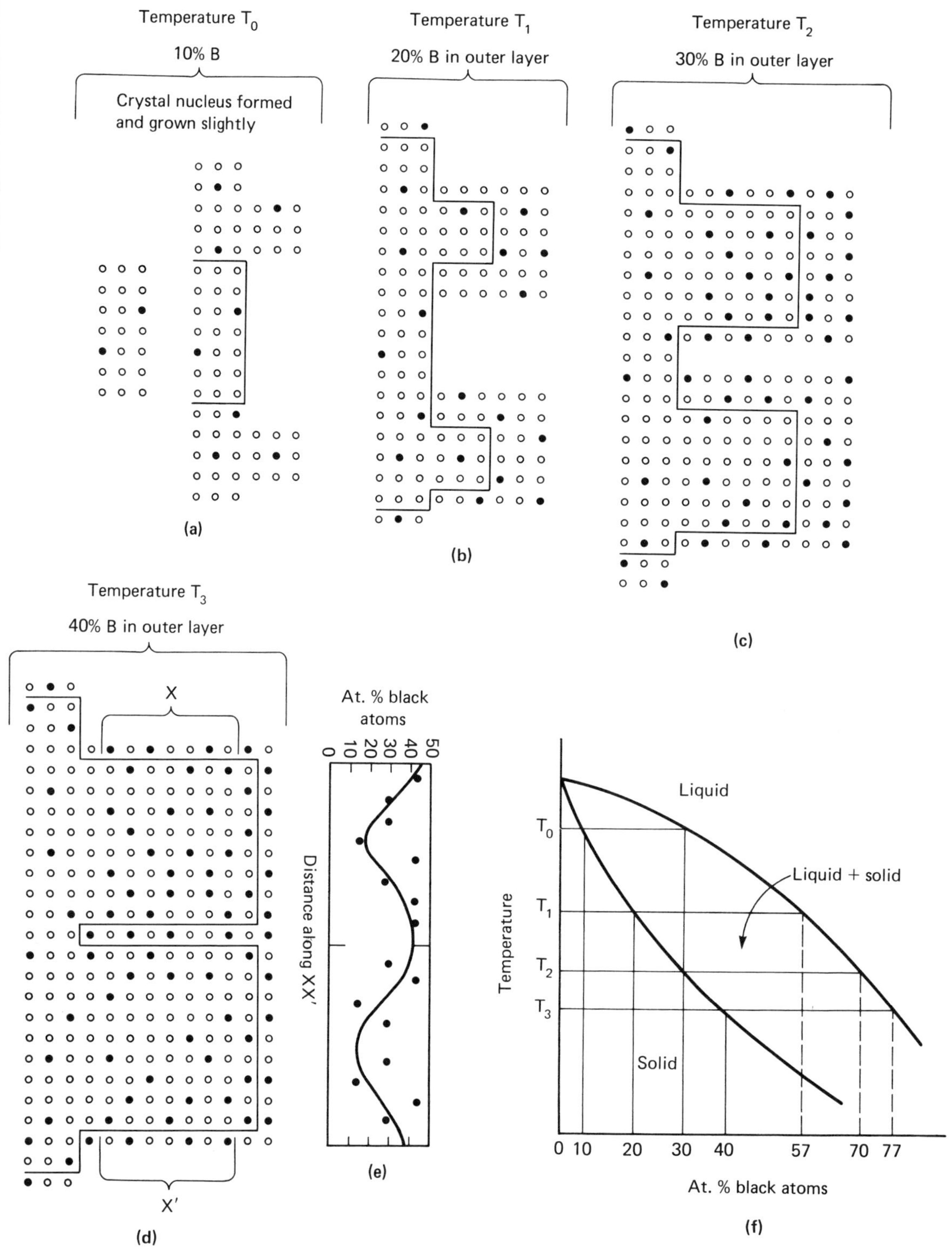

Fig. 2-12. Schematic illustration of the development of coring in a growing dendrite. See text for discussion.

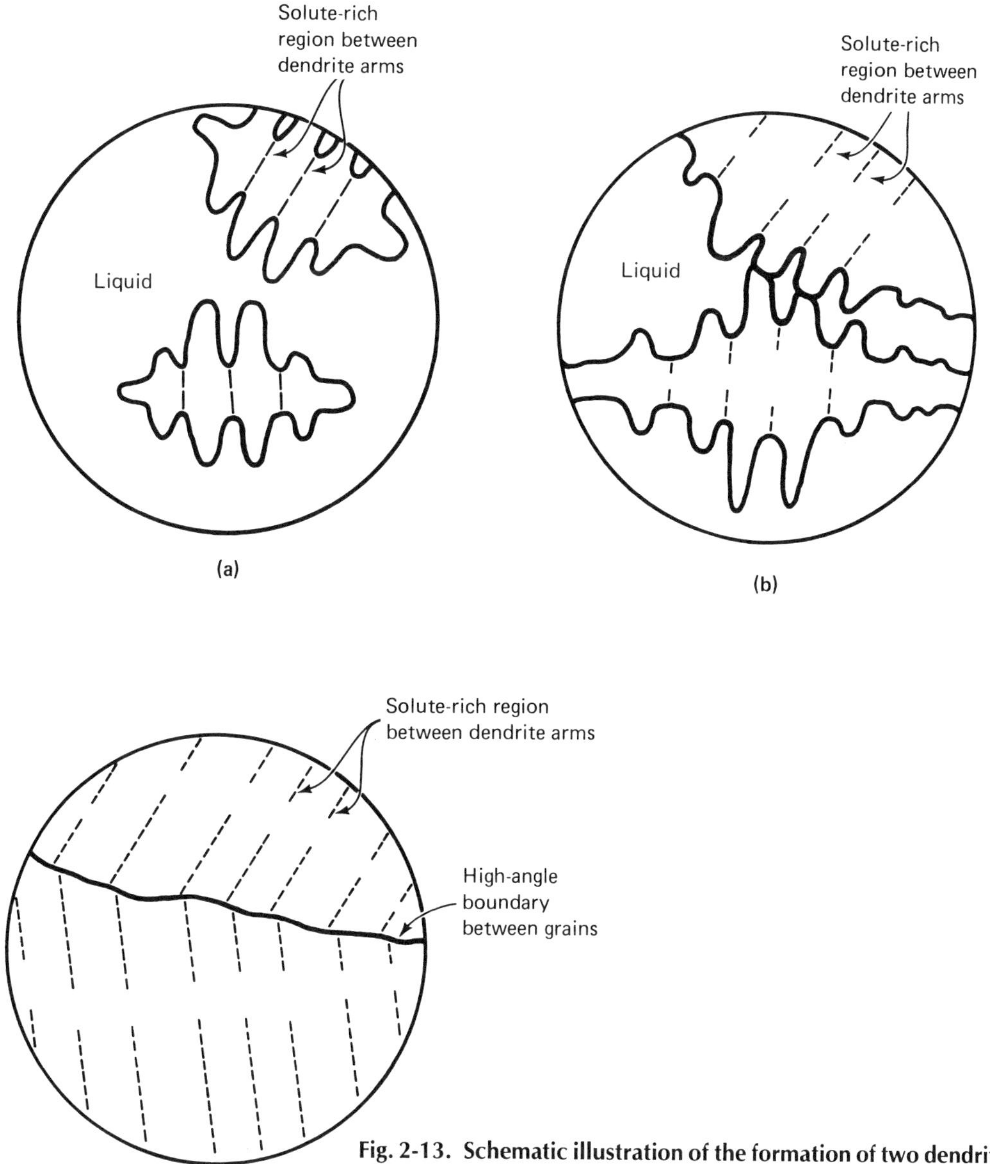

Fig. 2-13. Schematic illustration of the formation of two dendrites in an alloy system in which chemical segregation occurs. The boundaries between the dendrite arms are marked by the dashed lines; these regions are solute-rich if the alloy is 30% B (Fig. 2-12). The effect of etching a polished surface having the segregation shown in (c) is illustrated in Fig. 2-14.

composition of the alloy. Of course the heat treatment must be carried out below the solidus; in the case of the Ni-30Cu alloy in Fig. 2-15, this is below about 1340 °C. The higher the temperature, the shorter the time required. For example, the alloy whose as-cast structure is shown in Fig. 2-15 has the microstructure shown in Fig. 2-16 after heating for 5 h at 1000 °C, then cooling to 20 °C. The slight gradation in darkness is due to some slight

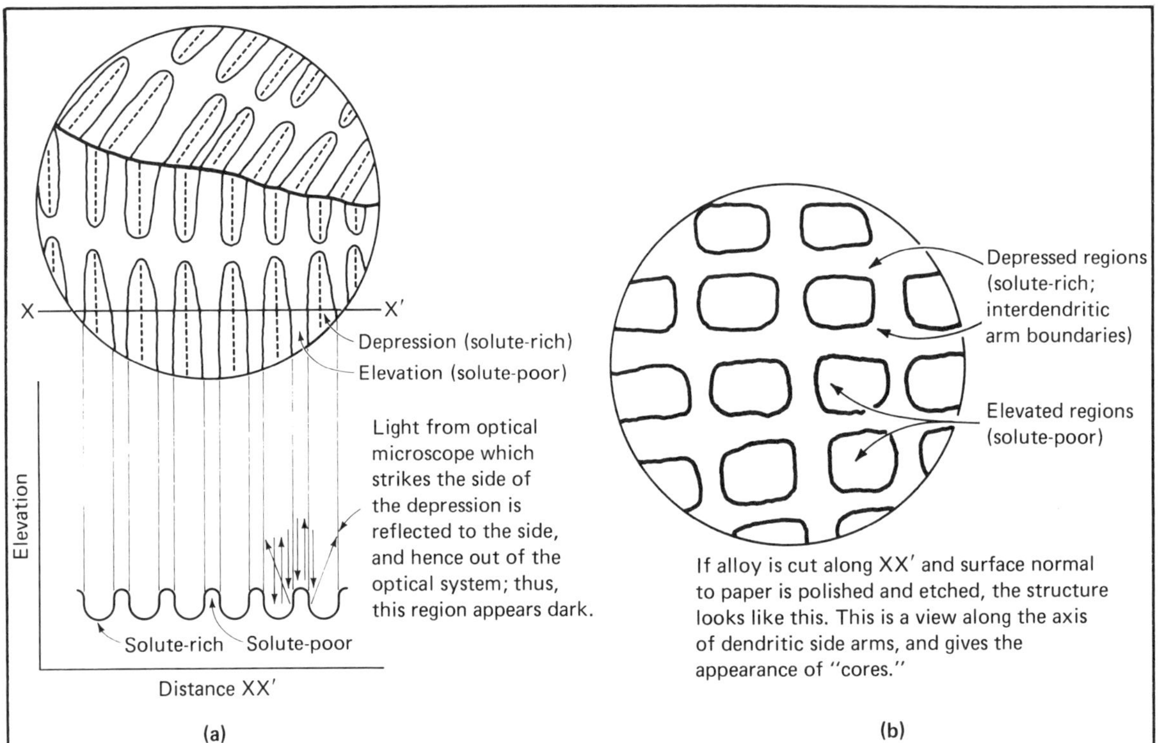

Fig. 2-14. Schematic illustration of the surface topology developed upon etching an alloy in which chemical segregation exists. If the plane of view in Fig. 2-13(c) is polished and etched, the solute-rich and solute-poor regions will be dissolved at different rates, leading to depressions (shown here along the solute-rich regions) and elevations. Upon viewing this surface with the optical microscope, the light striking the side of these regions will be reflected out of the optical system, and hence these regions appear dark. (b) Shows the appearance if the plane of cut is such that the ends of the dendrite arms are viewed.

coring remaining to cause an uneven surface, but it is seen that the alloy is more homogeneous than in the as-cast condition.

Upon solidification of some alloys, rapid cooling can cause not only coring but the appearance of other phases which would not be present upon slow cooling. It is important to understand this process, as it is responsible for phases in as-cast commercial material which are not predicted by the phase diagram. To illustrate this process, we will consider the microstructure developed in an Al-5Cu alloy which has been chill cast (cast into a cold mold, and hence rapidly cooled to 25 °C). The phase diagram is shown in Fig. 2-17. Note that upon cooling slowly from the liquid, this alloy passes into the liquid + α region, and then into the α region around 570 °C. This alloy remains single-phase α until the solubility limit is reached at about 525 °C. Below this temperature, the structure should consist of an α matrix with some θ precipitates. This description assumes that cooling is sufficiently slow that equilibrium cooling is attained.

Upon rapid cooling, coring is expected to develop. Thus the first α crystals to form will have about 1% Cu, and as the material cools the α

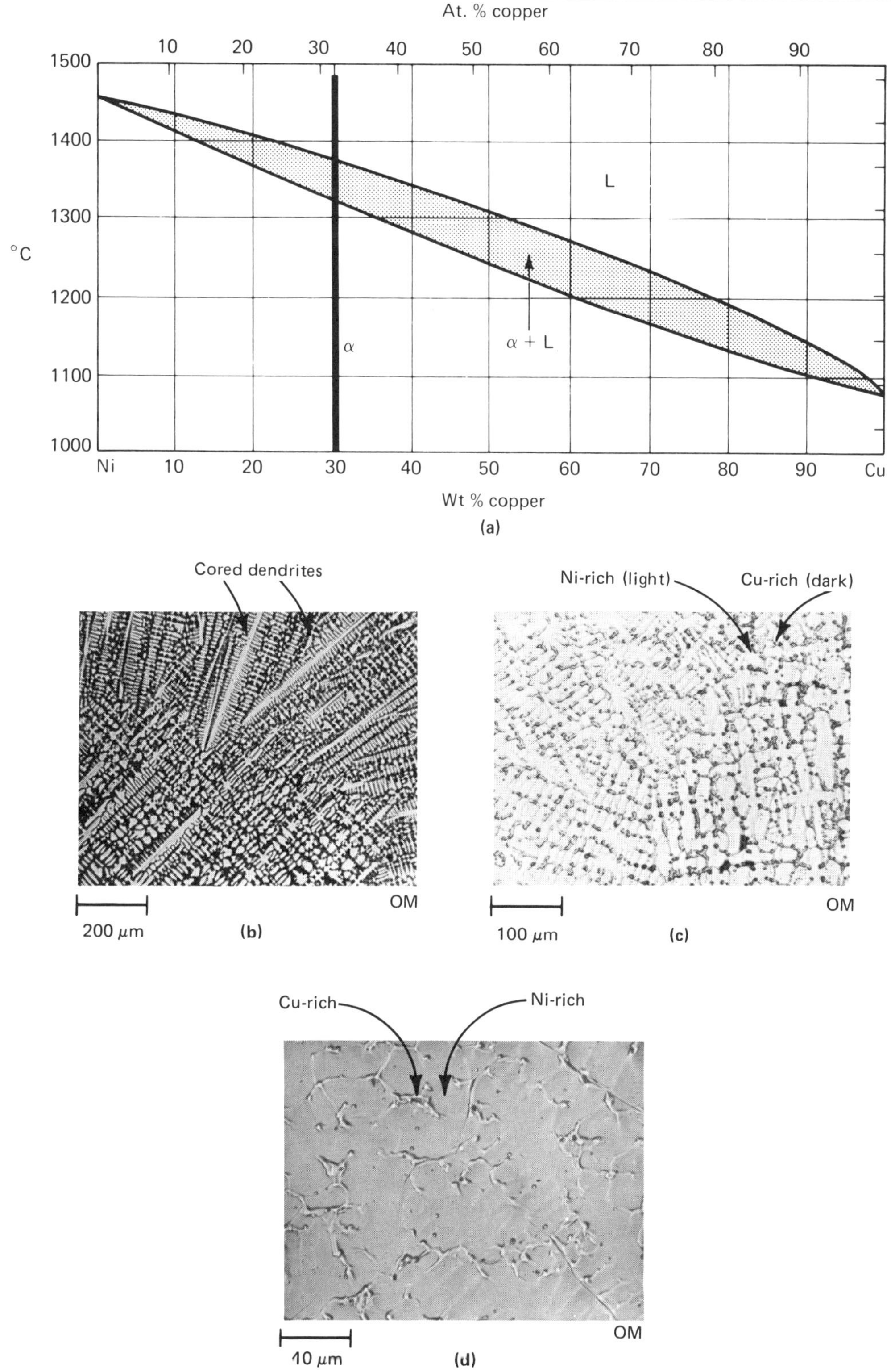

Fig. 2-15. Microstructure of a Ni-30Cu alloy which has been nonequilibrium cooled from the liquid to 20 °C. Coring is evident; this appearance originates by the mechanism described in Fig. 2-14 (especially compare with Fig. 2-14b). (Phase diagram adapted from Metals Handbook, 1948 Edition, American Society for Metals, Metals Park, Ohio)

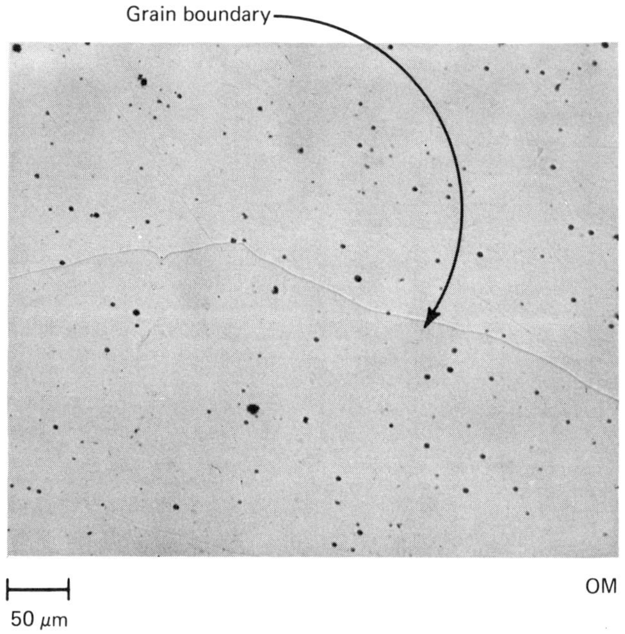

Fig. 2-16. Microstructure of a chill-cast Ni-30Cu alloy homogenized at 1000 °C for 5 h, then cooled to 20 °C. The microstructure before this heat treatment is shown in Fig. 2-15(b). The black dots are oxide inclusions and shrinkage cavities. Note the relative absence of coring.

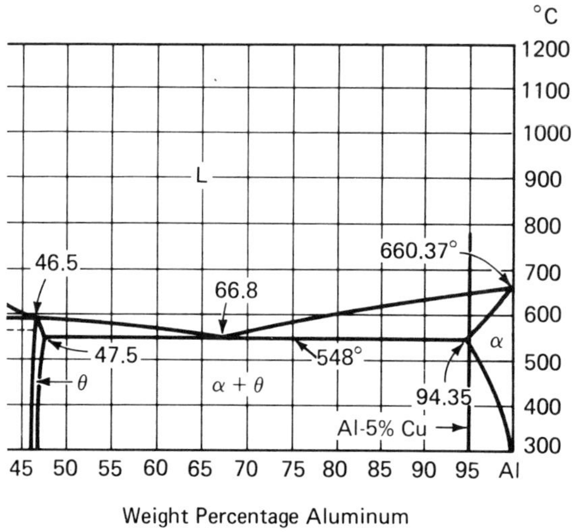

Fig. 2-17. The Al-Cu phase diagram, showing the region of Al-5Cu. (Adapted from same source as Fig. 2-5)

crystals increase in size, with each layer of atoms added to the crystal surface being richer in copper than the preceding layer. When the temperature of 570 °C is reached (at which freezing would be complete if the alloy were cooled slowly) the crystals have 1% Cu at the center but 5% Cu at the surface. Thus the average composition is somewhere between 1 and 5% Cu, and hence freezing is not yet complete. Further cooling increases the size of these α crystals, and the chemical composition of the outside of the crystals reaches about 5.6% Cu when the temperature reaches 548 °C, the eutectic temperature (see Fig. 2-18). At this temperature, the remaining liquid contains 66.8% Cu, which is the eutectic composition. Thus, this liquid should freeze to the eutectic solid mixture of $\alpha + \theta$. Figure 2-18 shows the process schematically.

Figure 2-19 shows microstructures of this alloy after chill casting. Note two features which reflect the above process. One is the light and gray areas, caused by the etch dissolving the α regions at different rates because of the difference in the Cu content caused by the coring. The other prominent feature is the rather dark gray areas which are clearly associated with the regions between the arms of the dendrites. It might be predicted that these regions are θ crystals, formed when the alloy cooled below 525 °C. However, the rapid cooling prevents this solid-state precipitation, so this process is not their origin. The uncertainty is resolved by high magnification, where the two-phase eutectic structure is clearly resolved (Fig. 2-19c), its origin being that described in Fig. 2-18.

Thus, it is seen that in some alloys rapid cooling may produce phases and structures which would not be present after slow cooling. The process

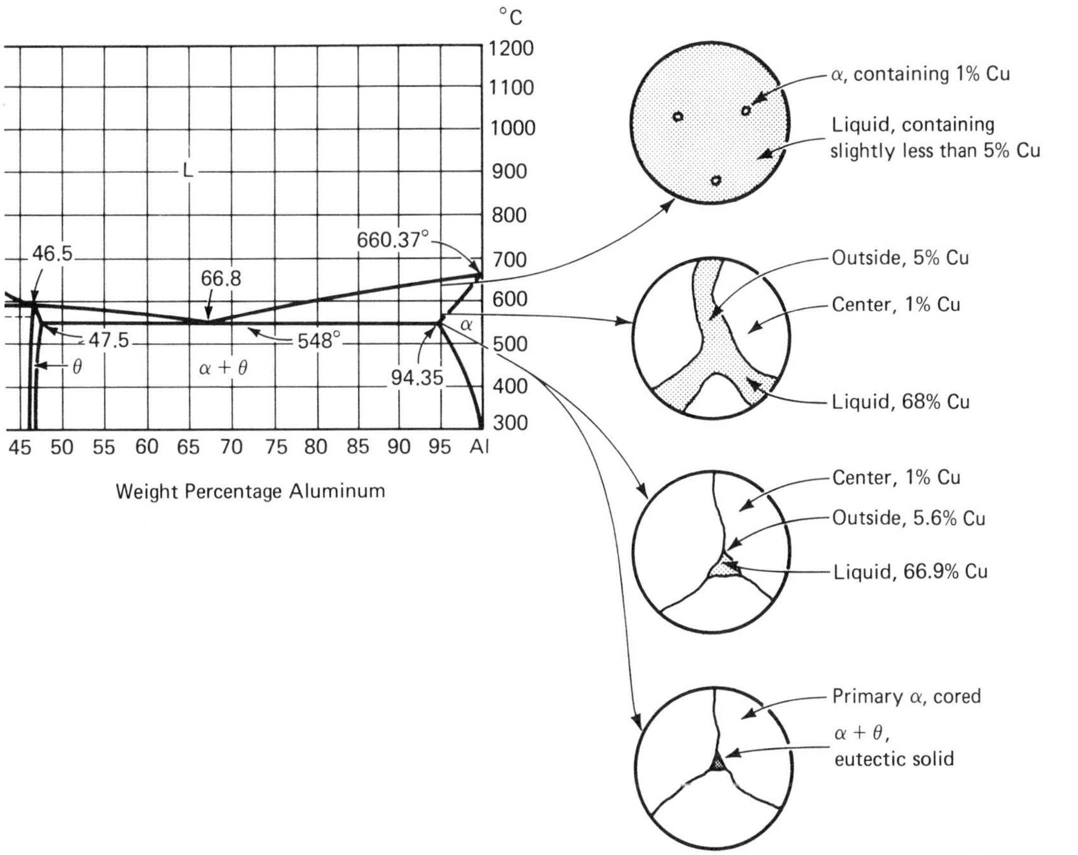

Fig. 2-18. Schematic representation of the development of a eutectic structure upon solidification by rapid cooling.

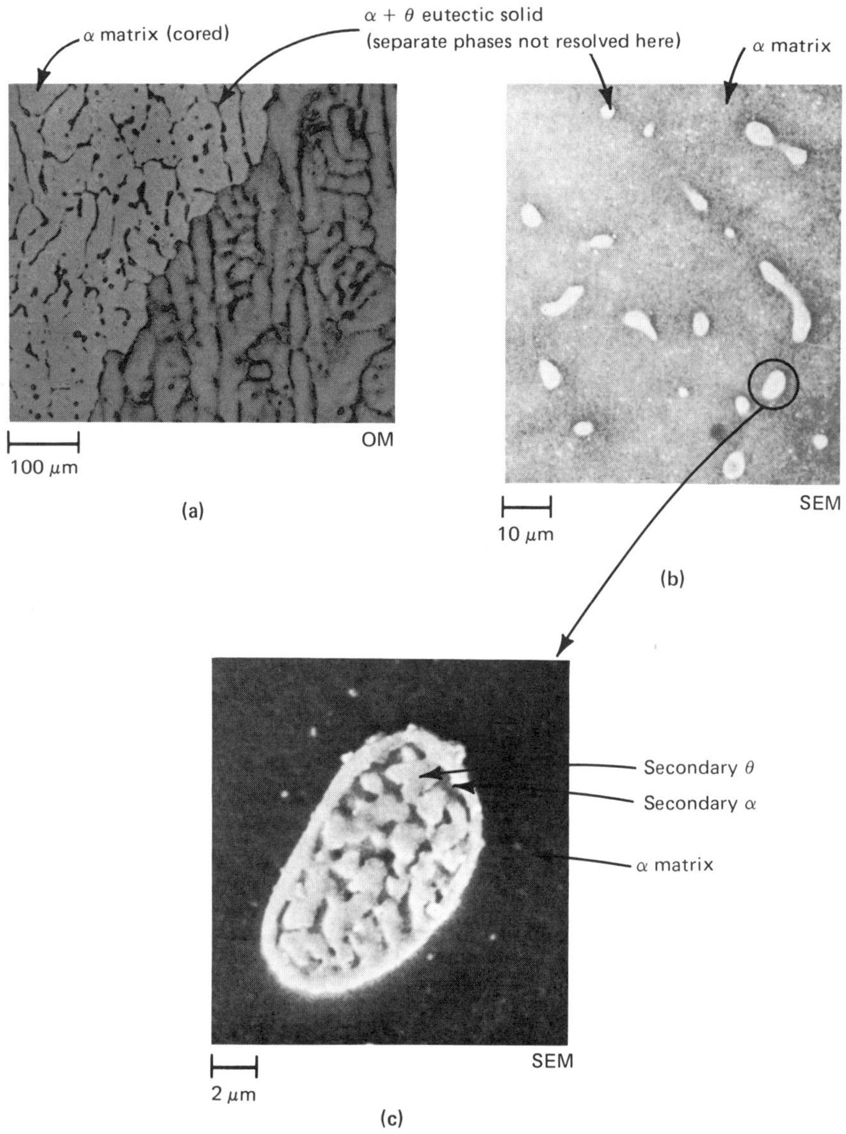

Fig. 2-19. Microstructures of an Al-5Cu alloy in the chill-cast condition, showing coring and eutectic solid formed through the coring process.

just described is responsible for many nonequilibrium structures in cast commercial alloys. This will be emphasized in subsequent chapters dealing with such alloys.

In many cases it is possible to dissolve the second phase by proper homogenization in the high-temperature single-phase region. For example, the Al-5Cu alloy is in the single-phase region at about 550 °C (see Fig. 2-17). Thus, if this alloy, after chill casting and hence with the two-phase microstructure shown in Fig. 2-19, is heated to 550 °C for one week, then cooled rapidly (to prevent θ precipitation) to 25 °C, the microstructure

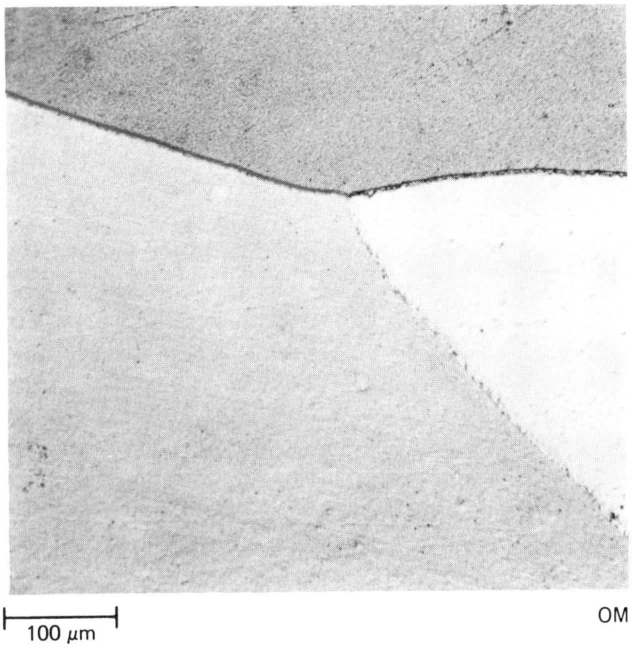

Fig. 2-20. Microstructure of an Al-5Cu alloy, originally chill cast, then homogenized at 550 °C for 1 wk and cooled rapidly to 25 °C. This heat treatment has removed the coring and dissolved the θ phase. Compare the as-cast structure in Fig. 2-19(a). (The slight grain contrast and decoration along the grain boundaries are caused by the etching, not by any precipitates present.)

shown in Fig. 2-20 is obtained. Note the absence of the θ phase and the lack of coring. Thus, this homogenization treatment has dissolved the θ and removed the coring, to obtain a well-homogenized, single-phase structure.

References

Liquid Metals and Solidification, American Society for Metals, Metals Park, Ohio (1958).

An Introduction to the Solidification of Metals, W. C. Winegard, Institute of Metals, London (1964).

Principles of Solidification, B. Chalmers, Wiley, New York (1964).

Transformations in Metals, P. G. Shewmon, McGraw-Hill, New York (1969).

Solidification, American Society for Metals, Metals Park, Ohio (1971).

Solidification Processing, M. C. Flemings, McGraw-Hill, New York (1974).

CHAPTER 3

Precipitation Hardening

A method of strengthening a metallic material is to add hard, finely dispersed particles. Although in principle this can be done by mixing, mixing is not always practicable. However, by proper choice of alloying additions and heat treatment, a dispersion can be obtained by solid-state precipitation of a second phase from the matrix in which it is dissolved. If strengthening occurs, this is precipitation hardening, a widely used method of strengthening metallic materials. In this chapter, the principles of precipitation hardening are reviewed, and in subsequent chapters this heat treatment will be examined as a method of strengthening aluminum-base, copper-base and nickel-base alloys.

3-1. Solid Solutions

If the chemical composition of a single-phase solid alloy can be varied, within limits, and yet maintain the single phase, the alloy is a solid solution. If the phase is based on the pure solvent, it is called a terminal solid solution. If the phase has a crystal structure different from that of any of the atoms (as pure elements) present in the solution, it is an intermediate solid solution. In both cases, it is important to realize that the atoms are arranged at random (or nearly so) on the lattice sites, and that any small grouping has the same chemical composition of any other if the alloy is at equilibrium. This is illustrated schematically in Fig. 3-1(a). In such an alloy, the microstructure may appear like that shown in Fig. 3-1(b), but here the dark grains and the light grains all have the same zinc content, 30%. The grain contrast is caused by the etch. (See Fig. 1-57.)

(a)

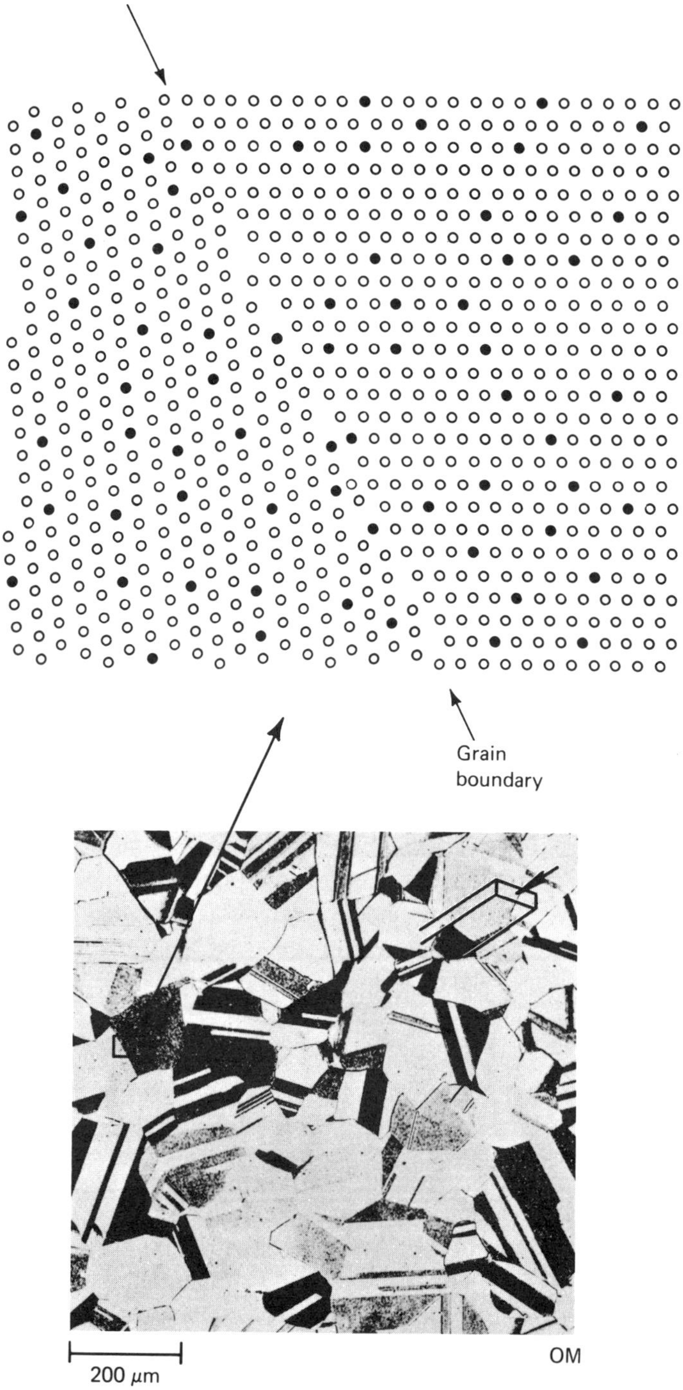

(b)

Fig. 3-1. (a) Schematic representation of two grains of a solid solution alloy containing 10% B–90% A atoms. (b) Microstructure of a Cu-30Zn solid solution alloy. The alloy has been etched to develop grain contrast, caused by the difference in orientation of each grain relative to the etched surface (see Fig. 1-57). Each grain contains 30% Zn (adapted from Metals Handbook, 8th Ed., Vol 7, American Society for Metals, Metals Park, Ohio, 1972).

Solid solutions (whether terminal or intermediate) can be formed by replacing a solvent atom at a lattice site by a solute atom; this forms a substitutional solid solution. However, if the solute atom is located among the solvent atoms, in their interstices, it is called an interstitial solid solution. The latter requires that the solute atom be relatively small, and usually occurs only for H, C, N, B and perhaps O as the solute atoms.

If we consider a property of a pure metal, such as density, it is to be expected that upon the addition of a solute the property will change. Since the crystal structure remains constant as the solute content increases, the density will increase if the solute atom has a higher mass than that of the solvent, and otherwise decrease. However, the effect usually is not linear, indicating an attraction or repulsion between the solute-solvent atoms. That is, the solution is not ideal. The property will, though, vary smoothly. If the solute is completely soluble in the solvent, then the property will change smoothly from that of one pure element to that of the other as the concentration of the latter is increased. This is illustrated in Fig. 3-2 for nickel-copper alloys. The phase diagram shows that only a single phase will exist at equilibrium in the solid state, and the strength data show a smooth transition from the values of pure nickel to those of pure copper.

3-2. Solubility Limit

The formation of terminal solid solutions will be favored by the presence of atoms of similar size, similar electronic structure and the same crystal structure. Nickel and copper differ in atom size by only 2.5%, they differ in the number of valence electrons by one, and they are both face-centered cubic. If these three factors are not closely similar, only limited solubility exists. This can be envisioned as follows. In a pure solvent, a few atoms are removed and in their place solute atoms substituted, and the mixture is allowed to come to equilibrium to form a solid solution. Then a few more solute atoms are substituted, to increase the solute concentration. Eventually, however, a level of solute concentration is attained such that, when a few atoms of the solvent are removed and solute atoms substituted, the matrix will not dissolve any more solute atoms, and these excess solute atoms are rejected. They may group together to form a small crystal of their own, perhaps with a few solvent atoms dissolved in it, or react with some of the solvent atoms to form a phase of a crystal structure different from that of either pure metal (i.e., an intermetallic compound).

The composition at which this precipitation will just occur is the solubility limit, and on the phase diagram bounds one side of the two-phase region. Such a case is illustrated in Fig. 3-3 for the copper-zinc system. For example, at 200 °C, the maximum amount of zinc which can be added to copper and still have at equilibrium a single phase which is face-centered cubic is 35% (point a). If more zinc than this is present in the alloy, the excess above 35% is present as an intermetallic compound γ containing 59% zinc (point b). The solubility limit is quite temperature dependent, as the

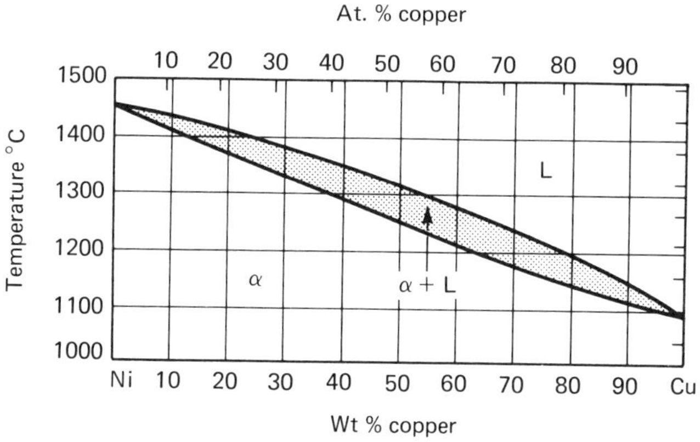

(a)

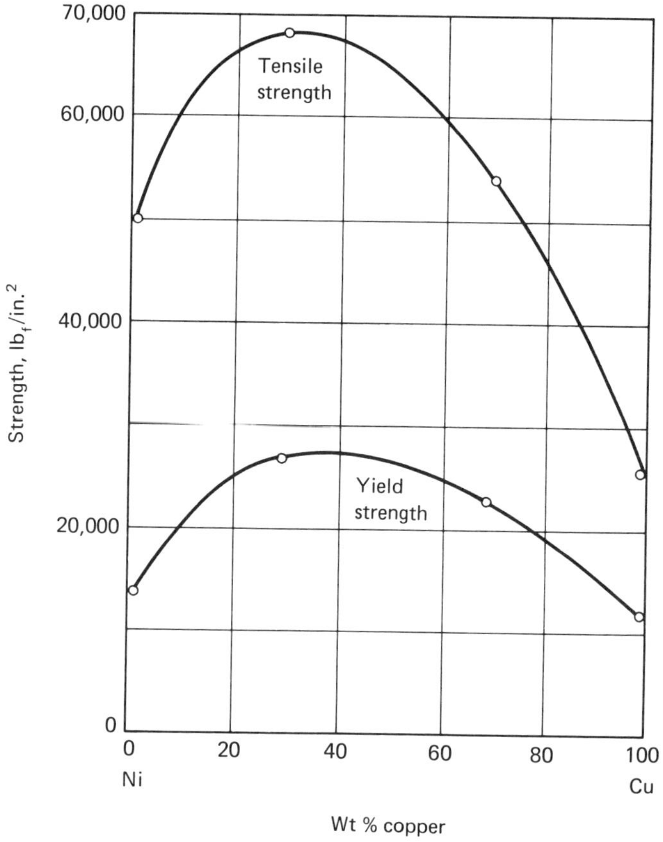

(b)

Fig. 3-2. (a) Phase diagram of the Ni-Cu system showing that Cu and Ni are completely soluble in each other in the solid state, so that only one phase is stable for any composition (adapted from Metals Handbook, 1948 Edition, American Society for Metals, Metals Park, Ohio). (b) The strength data, measured at 20 °C, vary smoothly from those of pure Ni to those of pure Cu as the Cu content increases but do exhibit a maximum (adapted from W. D. Jenkins, T. G. Digges and C. R. Johnson, J. Res. NBS, Vol 58, p 201, 1957).

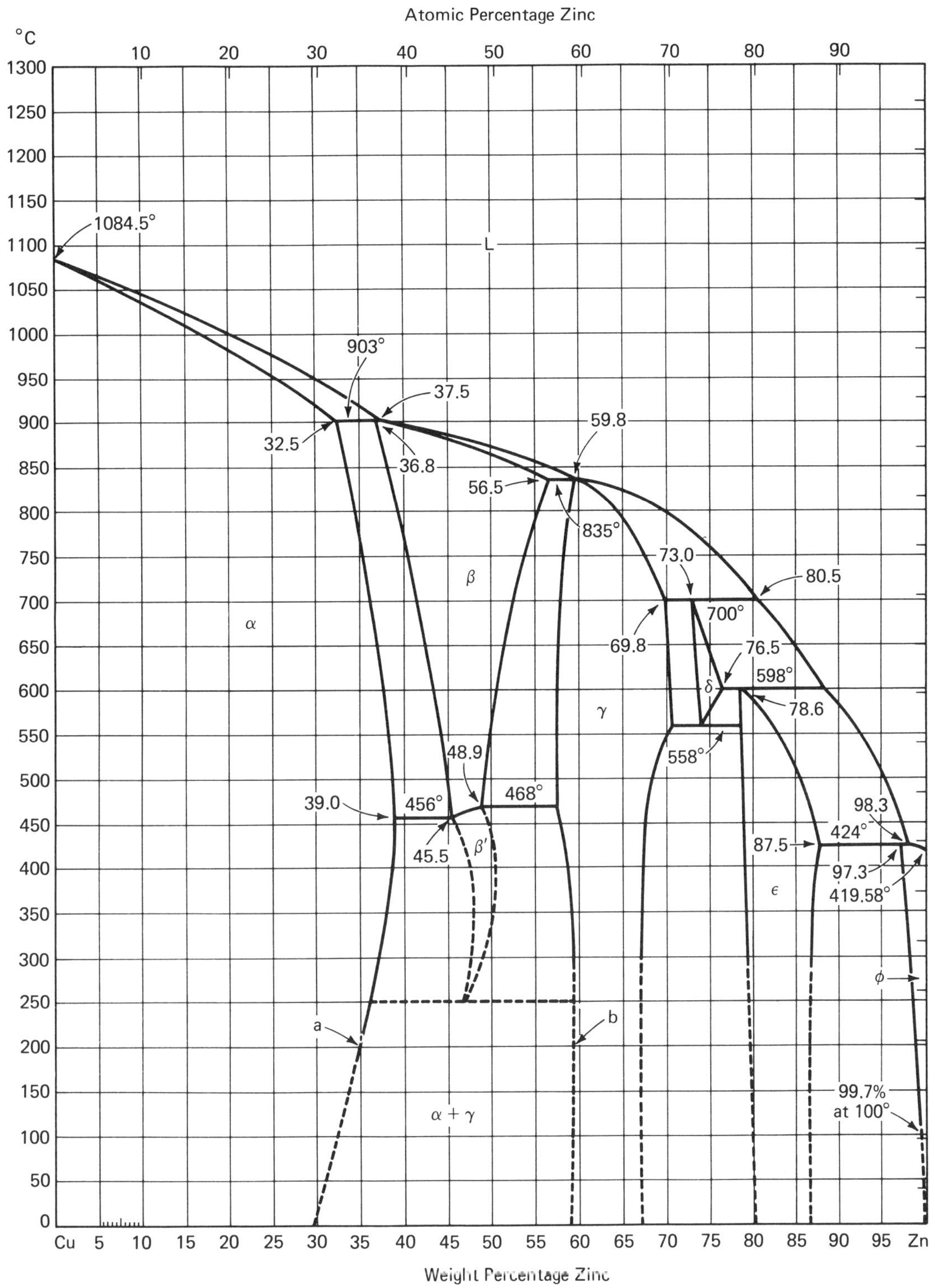

Fig. 3-3. The Cu-Zn phase diagram illustrating the solubility limit of Zn in Cu. (Adapted from Metals Handbook, 8th Ed., Vol 8, American Society for Metals, Metals Park, Ohio, 1973)

boundary shows. The composition of the γ in equilibrium with the α is also temperature dependent.

Even if the solubility is limited, in the solid solution range the properties will vary smoothly. For example, hardness as a function of the zinc content in the α range is shown in Fig. 3-4.

3.3. Precipitation From Solid Solution

Now we will consider the mechanism whereby precipitation occurs from a supersaturated solid solution. Assume that a system A-B exists such that part of the phase diagram is that shown in Fig. 3-5. An alloy containing 10% B is heated to a temperature T_0, in the single-phase, solid solution range, until equilibrium is attained. It is then cooled quickly to a temperature T_1, at which at equilibrium a two-phase mixture of α and θ exists. We assume that the cooling rate upon going from T_0 to T_1 is sufficiently rapid that no change occurs to the α other than the decrease in temperature to T_1. Thus, initially at T_1 the α is supersaturated, since it contains 10% B, but the solubility at T_1 is only 1% B. With time at T_1 some of the B atoms will react with some of the A atoms to form θ particles, causing the B content of α to be reduced. Eventually, at equilibrium, this becomes 1%.

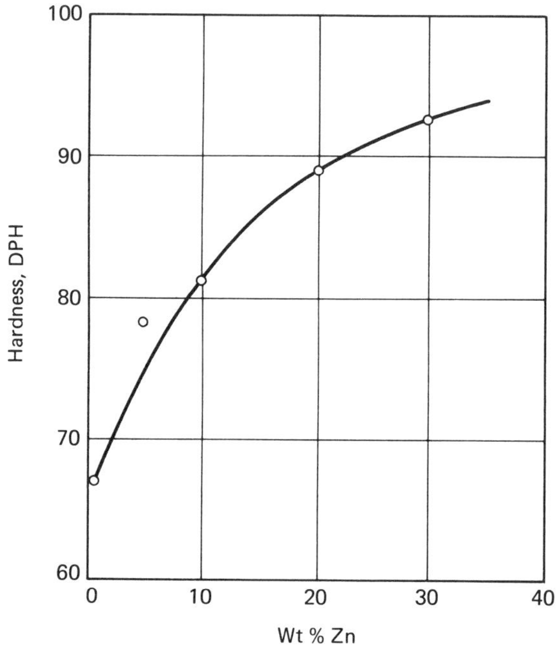

Fig. 3-4. Hardness as a function of Zn content for Cu-Zn solid solution alloys in the α terminal solid solution region.

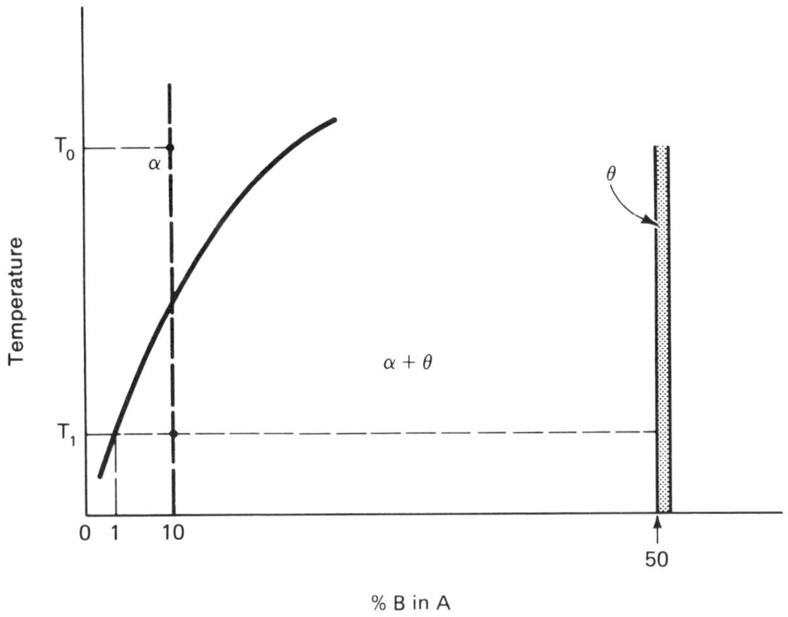

Fig. 3-5. Hypothetical phase diagram of system A-B. The decreasing solubility of B in α with decreasing temperature allows an alloy containing 10% B to be single-phase at high temperature (e.g., T_0) but two-phase at low temperature (T_1).

Figure 3-6 illustrates schematically the atom movements that might occur as a crystal of θ forms in the supersaturated α. (In actuality, the atoms move in the lattice by the vacancy mechanism, but for convenience of illustration the vacancies have not been shown.) It is seen that a small group of atoms statistically form a region having the lattice and composition of θ. This reduces the B content of the α surrounding the θ so that a concentration gradient of B exists in the α between regions away from the θ and regions close to θ. This gradient then acts to drive B atoms toward the α-θ interface, where the B atoms take on sites to increase the size of the θ crystal. The gradient is illustrated in Fig. 3-6(f), for the section shown in Fig. 3-6(e). (To actually obtain the chemical composition as a function of distance requires sampling a larger number of atoms than the group shown.) These figures show the quite early stages of formation of the precipitate, and at (e) equilibrium has not yet been reached.

Figure 3-7 shows schematically that the gradient in the α matrix is retained as the precipitate grows. This is a steady-state process, and the growth rate is controlled by the diffusion process. It is related to temperature by the expression exp $(-Q/RT)$, where Q is a constant, the activation energy for the diffusion process, R is the ideal gas constant and T is the absolute temperature. Eventually, the gradient zones begin to overlap with those of neighboring precipitates, the gradient decreases, the growth rate decreases, and finally chemical equilibrium is attained, with the matrix being uniformly 1% B.

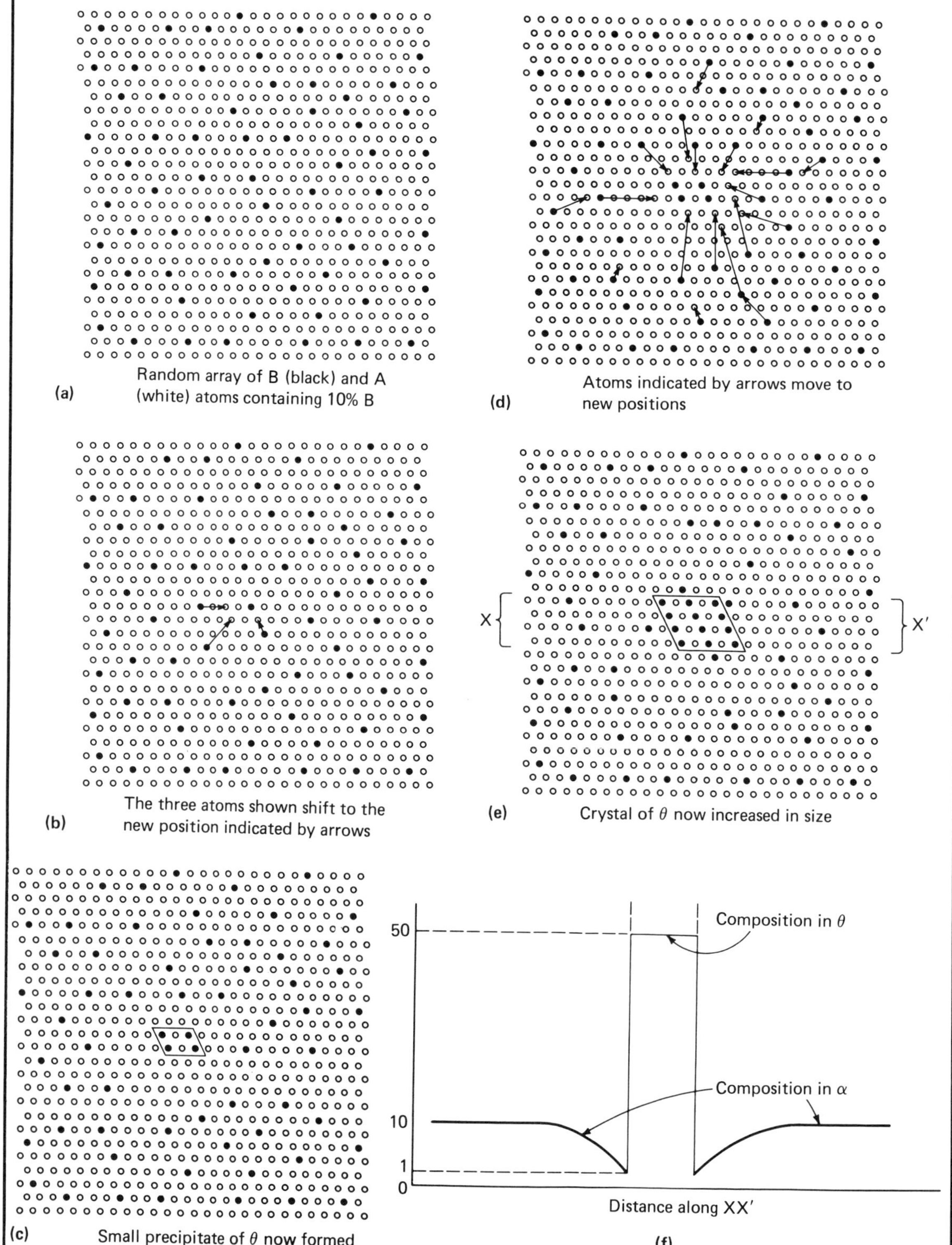

Fig. 3-6. Schematic illustration of a possible mechanism of precipitation from a supersaturated solid solution. Time is increasing from (a) to (e), but at (e) equilibrium is not attained.

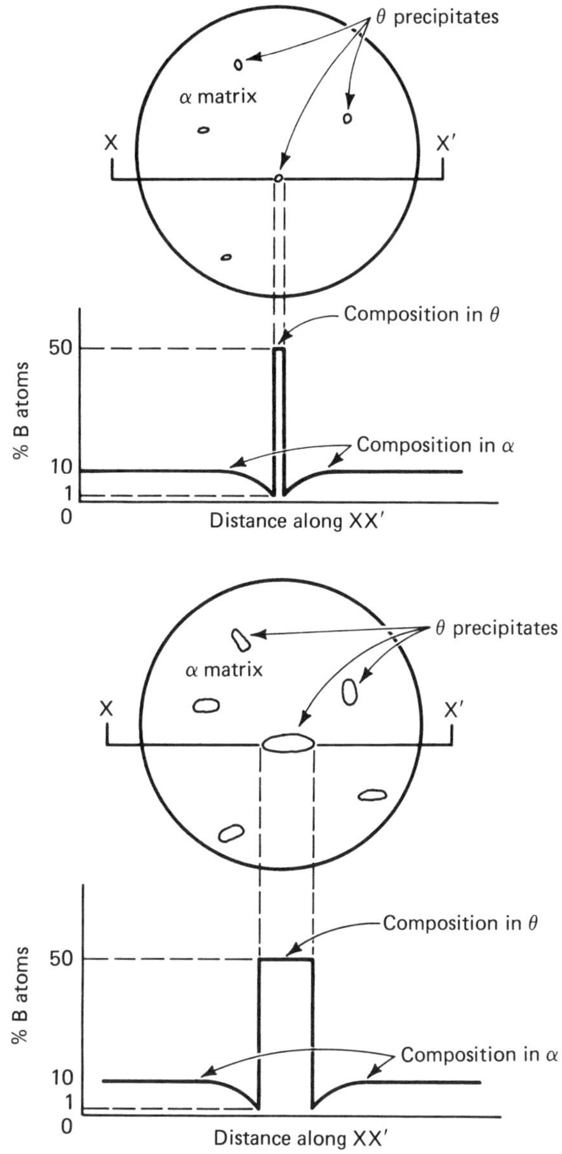

Fig. 3-7. Schematic illustration of the retention of the composition gradient in the α matrix as the precipitate grows.

3-4. Effect of Precipitation on Properties

In the previous section we examined possible atom movements during precipitation from a supersaturated solid solution. We will see that we will have to modify this picture somewhat, but before examining the details of the mechanism of the hardening associated with the precipitation process, we will first look at the effect of precipitation on properties.

To illustrate the changes which can occur, we will examine the behavior of an Al-4Cu alloy. The pertinent portion of the phase diagram is shown in

Fig. 3-8. If the alloy is heated for several hours at 550 °C, it will be a single-phase structure. If it then is cooled quickly (e.g., quenched in water) to 20 °C, this single-phase structure will be retained, and thus a supersaturated solid solution is obtained. The alloy is then heated to an inter-

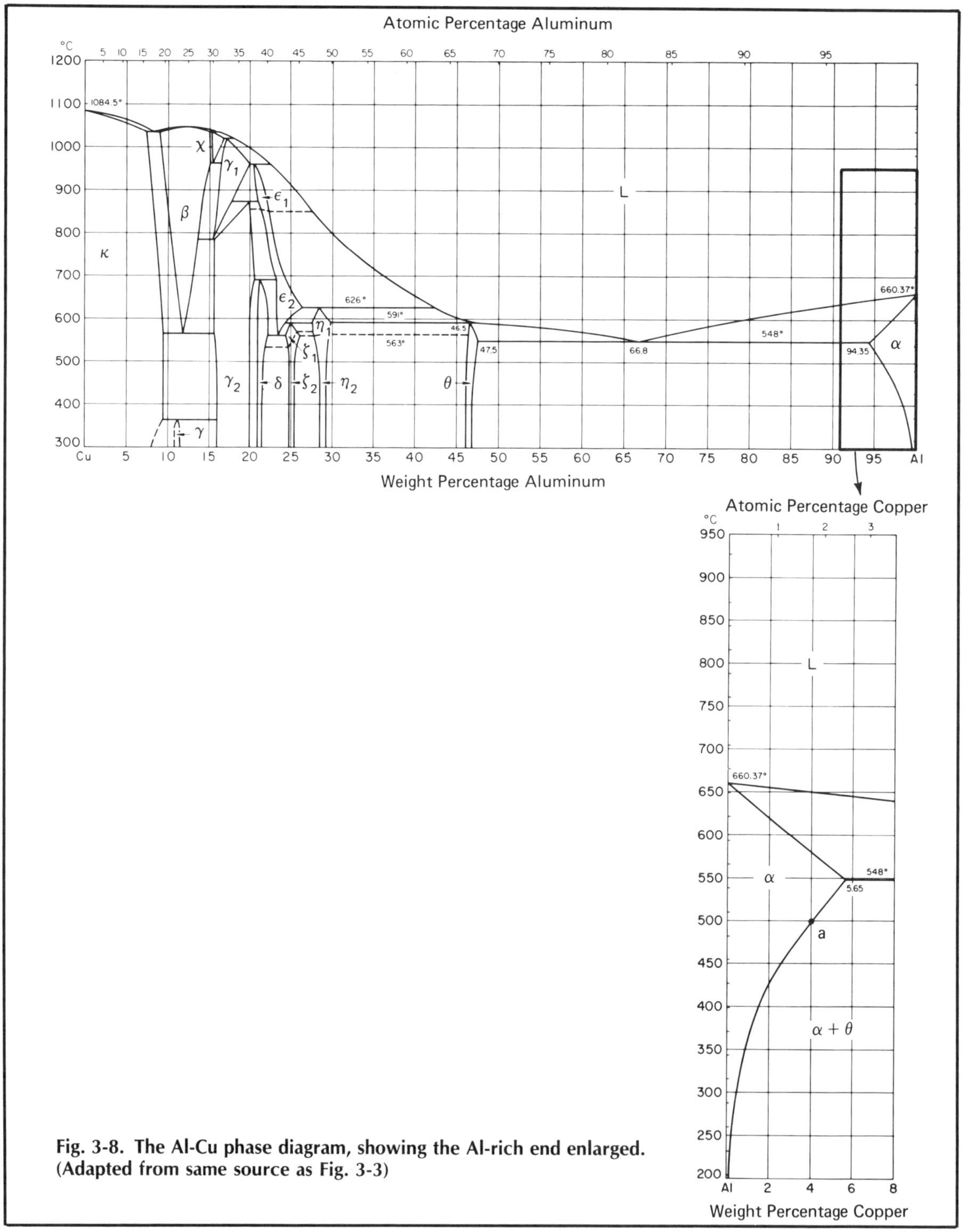

Fig. 3-8. The Al-Cu phase diagram, showing the Al-rich end enlarged. (Adapted from same source as Fig. 3-3)

mediate temperature (below 500 °C, point a in Fig. 3-8), and the properties are measured as a function of time. Figure 3-9 shows typical data obtained.

Note that the hardness and strength properties pass through a maximum, and that the curves are quite temperature dependent. As we will see, the maximum is associated with the precipitate attaining a critical size and distribution. When the particles become too coarse, they are too dispersed to be as effective in retarding dislocation motion, and hardness decreases. The formation of these particles to an optimum size and distribution will be more rapid at higher temperature, so that the maximum in the hardness curves occurs at shorter times as the precipitation temperature increases. However, the amount of precipitate which forms decreases (the degree of supersaturation decreases) with increasing temperature, and the magnitude of the hardness maximum decreases. Also, it is expected that at a given precipitation temperature the maximum will be attained faster the higher the solute content, because the supersaturation will be greater, and therefore diffusion will be faster.

The strengthening which occurs is called *precipitation hardening;* if precipitation occurs at room temperature, the strengthening is called age hardening.

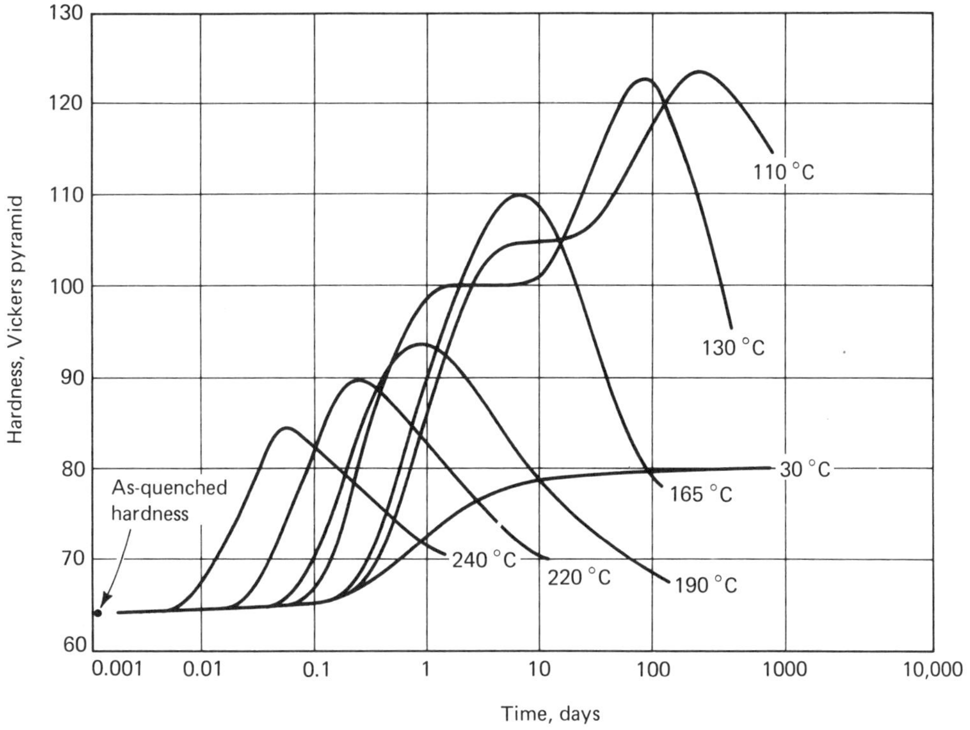

Fig. 3-9. Hardness as a function of time for an Al-4Cu alloy. The alloy was solution heat treated for at least 48 h at 520 °C, then cooled quickly (water quenched) to 20 °C. (Adapted from H. K. Hardy, J. Inst. Metals, Vol 79, p 321, 1951)

3-5. Precipitation Mechanism

In this section we will see that the schematic description (Fig. 3-6) of the precipitation process will have to be modified. This can be seen best by examining the microstructures developed upon precipitation; to do this, we consider an Al-5Cu alloy. If the alloy is heated for one week at 545 °C, it will equilibrate to the single-phase α (Fig. 3-8). If the alloy is then cooled slowly to 20 °C, precipitation occurs once the temperature decreases below the solubility limit (about 525 °C, Fig. 3-8), and the microstructure shows θ precipitates in an α matrix (Fig. 3-10). Note that the precipitates have formed on the α grain boundaries, and also in the grains. Some of those in the grains are parallel, and further examination would show that the precipitates are plates which are parallel to the {111} planes of the α. This type of geometric appearance is referred to as *Widmanstätten structure*. (See Section 9.6.)

Now if the alloy, after being heated for a week at 545 °C so that it is single-phase α, is cooled rapidly to 20 °C, the precipitation of θ is suppressed by the rapid cooling. Also the rate of precipitation at 20 °C is too slow for significant amounts of θ to form except for long aging times. Thus, the microstructure at 20 °C is single-phase α, as shown in Fig. 3-11.

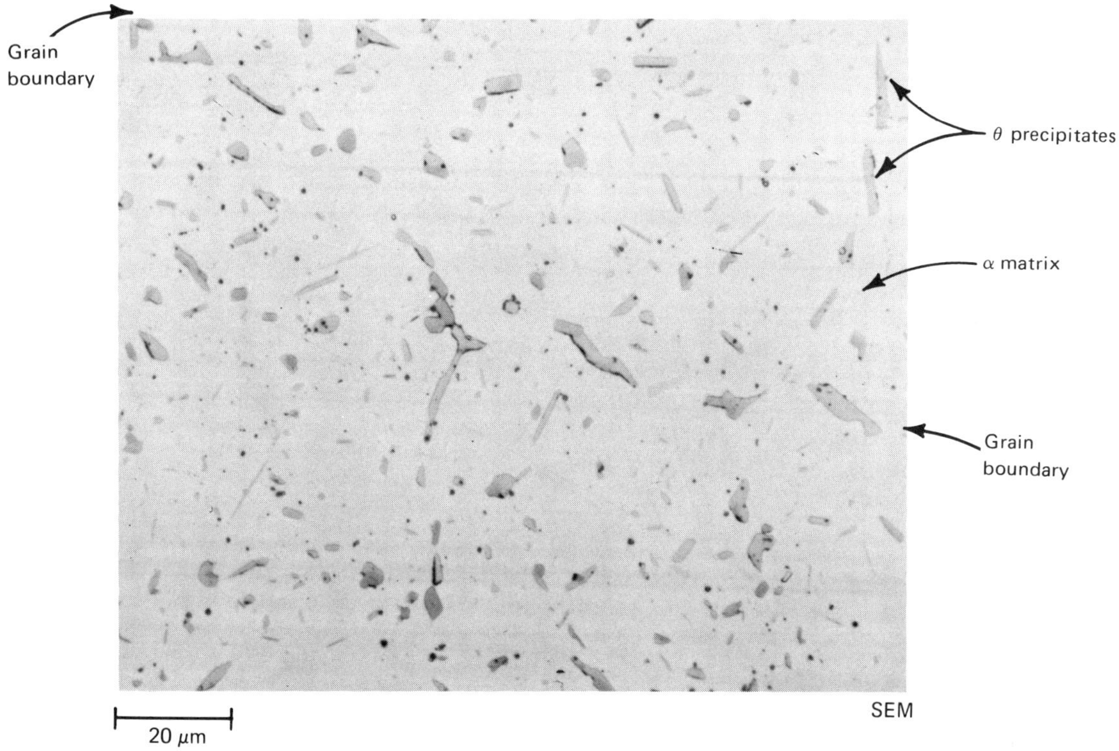

Fig. 3-10. Microstructure obtained after heating an Al-Cu alloy for 1 wk at 545 °C, then cooling slowly to 20 °C. The θ precipitates are quite prominent.

If, after rapid cooling to 20 °C, the alloy is heated to 300 °C, held for 12 h and then cooled rapidly to 20 °C, the microstructure developed is that shown in Fig. 3-12. The structure is quite fine. High magnification is required to resolve clearly the θ precipitates. If the alloy is heated at 400 °C for 12 h, the structure is coarser, the θ particles being clearly resolved in the optical microscope (Fig. 3-13).

In Fig. 3-13 it can be seen that larger particles are in the grain boundaries. This reflects the higher nucleation rate along the grain boundary, where diffusion is faster and atoms are more mobile. Thus, nucleation occurs first here, and later inside the grains. The early nucleation and growth of the particles in the α grain boundaries depletes the region adjacent to the boundary of solute (copper atoms), and thus this region has a matrix composition close to the equilibrium. This region then forms no precipitates when the matrix itself begins to nucleate θ. This sometimes (but not in Fig. 3-12) leaves a precipitate-free region on each side of the grain boundary, referred to as a denuded zone.

The hardness of the 5% copper alloy as a function of time at the precipitation temperature will pass through a maximum (similar to that shown in Fig. 3-9). However, this maximum occurs considerably before the precipitates become as coarse as those shown in Fig. 3-12 and 3-13. To reveal the early stage of precipitation requires electron microscopy, and Fig. 3-14 shows the type of microstructures that develop. In the quite early stages, thin plates of almost pure copper form, referred to as Guinier-Preston

Fig. 3-11. Microstructure of an Al-5Cu alloy heated for 1 wk at 545 °C, then cooled rapidly (water quenched) to 20 °C. The structure is single-phase α. Optical micrograph; 150×.

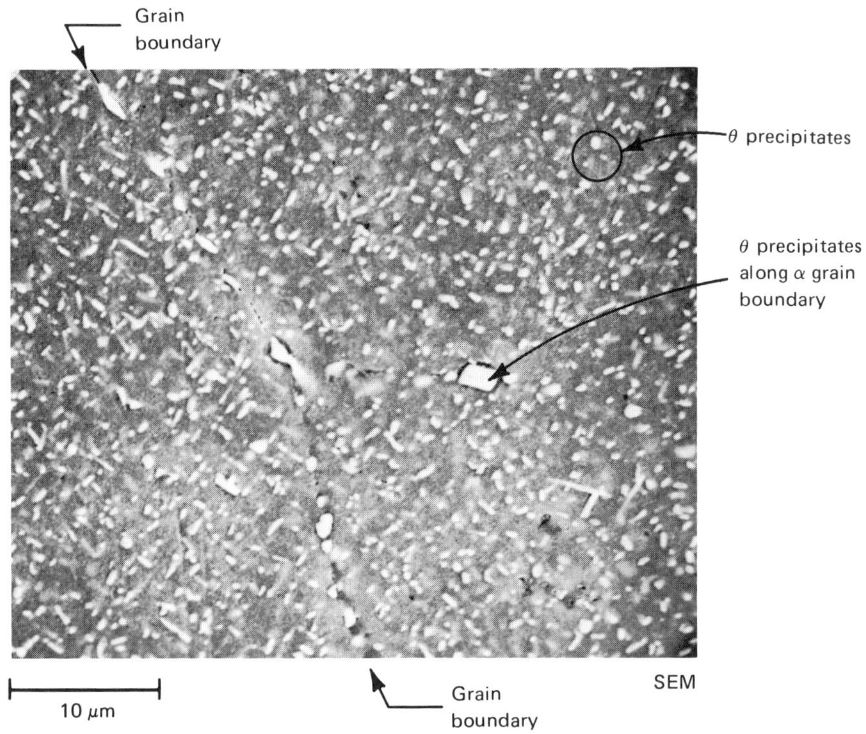

Fig. 3-12. Microstructure of Al-5Cu alloy heated for 1 wk at 545 °C, cooled rapidly to 20 °C, then held at 300 °C for 12 h, then cooled rapidly to 20 °C. The θ particles are clearly resolved.

(GP) zones. This formation is followed by the development of thicker regions, consisting of alternate layers of copper and an aluminum-copper mixture; this is called θ''. In these two stages the atoms of the precipitate remain approximately aligned with those of the matrix, but elastic strain occurs. In the next stage the θ'' changes into a transition crystal structure θ'. This stage usually corresponds to maximum hardness, and the precipitate becomes resolvable through optical microscopy. There is still atom alignment along some interfaces between the θ' and the matrix, and elastic strain is still present. Finally, the precipitate grows sufficiently large that the elastic strain is relieved by the formation of interface dislocations, and the crystal structure of θ' is converted to that of the equilibrium phase, θ. From this stage on, the precipitates coarsen, with the smaller particles dissolving and the larger ones increasing in size. The quantity of θ remains essentially at the equilibrium amount, but the number of θ particles decreases and the average distance between neighboring particles increases. This causes the strength to decrease. The microstructures in Fig. 3-12 and 3-13 are in this coarsening stage.

This process of precipitation is shown schematically in Fig. 3-15. Initially, B atoms segregate to form thin plates, GP zones. These thicken into plates in which B and A layers alternate, the θ'' structure. As the θ'' precipitates increase in size, their crystal structure changes to give the θ' precipitate

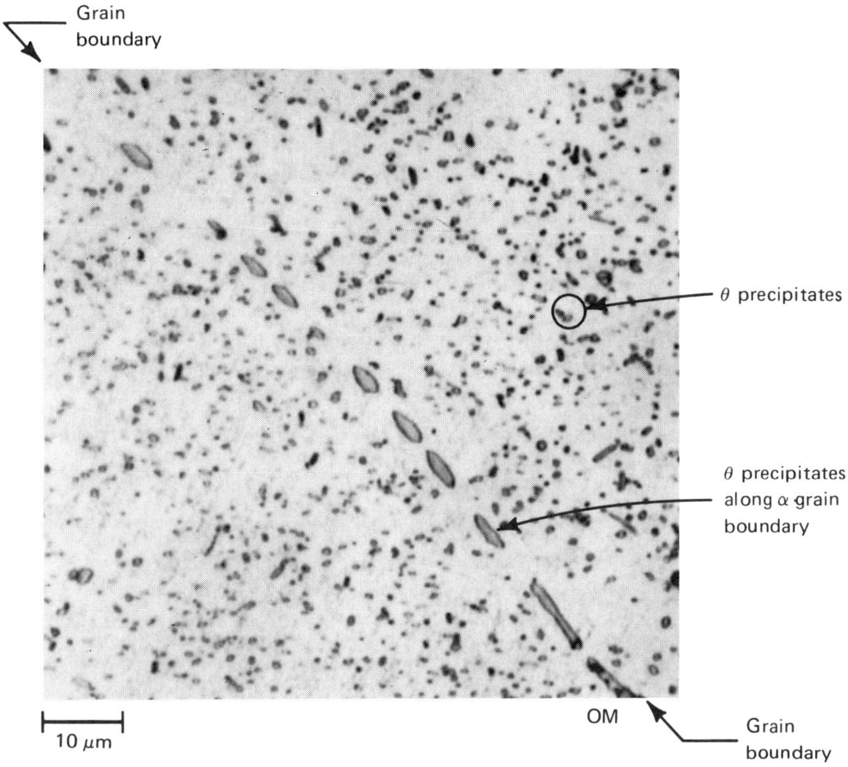

Fig. 3-13. Microstructure of the same alloy and heat treatment as in Fig. 3-12, except that the alloy was heated for 12 h at 400 °C. The θ particles are coarser and farther apart than those formed upon heating at 300 °C.

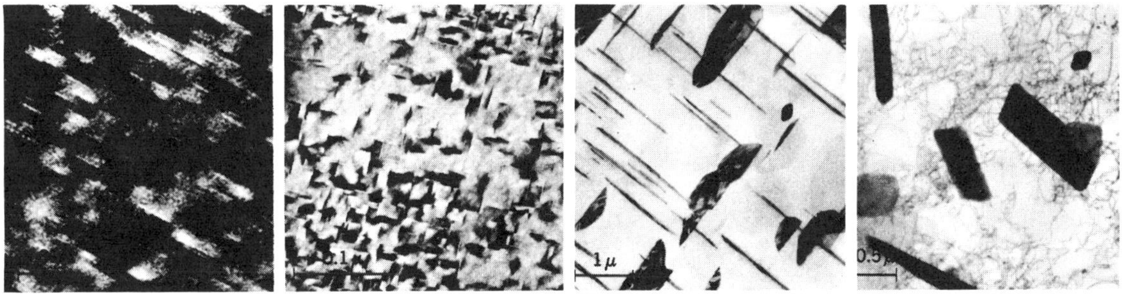

Fig. 3-14. Transmission electron micrographs of precipitates formed in an Al-4.6Cu alloy with increased aging time, shown here from left to right. (Adapted from Introduction to Materials Science, by A. G. Guy. Copyright © 1972, McGraw-Hill, New York. Used with the permission of McGraw-Hill Book Company)

The four products of reaction in an Al-2 *A/o* Cu alloy.

(shown by the square array in Fig. 3-15d). Note that this precipitate still has some interfaces on which atom positions match those in the matrix, giving a *coherent* interface. The match is not perfect, however, and as the θ' increases in size, the elastic strains at the interface increase, eventually attaining a magnitude to generate a complex dislocation network (not shown

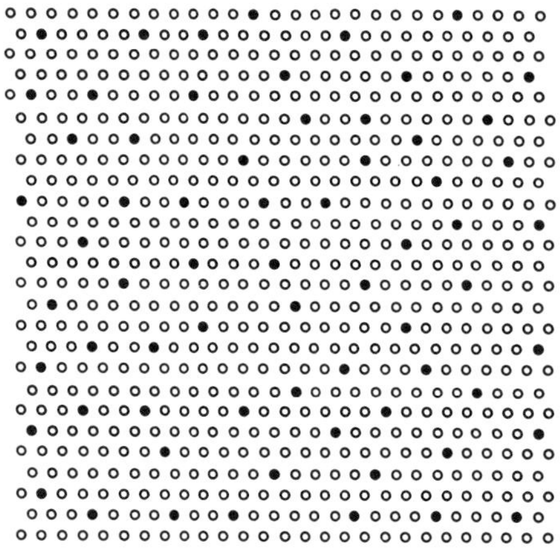

(a) Random solid solution of 10% B

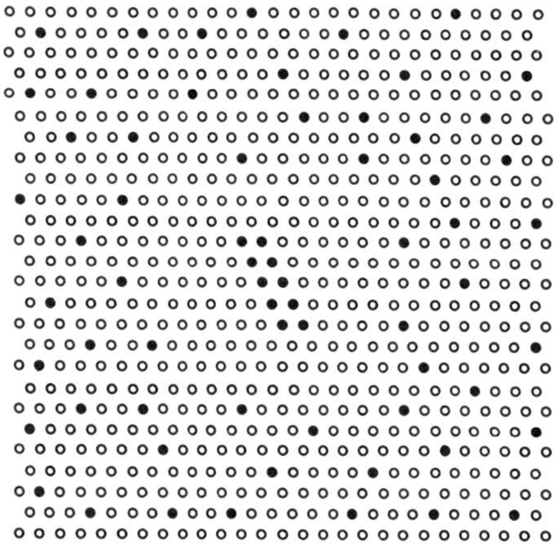

(b) Nuclei of GP-zone formed, consisting of B atoms
B atoms diffuse to G-P zone region to form

(c) B atoms diffuse to G-P zone region to form alternate A-B layer, forming θ''

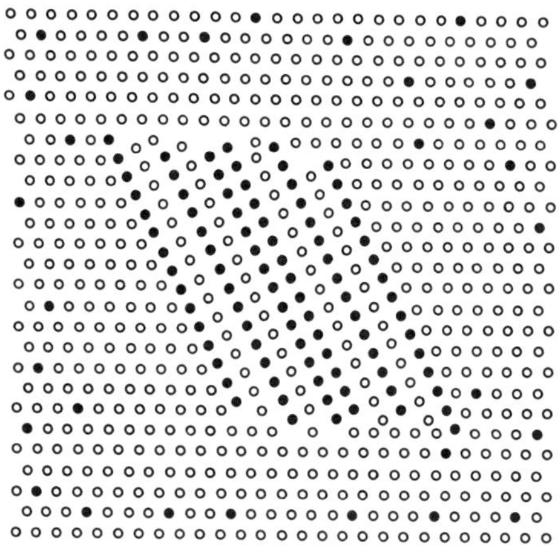

(d) More B atoms diffuse to θ'' region, and a change in crystal structure occurs (shown here as a square lattice), forming θ'

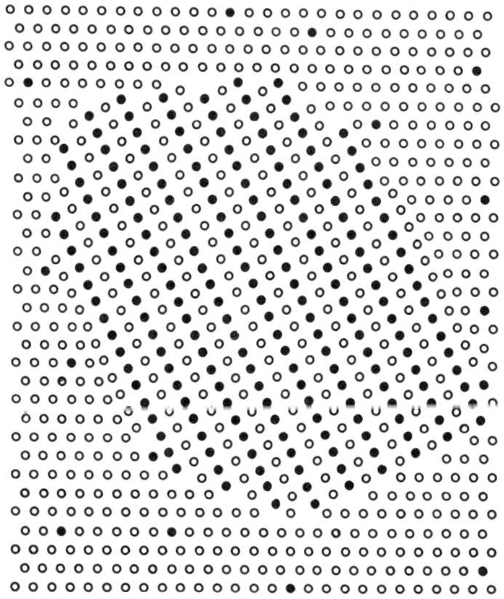

(e) More B atoms diffuse to region of θ', and rearrangement of structure forms equilibrium phase θ, with incoherent interface.

Fig. 3-15. Schematic illustration of a possible precipitation sequence.

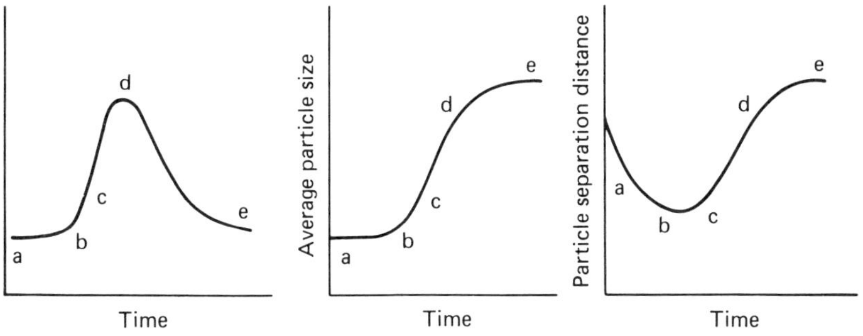

Fig. 3-15 (continued).

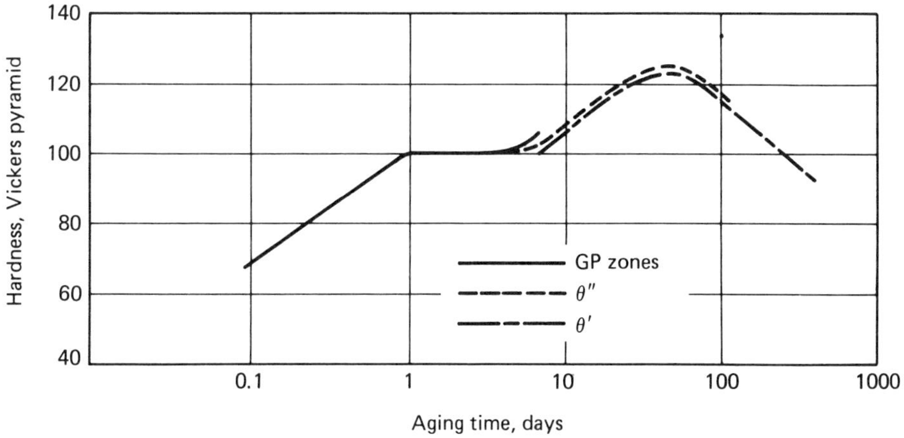

Fig. 3-16. Hardness curve for an Al-4Cu alloy showing the relationship between the various precipitates formed and the hardness upon aging at 130 °C. (Adapted from J. M. Silcock, T. J. Heal and H. K. Hardy, J. Inst. Metals, Vol 83, p 239, 1953-54)

in the figures). The strain is relieved and the interface becomes *incoherent*. Also, the crystal structure and chemical composition of θ' change to that of θ. From here on, the larger θ particles grow in size and the smaller θ particles dissolve, and the structure coarsens.

The relation of these structural changes to the hardness curve for an Al-Cu alloy is illustrated in Fig. 3-16. Note that the various stages overlap somewhat. The early stage of hardening is due to GP zone formation, and then additional hardening occurs as θ'' forms. The maximum corresponds to the presence of both θ' and θ''. The hardness decreases as θ' coarsens, and finally θ would be observed. (The aging time was not long enough to form θ.)

It can be shown by thermodynamic arguments that the chemical composition of the matrix in metastable equilibrium with GP zones, θ'' and θ' is greater than that in equilibrium with θ. This is shown in Fig. 3-17. Note that these curves predict at high temperature that the α solid solution should decompose without the formation of GP zones or θ''. In Fig. 3-9 the higher aging temperatures show a lower maximum and do not have the plateau

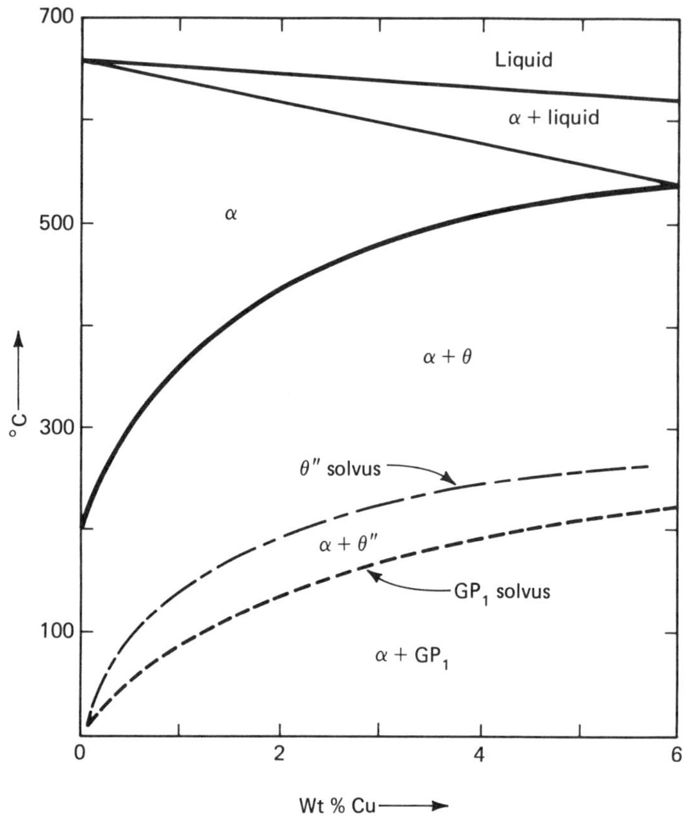

Fig. 3-17. The Al-rich section of the Al-Cu phase diagram showing the equilibrium and metastable equilibrium solvus. (From R. H. Beton and E. C. Rollason, J. Inst. Metals, Vol 86, p 77, 1957)

typical of the formation of GP zones at the lower temperatures. Also, if the alloy is aged to form GP zones and then heated above the GP solvus but below the equilibrium solvus (e.g., for an Al-4Cu alloy, heating above about 250 °C but below about 500 °C), the GP zones dissolve, and the alloy softens. This process is called *reversion*. The alloy will eventually reharden as θ' or θ precipitates.

3-6. Strengthening Mechanisms

Not only are the fine details of dislocation theory still debated, but the interaction of dislocations with structural details such as complex precipitates is complicated. However, some comments and observations are useful in terms of what possible strengthening mechanisms occur in precipitation hardening.

The precipitates present as GP zones, θ'' and θ' have sufficient mismatch with the matrix that elastic strains are present in the matrix and

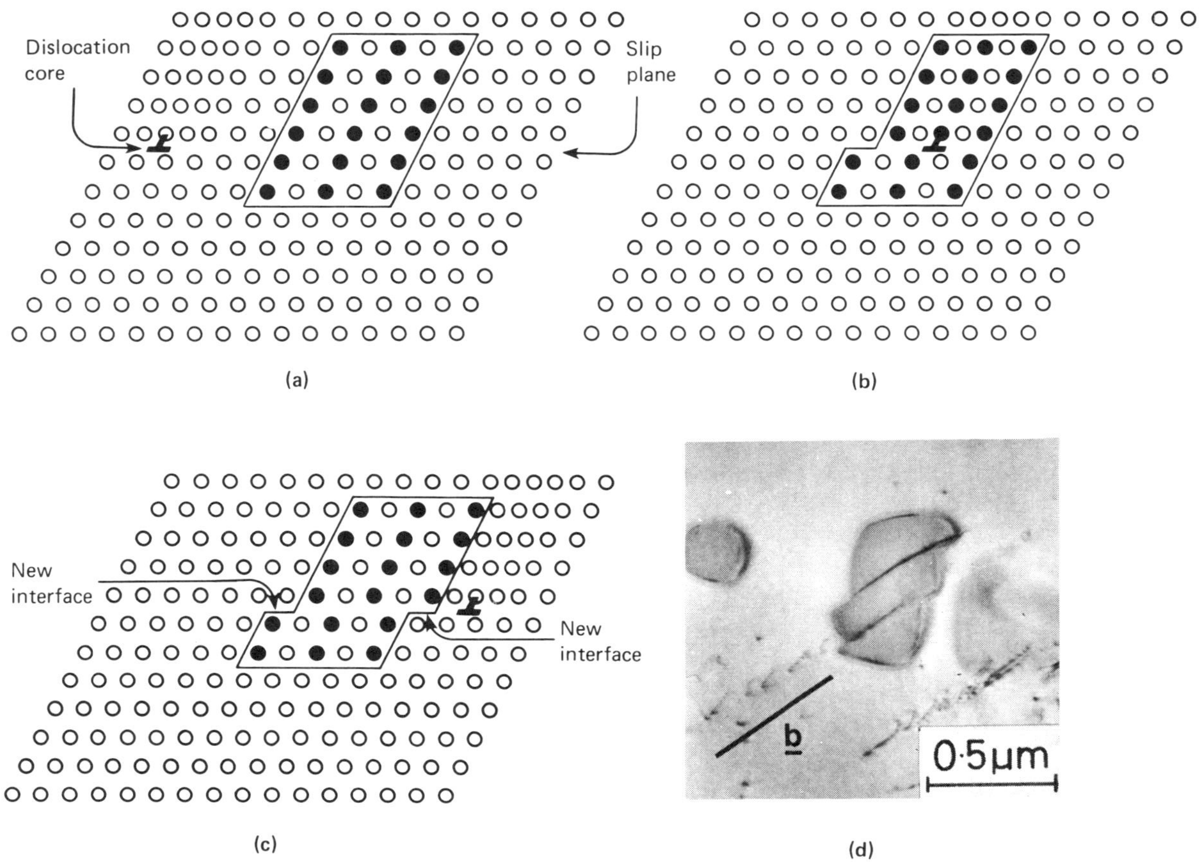

Fig. 3-18. (a)-(c) Schematic illustration of a dislocation cutting through a coherent precipitate. Note that the passage of the dislocation creates a new particle-matrix interface and changes the atomic alignment across the slip plane in the particle. (d) Transmission electron micrograph of an ordered Ni_3Si particle in a Ni-6.5Si alloy, showing dislocations cutting the particle (from F. J. Humphreys and V. Ramaswamy, in High Voltage Electron Microscopy, ed. P. R. Swann, C. J. Humphreys and M. J. Goringe, p 268, Academic Press, New York, 1974).

particle. The strain is not shown in the schematic illustration of precipitation (Fig. 3-15), but it is clear in the electron micrographs. The interatomic distance in the matrix of the regions adjacent to the particles is different from that in the regions far removed from the precipitates, and thus the regions adjacent to the particle may diffract electrons when the other regions do not. Since the diffracted electrons are prevented by an aperture from contributing to the image, these regions appear dark. This is the origin of the contrast (strain contrast) especially prominent in Fig. 3-14(a). This strain field interacts with that of an approaching dislocation, and this may (depending on the net strain field) require an increase in force to move the dislocation through the strain field. If the dislocation moves to the particle-matrix interface, it must then be forced through the particle. Because of the offset generated, an increase in surface energy is involved where the dislocation enters and leaves the particle. Also, if the precipitate is ordered, the dislocation moves atoms into positions in which they repel each other, requiring work to do so. These last two effects are illustrated schematically in Fig. 3-18.

The dislocation can bow around the particles (see Fig. 1-19), and thus not have to cut through them. Recall that the force to do this is inversely proportional to the interparticle spacing. As precipitation progresses, the particles grow in size, and the stress required to force dislocations through them increases because the new matrix-particle interface increases and the number of mismatched atoms increases (Fig. 3-15f). However, the particles begin to coarsen and the average distance between them increases, so that the dislocation can move around the particles by bowing (Fig. 1-19). This causes the strength to decrease. (See Fig. 3-15f.) Thus, the hardness–aging time curve passes through a maximum. In addition, as precipitation progresses the matrix solute concentration is decreasing, so that solid solution strengthening of the matrix is being reduced. This effect is usually small compared to that of the precipitation hardening, but it does influence the hardness after long aging times.

References

Transformations in Metals, P. G. Shewmon, McGraw-Hill, New York (1969).

Precipitation From Solid Solution, American Society for Metals, Metals Park, Ohio (1958).

Precipitation Hardening, J. W. Martin, Pergamon, New York (1968).

CHAPTER 4

Aluminum-Base Alloys

Aluminum, with its low density, was early an attractive candidate as the base for the development of engineering alloys. Although the strengths attained in these alloys are not as great as those attainable in steels, the strength-to-weight ratio is very favorable, and hence many commercial aluminum-base alloys are available. Most of these alloys have their strength increased by precipitation hardening, a process reviewed in Chapter 3.

In this chapter, the physical metallurgy underlying the heat treatment of one commercial aluminum-base alloy, 2024, is examined in detail. The pertinent ternary and quaternary phase diagrams are used to describe the phase relations and to show why specific heat treatments are used.

4-1. Ternary Phase Diagrams

The phase relations in ternary systems are usually presented in terms of isothermal sections through triangular composition-temperature phase diagrams. We will briefly examine the meaning of such sections here. Consider three components, A, B and C, which in pairs have the binary phase diagrams shown in Fig. 4-1, each being a simple eutectic system. To depict the phase relations for a ternary alloy, it is most common to give the composition on a ternary plot and show temperature as the variable in the vertical direction. We will assume here that the ternary system of A, B and C has a eutectic diagram. That is, there is a ternary alloy which has a lowest melting point. The assumed diagram is shown in Fig. 4-2. Each of the terminal solid solutions has some solubility.

If a constant temperature plane is passed through the phase diagram, the intersection of this plane with the phase equilibria surfaces (e.g., solid-

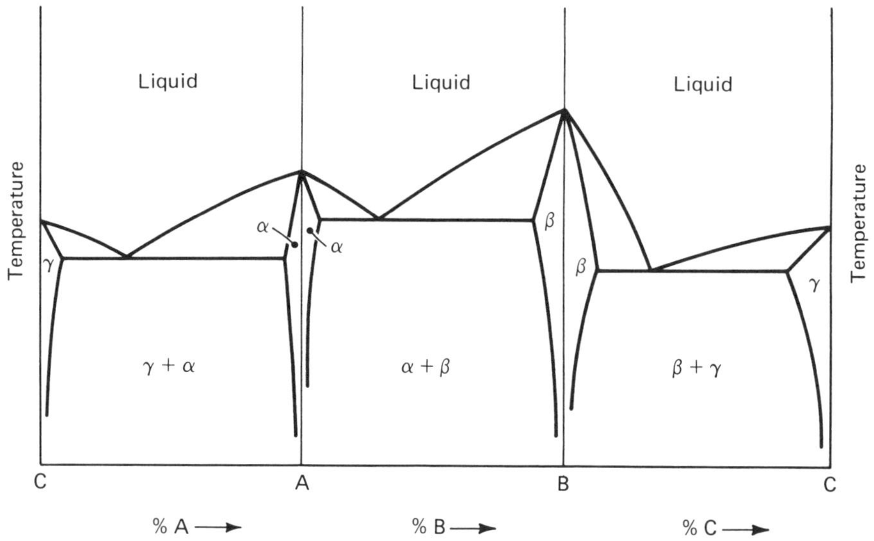

Fig. 4-1. Three eutectic binary phase diagrams of the systems A-B, B-C and C-A.

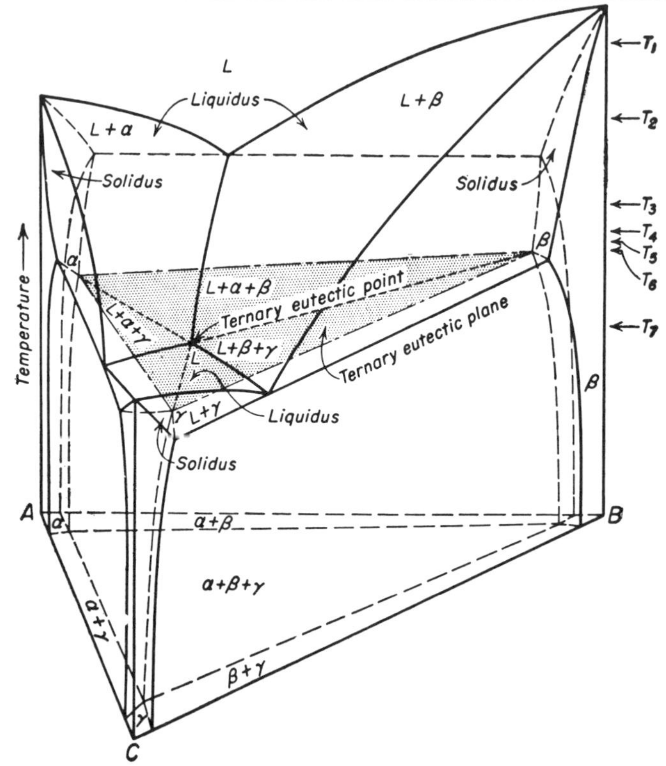

Fig. 4-2. Assumed ternary diagram for the system A-B-C. (From Phase Diagrams in Metallurgy, by F. N. Rhines. Copyright © 1956, McGraw-Hill, New York. Used with the permission of McGraw-Hill Book Company)

liquid boundaries) generates an isothermal section. Those given by the temperatures shown in Fig. 4-2 are shown in Fig. 4-3. Note that three-phase equilibrium can exist over a range of temperatures in a ternary system, whereas it is limited to a fixed temperature in a binary system (e.g., two solids and a liquid at the eutectic temperature).

If the composition of an alloy is in the three-phase region, the chemical composition of each phase is given by the apex at each corner. This is

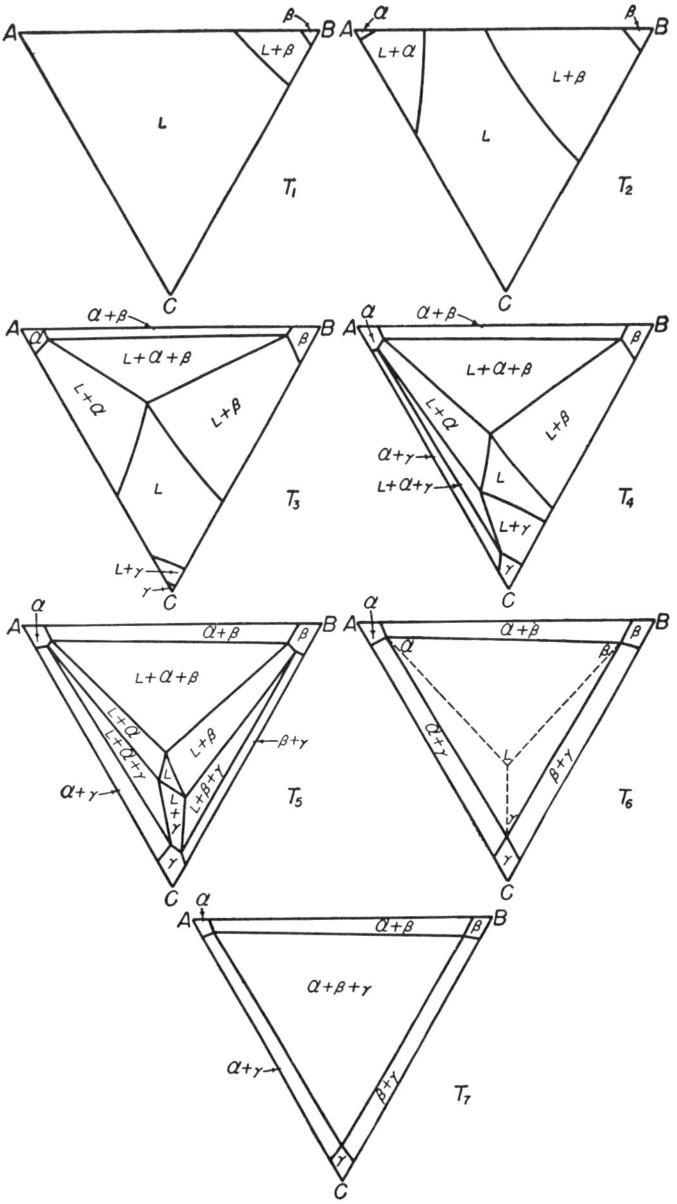

Fig. 4-3. Isothermal sections from the ternary A-B-C phase diagram, at the temperatures indicated in Fig. 4-2. (From same source as Fig. 4-2; used with the permission of McGraw-Hill Book Company)

illustrated in Fig. 4-4. In the two-phase region the composition is given by a tie line, which must be experimentally located. This is illustrated also in Fig. 4-4. For example, the alloy which has the composition given by point 1 is in the α + L region. The composition of the liquid (point 3) and of the α (point 2) must be connected by a straight line through point 1. From a mass balance the amount of each phase can be calculated. If the composition is at point 4, β, L and α are present, and the chemical composition of each phase is given by point 5, 6 and 7, respectively. Again, the amount of each phase can be calculated.

It is also useful to examine the projections of the liquidus, solidus and solvus, and these are illustrated in Fig. 4-5 and 4-6, using the B-rich corner of the ternary diagram in Fig. 4-2.

If there are four components present, the presentation of phase equilibria is more complicated. One method is to present the composition-phase relations at a given temperature in a tetrahedron. This is illustrated in Fig. 4-7 for a hypothetical system in which the four components are completely soluble in each other. At the temperature for which the diagram applies, there is a single-phase α region at the A-rich corner of the diagrams. A two-phase region (α + liquid) separates this α region from the liquid region.

It is difficult to analyze quantitatively data in such a figure. Instead it is more useful to present constant-composition sections from the tetrahedron, as illustrated in Fig. 4-8. It is seen that these isothermal sections are similar in appearance to those of ternary systems. However, most binary systems are

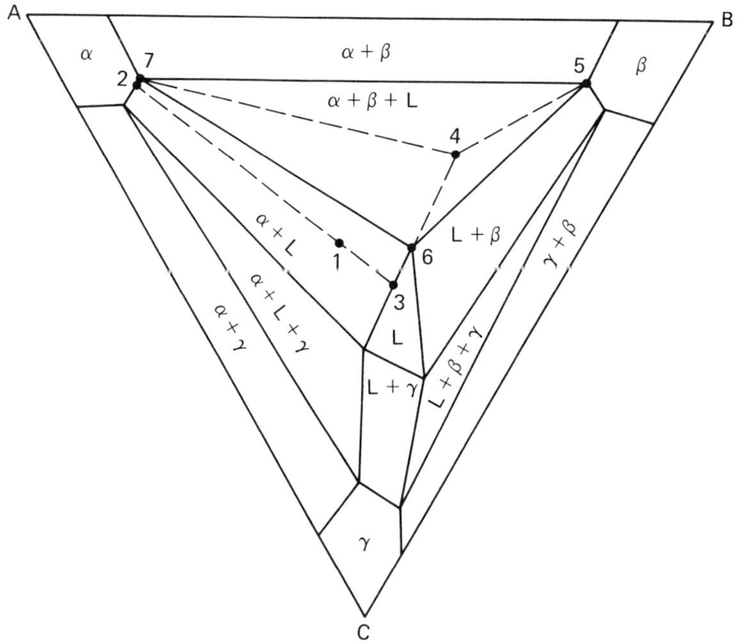

Fig. 4-4. Isothermal section at temperature T_5, used to illustrate the chemical composition of the phases in a two-phase and three-phase region.

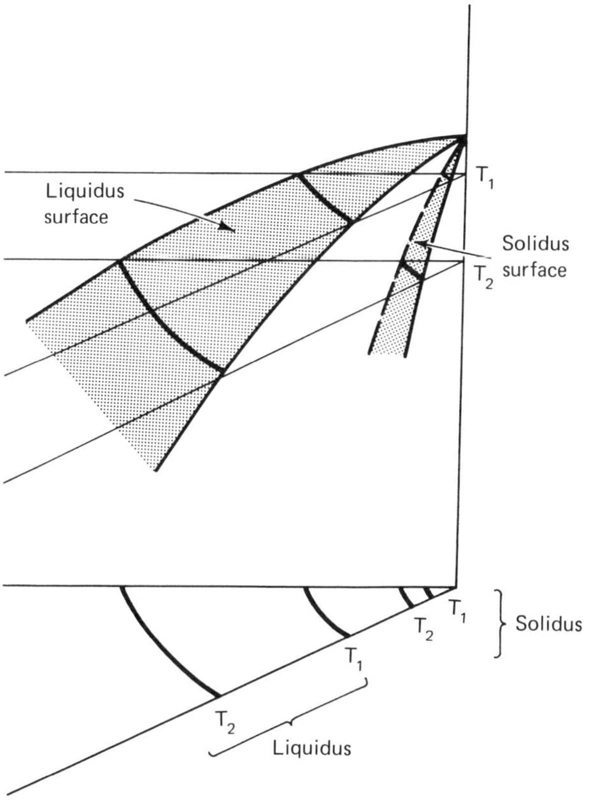

Fig. 4-5. Illustration of the relation between the liquidus and solidus surfaces, and projection of their isotherms onto a plane.

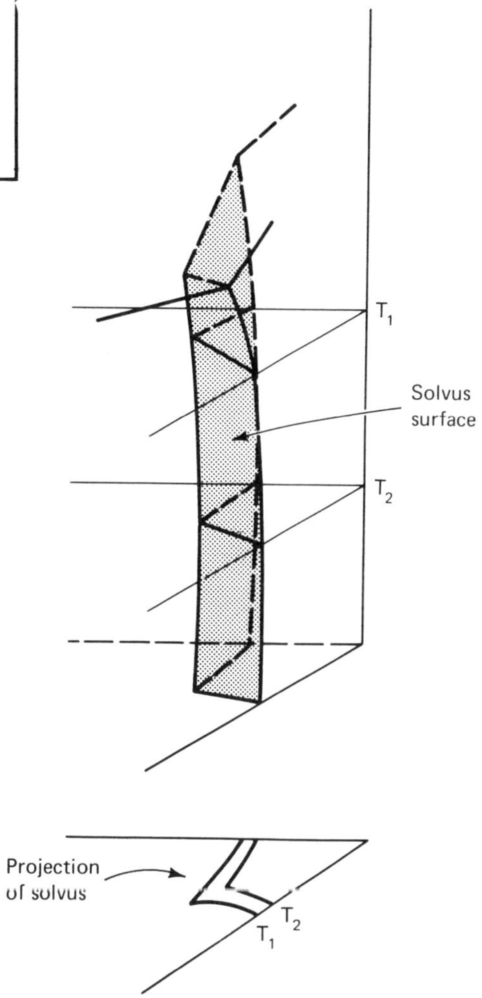

Fig. 4-6. Illustration of the relation between the solvus and its isotherms projected onto a plane.

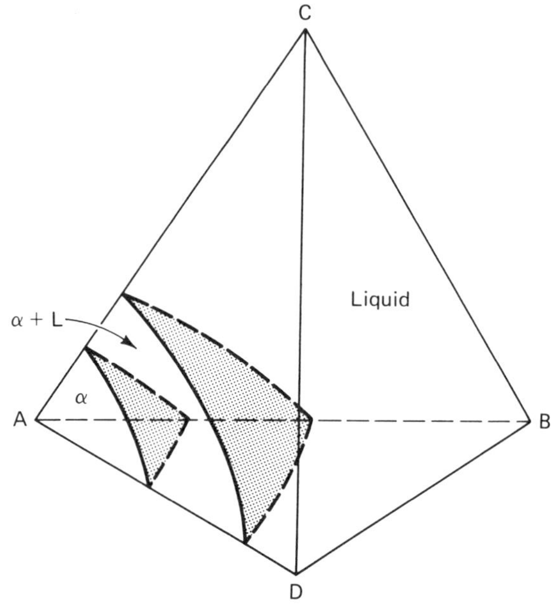

Fig. 4-7. Schematic quaternary isothermal phase diagram for the system A-B-C-D where the components are completely soluble.

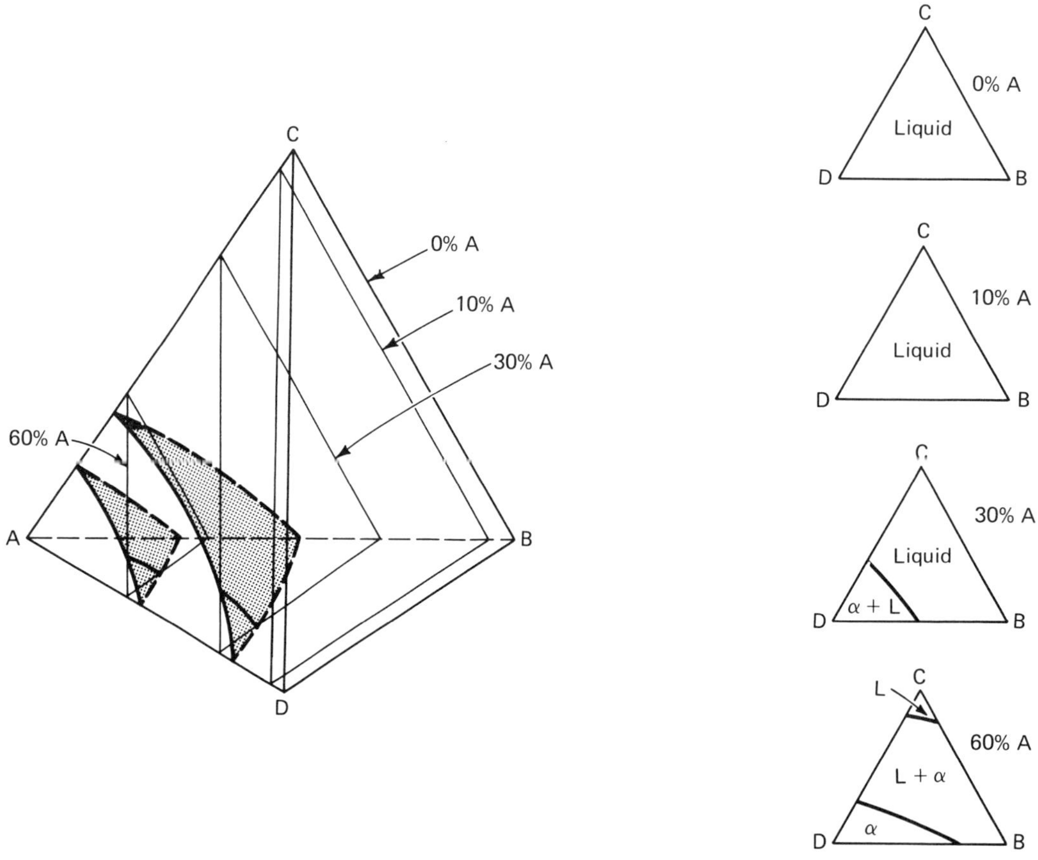

Fig. 4-8. Constant-composition sections from the quaternary diagram shown in Fig. 4-7.

sufficiently complicated that the quaternary diagram is difficult to present in any form other than the constant-composition sections.

4-2. Aluminum-Copper-Magnesium Alloys

In the binary aluminum-copper system, the aluminum-rich terminal solid solution is in equilibrium with the intermetallic compound θ, which has approximately the formula $CuAl_2$, although some solid solubility exists. (See Fig. 3-8). The addition of magnesium allows the formation of more intermetallic compounds, as illustrated in Fig. 4-9 by the isothermal section at 430 °C for the ternary system. Since the alloys of interest are aluminum-rich, only those phases in equilibrium with α are of interest, and there are four possibilities. Although these are represented by compound formulas, in actuality there is some solid solubility.

The liquidus, solidus and solvus isotherms are shown in Fig. 4-10. To illustrate the heat treatment of these alloys, we will consider an alloy containing 4% Cu and 1% Mg. This composition is located at the circle in each of the isothermal projections. The interpolated isotherms through this composition are indicated on the figures. To visualize the relation of this composition to the isotherms, Fig. 4-11 shows the relation of the liquidus,

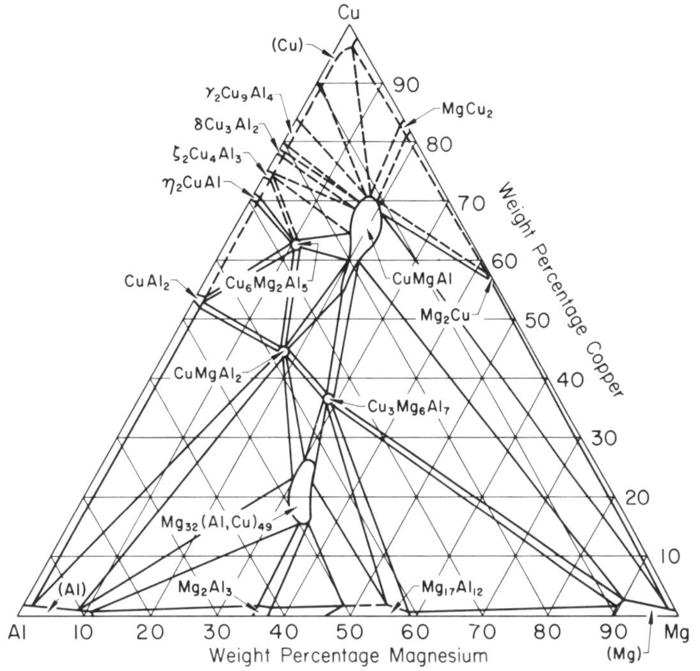

Fig. 4-9. Isothermal section at 430 °C for the Al-Cu-Mg system. Note that there are several intermetallic compounds present. Four of these — Mg_2Al_3, $Mg_{32}(Al, Cu)_{49}$, $CuMgAl_2$ and $CuAl_2$ — can be in equilibrium with the α phase. (Adapted from Metals Handbook, 8th Ed., Vol 8, American Society for Metals, Metals Park, Ohio, 1973)

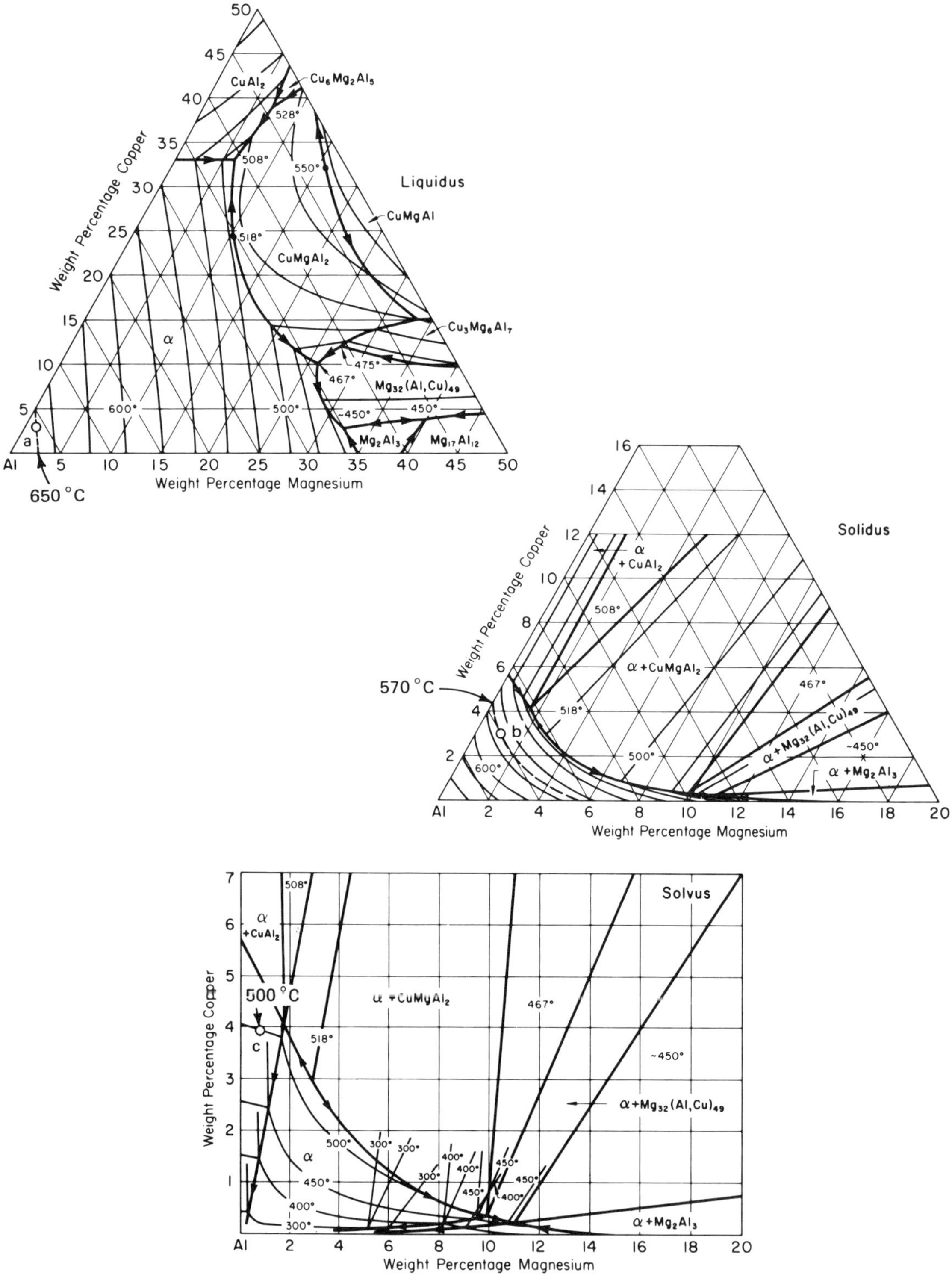

Fig. 4-10. Projections of the liquidus, solidus and solvus isotherms for the Al-rich end of the Al-Cu-Mg system. The circle in each diagram denotes the alloy containing 4% Cu and 1% Mg, and the temperatures noted are for this composition. The letters a, b and c refer to locations in Fig. 4-11. (Adapted from same source as Fig. 4-9)

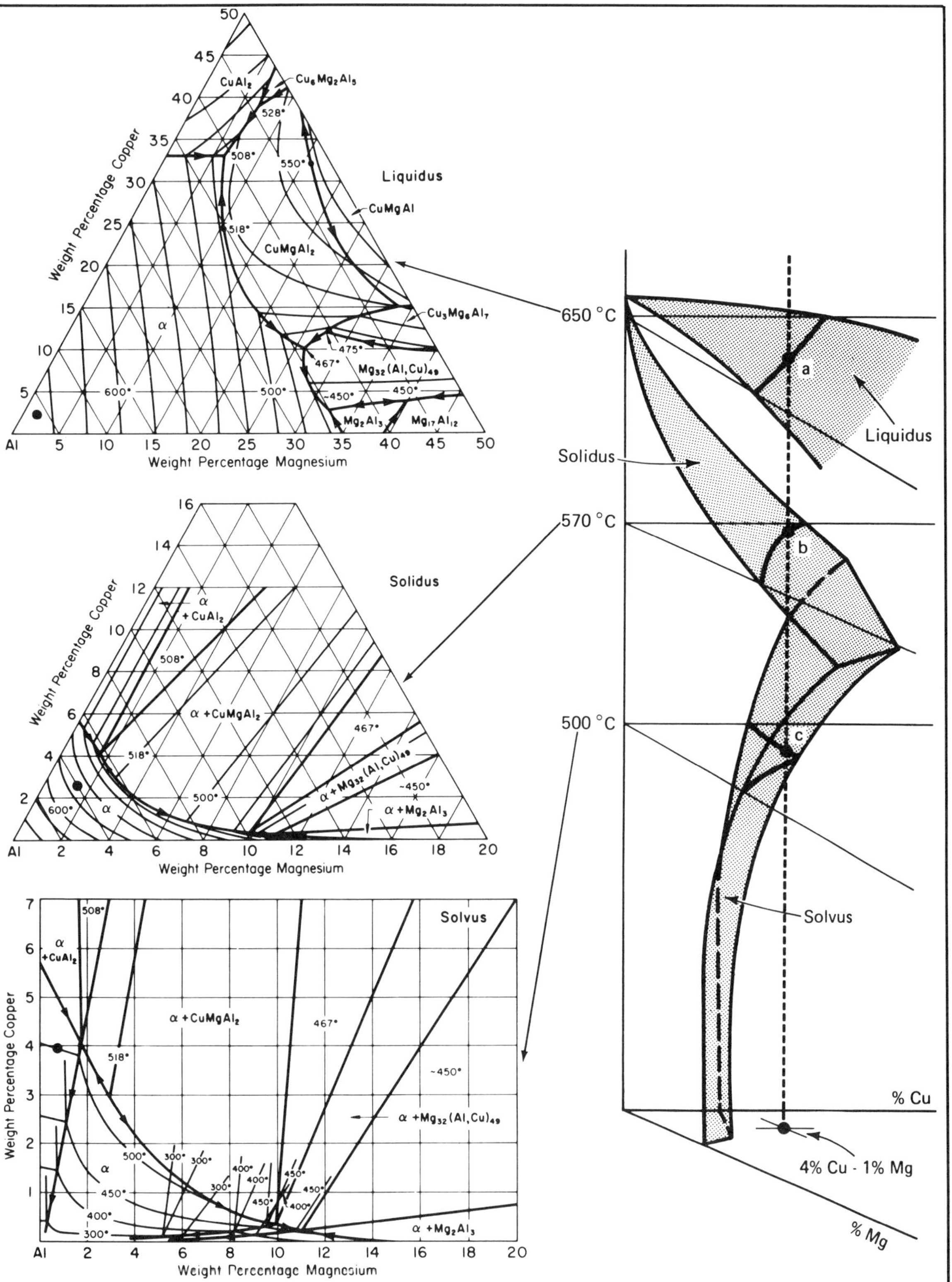

Fig. 4-11. Illustration of the relation between the projection of the liquidus, solidus and solvus isotherms and the three-dimensional phase diagram. The temperatures noted are for an aluminum alloy containing 4% Cu and 1% Mg.

solidus and solvus projections to the three-dimensional ternary phase diagram. It is seen that, for this composition, the alloy is liquid above 650 °C (point a); in the two-phase liquid-solid region between 570 °C (point b) and 650 °C; in the single-phase α region between 570 °C and 500 °C (point c); and in the three-phase region below 500 °C. The three-phase relation is not shown in Fig. 4-11, but it is in Fig. 4-12, the isotherm at 430 °C. It is seen that at this temperature this alloy has α in equilibrium with $CuAl_2$ and $CuMgAl_2$.

Now for precipitation hardening, the first step in the heat treatment is to equilibrate the alloy in the α region (Fig. 4-11); that is, to solution heat treat the alloy. This alloy must be heated between 500 and 570 °C for sufficient time to homogenize the structure, which by experimentation would be found to be a few hours. [However, the original casting may require up to 100 h to remove dendritic segregation (coring) and dissolve second-phase particles. See Chapter 2.] After the solution heat treatment, the alloy must

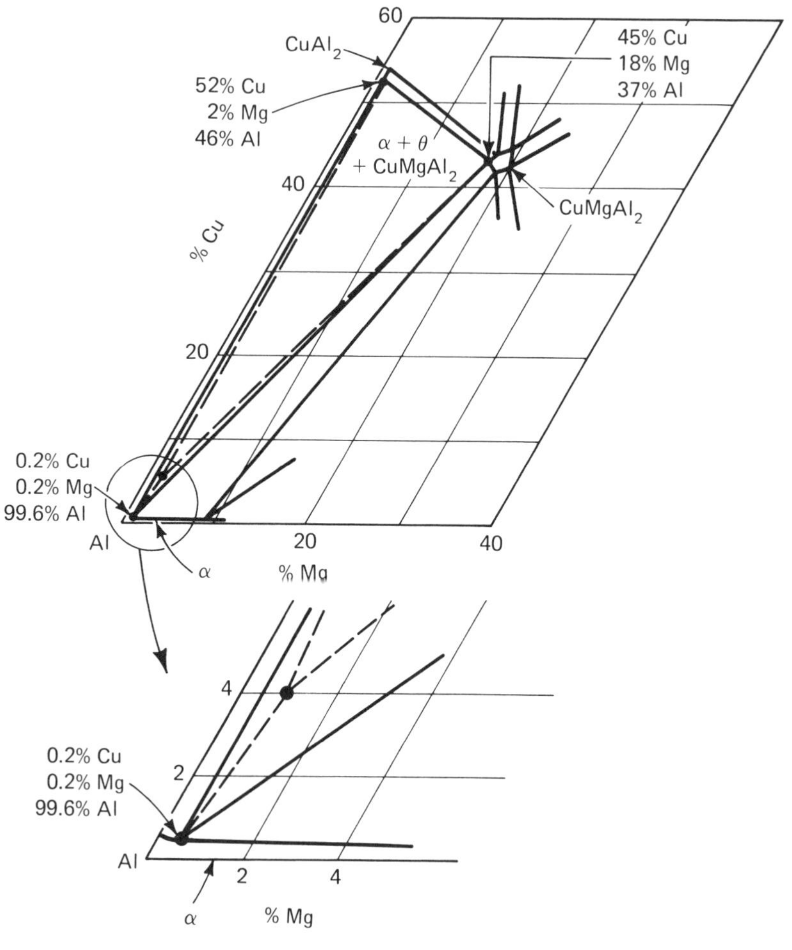

Fig. 4-12. Enlargement of the Al-rich portion of the Al-Cu-Mg isothermal section at 430 °C. The alloy containing 4% Cu and 1% Mg is seen to be in the three-phase region in which α, $CuAl_2$ and $CuMgAl_2$ are in equilibrium.

be cooled to 20 °C at a rate sufficient to suppress any precipitation. For example, plate about 0.5 in. thick (or less) can be quenched into cold water. The precipitation process begins immediately, and the rate depends upon the temperature at which the alloy is held after quenching from the solution treatment temperature. In this Al-Cu-Mg alloy, the rate at 20 °C is sufficiently great to cause prominent mechanical property changes, as shown in Fig. 4-13. (Precipitation hardening at room temperature is referred to as age hardening, or natural hardening.) Note that the yield strength increases from about 12,000 psi for the solution-treated condition to about 38,000 psi after two years. As overaging has not been completed, equilibrium at 20 °C is far from established after two years.

At 430 °C, the alloy will contain at equilibrium about 3% $CuAl_2$ and 5% $CuMgAl_2$. (This is calculated by a mass balance from the data in Fig. 4-12.) Assuming that at equilibrium at 20 °C the Cu and Mg content of the α is negligible, the amount of $CuAl_2$ will be about 4% and $CuMgAl_2$ about 4%. This is about the same as at 430 °C. Thus, the calculation of the amount of the phases present at equilibrium at 430 °C can be used as an approximation of the amount present at 20 °C. Examination of Fig. 4-12 shows that as the Mg content is increased to 1.5% Mg but the Cu content kept at 4%, the amount of $CuAl_2$ decreases to less than 1% and that of $CuMgAl_2$ increases to 8%. Thus, as the Cu/Mg ratio decreases, the $CuMgAl_2$ formation becomes more important in determining the precipitation-hardening process. Note that the steps leading to the formation of this phase are more effective in precipitation hardening than those associated with the formation of $CuAl_2$. This is seen in Fig. 4-13. Even though the

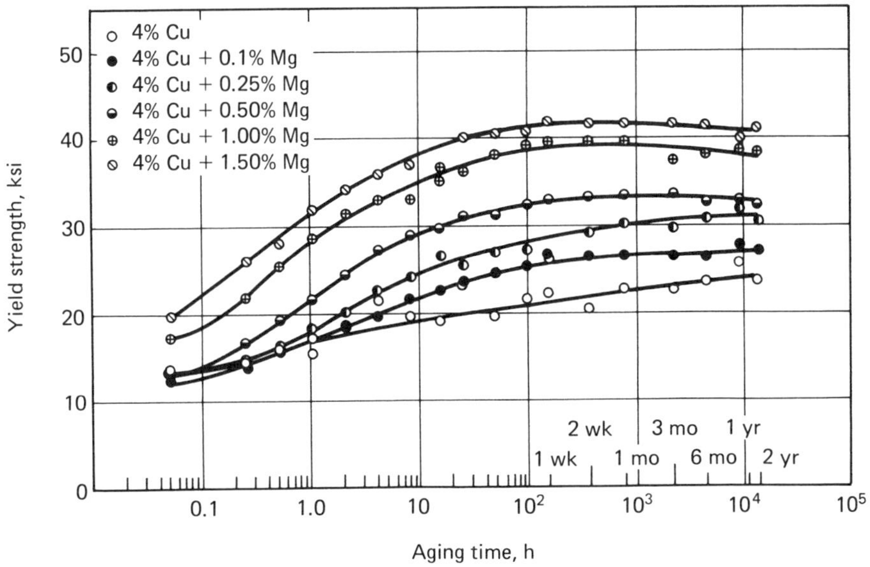

Fig. 4-13. Effect of aging at 20 °C on the yield strength of Al-Cu-Mg alloys. (Adapted from W. L. Fink, D. W. Smith and L. A. Willey, in Age Hardening of Metals, American Society for Metals, Metals Park, Ohio, 1940)

binary alloy, in which only $CuAl_2$ is involved, does harden appreciably, the addition of Mg causes significantly greater hardening.

In these ternary alloys, the precipitation mechanism is similar in scheme to that described in Chapter 3 for copper-aluminum alloys. First to form are GP zones, followed by a coherent transition phase. Eventually the equilibrium $CuMgAl_2$ phase forms, then coarsens. If the Cu/Mg ratio is high, the formation of $CuAl_2$ can occur, and its formation mechanism should be the same as described for the Al-Cu binary alloy. If the Cu/Mg ratio is such as to allow comparable amounts of both phases to form, the mechanism may involve parallel formation of each phase by individual mechanism or formation of only a single GP zone, with the separate phases appearing later in the sequence. The precipitation details are not well established, even though the process involves the type of steps described in Chapter 3.

4-3. Aluminum-Copper-Magnesium Commercial Alloys

A commonly used commercial alloy based on the Al-Cu-Mg ternary system is 2024, and we will examine its heat treatment in detail. The chemical composition is within the following specification: 3.8-4.9% Cu, 1.2-1.8% Mg, 0.4-0.9% Mn, 0.50% max Si, 0.50% max Fe, 0.10% max Cr, 0.25% max Zn.

Let us first examine this alloy as if it contained only Al, Cu and Mg, using an average composition of 4.5% Cu and 1.5% Mg. For such a ternary alloy, Fig. 4-10 shows a solidus of about 560 °C and a solvus of about 520 °C. For the 2024 alloy, the recommended solution treatment range is 488-499 °C. This seems somewhat low, but the range of composition must be considered in specifying a temperature range for use with the alloy. Figure 4-14 shows the Cu and Mg ranges blocked out for the solidus and solvus. Note that if the Cu and Mg are both on the low side, the solidus will be about 560 °C and the solvus about 500 °C. However, if the Cu and Mg are both on the high side, the situation is different. To see this, we examine Fig. 4-15. The aluminum-rich section of the phase diagram is shown at 510 °C, where the blocked-out area covers the composition limits of Mg and Cu. Note that for high Cu and Mg, liquid will be present with the α, and the solid phase $CuMgAl_2$ may also be present. The ternary eutectic temperature for equilibrium between liquid, α, $CuMgAl_2$ and $CuAl_2$ is 508 °C. The diagram for 508 °C in Fig. 4-15(b) was derived from that for 510 °C in Fig. 4-15(a). Note that the liquid region has shrunk to a point corresponding to the ternary eutectic. If the temperature falls below 508 °C, this liquid phase disappears. Thus, at 508 °C in this general composition range the solidus and solvus are identical. This can be seen in Fig. 4-14, which shows the composition range blocked out on the solidus and solvus projections. For high Cu and Mg content, at 508 °C the alloy would fall within the phase fields involving α, $CuAl_2$, $CuMgAl_2$ and liquid.

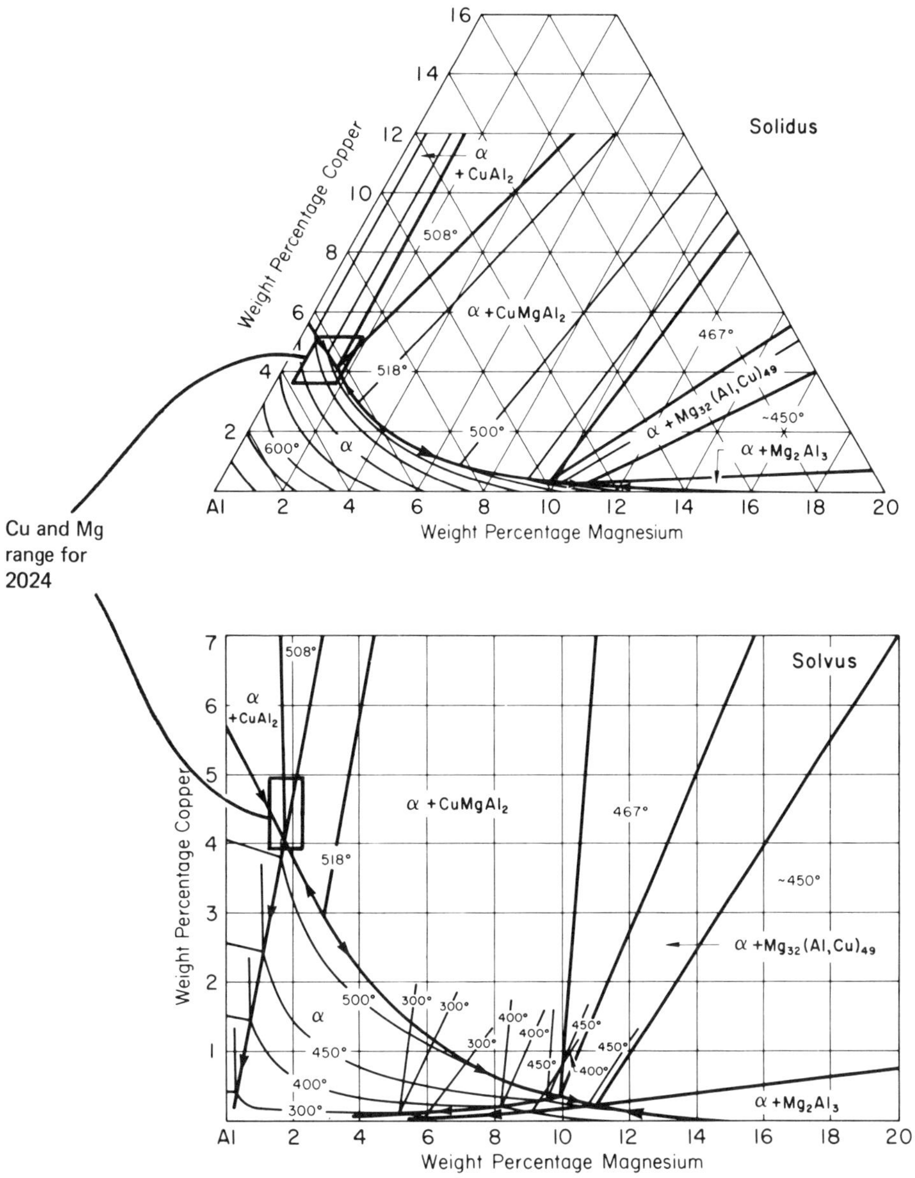

Fig. 4-14. Solidus and solvus projections with the Cu and Mg ranges for 2024 shown blocked out. (Adapted from same source as Fig. 4-9)

For the 2024 alloy, two important facts are derived from this analysis. First, to avoid the possibility of localized melting in the alloy if the Cu and Mg are on the high side, the solution treatment temperature is below 508 °C (e.g., 495 °C), and if the Cu and Mg content are on the high side of the allowable range, then the alloy will not be in a single-phase region but in a multiphase solid field. To visualize this, examine the isotherm at 430 °C in Fig. 4-16. It is seen that if Cu and Mg content are high the alloy will be in

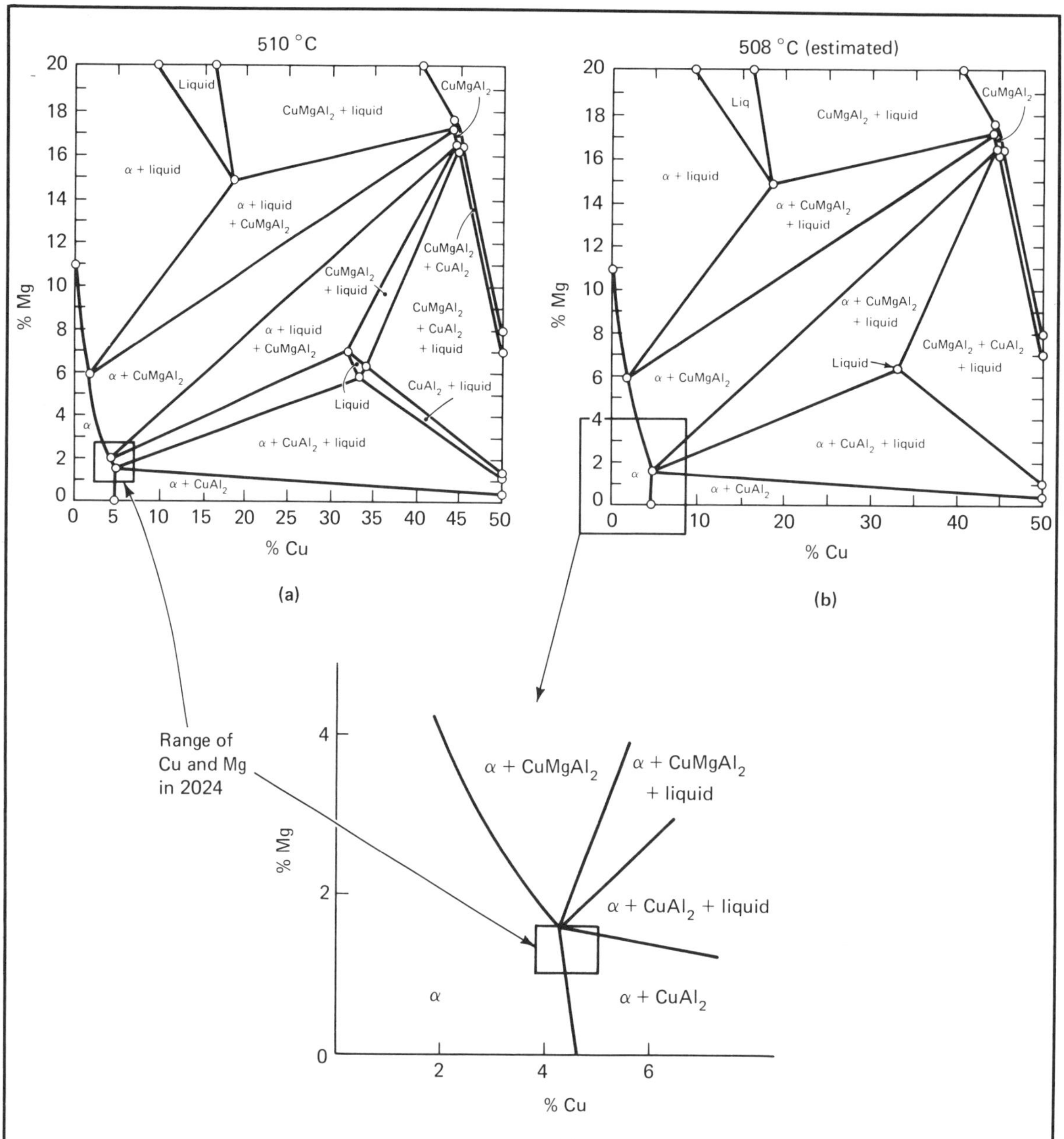

Fig. 4-15. (a) Isotherm at 510 °C (adapted from H. W. L. Phillips, Annotated Equilibrium Diagrams of Some Aluminum Alloy Systems, The Institute of Metals, London, 1959). (b) Isotherm at 508 °C, the ternary eutectic temperature involving liquid, $CuAl_2$ and $CuMgAl_2$. Just below this temperature, the liquid shown by the point disappears. (c) Enlargement of the region shown in (b).

the α-$CuAl_2$ region or in the α-$CuAl_2$-$CuMgAl_2$ region, depending upon the exact chemical composition. Thus, solution treating the alloy between 430 and 508 °C will result in some solid phases present along with the α. That is, the solution-treated condition may contain some $CuAl_2$ or $CuMgAl_2$

along with the α. (Also, in the aged condition, both of these solid phases may form.)

The lower solution temperature of 488 °C recommended for 2024 is based on the highest temperature consistent with minimizing the possibility of melting and yet obtaining maximum solution of solute. The recommended range also takes into account the uncertainty in temperature control and the temperature gradients that exist in commercial heat treating furnaces.

Now the effect of the presence of Si, Mn and Fe must be considered. Figure 4-17 shows the projection of the aluminum-rich corner of the quaternary Al-Cu-Mg-Si phase diagram in the temperature range where only solid phases are present. The addition of Si will favor the formation of the quaternary compound $Cu_2Mg_8Si_5Al_4$. Figure 4-18 shows a section from the quaternary diagram at 502 °C for a fixed Cu content of 4%. The range of the allowable variation in Mg and Si is blocked out, and it is seen that, depending upon the exact Mg and Si content, α could be in equilibrium with Mg_2Si, $CuMgAl_2$, $CuAl_2$ and $Cu_2Mg_8Si_5Al_4$. Thus the addition of Si allows the formation of two more solid phases which were not present in the Al-Cu-Mg ternary system: Mg_2Si and $Cu_2Mg_8Si_5Al_4$.

The presence of Mn and Fe also allows the formation of additional intermetallic compounds. Figure 4-19(a) and (b) show the projections of the aluminum-rich corner of the Al-Cu-Mg-Mn and Al-Cu-Mg-Fe diagrams in the solid range. Adding Mn promotes the formation of $Cu_3Mn_2Al_{20}$, and adding Fe promotes Cu_2FeAl_7. Figure 4-19(c) shows the Al-Cu-Mn-Fe diagram. Increased Mn *and* Fe promote the formation of the phase which is a solid solution compound based on (CuFe) Al_6 and $MnAl_6$; it is usually designated as (Cu, Fe, Mn)Al_6.

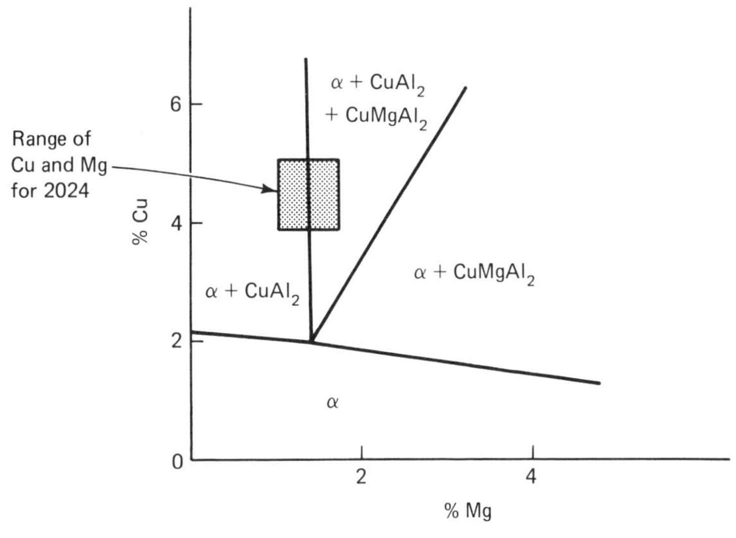

Fig. 4-16. Isotherm at 430 °C. Enlargement of the Al-rich corner from Fig. 4-9.

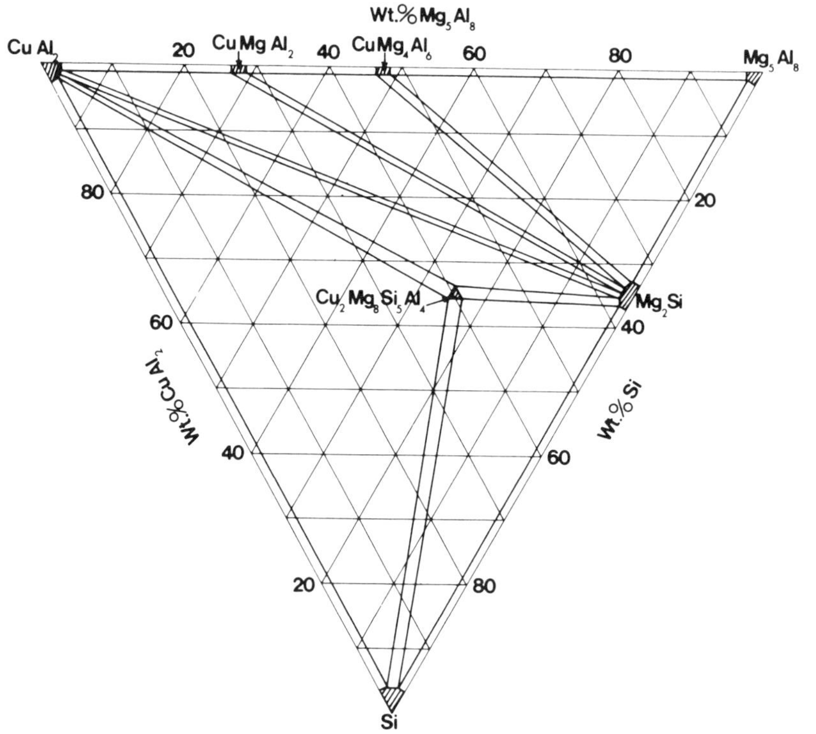

Fig. 4-17. Projection of the Al-rich corner of the Al-Cu-Mg-Si phase diagram for the solid condition. (From L. F. Mondolfo, Aluminum Alloys: Structure and Properties, Butterworths, London, 1976)

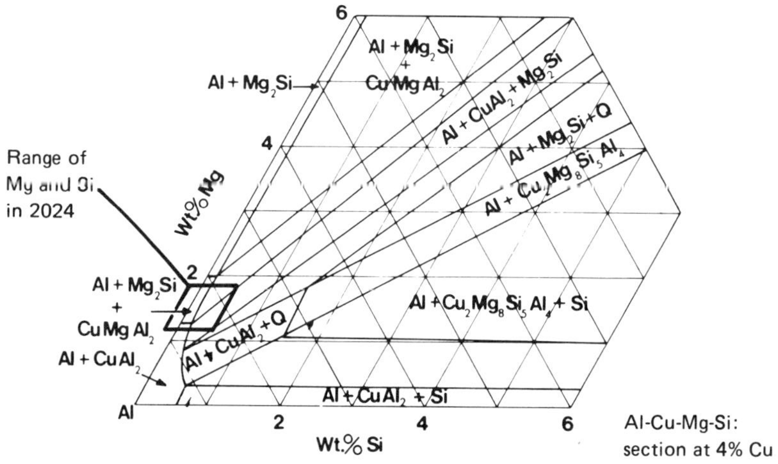

Fig. 4-18. Section of the quaternary phase diagram for the Al-Cu-Mg-Si system at 502 °C and 4% Cu. The phase called Q is $Cu_2Mg_8Si_5Al_4$. (From same source as Fig. 17)

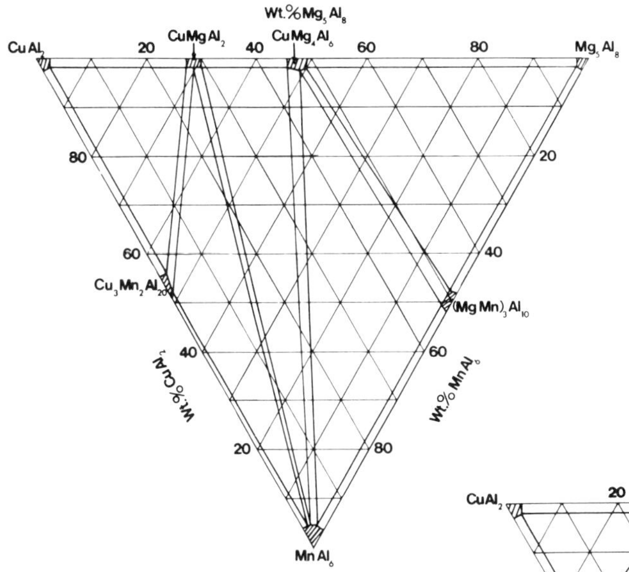

(a) Al-Cu-Mg-Mn

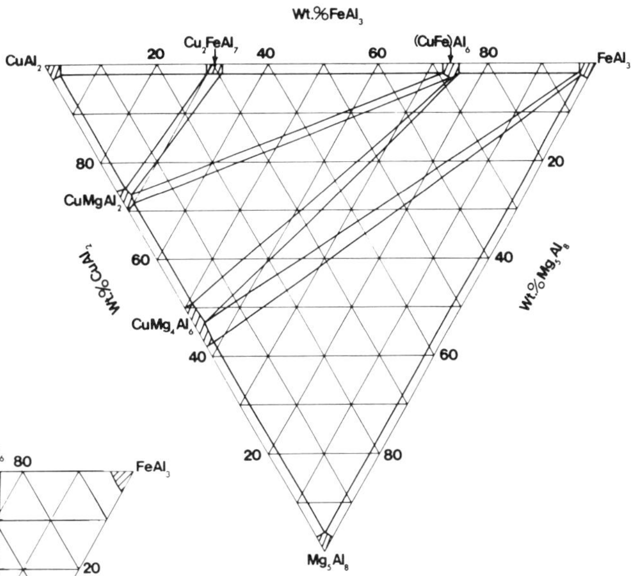

(b) Al-Cu-Mg-Fe

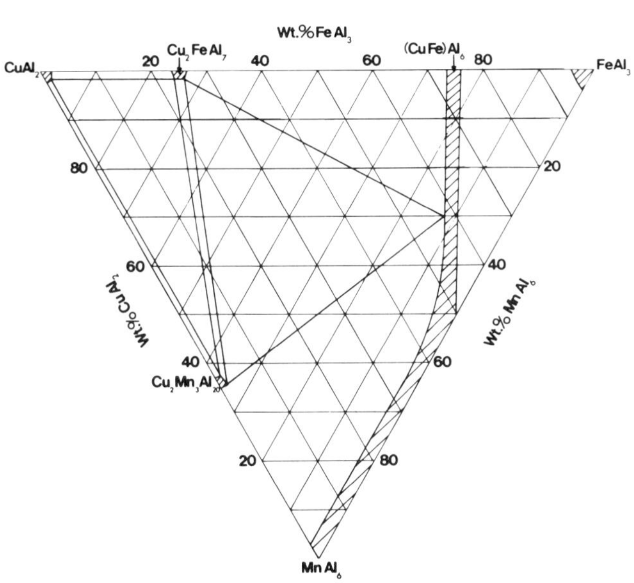

(c) Al-Cu-Mn-Fe

Fig. 4-19. Projections of the Al-rich corner of the (a) Al-Cu-Mg-Mn, (b) Al-Cu-Mg-Fe and (c) Al-Cu-Mn-Fe phase diagrams for the solid condition. (From same source as Fig. 4-17)

Thus we see that in the 2024 commercial alloy several intermetallic phases may be present, depending upon the exact chemical composition and the heat treatment. It also depends quite sensitively upon how the alloy was cast, because the cooling rate from the liquid affects the phases formed in the eutectic reactions. In setting the upper limit of the solution treatment temperature, the possibility of a eutectic reaction between the α and some of these other phases must be considered. However, it is found that the lowest eutectic temperature is 505 °C (involving $CuAl_2$, $CuMgAl_2$ and Cu_2FeAl_7 and α), so that the upper limit of 499 °C is still valid.

The microstructure of a 2024 alloy which has been annealed (cooled slowly from the solution treatment temperature) is shown in Fig. 4-20. The large black particles are $CuMgAl_2$, which were present at the solution treatment temperature. Also present in that condition were a few large gray particles of (Cu, Fe, Mn)Al_6, due to the Fe and Mn in the alloy. However, the majority of the particles are $CuMgAl_2$. The fine black particles are also $CuMgAl_2$, but they formed upon cooling slowly from the solution treatment temperature. (Note that the particular etch used did not reveal the α grain boundaries.)

The microstructure in the solution-treated condition is shown in Fig. 4-21. The etch used revealed the α grain boundaries, as well as some particles present at the solution treatment temperature. These were identified as $CuMgAl_2$ along with some $Cu_2Mn_3Al_{20}$ and Cu_2FeAl_7. Note that the grains are elongated, appearing similar to that of cold worked structure. This is caused by a nonuniform distribution of undissolved phases present at the solution temperature during annealing of the cold worked material. Upon completion of recrystallization, the grains begin to grow, but their boundary movement is inhibited by the particles present, and since their distribution is nonuniform, the annealed grain structure is still elongated.

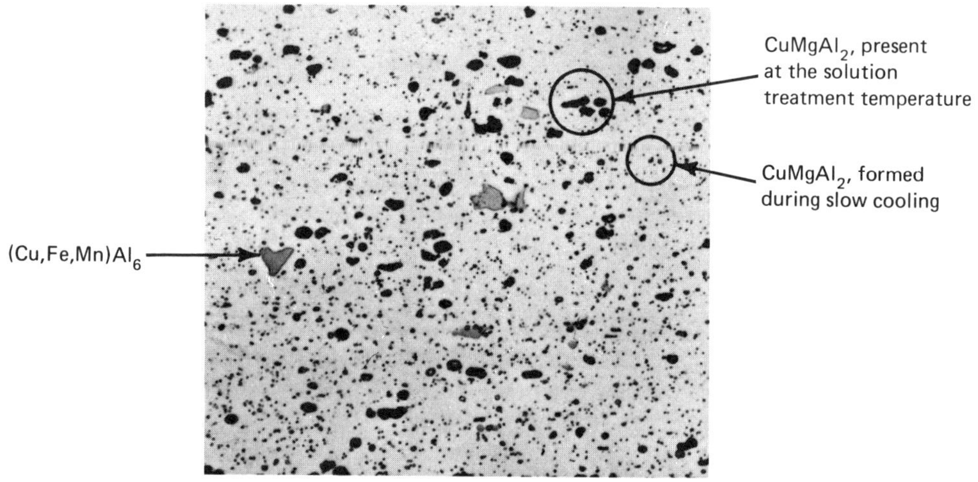

Fig. 4-20. Microstructure of a 2024 alloy in the annealed condition (condition TO). Etched in 25% nitric acid. Optical micrograph; 500×. (Adapted from Metals Handbook, 8th Ed., Vol 7, American Society for Metals, Metals Park, Ohio, 1972)

During precipitation, the structure is too fine to properly resolve with optical microscopy. Electron microscopy reveals that the precipitation process is similar to that described in Chapter 3 for Al-Cu alloys; however, the details of the precipitation process do depend upon the exact chemical composition of the alloy. In all cases, though, GP zones and transition phases form which are responsible for hardening. Eventually, the equilibrium phases appear.

The effect of the precipitation process on the mechanical properties of a 2024 alloy is illustrated in Fig. 4-22 and 4-23. Aging at low temperature

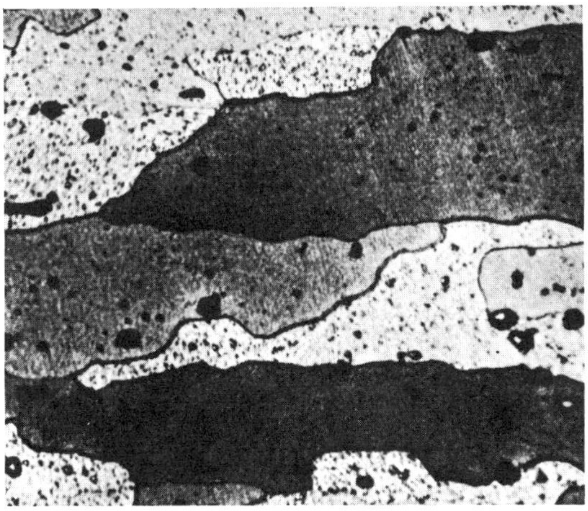

Fig. 4-21. Microstructure of a 2024 alloy in the solution heat treated condition (condition T3). Optical micrograph; 500×. The dark particles are $CuMgAl_2$, $Cu_2Mn_3Al_{20}$ and Cu_2FeAl_7. Etched in Keller's reagent. (From same source as Fig. 4-20)

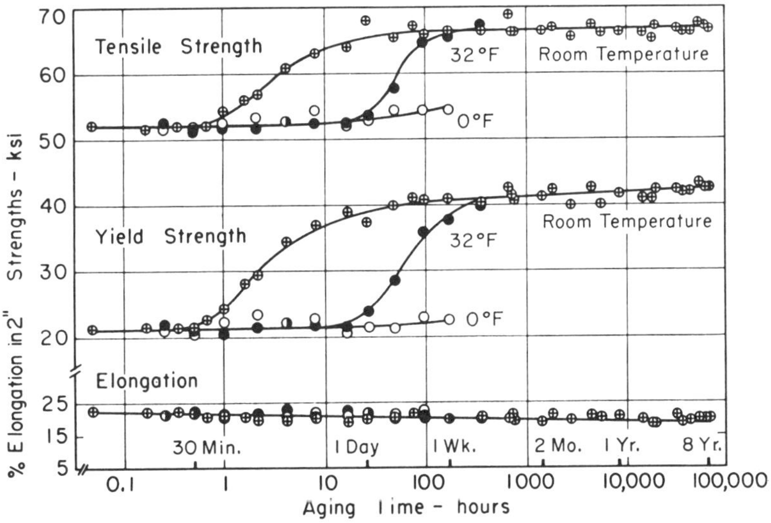

Fig. 4-22. Effect of aging time at 20, 0 and −17 °C on the mechanical properties of 2024 alloy. (From W. A. Anderson, in Precipitation From Solid Solution, American Society for Metals, Metals Park, Ohio, 1958)

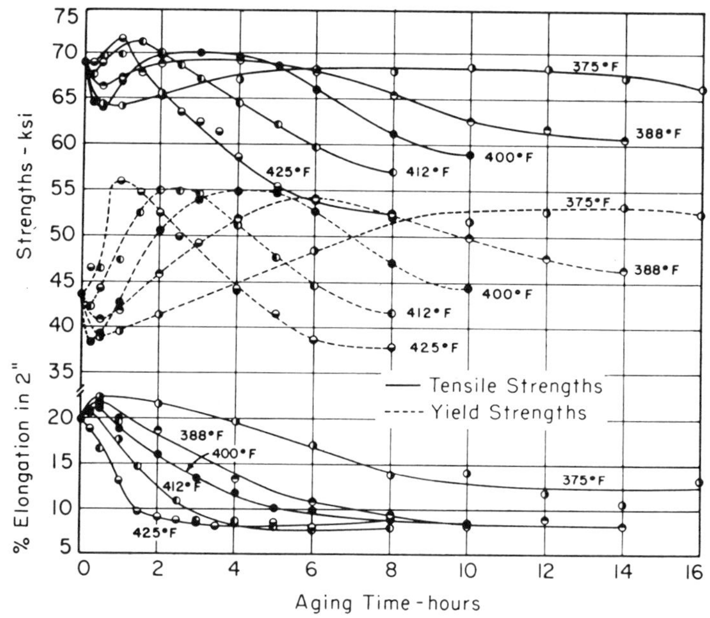

Fig. 4-23. Effect of aging time and temperature on the mechanical properties of 2024 alloy. The initial condition was the natural aged state (temper T4). (From same source as Fig. 4-22)

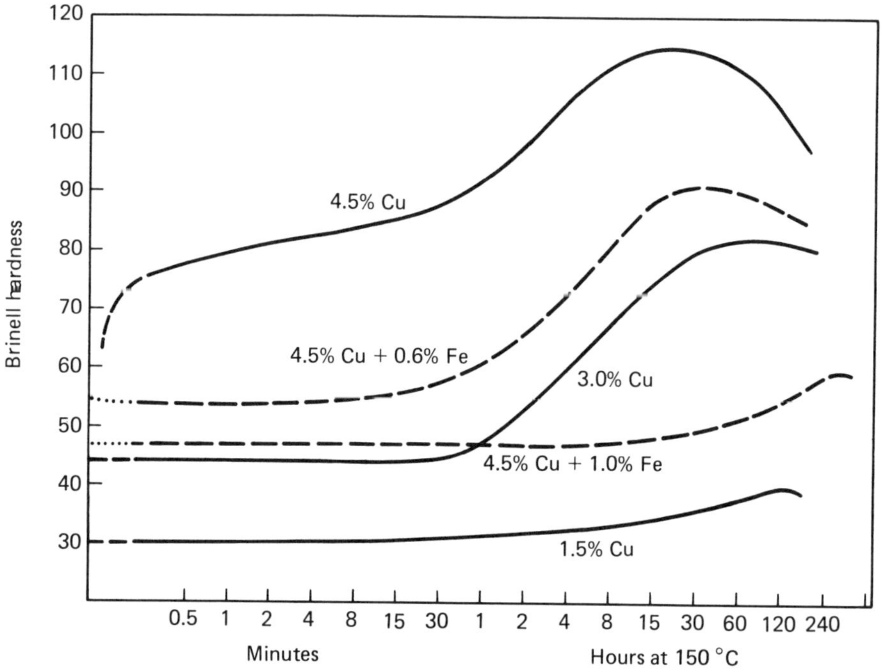

Fig. 4-24. Effect of Fe additions on the age hardening at 150 °C of Al-Cu alloys. (From H. Y. Hunsicker, in Age Hardening of Metals, American Society for Metals, Metals Park, Ohio, 1940)

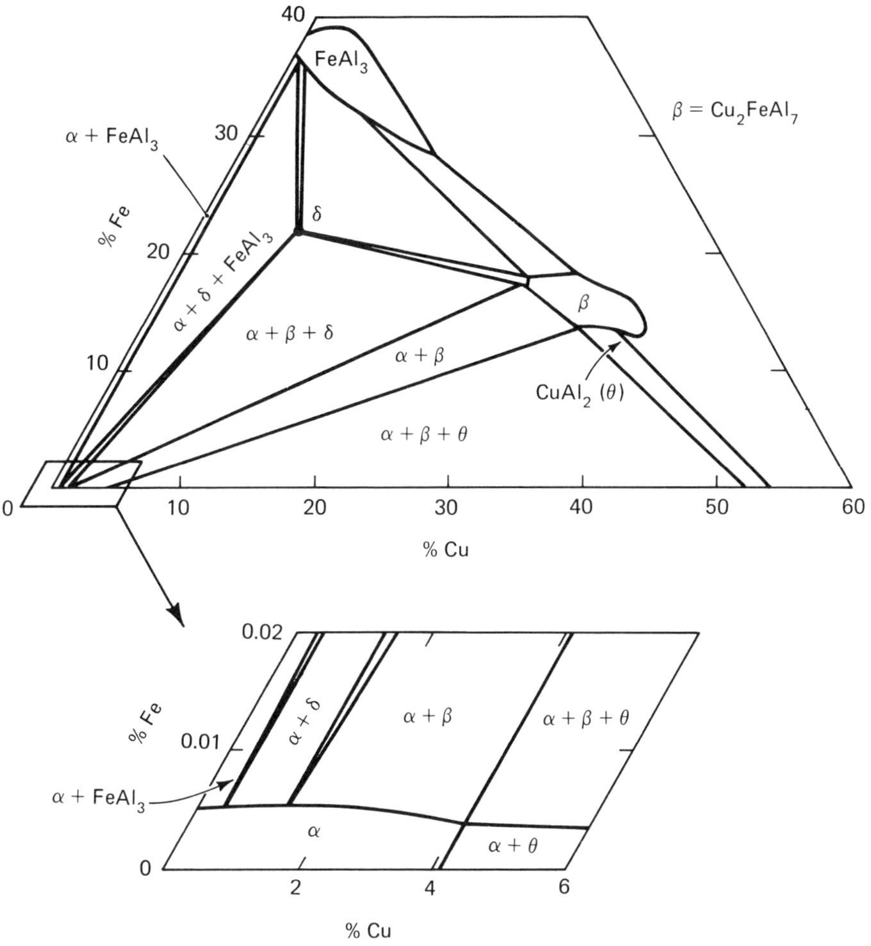

Fig. 4-25. Isothermal section at 500 °C for the Al-Cu-Fe system. (Adapted from H. W. L. Phillips, J. Inst. Metals, Vol 82, p 197, 1953/54)

hardens the alloy in a manner very similar to that for Al-Cu-Mg ternary alloy (Fig. 4-13). Elevated temperature precipitation is much more effective in hardening this alloy. The data in Fig. 4-23 show that the strength can be increased from about 40,000 to about 55,000 psi by aging at about 200 °C (400 °F) for a few hours. For the 2024 alloy, a recommended aging treatment is to heat for 11 to 13 h in the range 185 to 196 °C (365 to 385 °F). Figure 4-23 shows that this time is in the range of the maximum in strength.

The mechanical properties upon aging do not vary greatly with variation in chemical composition within the allowable range for 2024 alloy from those shown in Fig. 4-22 and 23. However, the effect of Fe deserves specific attention since its presence can lower the hardening response. Figure 4-24 shows that the addition of 1.0% Fe to an Al-4.5 Cu alloy almost eliminates any hardening. Examination of the Al-Cu-Fe ternary 500 °C isotherm (Fig. 4-25) shows why the iron has such an effect: the solubility of Fe in α is quite low, and the intermetallic compound based on the approximate formula Cu_2FeAl_7 forms. Figure 4-26 is an enlargement of the aluminum-rich corner. The line at constant Cu content of 4.5% is shown, and the composition of 4.5% Cu and 1.0% Fe is in the $\alpha + \beta$ region. The tie line through this composition, extending from the α phase to the compound $(Cu_2FeAl_7)_3$, reveals that the Cu content of the α is 2.2%. Thus, although the alloy contains 4.5% Cu, the compound Cu_2FeAl_7 takes about half of the copper out of the solution.

Upon rapid cooling to 20 °C, the alloy consists of particles of the compound in an α matrix, but the α contains only 2.2% Cu. Upon aging, very little additional Cu_2FeAl_7 will form since its solubility in α is very low at 500 °C. The phase boundaries shift to favor the precipitation of $CuAl_2$, which is the phase which forms in the binary Al-Cu alloy. However, even though the ternary alloy contains 4.5% Cu, upon aging it behaves like a binary alloy containing about 2.2% Cu. This is responsible for the very low age hardening of the iron-containing alloys shown in Fig. 4-25, and for the iron content of the 2024 alloy being held to 0.5% or less.

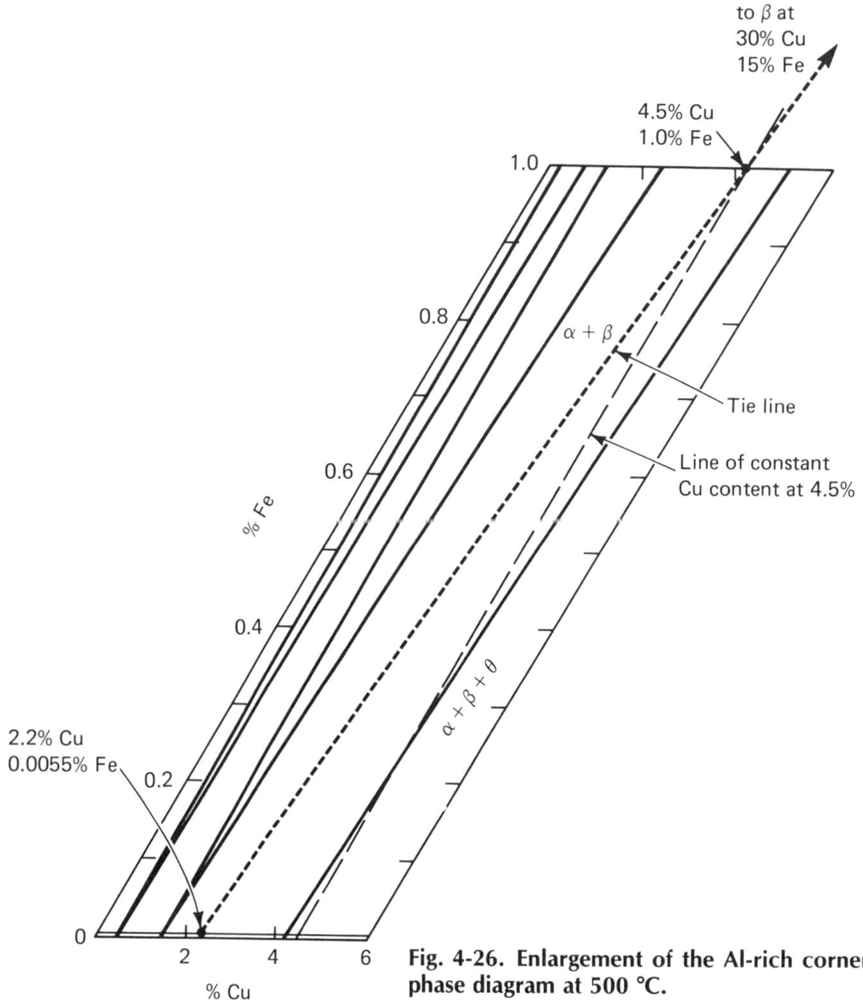

Fig. 4-26. Enlargement of the Al-rich corner of the Al-Cu-Fe ternary phase diagram at 500 °C.

References

Metallography of Aluminum and Its Alloys, V. Fuss, The Sherwood Press, Cleveland (1936).

Age Hardening of Metals, American Society for Metals, Metals Park, Ohio (1940).

Metallography of Aluminum Alloys, L. F. Mondolfo, Wiley, New York (1943).

Physical Metallurgy of Aluminum Alloys, American Society for Metals, Metals Park, Ohio (1949).

Precipitation from Solid Solution, American Society for Metals, Metals Park, Ohio (1958).

Aluminum, Vol I: Properties, Physical Metallurgy and Phase Diagrams, ed. K. R. Van Horn, American Society for Metals, Metals Park, Ohio (1967).

Metallurgy of Aluminum Alloys, M. Van Lancker, Wiley, New York (1967).

Aluminum Alloys: Structure and Properties, L. F. Mondolfo, Butterworths, Boston (1976).

Aluminum Transformation Technology and Applications, ed. C. A. Pampillo, H. Biloni and D. E. Embury, American Society for Metals, Metals Park, Ohio (1980).

CHAPTER 5

Nickel-Base Alloys

A critical application of materials is in high-temperature systems such as jet aircraft engines and gas turbines in which corrosion and oxidation resistance and strength must be maintained. The alloys being used for these applications are iron-, cobalt- or nickel-base, and are called superalloys. The nickel-base alloys are the most widely used. Their applicability is based on the presence of chromium, especially to impart oxidation resistance, and other alloying elements to impart high-temperature strength, especially creep resistance. In this chapter, we will examine the physical metallurgy of some of the nickel-base alloys.

Appendix 3 lists a number of nickel-base superalloys and gives typical applications. The chemical composition is so varied that a very wide range of properties is obtained from this class of alloys, and it is impossible to describe any general heat treatment, based on physical metallurgy principles, which is applicable to all of these alloys. Instead, this chapter will give some description of simpler nickel-base alloys to illustrate the reasons for inclusion of many of the alloying elements present, and then a few of these alloys will be examined in detail.

Most of these alloys depend for their oxidation and hot corrosion resistance upon chromium and for their strength upon aluminum and titanium to allow the precipitation of an ordered compound based on the formula $Ni_3(Al, Ti)$. This phase is called γ' to distinguish it from the face-centered cubic matrix in which it forms, denoted γ. The mechanical properties of superalloys, then, depend upon the control of such factors as the amount of γ, its morphology and the elastic strain induced when it precipitates because of the lattice mismatch between it and the γ matrix, and other factors to be explored in this chapter.

The γ' precipitate is usually quite fine, requiring electron microscopy to reveal it properly. This is illustrated in Fig. 5-1 for a specific superalloy. Note

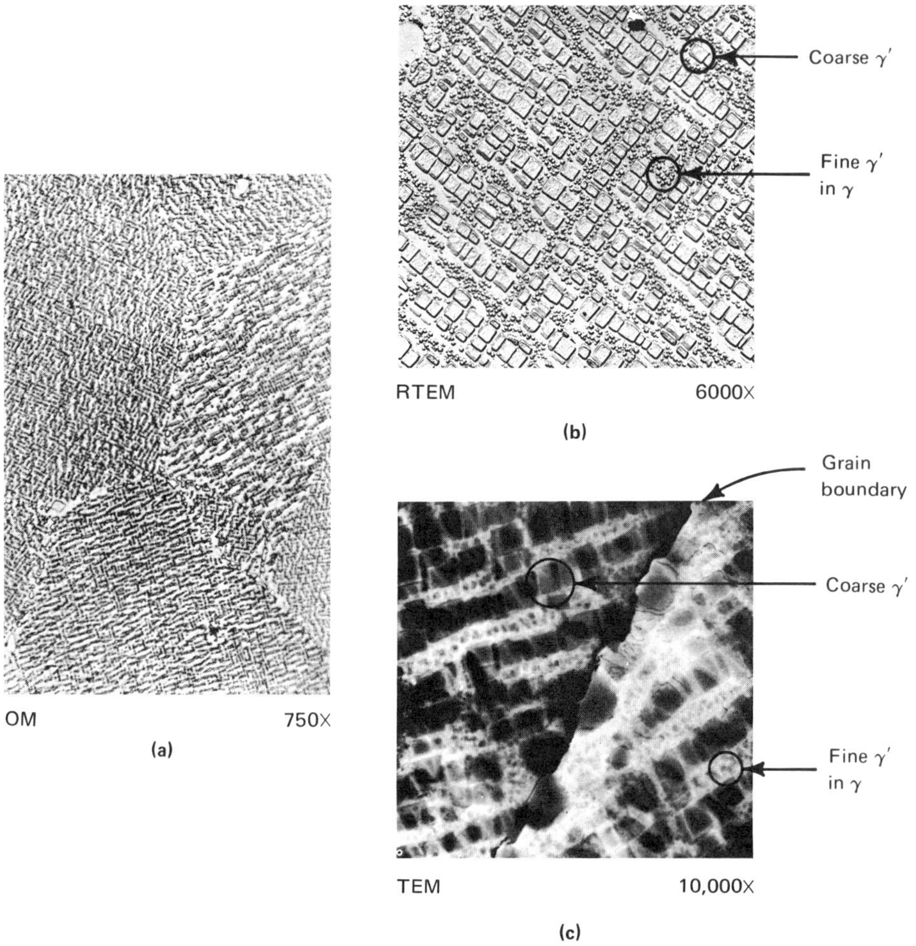

Fig. 5-1. Microstructure typical of Nimonic 115, heated for 1.5 h at 1190 °C, air cooled to 1100 °C, held 6 h, air cooled to 25 °C. This alloy contains approximately (wt %) 15Cr-5Al-4Ti-15Co-4Mo-0.18B-0.16C. (Adapted from C. H. White, in The Nimonic Alloys, 2d Ed., ed. W. Betteridge and J. Heslop, Edward Arnold Ltd, London, 1974)

that in the optical micrograph a pattern is revealed (a), but higher resolution and magnification (b) reveal that it consists of cubes of γ' of two sizes, one considerably smaller than the other. This is also shown in the direct transmission electron micrograph (c) of a very thin sample prepared from the alloy. Here the γ' precipitates are being revealed directly, whereas in (a) and (b) the surface topology of a polished and etched surface is being observed.

Because the structure is so fine, most of the micrographs used in this chapter are electron micrographs, either scanning electron micrographs (SEM) or replica transmission electron micrographs (RTEM). For the latter, a plastic replica is made of the etched surface, coated with a very thin layer of carbon and heavy metal (e.g., chromium) and then the plastic is dissolved. The remaining replica is observed in the electron microscope. Note that both RTEM and SEM present a view of the surface topology.

5-1. Nickel-Chromium Alloys: Oxidation Resistance

The gaseous oxidation of metals involves the contact of oxygen molecules of the gas with the metallic atoms at the metal surface, and their reaction to form the oxide. (Oxidation can also occur in other gases, such as CO_2 and water vapor; here we will consider only oxidation with oxygen molecules.) This reaction forms an oxide layer on the surface, and further oxidation may require oxygen atoms (from the dissociation of the oxygen molecule at the surface) to diffuse through the oxide layer to the oxide-metal interface, where they can encounter the metal atoms (Fig. 5-2). Or the metal atoms may diffuse through the oxide layer to the oxide-gas interface, where they can react with oxygen. In either case, it is reasonable to assume that the rate of oxidation will decrease with time, as the diffusing species must transverse a thicker and thicker layer for more oxide to form.

A relation which shows this behavior is

$$dm/dt = kt^{+n},$$

where m is the mass of oxide formed, t is time, k is a constant (temperature dependent) and n is a positive number less than one. It is frequently found that $n = 1/2$, so that integration of the above equation obtains for the weight gain Δm

$$(\Delta m)^2 = k_p t,$$

where k_p is another constant, called the parabolic rate constant. This relation is referred to as the parabolic oxidation law. The oxidation rate can be characterized by the value of k_p.

Figure 5-3 shows the effect of chromium additions to nickel for oxidation under a certain condition. The rate of oxidation increases for low chromium content, and this has been associated with a "defect" oxide (i.e., non-stoichiometric) forming in which oxygen can diffuse more easily than in pure NiO. When the chromium content reaches 20% or above, the oxidation rate decreases, and this corresponds to the formation of a very adherent, dense Cr_2O_3 oxide. Especially important is the adherence of this oxide, and the lower diffusion rate of oxygen through it. Thus, in the range 20-30% Cr, the oxidation rate constant is about one-half of that for pure nickel for the oxidation conditions of the data in Fig. 5-3.

It is not surprising that elements such as nickel and chromium easily oxidize. After all, many of the ores of metals are found as oxides and must be reduced to obtain the pure metal. A prominent example is aluminum, where the oxide is so stable that it is used as a high-temperature refractory material. Thus, thermodynamically, aluminum oxide is extremely stable relative to the pure metal. However, the oxide is very dense and adherent and the diffusion rate of oxygen through it is quite low, and once a thin layer forms on the metal, the rate of oxidation is so low that, for many applications, oxidation is of no consequence. The Ni-Cr alloys have this

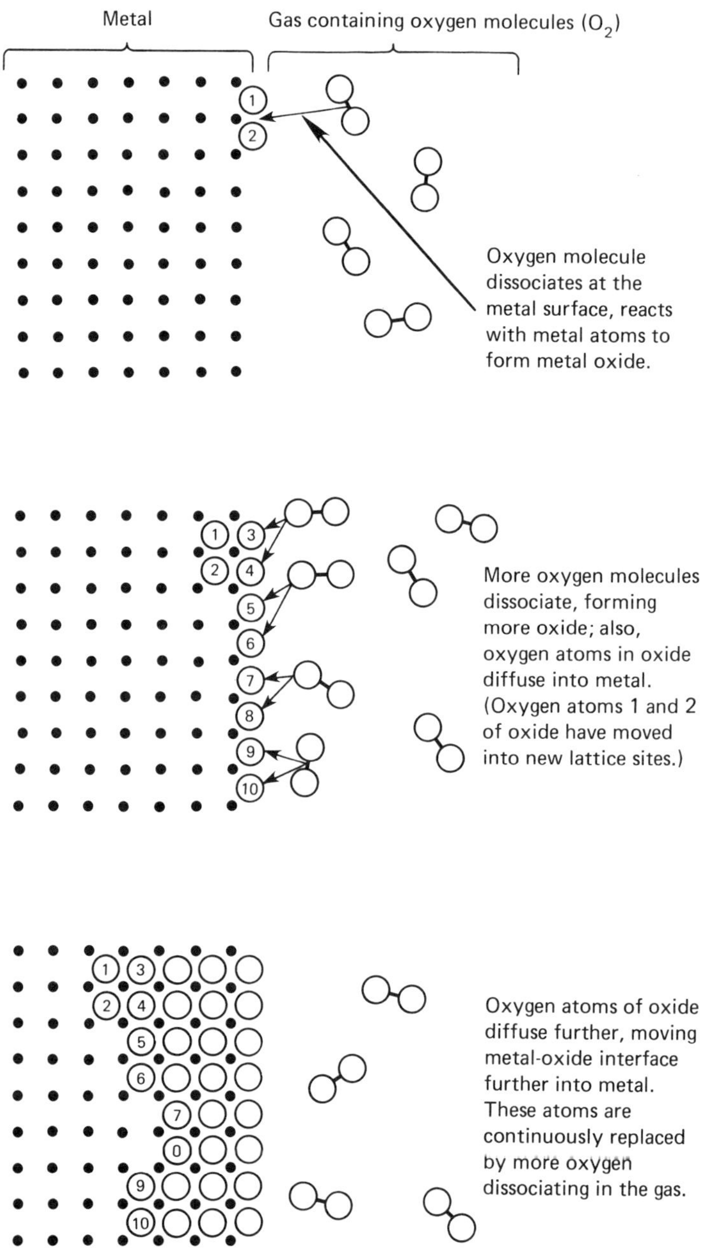

Fig. 5-2. Schematic illustration of a possible mechanism of oxidation of a metal.

same behavior, where thermodynamically Cr_2O_3 is stable, but the rate of formation is sufficiently low that they can be used in many high-temperature oxidizing conditions.

Another important consideration is the adherence of the protective oxide film during thermal cycling. The underlying metal and the oxide have differ-

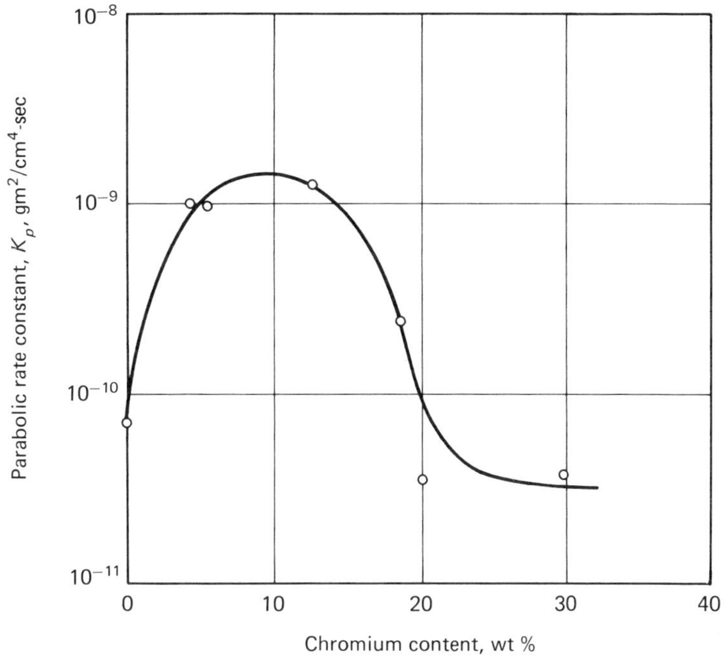

Fig. 5-3. Parabolic rate constant for oxidation of Ni-Cr alloys. Specimens were oxidized at 1000 °C in oxygen at 0.1 atm pressure for up to 70 h. From plots of $(\Delta m)^2$ versus t the constant K_p was obtained. (Adapted from C. S. Giggins and F. S. Pettit, Trans. Met. Soc. AIME, Vol 245, p 2495, 1969)

ent coefficients of thermal expansion, and the possibility must be examined of the brittle oxide cracking from the thermal stresses and separating from the metal (spalling) if the temperature changes from that at which the oxide formed. Such spalling exposes fresh metal to oxidation, and parabolic oxidation would not be followed but instead a higher rate of oxidation (e.g., linear). It has been found that the Cr_2O_3 scale is very adherent, and thus the oxidation resistance remains even under cyclic conditions. Therefore, the element Cr plays a key role in developing resistance to oxidation in Fe-base (e.g., stainless steels), Ni-base and Co-base alloys, and another element with such characteristics which is as economical to use has not been found.

In this section it has been shown that chromium is a key element in imparting oxidation resistance to nickel, if present in the amount of 20% or above. Other potential alloy additions do not impart such resistance, and thus if they are added for strengthening, the chromium content must be regulated carefully to retain oxidation resistance. Now, the Ni-Cr alloys considered in this section are single-phase (up to about 35% Cr), whereas the commercial superalloys are multiphase structures, and thus the resistance of this more complex structure to oxidation must be considered. However, before doing so, we will first examine the strengthening mechanisms in superalloys, then return to a consideration of oxidation resistance.

5-2. Nickel-Aluminum Alloys

In considering methods to strengthen nickel, two probably would come to mind first: solid solution strengthening and precipitation hardening. Generally, it would be expected that the addition of any solute would cause strengthening (see Chapter 1). However, if high-temperature strength is considered, such as creep strength, solute addition may not increase the strength. Since creep strength is controlled by atom mobility (diffusion), this decreases with decreasing melting point. The Ni-Cr phase diagram (Fig. 5-4) shows that chromium lowers the melting point of nickel in the composition range where the alloy is a solid solution (0-35% Cr), and thus chromium may not be an effective strengthener for high-temperature properties.

It was pointed out in Chapter 3 that extensive solubility is favored by similar atom size, similar electronic structure and similar crystal structure of the solute and solvent. Thus an element that has a high solubility in nickel will probably be similar in size. Solid solution strengthening is related to the size difference, and hence those solutes having a large size difference from nickel will be the most potent strengtheners, but the solubility will be low. This is illustrated in Table 5-1, where the size difference and solubility at 1000 °C are listed for several solutes. Note that carbon should be a potent solid solution strengthener, but the solubility is very low. Solutes that appear to be most promising from both the size difference and the solubility are Al, Ti, Mn, Nb, Mo, Ta and W. Figure 5-5 shows the effect of some of these solutes on the yield strength at 25 °C. There should be an approximate

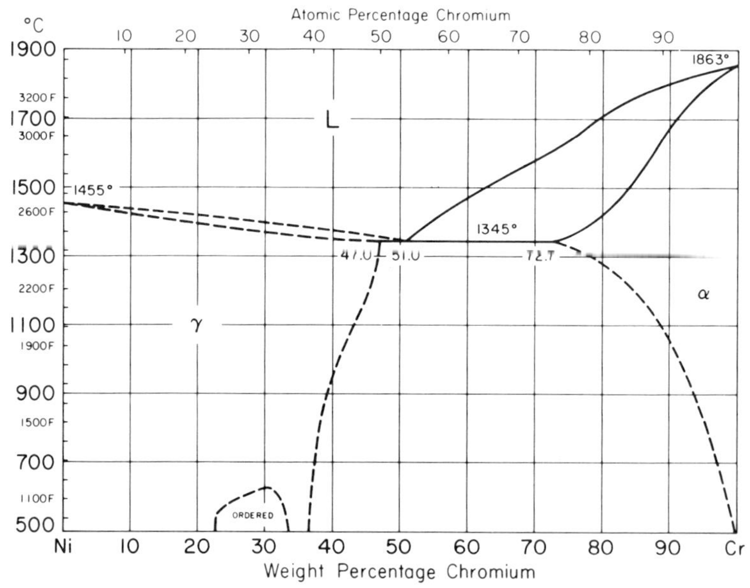

Fig. 5-4. The Ni-Cr phase diagram. The ordered structure occurs at Ni_2Cr, but its formation rate is very low. (From Metals Handbook, 8th Ed., Vol 8, American Society for Metals, Metals Park, Ohio, 1973)

Table 5-1. Approximate Atom Diameter Size Difference and Approximate Solubility in Ni of Several Solutes

Solute	Approximate atom diameter size difference, % $(d_{Ni} - d_M)/d_{Ni}$	Approximate solubility in Ni at 1000 °C, wt %	Solute	Approximate atom diameter size difference, % $(d_{Ni} - d_M)/d_{Ni}$	Approximate solubility in Ni at 1000 °C, wt %
C	+43	0.2	Fe	+0.3	100
Al*	−15	7	Co	−0.2	100
Si	+6	8	Cu	−3	100
Ti*	−17	10	Nb*	−15	6
V	−6	20	Mo*	−9	34
Cr	−0.3	40	Ta*	−15	14
Mn*	+10	20 (at 500 °C)	W*	−10	38

Note. Solutes marked with an asterisk (*) show the best combination of large size difference and high solubility and should be the most potent solid solution strengtheners.

correlation between the slope of these curves and the atom size difference (Table 5-1), and this exists as seen by comparing the effect of Ti with that of the other elements. On this basis, Ta and Nb should be potent solid solution strengtheners. Also, Mo, W and Cr are all effective strengtheners.

Figure 5-6 show some high-temperature creep data. Both Mo and W are more potent than Cr in high-temperature strengthening.

Thus it is seen that several elements have sufficient solubility to be effective in strengthening nickel. Of course, their effect may be modified when several solutes are present, and this will be examined when commercial, Ni-base alloys are discussed.

The significant solid solution strengthening caused by the addition of several solutes to nickel naturally leads to an examination of the possibility of increasing the amount of the element to exceed the solubility limit, and determining the precipitation-hardening response of the alloys. We will look at precipitation in Ni-Al alloys first.

The Ni-Al phase diagram (Fig. 5-7) shows that compositions in the range of 5-12% Al might show precipitation hardening. The aluminum solute can be dissolved at high temperature, and if precipitation of γ' upon cooling can be suppressed, then precipitation can be controlled by proper choice of the aging temperature and time. Whether precipitation causes strengthening will have to be determined empirically.

The γ phase is face-centered cubic, and the γ' has the approximate formula Ni_3Al. It is an ordered face-centered cubic compound, with the Al atoms at the cube corners and the Ni at the face centers. Since both γ and γ' are face-centered cubic and the lattice parameters are quite similar, differing by only about 0.5%, it is expected that γ' will form coherent with the matrix, with the unit-cell cube faces and edges of γ' parallel to those of γ. Figure 5-8 illustrates schematically the precipitation sequence and the crystallography involved. Shown is a (100) plane, with the Al atoms (solid circles) placed at random on the face-centered square array of sites. This is typical of the structure in the solution-treated condition, where the structure

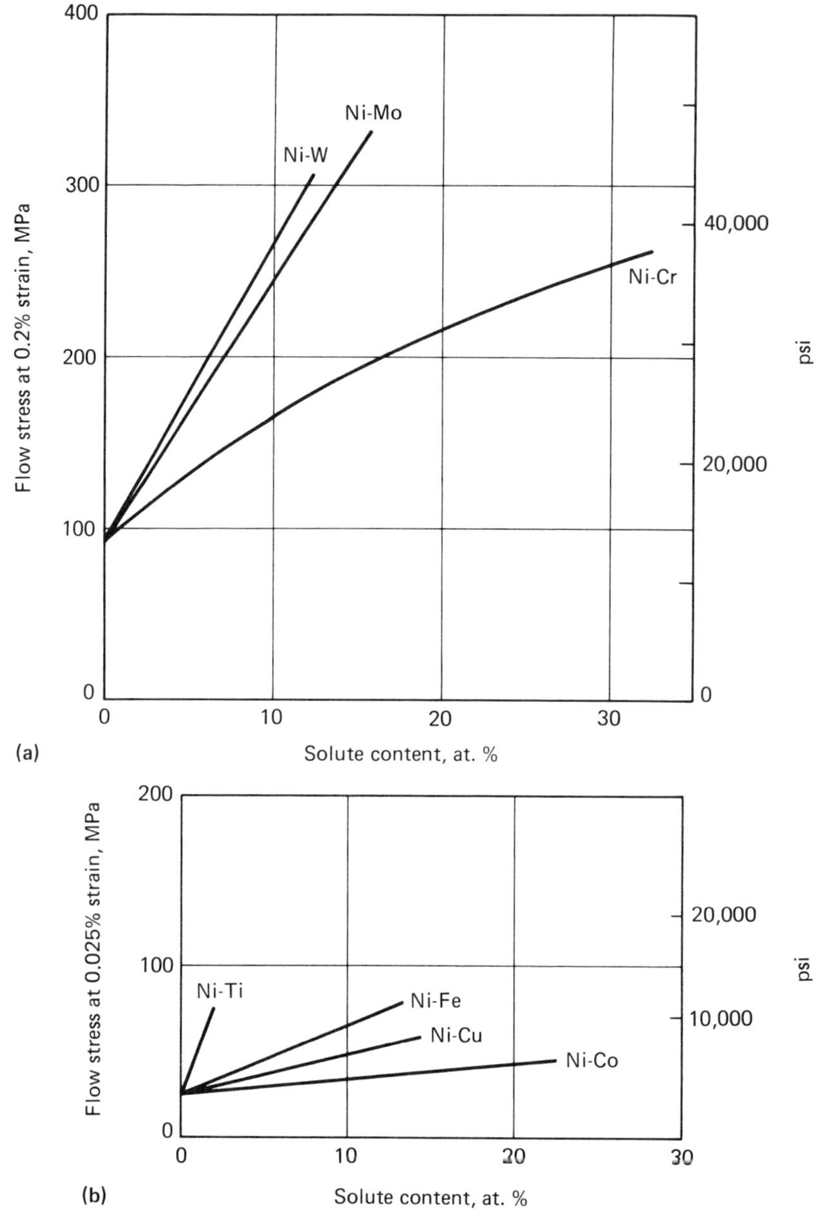

Fig. 5-5. Effect of solute content of Ni on yield strength (flow stress) at 25 °C. The alloys were all solid solution, single-phase. (a) adapted from R. M. N. Pelloux and N. J. Grant, Trans. Met. Soc. AIME, Vol 218, p 232, 1960; (b) adapted from E. R. Parker, in Relation of Properties to Microstructure, American Society for Metals, Metals Park, Ohio, 1954.

is single-phase γ. Upon reheating in the two-phase region, the bonding between Ni and Al atoms is strong enough to overcome their tendency to randomize on the lattice sites as a result of thermal agitation. Thus, when an Al atom moves into a position relative to a Ni atom, so that this small group of atoms attains the ordered structure, this region is relatively stable. Since, in the case shown here, the solute content is high enough for this to occur

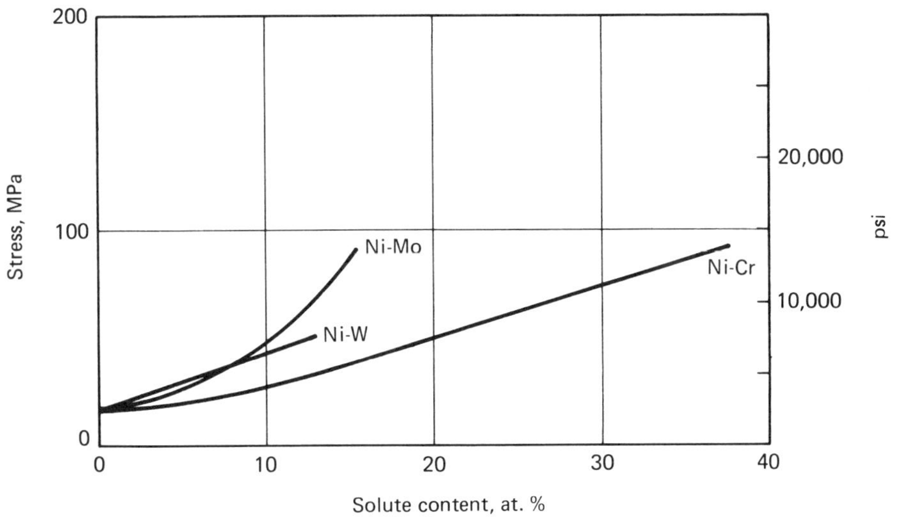

Fig. 5-6. Stress for 100-h rupture life at 820 °C for Ni-base solid solutions. (From same source as Fig. 5-5a)

easily without Al atoms moving long distances, the nucleation rate is high and many fine precipitates form initially.

The precipitation of γ' causes strengthening, as shown in Fig. 5-9. The aging curve has the shape typical of that associated with precipitation hardening, passing through a maximum with time at the aging temperature. In Chapter 3 it was pointed out that in many precipitation processes the initial precipitate is a nonequilibrium, transition phase, eventually disappearing as the equilibrium precipitate appears. In these Ni-Al alloys, however, the equilibrium γ' phase is the only precipitate.

Pure γ' (Ni_3Al) is weak at 25 °C, having a yield strength of about 20,000 psi. For a Ni-14Al alloy, the solution heat treated alloy will have a yield strength at 25 °C of about 40,000 psi. Thus, the extensive hardening upon precipitation of γ' is not due to its intrinsic strength. The lattice parameter mismatch between γ and γ' is about 1%, so that there are elastic interface strains which may impede dislocation motion. Also, if the dislocations cut the γ' particles, they create more γ-γ' interface, and this interface energy may be sufficiently high to resist cutting. When precipitation commences, the particle density is low and the particles are widely separated. Thus, dislocations can bypass the particles by bowing. (See Chapter 3.) As precipitation progresses, the density of the particles increases, bowing becomes more difficult, and the strength increases. In the Ni-12.7 Al alloy aged at 700 °C (Fig. 5-9), the γ' particles at the stage of maximum strength are quite small, about 150 Å. Upon aging this alloy beyond 5 h, it is observed that the smaller γ' particles begin to dissolve and the larger ones increase in size, so that the average distance between γ' particles increases. Thus, bowing becomes an active method of dislocations bypassing the particles, and the strength begins to decrease. (See Chapter 1.)

The coarsening is seen more clearly in Fig. 5-10. The γ' particles are cubes with sides parallel to the (100) planes of the γ matrix. The average size

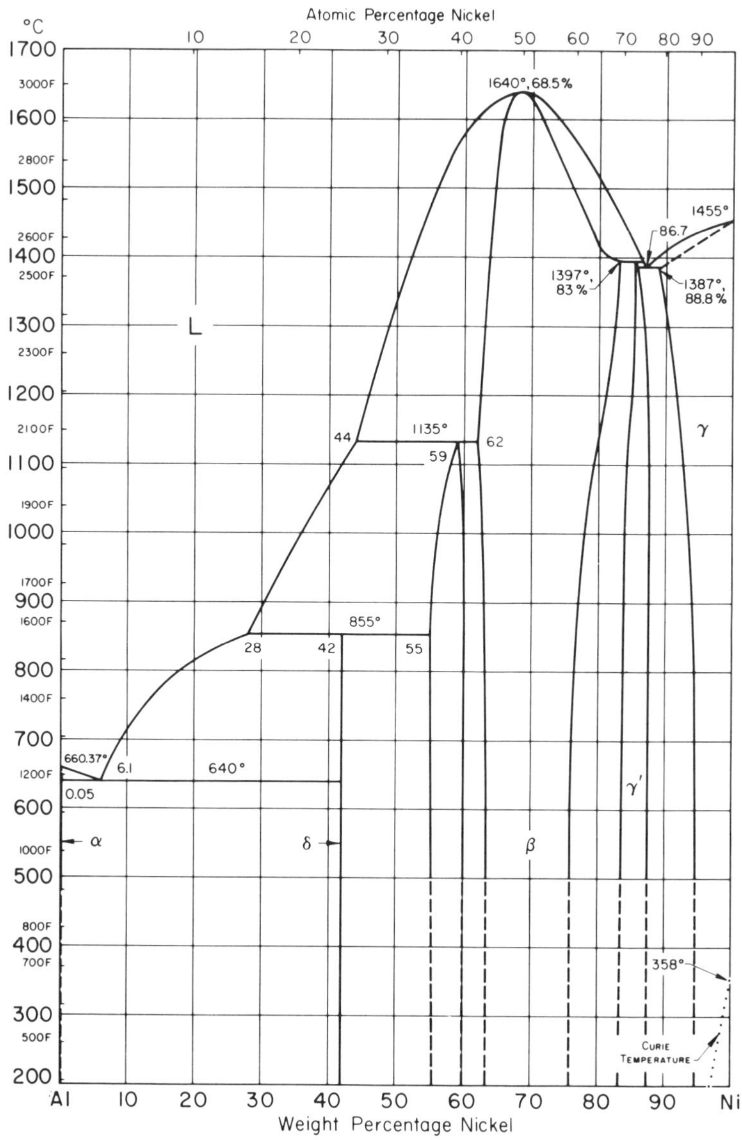

Fig. 5-7. The Ni-Al phase diagram. (Adapted from same source as Fig. 5-4)

of the γ' particles and the average distance separating the particles increase with time. (Note that the four micrographs are at different magnifications.) It is emphasized that the total amount of γ' remains constant; the only microstructural effect is the coarsening.

In Ni-Al alloys (and many superalloys) the crystallography predicted in Fig. 5-8 is correct. However, the morphology (shape) of the precipitate is determined by the γ'-γ surface energy and the elastic strain generated by the lattice misfit between γ and γ'. These are orientation dependent, especially the lattice strain. If the strain is low, the morphology which minimized the surface and strain energy per volume of precipitate would be spherical.

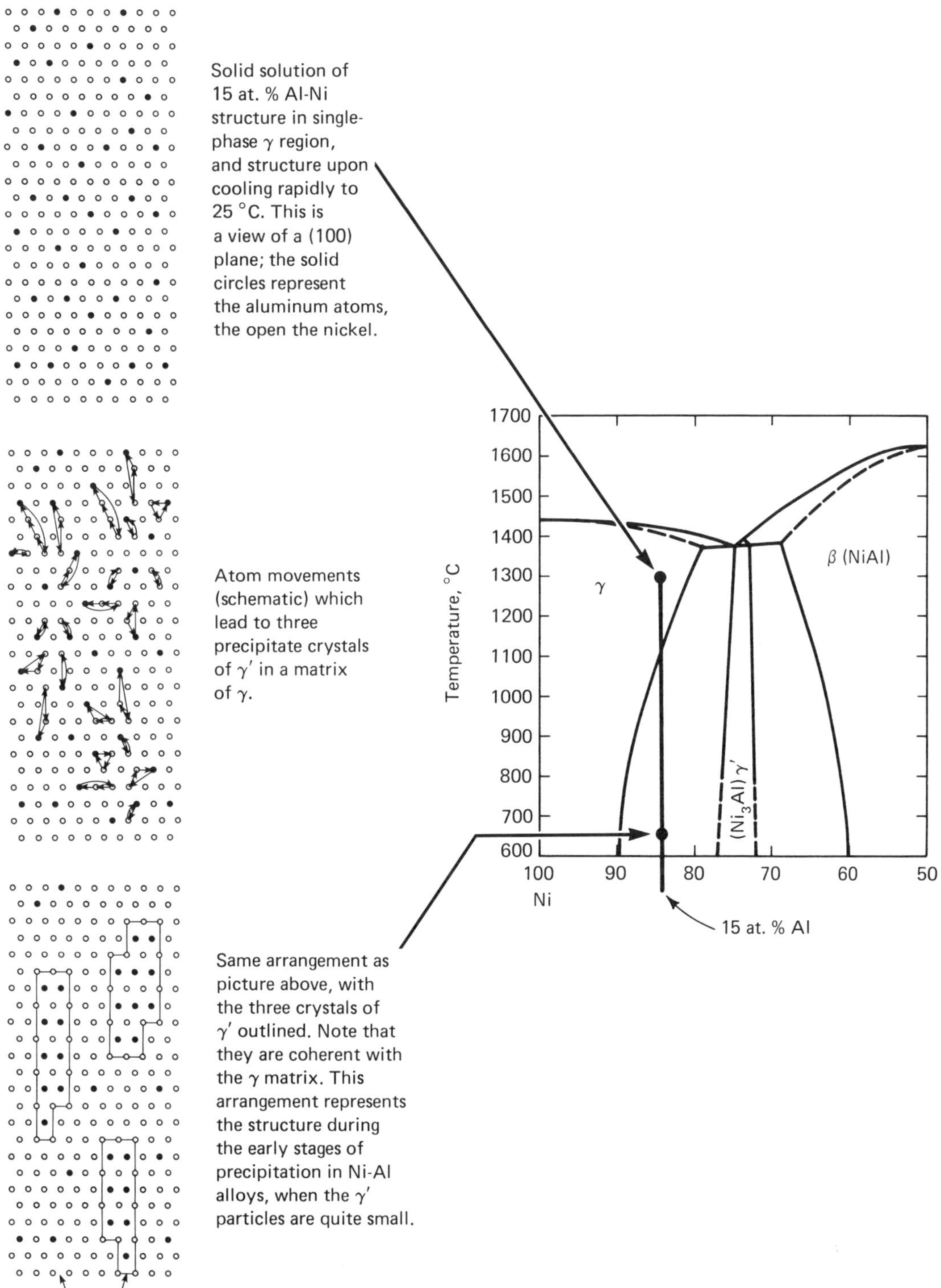

Fig. 5-8. Schematic arrangement of the precipitation of γ' crystals from γ in a Ni-Al alloy containing 15 at.% Al. The atom movements show possible atom shifts to create the three crystals of γ'. (The exact mechanism would involve vacancy movement, not depicted here.)

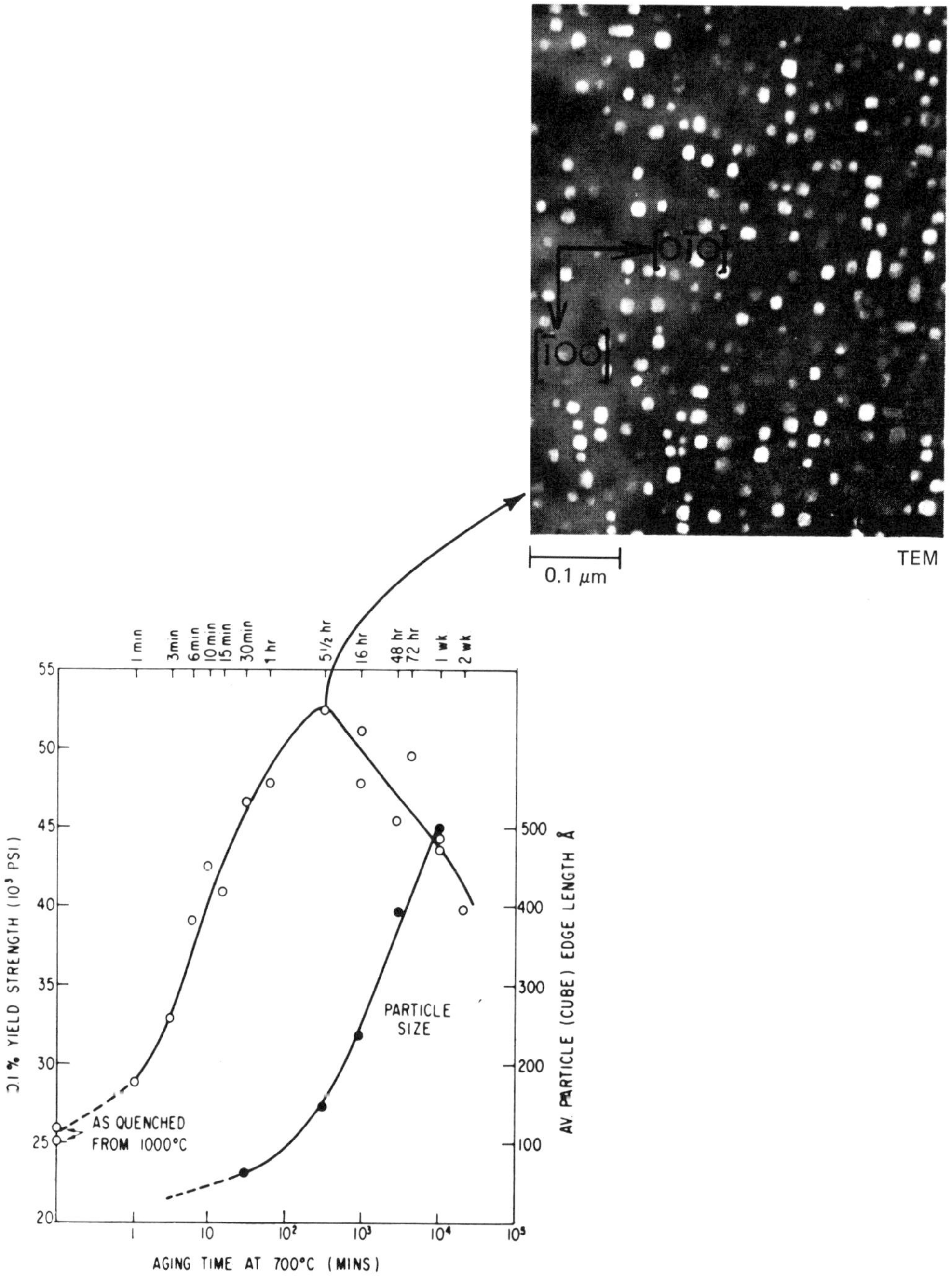

Fig. 5-9. Effect of aging time at 700 °C on the yield strength (at 25 °C) and γ' particle size of a Ni−12.7 at.% Al alloy. The microstructure (dark-field electron micrograph) shows the cuboidal γ' particles in the γ matrix. (Adapted from V. A. Phillips, Acta Met., Vol 14, p 1535, 1966)

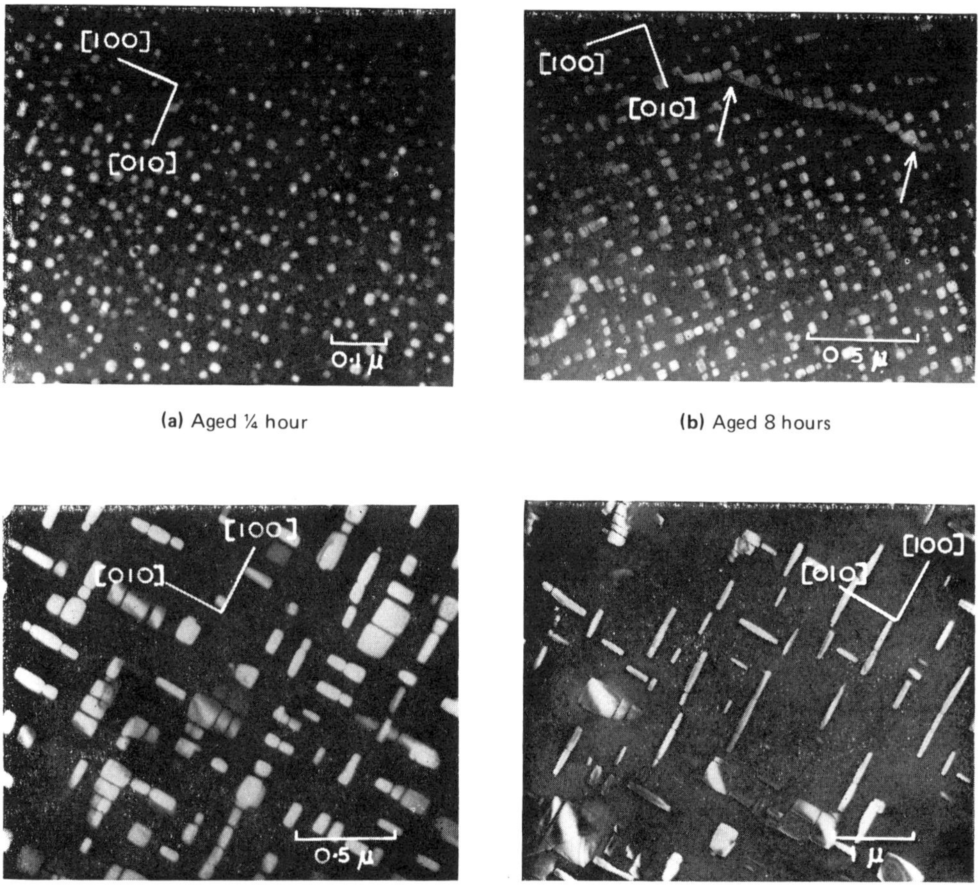

(a) Aged ¼ hour (b) Aged 8 hours

(c) Aged 96 hours (d) Aged 450 hours

Fig. 5-10. Transmission electron micrographs (dark-field) of a Ni – 6.71 at.% Al alloy aged at 750 °C for various times. (Adapted from A. J. Ardell and R. B. Nicholson, Acta Met., Vol 14, p 1299, 1966)

However, if the strain is sufficiently high, the morphology will not be spherical, and in the Ni-base alloys it is usually cubic. The origin of the lattice strain due to lattice mismatch is illustrated in Fig. 5-11. Note that the lattice is strained if the lattice parameter of the γ' is either larger or smaller than that of the γ in which it precipitates. The high-temperature creep properties are dependent upon the stability of the γ' precipitate. That is, for long time applications, by heat treatment a γ' structure is developed which has the desired creep properties, and these properties are best when the γ' particle size is stable, or increases only slowly. We shall see that the lattice mismatch gives an indication of the stability of the precipitate, as reflected in the growth rate. Figure 5-12 shows growth data of the Ni – 12.7 at.% Al alloy when aged at 700 °C. Note that the data are fitted approximately by a

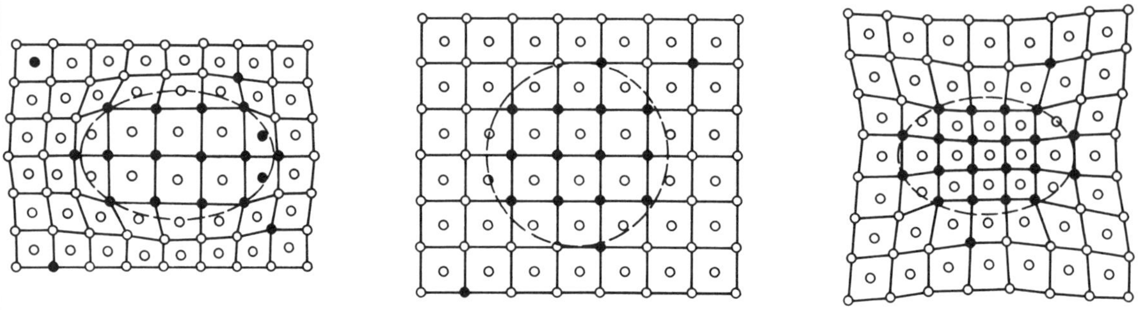

Fig. 5-11. Schematic illustration of lattice strain in γ matrix due to lattice mismatch between γ' and γ. (Adapted from H. Warlimont, in Electron Microscopy and Structure of Materials, ed. G. Thomas, University of California Press, Los Angeles, 1971)

straight line (on the log-log plot) with a slope of 1/3, which implies a relation of type

$$d = Kt^{1/3}$$

where K is a constant. Also note that the data pass through the origin on a plot of d versus $t^{1/3}$. In many Ni-base alloys the growth of γ' approximately follows this relation.

The equilibrium chemical composition of a phase dispersed in another turns out to be dependent upon the size of the precipitate. Thus, if there is a precipitate size distribution, then a composition gradient will exist in the matrix from one particle to another. This gradient is such that the element movement will be from the smaller particles to the larger, so that the smaller particles dissolve and the larger ones increase in size. This "ripening" or "coarsening" then is a diffusion-controlled process, and it is reasonable to expect some type of diffusion model to be used to predict the growth (coarsening) rate. Such a model has been developed, and it predicts the constant K to be given by

$$K = \left[\frac{64\Gamma D C_0 V_m^2}{9RT}\right]^{1/3}$$

where Γ is the γ-γ' interfacial energy, D is the diffusion coefficient of the solute in the γ matrix, C_0 is the solubility of the solute in equilibrium with a particle of infinite radius, V_m is the molar volume of the γ', R is the ideal gas constant and T is the absolute temperature. The 1/3 power in this equation is the reason for emphasizing the 1/3 slope in Fig. 5-12. Γ, D, C_0

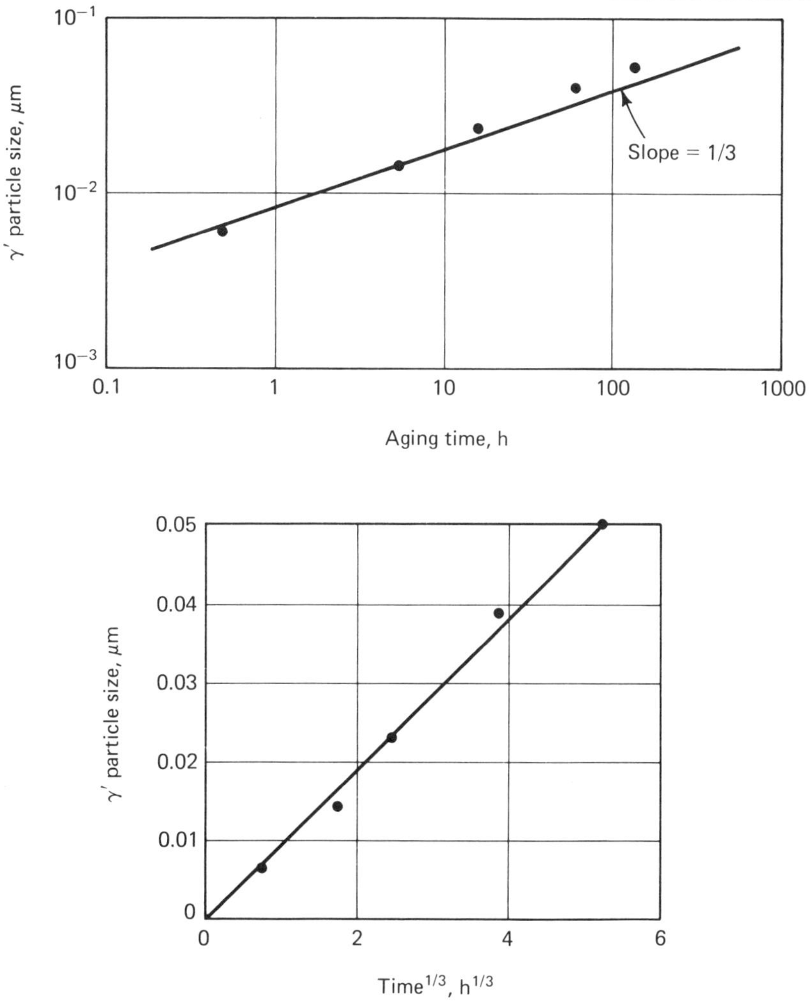

Fig. 5-12. γ′ particle size as function of time at 700 °C for a Ni−12.7 at.% Al alloy. Note that on the log-log plot the data can be represented reasonably well by a straight line with a slope of 1/3, and that on a linear plot against time the straight line passes through the origin. (Adapted from same source as Fig. 5-9)

and V_m are all functions of temperature, but especially D is temperature sensitive, being given by

$$D = D_0 e^{-Q/RT}$$

where D_0 and Q (the activation energy for the process) are both constants. This exponential temperature dependence makes the coarsening rate quite temperature dependent. We will examine these relations in a following section, because it is important to stabilize the γ' particles, or at least to minimize the coarsening. These equations predict that we should especially examine the effect of solute on the diffusion coefficient.

The variation of the yield strength of Ni_3Al with temperature is shown in Fig. 5-13. At 25 °C, the strength is about 20,000 psi, similar to the solution heat treated Ni – 12.7% Al alloy (Fig. 5-9). However, Ni_3Al shows the remarkable behavior of increasing strength (up to 800 °C) with increasing temperature. This is opposite to the behavior of most stable structures (i.e., structure in which at temperature no precipitation or other structural changes are occurring), such as shown in Fig. 5-13 for the quenched Ni – 14 at.% Al alloy. In this case the structure is single-phase, and the strength decreases with temperature; measurements were not made above about 400 °C because precipitation could not be suppressed. For the alloy aged at 700 °C for one hour, the yield strength is independent of temperature (up to about 800 °C). Thus, in the two-phase alloy, with γ' precipitates in a matrix of γ, the matrix decreases in strength with temperature but the γ' increases, so that the strength of an alloy containing both phases remains approximately independent of temperature, retaining the strength induced by the aging treatment.

It would appear that for high-temperature applications alloys which are completely γ' would be desired. However, the ductility of γ' can be quite low, so that it is necessary to use alloys in which γ' is in a more ductile matrix.

The information in this section has shown that in Ni-Al alloys, although the Ni_3Al phase (γ') itself is not necessarily stronger than the γ matrix, when properly dispersed it has a potent strengthening effect. Also, the Ni_3Al has

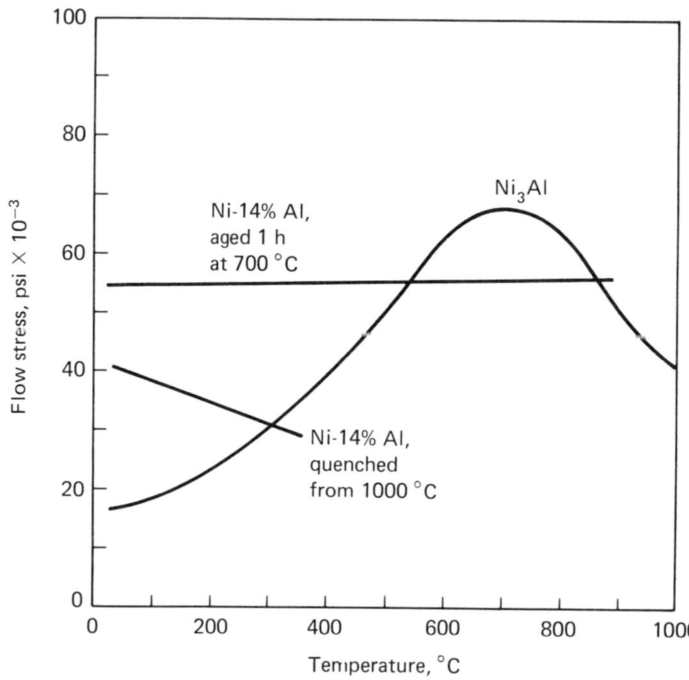

Fig. 5-13. Variation of the 0.2% yield strength with temperature for Ni_3Al, and for a Ni – 14 at.% Al alloy, quenched and also aged. (Adapted from R. G. Davies and M. S. Stoloff, Trans. Met. Soc. AIME, Vol 233, p 714, 1965)

the unusual behavior (not well understood) of having a greatly increasing yield strength with temperature so that its influence on high-temperature properties is very significant.

Since we have already seen that chromium is necessary for high-temperature oxidation resistance, it is a logical step to examine the possibility of obtaining γ' in ternary Ni-Cr-Al alloys. We will do this in the following section. Then we will be in a position to examine the more complicated commercial alloys.

5-3. Nickel-Chromium-Aluminum Alloys

The Ni-Cr and Ni-Al phase diagrams (Fig. 5-14) show that both Cr and Al are soluble to a considerable degree in nickel. However, the Al-Cr phase diagram shows that several intermetallic compounds form in this system, so that it is difficult to predict the influence of both Cr and Al in Ni on the formation of γ'. The ternary phase diagrams (Fig. 5-15) show that both Cr and Al are relatively soluble in combination in Ni and that the γ' phase (Ni_3Al) will dissolve up to about 20 at.% Cr at 750 °C. That is, in the binary compound Ni_3Al, Cr atoms can be substituted for Al atoms up to about 20% Cr.

Now the question is whether high strength can still be retained in the ternary alloys. Figure 5-16 shows an isopleth at 75% Ni, and isothermal sections at 750 and 1150 °C are also illustrated. Several alloys whose composition is shown at 900 °C by the dots were held at this temperature for 16 h (until equilibrium attained), then cooled to 25 °C. The cooling did not suppress the precipitation of γ' in the γ present at 900 °C, and a fine (75 Å) precipitate was present along with any γ' formed at 900 °C. The amount of this fine γ' was only about 3%. Figure 5-17 shows the effect of the amount of γ' on the strength at 25 and 900 °C. Note that the alloy which was all γ at 900 °C, and hence γ with about 3% γ' at 25 °C, has a composition of about 7% Al – 18% Cr, and a yield strength of about 40,000 psi. As shown in Fig. 5-9 a binary alloy of about 13% Al aged for two weeks at 700 °C to a γ' particle size of about 400 Å has a strength of about 40,000 psi. From the phase diagram, the equilibrium amount of γ' is calculated to be about 13% in this case. Thus, the binary alloy and the ternary alloy have about the same strength, but with less γ' present in the ternary alloy. Hence, the presence of Cr in Ni-Al alloys retains the high strength capabilities and increases oxidation resistance.

Figure 5-17 shows that the ternary alloy retains the excellent high-temperature strength associated with the presence of γ' (see Fig. 5-13). Further, at 900 °C the Cr strengthens γ' considerably compared with γ' in Ni-Al binary alloys. In the ternary alloy, a structure of all γ' has a yield strength at 900 °C of about 80,000 psi, whereas in the binary alloy, Ni_3Al (γ') has a strength of about 50,000 psi at 900 °C (Fig. 5-13). However, the γ' phase is generally brittle, so for the strength of this phase to be utilized

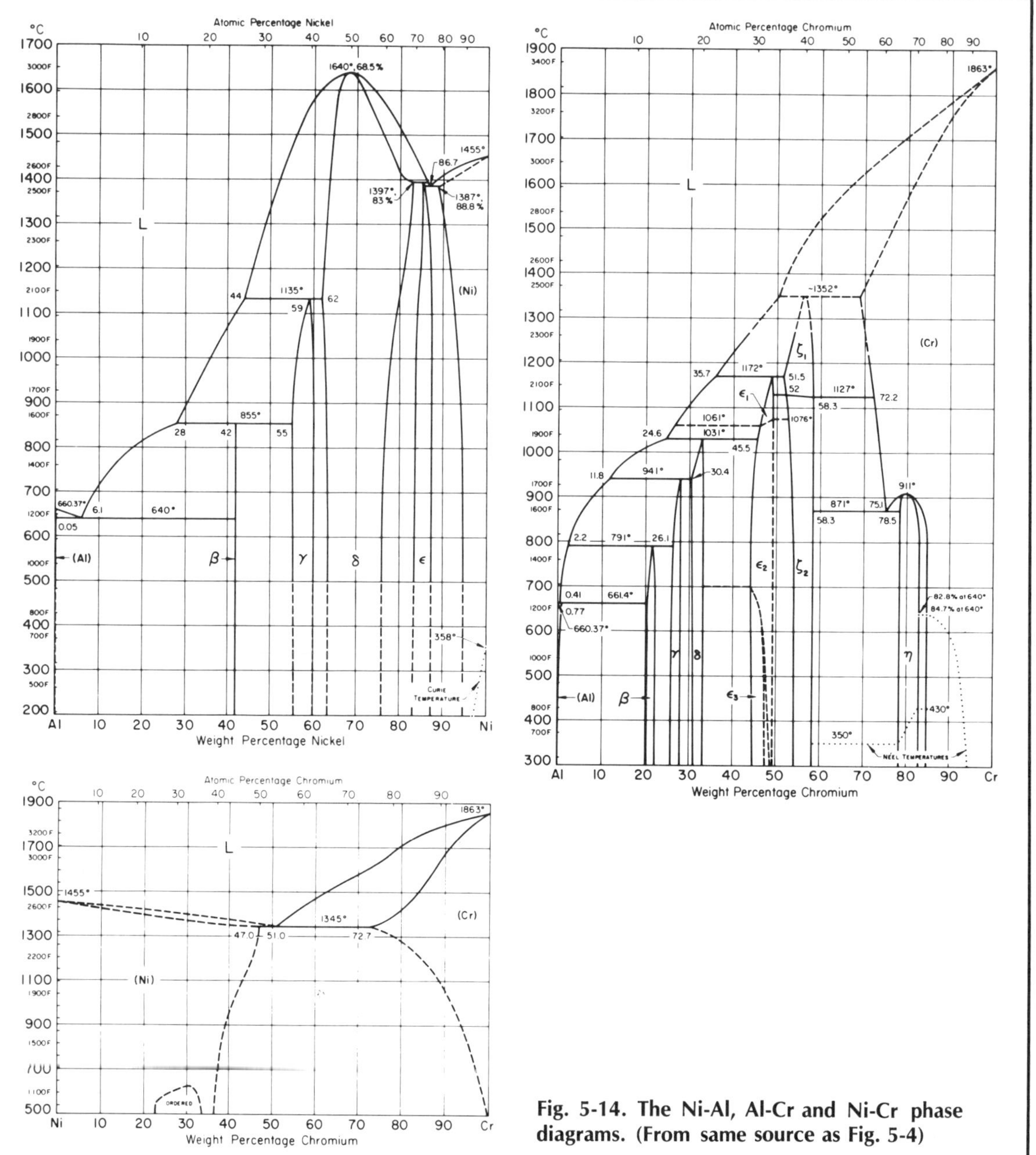

Fig. 5-14. The Ni-Al, Al-Cr and Ni-Cr phase diagrams. (From same source as Fig. 5-4)

it must be embedded in a more ductile matrix, and hence the useful alloys will usually be in the two-phase γ-γ' range.

Thus we see that the Ni-Cr-Al alloys show excellent low-temperature and short-time high-temperature strength. This increased strength of the ternary alloys over the Ni-Al alloys is associated with the increased solid solution strengthening of the γ matrix and of the γ'. However, it is emphasized that a major design criterion in developing superalloys is to control the amount of γ', as its presence is a major strengthening variable (Fig. 5-17).

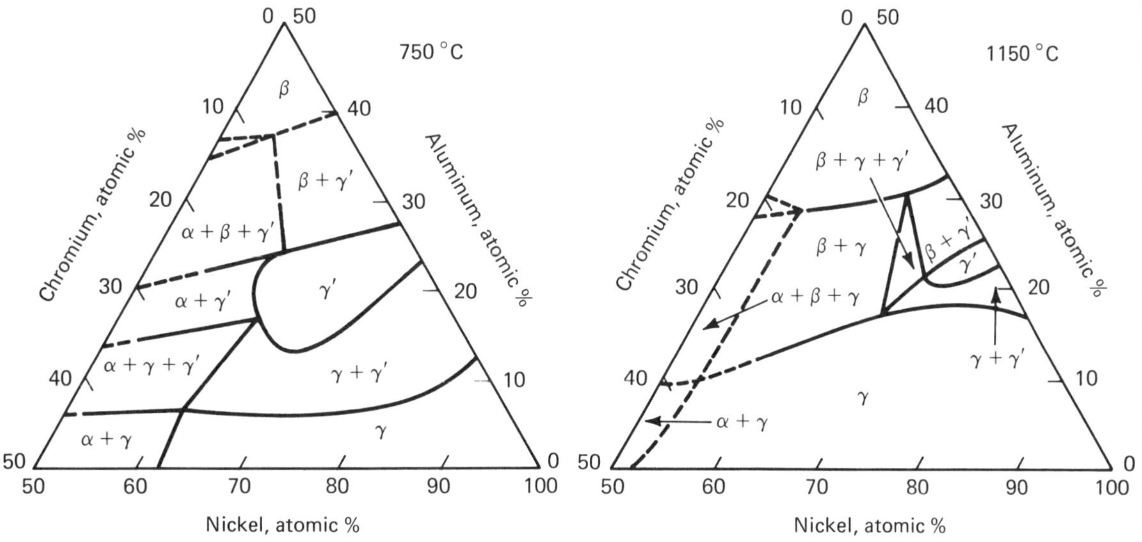

Fig. 5-15. Isothermal sections at 750 and 1150 °C from the Ni-Cr-Al phase diagram. (Adapted from A. Taylor and R. W. Floyd, J. Inst. Metals, Vol 81, p 455 and 460, 1952-53)

The maintenance of the high-temperature properties with time at temperature will depend upon the stability of the microstructure. Especially important in creep behavior is the ability to retard the growth of γ'. This can be achieved by decreasing the diffusion rate through proper choice of alloying elements, although the required information (e.g., diffusion coefficients) to design on this basis is not well known. Also, the γ-γ' interface is more stable the smaller the mismatch strain. This also can be controlled by proper alloying additions, and Fig. 5-18 illustrates this for Ni-Cr-Al alloys. The rupture life is a maximum at zero mismatch. Note that this presents a dilemma: greater age-hardening effect occurs for greater mismatch strain (and hence greater low-temperature strength and greater short-time high-temperature strength), but better creep properties are obtained for less mismatch strain. Thus, in alloy design, the specific application will dictate which criteria must be followed.

The addition of Cr to Ni-Al alloys markedly increases the static oxidation resistance (Fig. 5-19). More important, the cyclic oxidation resistance is retained (Table 5-2).

5-4. Nickel-Chromium-Aluminum-Titanium Alloys

The possibility of adding Ti to Ni-Cr-Al alloys for additional strengthening is worth examining, as the stability of titanium oxide indicates that this element may retain or improve the oxidation resistance of an alloy to which

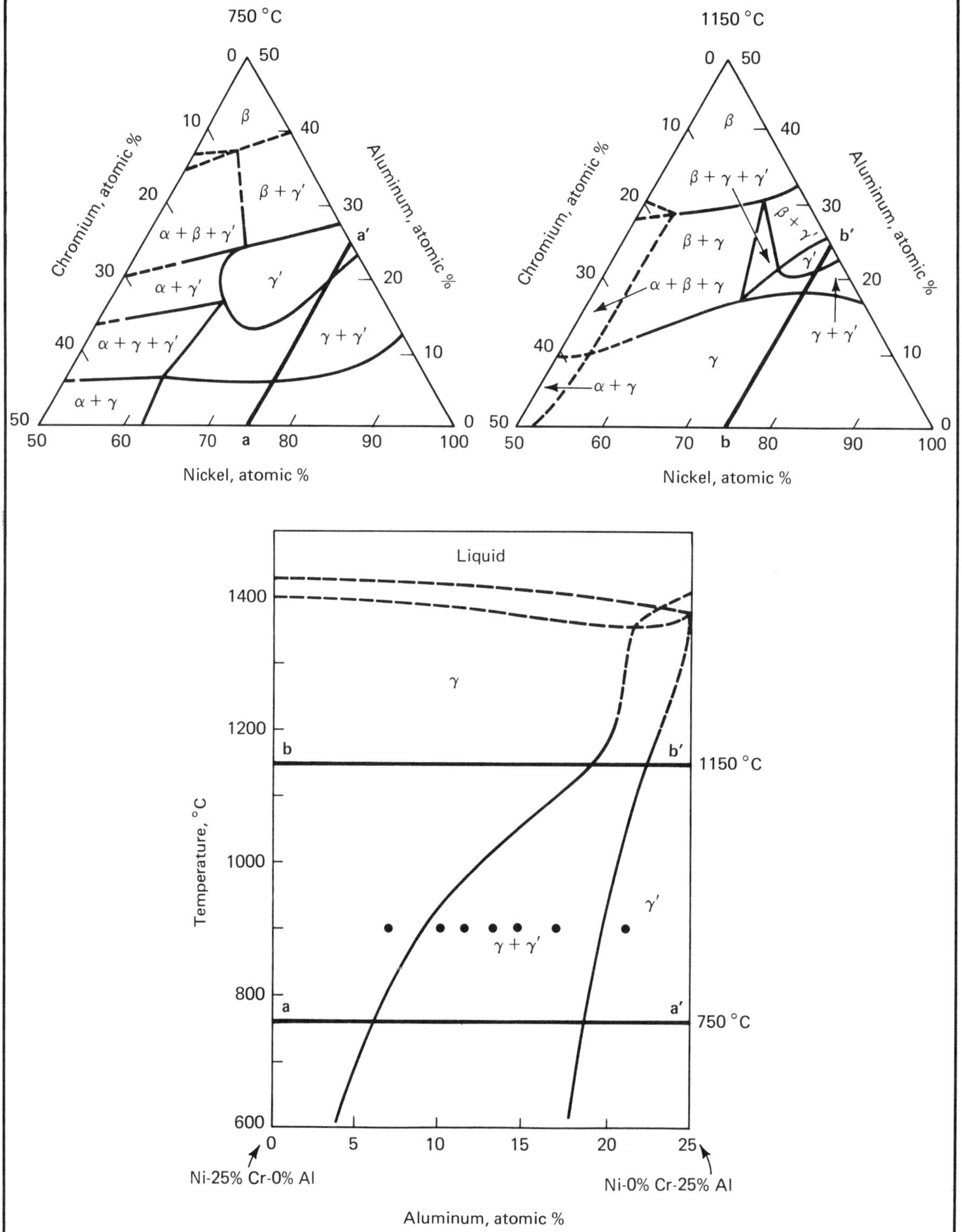

Fig. 5-16. Isopleth at 75% Ni in the Ni-Al-Cr phase diagram. The relation between this section and the isothermal sections at 750 and 1150 °C is shown. (Top diagrams from same source as Fig. 5-15; bottom diagram adapted from P. Beardmore, R. G. Davies and T. L. Johnston, Trans. Met. Soc. AIME, Vol 245, p 1538, 1969)

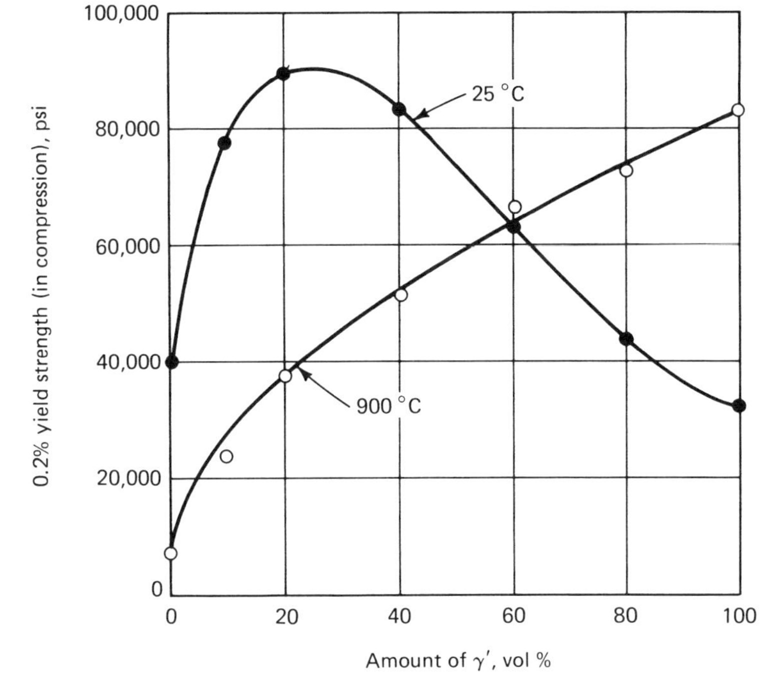

Fig. 5-17. Effect of the amount of γ' in Ni-Cr-Al alloys of fixed Ni content (75%) on the yield strength (in compression tests) measured at 25 °C and at 900 °C. All alloys were solution heat treated at 1150 °C for 2 h, air cooled at 25 °C, then aged for 16 h at 900 °C to develop the equilibrium phases. (Adapted from same source as Fig. 5-16, bottom diagram)

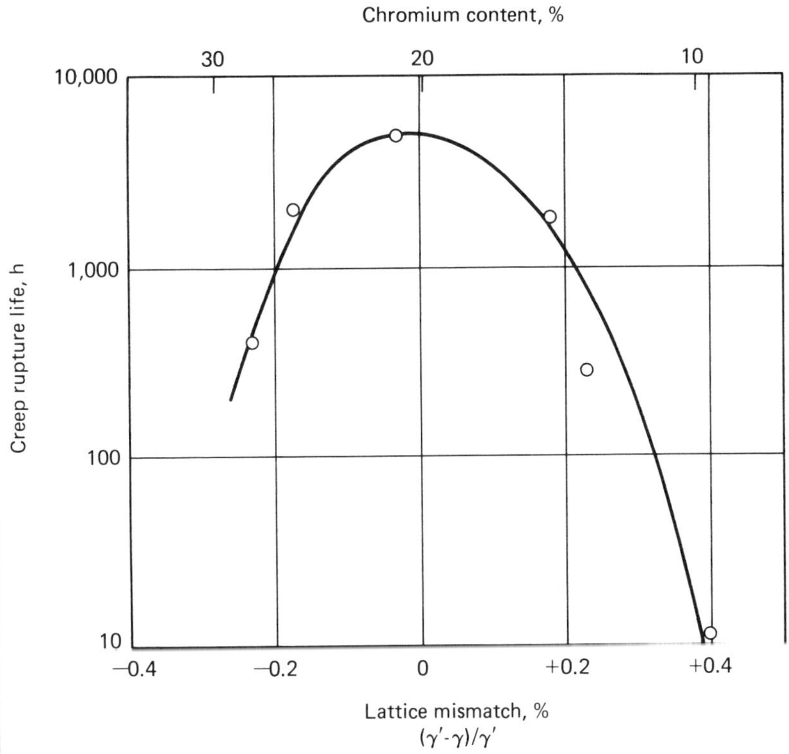

Fig. 5-18. Effect of lattice mismatch between γ and γ' on the creep rupture life at 21,200 psi and 700 °C for Ni-Cr-Al alloys. The Al content was approximately constant at 7 at.%. (Adapted from I. L. Mirkin and O. D. Kancheev, Metal Sci. and Heat Treat., No. 1, p 10, 1967)

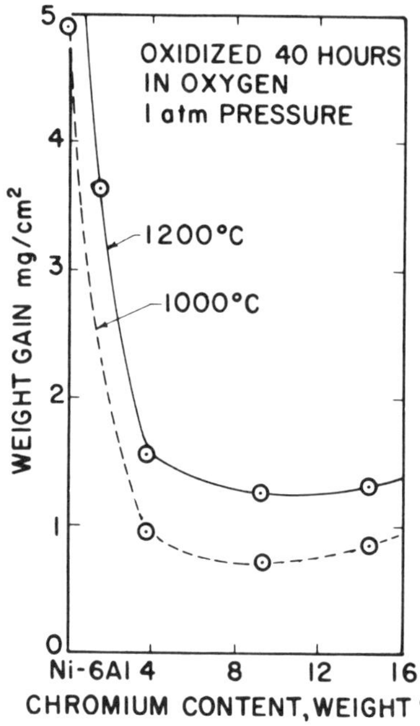

Fig. 5-19. Effect of Cr content on the oxidation of Ni-Cr-Al alloys of fixed Al content (6%). (From I. A. Kvernes and P. Kofstad, Met. Trans., Vol 3, p 1518, 1972)

Table 5-2. Comparison of the Effect of Cr and Al Content on the Oxidation Resistance of Ni After Exposure to Still Air at 1 Atm Pressure for 40 Hours

Alloy composition, at.%	Temperature, °C	Approximate weight gain, gm/cm²	Alloy composition, at.%	Temperature, °C	Approximate weight gain, gm/cm²
Static			**Cyclic**		
20Cr........	1000	0.002 (in 0.1 atm O_2)	6Al-15Cr...	1100	0.002
12Al........	1000	0.005	11Al-18Cr..	1100	0.002
12Al-16Cr...	1000	0.001	12Al-13Cr..	1100	0.0003

Calculated from data of C. A. Barrett and C. E. Lowell, Oxid. Metals, Vol 11, p 199, 1977, and from data in Fig. 5-3 and 5-19.

it is added. The Ni-Ti phase diagram (Fig. 5-20) shows that the solubility of Ti in Ni is rather limited, but hardening in the binary system may be feasible upon precipitation of the η phase. This phase is the compound Ni_3Ti, but it is hexagonal, not face-centered cubic as Ni_3Al is. Also, this intermetallic compound is quite stoichiometric, showing little solubility of Al or Ti from the compound formula, unlike γ'. As shown in Fig. 5-21, at 1150 °C for the Ni-Ti binary system the solubility in γ is about 13 at.% Ti; for an alloy containing 20 at.% Cr, the solubility is about 8 at.% Ti. This is still sufficient to give significant solid solution strengthening (Fig. 5-5), and the solubility decreases with temperature so that Ni_3Ti may be precipitated.

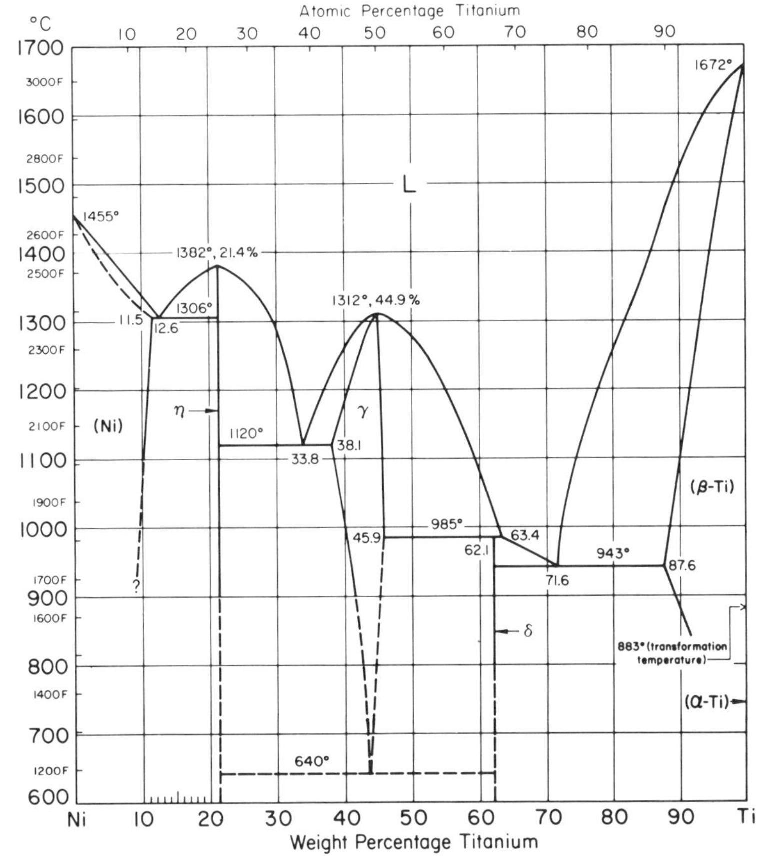

Fig. 5-20. The Ni-Ti phase diagram. η is Ni_3Ti, a hexagonal close-packed, ordered phase. (Adapted from same source as Fig. 5-4)

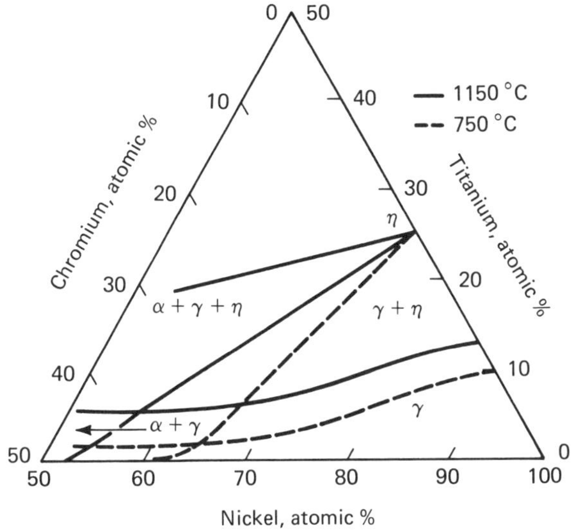

Fig. 5-21. The Ni-Cr-Ti phase diagram at 750 and 1150 °C. η is Ni_3Ti. (Adapted from A. Taylor and R. W. Floyd, J. Inst. Metals, Vol 80, p 577, 1951-52)

Even though Ni_3Al (γ') and Ni_3Ti (η) have the same stoichiometry, they have different crystal structures, so that Al cannot be substituted completely for Ti in Ni_3Ti, or Ti for Al in Ni_3Al. The stoichiometry of Ni_3Ti indicates that substitution of Al for Ti may be difficult, whereas the rather wide solubility range of Ni_3Al indicates that Ti may be substituted extensively for Al in this phase. These possibilities are found to be true, as shown in Fig. 5-22. Note that by lowering the Al content to about 9% the Ti content of γ' can be increased to about 16% (see point in Fig. 5-22). However, the solubility of Al in Ni_3Ti is negligible.

Although the equilibrium precipitate in Ni-Al-Ti alloys of high Ti – low Al content is η, its formation is preceded by a Ni_3(Ti, Al) γ' phase. Thus, even binary Ni-Ti alloys will form a metastable γ' (Ni_3Ti) precipitate. The effect of the Ti and Al content on the hardness of aged alloys is shown in Fig. 5-23. Note that the maximum hardness is obtained for alloys containing both Ti and Al. The lattice mismatch between γ' and γ is also a function of the Al and Ti content (Fig. 5-23): the greater the mismatch the greater the hardness. However, as pointed out with regard to Fig. 5-18, the best creep properties are obtained for lowest lattice mismatch.

The addition of Ti to γ' will cause solid solution hardening (Fig. 5-24) at 25 °C, and this effect is retained at high temperature (Fig. 5-25). Also, Cr will solid-solution strengthen γ' (Fig. 5-24), and the solubility of both Cr and Ti is sufficiently high, as reflected in the size of the γ' region of the Ni-Cr-Al-Ti phase diagram at 750 °C (Fig. 5-26), that the strengthening effect of both Cr and Ti can be utilized.

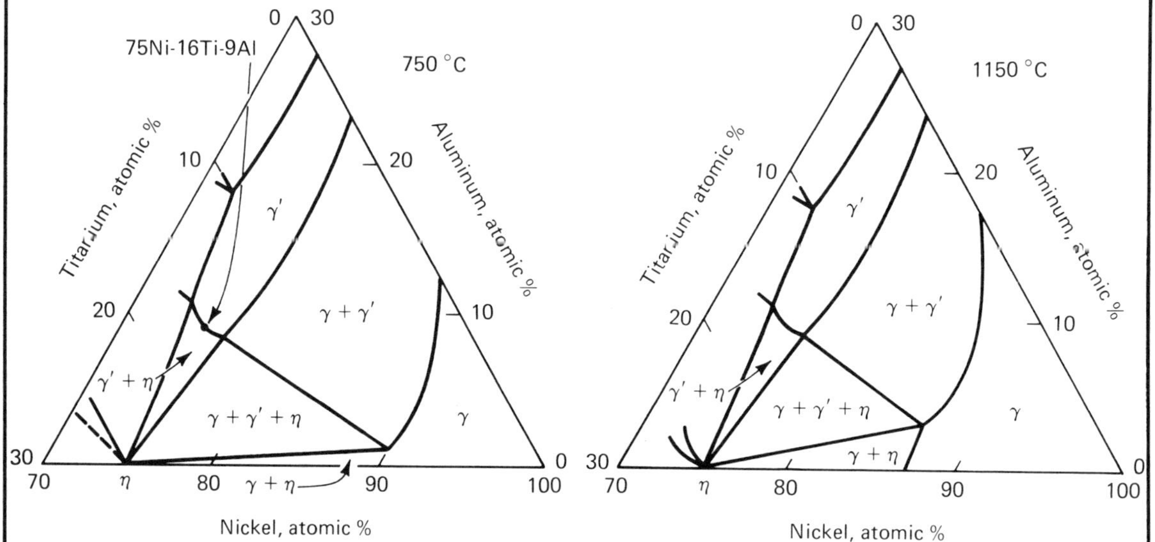

Fig. 5-22. The Ni-Al-Ti phase relations at 750 and 1150 °C. (Adapted from same source, p 25, as Fig. 5-15)

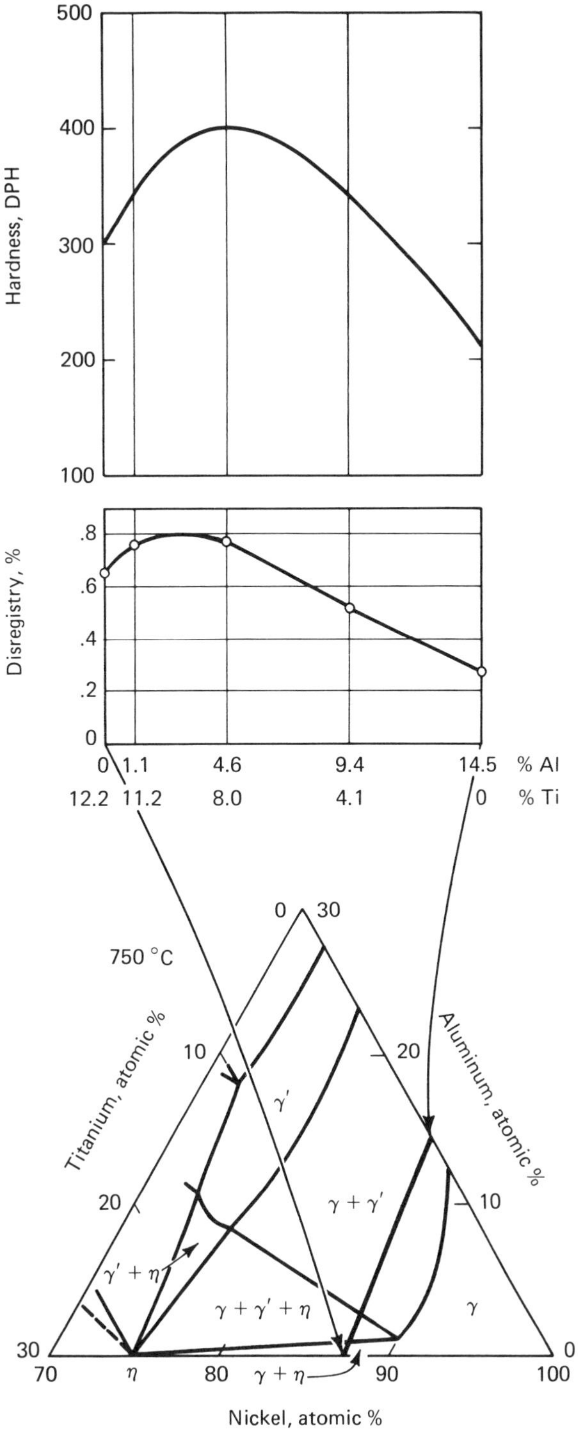

Fig. 5-23. Effect of Ti and Al content of Ni-Al-Ti alloys on the hardness (at 25 °C) obtained by aging at 760 °C to maximum hardness, and on the lattice mismatch [γ'-γ'/γ'] after aging for 1000 h at 760 °C. (Adapted from J. R. Mihalisin and R. F. Parker, Trans. Met. Soc. AIME, Vol 218, p 507, 1960)

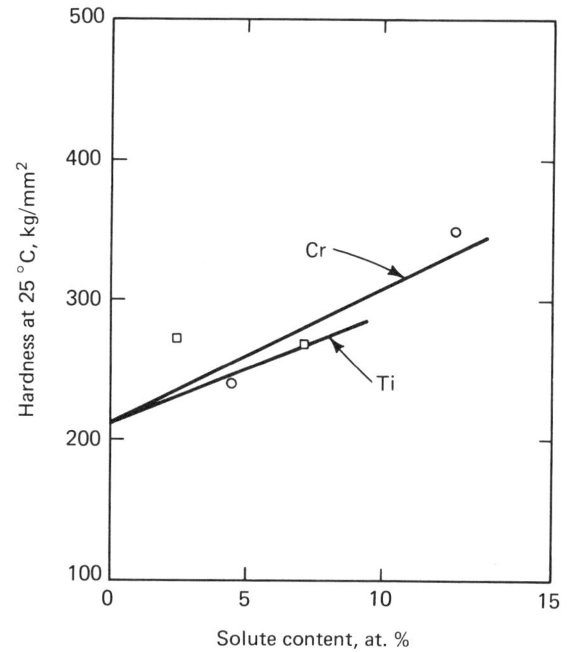

Fig. 5-24. Effect of Ti and Cr dissolved in Ni_3Al (γ') on the hardness at 25 °C. (Adapted from R.W. Guard and J.H. Westbrook, Trans. Met. Soc. AIME, Vol 215, p 807, 1959)

5-5. Complex Nickel-Base Alloys

The preceding sections have laid the foundation to understanding the physical metallurgy of complex, commercial nickel-base superalloys. This was done by examining in simple systems the effects of chemical composition and heat treatment on the phase relations and hence on properties. In this section, then, we will examine the chemical composition of some commercial superalloys in terms of the principles determined from the behavior of the simpler systems.

The main factors in alloy design which will be considered right now are listed in Table 5-3. Some data will be presented to justify the effects of the elements listed, and then these effects will be correlated with the chemical composition of some commercial alloys.

1. Solid Solution Strengthening of Gamma

The effect of solutes in Ni-base binary alloys on the yield strength at 25 °C of the face-centered cubic γ phase was illustrated in Fig. 5-5. The slopes of the lines indicate the relative potency of the elements in strengthening, and Ti, W and Mo are seen to be much more effective than Cr, Fe, Cu or Co. Based on its effect on the lattice parameter, Al should also be a potent strengthener.

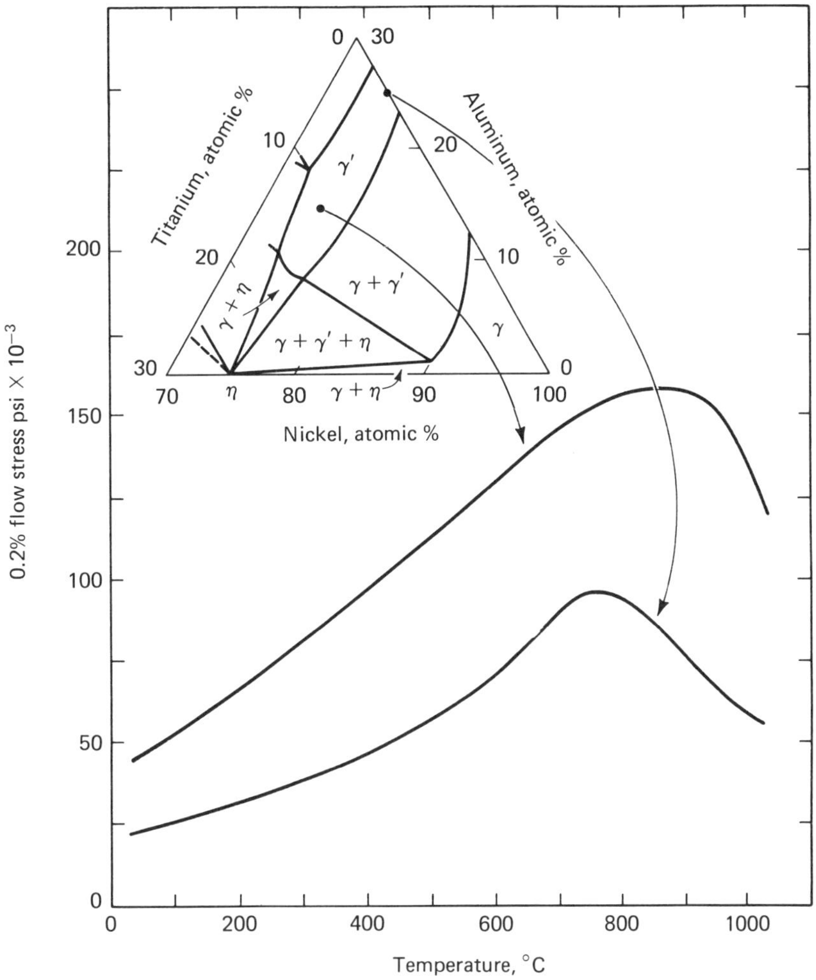

Fig. 5-25. Yield strength of the γ′ phase as a function of temperature for the binary Ni_3Al and for the ternary Ni_3 (Al, Ti). The two compositions are located on the ternary phase diagram at 750 °C. (Strength data from P.H. Thornton, R.G. Davies and T.L. Johnston, Met. Trans., Vol 1, p 207, 1970)

2. Solid Solution Strengthening of Gamma Prime

We have already seen (Fig. 5-24) that Cr and Ti increase the strength of γ′. The effect of other elements is shown in Fig. 5-27. This is the basis for listing Cr, Mo, Ti and Si as strengtheners of γ′. Based on the effect of Ti (Fig. 5-25), it would be expected that these elements would retain their influence on the strength at elevated temperature.

The effect of the solid solution strengthening depends, of course, on how much solute can be added (i.e., the solubility limit). Figure 5-28 shows the γ′ area at 1150 °C for Ni-Al-X ternary alloys. The straight line represents a constant Ni content of 75%. Note that Ti is highly soluble. So are Co and Cu, but these elements do not impart strengthening. Also, note that the

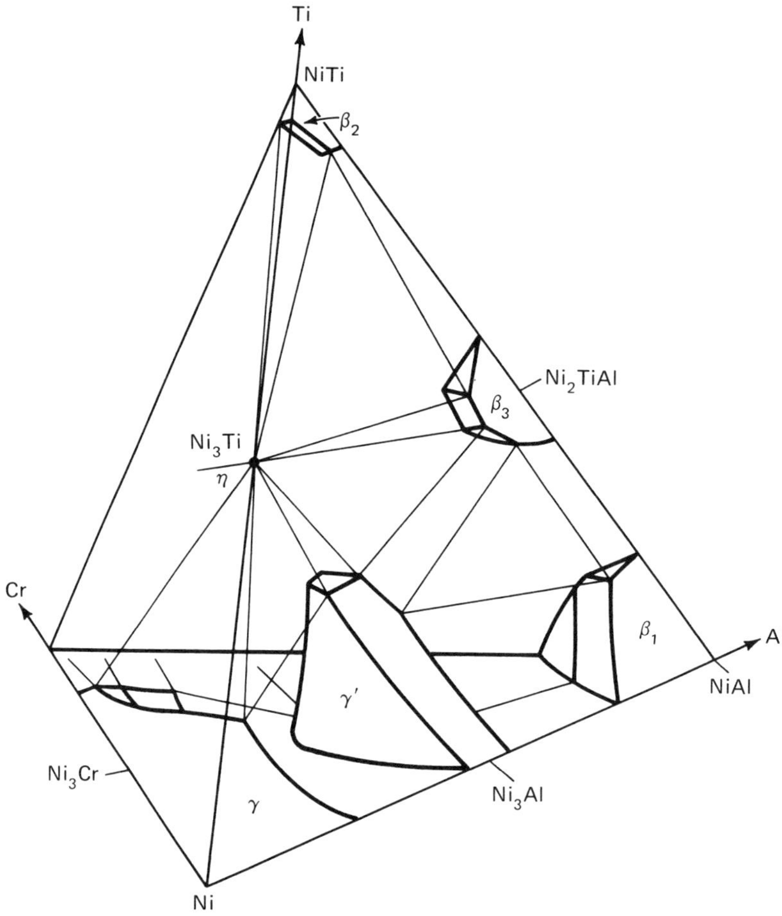

Fig. 5-26. The Ni-Cr-Al-Ti phase diagram at 750 °C. Note that both Cr and Ti dissolve extensively in γ'. (From A. Taylor, Trans. AIME, Vol 206, p 1356, 1956)

solubility of Mo is low, so that its high strengthening effect (Fig. 5-27) cannot be utilized as much as that of, say, Ti, which has a much higher solubility in γ'.

3. Amount of Gamma Prime

The data on Ni-Cr-Al alloys showed the strength to be dependent upon the amount of γ' (Fig. 5-17). At 25 °C, the yield strength is a maximum around 30 vol % γ', but at 900 °C these alloys are strongest for 100% γ'. This implies, for the commercial alloys, that the amount of γ' desired depends upon the temperature range of applications.

Figure 5-29 shows the relation between the amount of γ' in commercial alloys and the high-temperature creep strength. The amount of γ' is limited, however, by the influence of the lack of ductility of γ' on the ductility of the alloy.

Table 5-3. List of Important Factors To Consider in the Design of Ni-Base Superalloys

Phase relation/property	Effect	Reference
1. Solid solution strengthening of γ	W, Mo, Ti, Al, Cr strengthen; Fe, Co, Cu strengthen slightly	Fig. 5-5, Table 5-1
2. Solid solution strengthening of γ'	Mo, W, Si, Ti, Cr strengthen; V, Co weaken; Mn, Fe, Cu strengthen slightly	Fig. 5-27
3. Amount of γ'	Cr, Ti, Al, Nb, Mo, Co, Ta, V, Fe increase	Fig. 5-31 to 5-36
4. Antiphase boundary energy	Ti, Co, Mo, Fe increase; Al, Cr decrease	Table 5-4
5. γ-γ' lattice mismatch	Ta, Nb, C, Ti increase; Cr, Mo, W, Cu, Mn, Si, V decrease; Al, Fe, Co negligible effect	Fig. 5-39 to 5-44
6. Coarsening rate of γ'	Ti, Mo, Nb, Co, Fe decrease; Cr increases; Zr, B no effect	Fig. 5-47
7. Oxidation and hot corrosion resistance	Cr improves; other elements variable	Fig. 5-47 and 5-48

Note. Numbers in first column relate to text discussion. Information after R. F. Decker, in Steel Strengthening Mechanisms, Climax Molybdenum Company, Greenwich, Conn., 1969.

The effect of alloying additions on the amount of γ' will depend on the temperature and the specific element. It is not sufficient to examine just the size of the γ' region in the phase diagram; the γ-γ' field dictates whether increasing the amount of a component will increase the amount of γ'. To illustrate, we will look at a Ni – 15 at.% Al alloy. At 1150 °C, this is in the single-phase γ region. If Ti is added to the binary alloy, keeping the Ni fixed at 85 at.%, then the composition moves along the line a-b in Fig. 5-30(a). At about 4% Ti, the solubility limit is reached. As the amount of Ti is increased and the Al lowered (keeping the Ni content constant) from *a* toward *b,* the amount of γ' will increase. Remember (see Chapter 4) that for a given composition, the amount of γ and γ' present can be calculated if the tie line is known, such as the dashed line in Fig. 5-30(a). Now at 750 °C, the same Ni-Al binary alloy is in the two-phase region (point a). Increasing the Ti and lowering the Al content will increase the amount of γ'. If we have a Ni—15 at.% Al alloy at 1150 °C, and allow the Ni content to be lowered as the Ti content is increased, such as along line a-c, then the amount of γ' can be increased more effectively than moving along

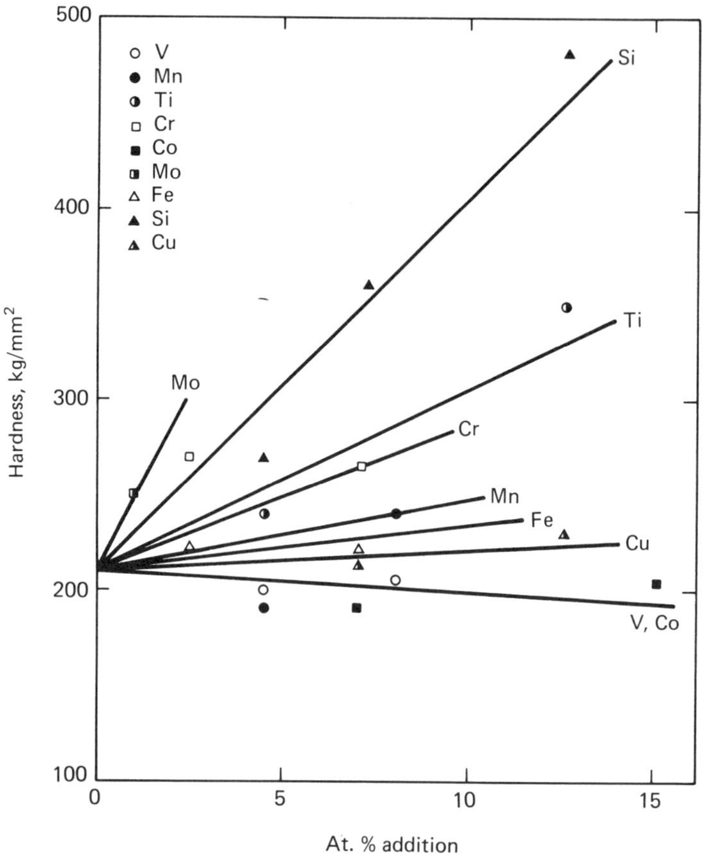

Fig. 5-27. Effect of alloying elements dissolved in γ' on the hardness at 25 °C. (From same source as Fig. 5-24)

line a-b. Indeed, moving from *a* to *c* will move the alloy from the single-phase γ region into the single-phase γ' region, or 100% γ'.

In the Ni-Al-Cr system, the behavior is somewhat different. At 1150 °C, a Ni–15 at.% Al alloy is in the single-phase region, and substituting Cr for Al (moving along line a-b in Fig. 5-30b) will retain this single phase. At 750 °C, however, the Ni–15 at.% Al alloy is in the two-phase γ-γ' region, and substituting Cr for Al will move the alloy along a-b into the single-phase γ region. One way to increase the amount of γ' at this temperature is to keep the Al content constant and decrease the Ni content as the Cr is increased, moving along line a-c. Figure 5-31 shows the agreement between the phase diagram and some experimentally measured amounts of γ'.

Unfortunately, phase diagrams necessary to analyze the effect of all of the important solutes on the γ'-γ phase boundaries are not available. Some studies in which an element is systematically varied and the boundaries are determined have been made, though, and the results of one are presented in Fig. 5-32(a) showing the influence of Mo. The solubility limit is shown for

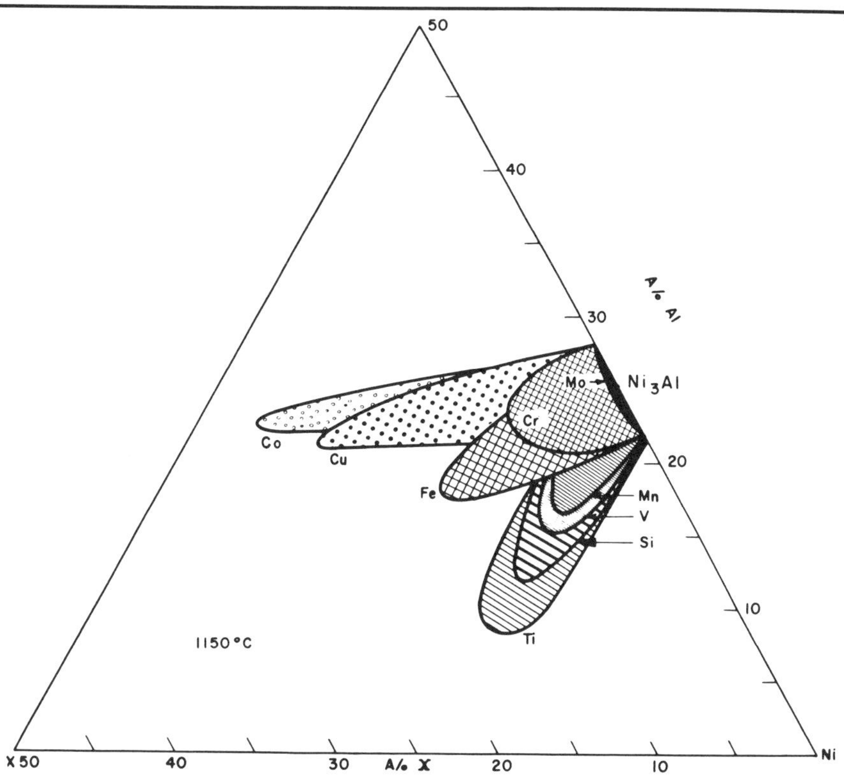

Fig. 5-28. Ternary phase diagrams of Ni-Al-X showing the effect of the element X on the size of the γ' field at 1150 °C. (From same source, p 810, as Fig. 5-24)

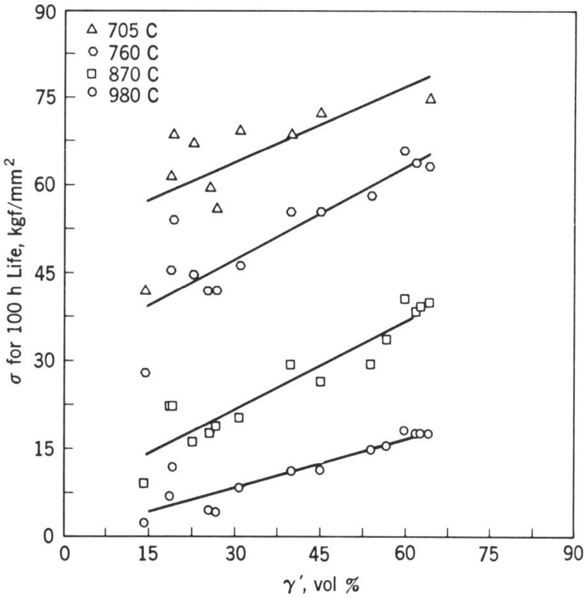

Fig. 5 29. Effect of the amount of γ' on the creep strength. The data are from commercial alloys. (From R. F. Decker, in Steel-Strengthening Mechanisms, Climax Molybdenum, Greenwich, Conn., 1969)

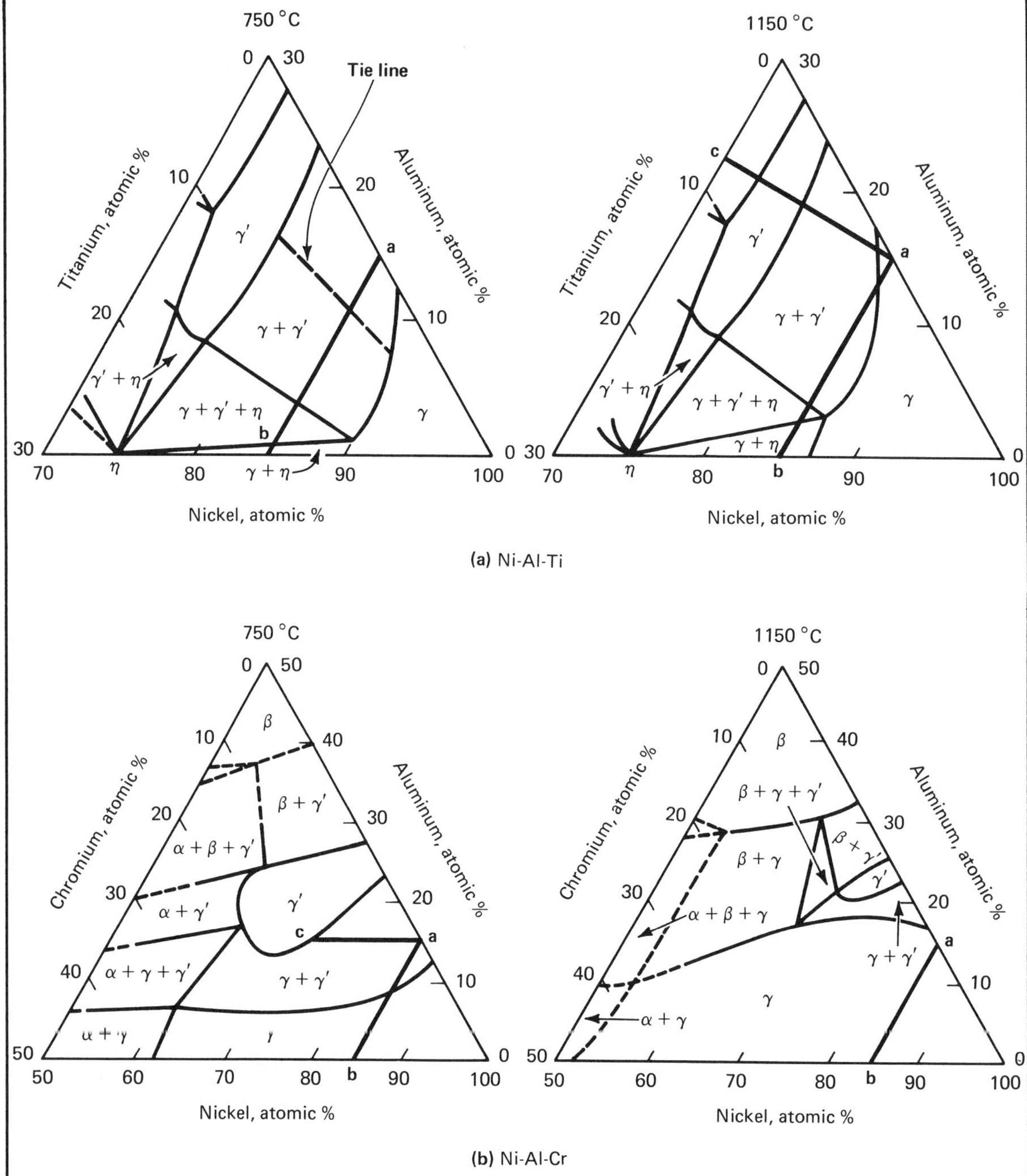

Fig. 5-30. The Ni-Al-Ti and Ni-Al-Cr phase diagrams at 750 and 1150 °C, illustrating the extent of the γ-γ' region. (a) from same source as Fig. 5-22; (b) adapted from same source, p 451, as Fig. 5-22.

alloys of constant Cr content at about 16 at.%. Note that in Ni-Cr-Al-Mo alloys the phase boundary is shifted up in temperature as the Mo content increases. Although the effect on the γ' boundary is not shown, this implies that the equilibrium amount of γ' at a given temperature will increase as the Mo content increases (as can be seen by applying the mass balance equa-

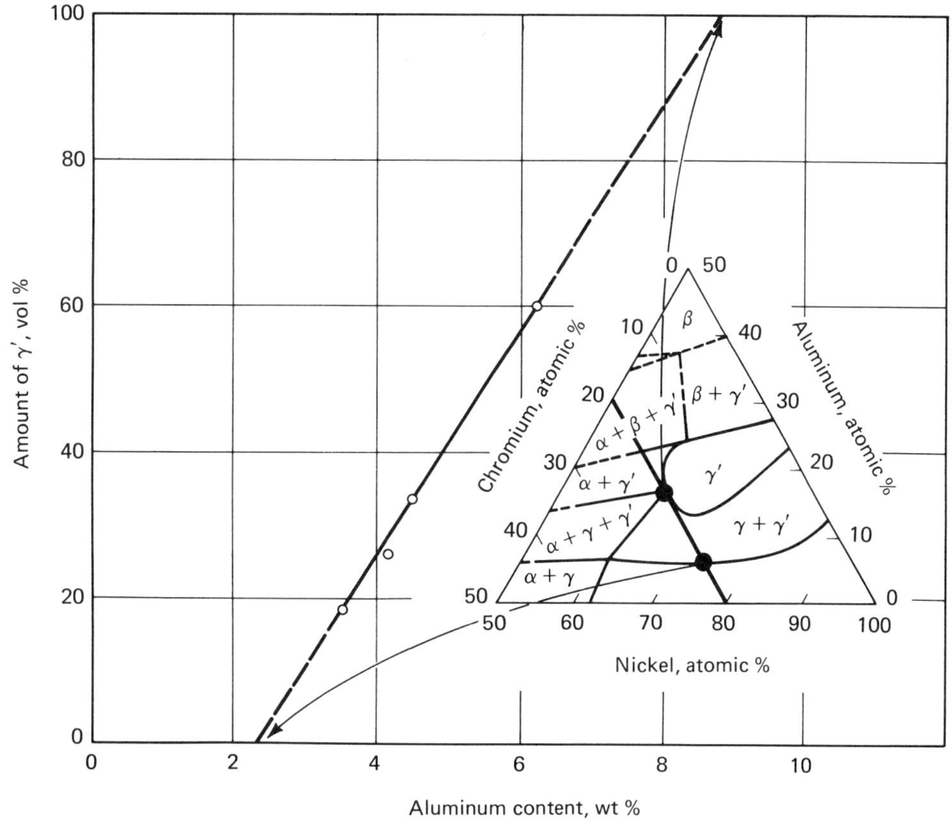

Fig. 5-31. The amount of γ' at 750 °C as a function of Al content for a Cr content of 20%. (From data of T. B. Gibbons and B. E. Hopkins, Metal Sci. J., Vol 5, p 233, 1971)

tion), and this is shown in Fig. 5-32(b). Also note in Fig. 5-32 that adding Ti to these alloys further displaces the phase boundary to high temperature. Figure 5-33 shows that the effect of Mo is retained if the Cr content is increased to 20%. Thus, Mo additions to Ni-base superalloys would be expected to increase the amount of γ' *and* to increase the temperature at which an alloy would have to be heated for solution heat treatment.

Figure 5-34 shows some data for Ni-Cr-Al-Ti and Ni-Cr-Al-Nb alloys in which the Cr content was fixed at 20%. These data predict that Ti and Nb should increase the amount of γ' in commercial alloys, and this is further supported by the data in Fig. 5-35 for Ni-Cr-Al-Ti-Mo alloys.

The effect of Co on the solubility of Ti in γ is shown in Fig. 5-36 for Ni-Cr-Ti and for Ni-Cr-Co-Ti alloys. These are isopleth sections at a constant Cr content of 22%. It is seen that, for a given Ti content, the addition of Co raises the solubility curve in temperature. Although the effect on the composition of γ' is not known, it would be expected (from the mass balance calculations) that the amount of γ' at equilibrium at a given temperature would be greater for the alloy containing Co.

The results presented in this section are the basis for listing Cr, Ti, Al, Nb, Mo and Co in Table 5-3 as increasing significantly the amount of γ'.

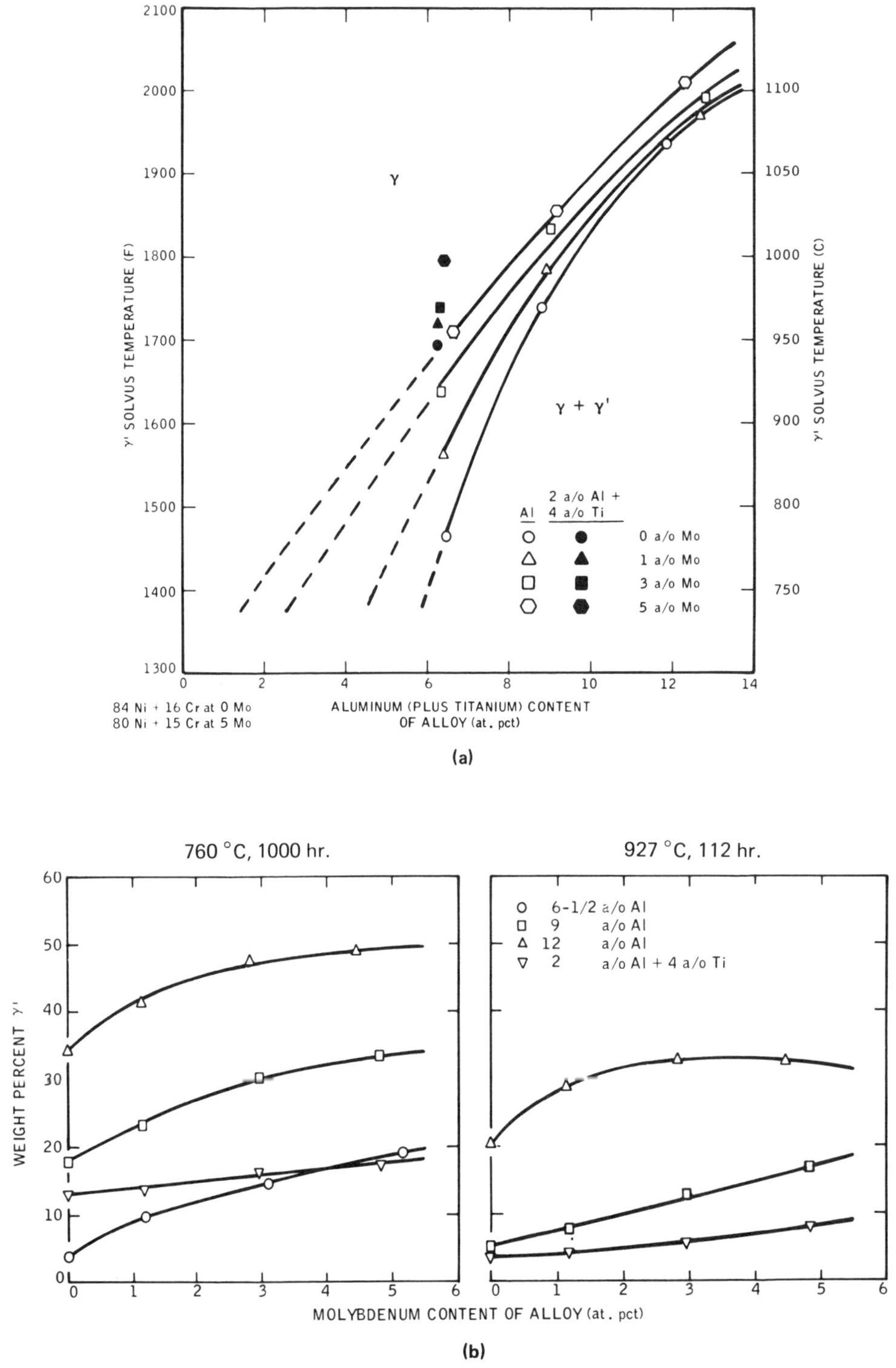

Fig. 5-32. Effect of Mo content on (a) the $\gamma/\gamma' + \gamma$ phase boundary and (b) the amount of γ' present at 760 and 927 °C in Ni-Mo-Al-Ti alloys in which the Cr content was approximately constant at 16 at.%. (After W. T. Loomis, J. W. Freeman and D. L. Sponseller, Met. Trans., Vol 3, p 992, 1972)

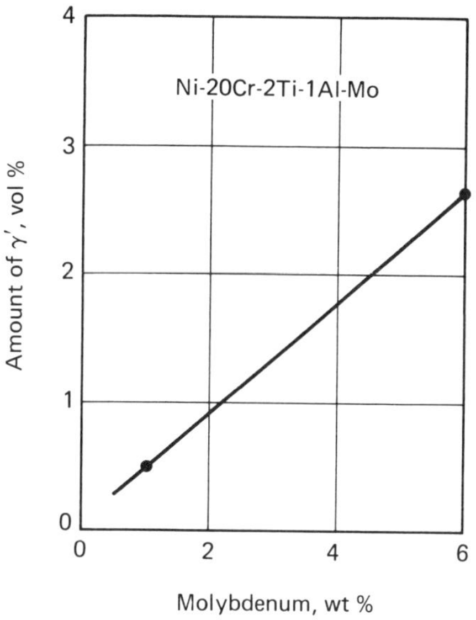

Fig. 5-33. Effect of Mo content on the amount of γ' in Ni-Cr-Ti-Al-Mo alloys. The alloys were aged for 50 h at 900 °C after solution treatment at 1120 °C. (Adapted from G. N. Maniar and J. E. Bridge, Met. Trans., Vol 2, p 95, 1971)

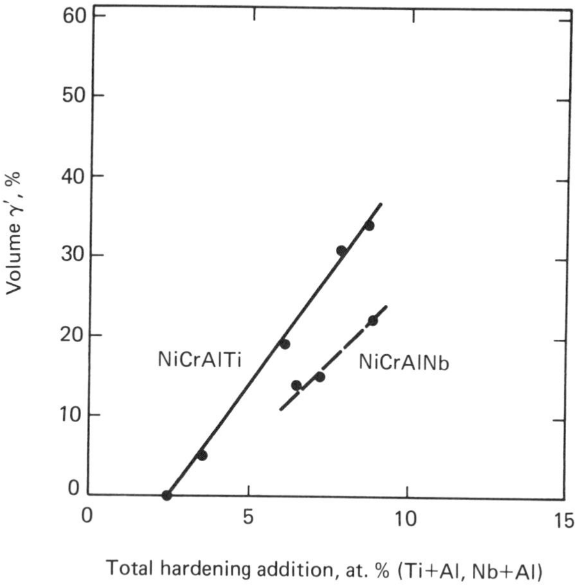

Fig. 5-34. Effect of Al, Ti and Nb on the amount of γ' formed at 750 °C in Ni-Cr-Al-Ti and Ni-Cr-Al-Nb alloys in which the Cr content was 20%. (Adapted from same source as Fig. 5-31)

Observations in specific commercial alloys are the basis for listing Ta, V and Fe.

4. Antiphase Boundary Energy

The shear stress required to move a dislocation through an array of ordered precipitates is related to the effect a dislocation has on the bonding sequence

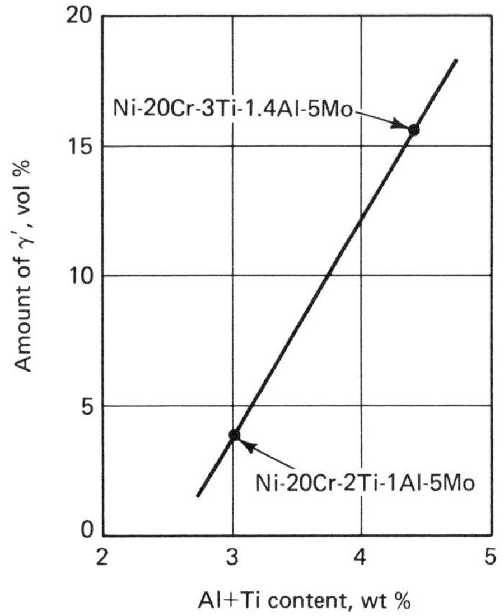

Fig. 5-35. Effect of Al and Ti content on the amount of γ' in Ni-Cr-Al-Ti-Mo alloys. The alloys were aged for 50 h at 900 °C after solution treatment at 1120 °C. (Adapted from same source as Fig. 5-33)

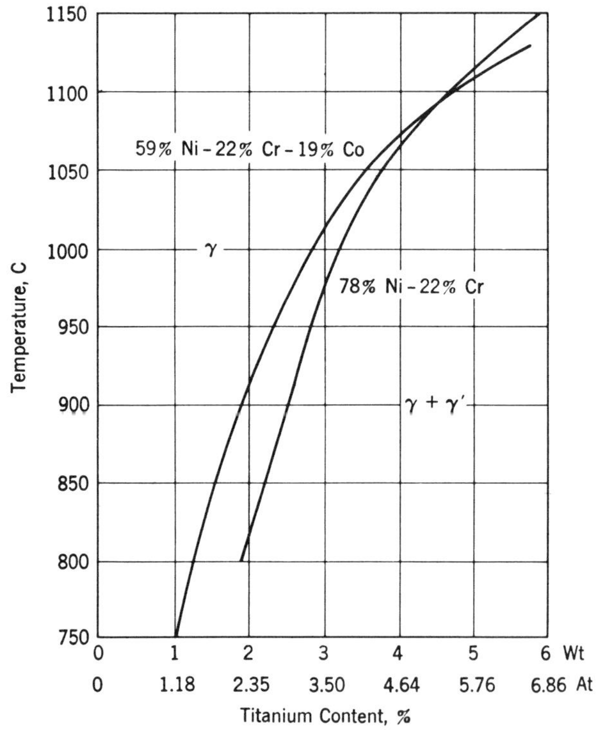

Fig. 5-36. Isopleth sections of the ternary Ni-Cr-Ti and the quaternary Ni-Cr-Co-Ti phase diagrams, at 22% Cr. (From J. Heslop, Cobalt, Vol 24, p 128, 1964)

in the precipitate (Fig. 5-37). Once a dislocation has reached the matrix-precipitate boundary, the shear stress must be increased to force the dislocation to enter the precipitate. This is because the atoms across the slip plane (see Fig. 5-37a) behind the dislocation are now like atoms, which is not the equilibrium, lowest energy arrangement of the precipitate crystal. For the second dislocation to enter the crystal requires less energy, as this restores the proper arrangement across the interface (see 5-37c). It can be seen that if two dislocations are on the same slip plane, their movement as an adjacent pair through the precipitate would cause no rearrangement. However, as the two dislocations approach each other, the elastic strain energy increases. Thus, their separation is a balance between this strain energy and the antiphase boundary energy of the region between them, and the dislocations will tend to move in pairs through an ordered precipitate, or through a region in which there are many precipitates closely spaced. The mechanism depicted in Fig. 5-37 is illustrated in Fig. 5-38 for a specific Ni-base alloy in which γ' has precipitated.

The arrangement marked by *a*'s in Fig. 5-37, corresponding to like atoms being neighbors across a plane (the slip plane in this case), is referred to as an antiphase boundary (APB), as the atoms are out of phase with their equilibrium sequence. The energy of such a boundary is the antiphase boundary energy, and for best strengthening of the structure by the γ' precipitates it is desired to have this energy high. There is evidence that this might be the most important factor impeding the movement dislocations through a matrix containing γ' particles. (The other important factor is lattice strain from the lattice mismatch between the γ and γ'.) Few data are available on the APB energy; some values are given in Table 5-4. The addition of Cr seems to lower the energy. A higher Ti-Al ratio increases it; thus increased Ti and decreased Al content is desired. These data also indicate that Co, Mo and Fe may increase the APB energy.

5. Gamma – Gamma-Prime Lattice Mismatch

As an element is added to a two-phase γ-γ' alloy, it usually is not distributed equally between the two phases but partitions or segregates between them. The effect this has on the lattice parameters of γ and γ' determines the lattice mismatch, which is related to the lattice strain. Even though an element expands (or contracts) the γ lattice, it also may expand the γ' lattice, so that increased mismatch may not occur. The mismatch depends upon how much the element affects each phase, and the amount dissolved in each. This effect is difficult to predict, but a satisfactory picture for alloy-design purposes is obtained from experimental studies.

Figure 5-39 shows that Ta, Nb, C, V, Si and Mn will increase the mismatch, and hence the hardness at 25 °C. The Ni-Al-Ga alloy serves as a reference for minimum effect, as Ga is similar to Al. Note that Si and Mn have a measurable effect, but that V, C, Nb (Cb) and Ta are quite potent in increasing the mismatch. The effect of Cr on the mismatch is shown in Fig. 5-40 for Ni-Cr-Al-Ti alloys. These data indicate that increasing Cr

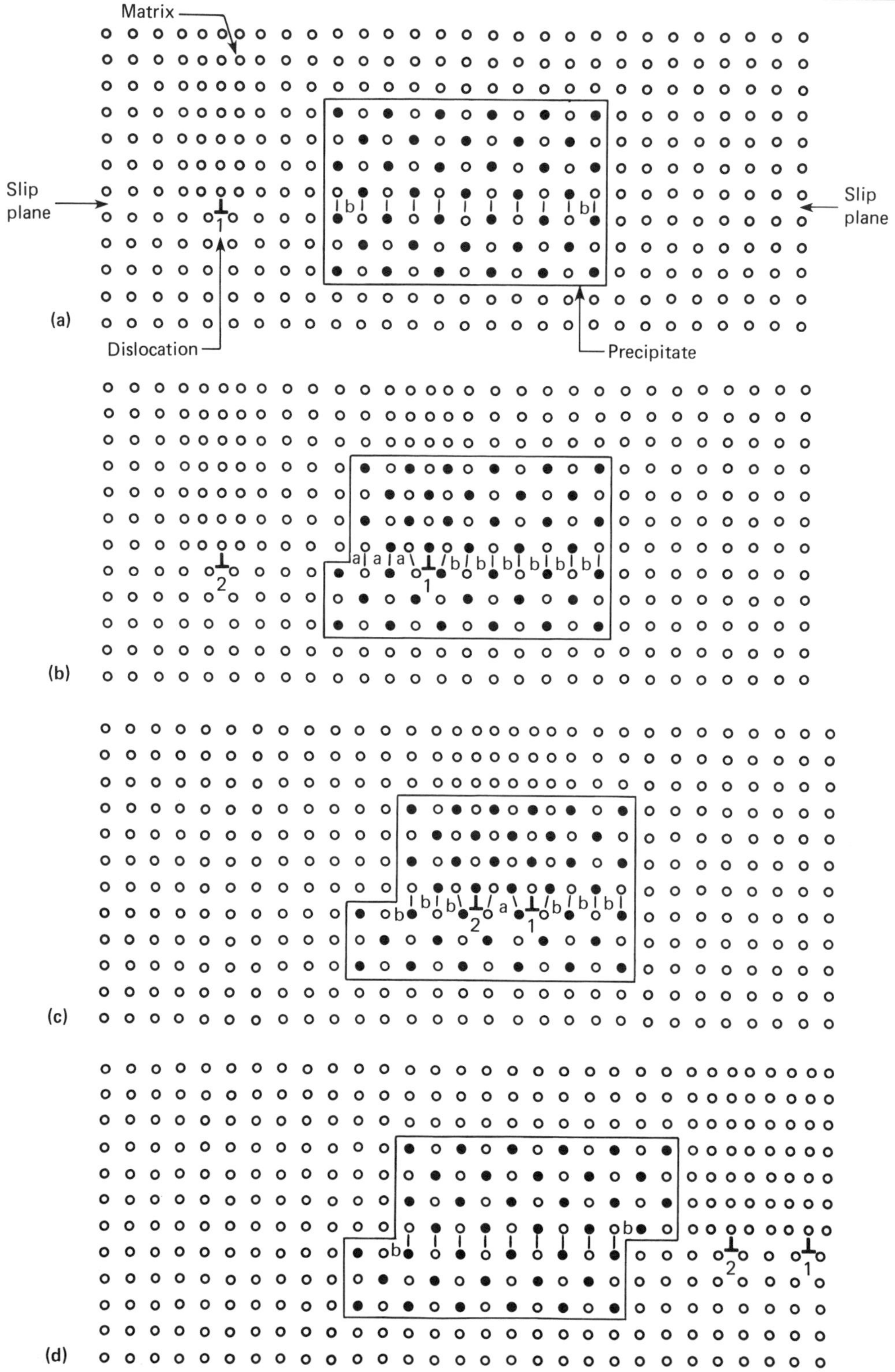

Fig. 5-37. Schematic illustration of the development of repulsive bonding (marked a) if a dislocation enters an ordered precipitate particle, and the minimization of the number of such bonds if the dislocations move in pairs through the precipitate.

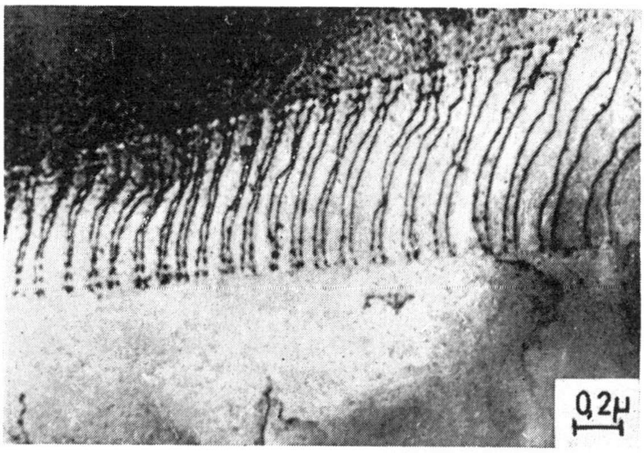

Aged 1 h at 750 °C, deformed 2%.
The fine γ' is not resolved here.

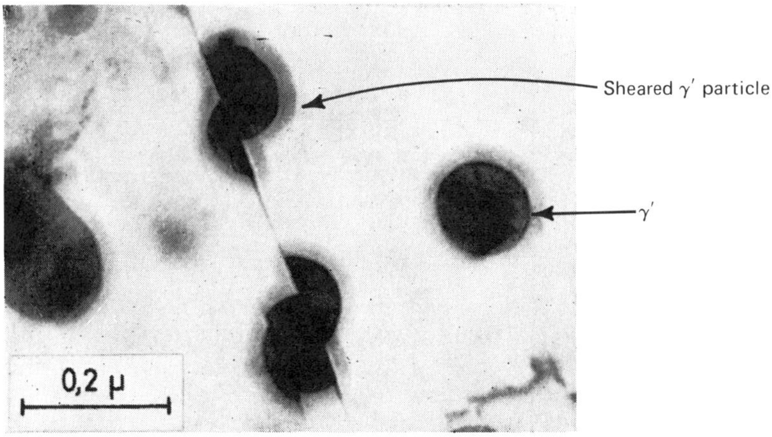

Aged 540 h at 750 °C, deformed 3%.

Fig. 5-38. Transmission electron micrographs illustrating the interaction of ordered precipitates and dislocations as depicted in Fig. 5-37. In (a) the dislocations are moving as pairs through a very dense, fine, ordered γ' precipitate dispersion. (b) illustrates the cutting of the particles by dislocations. The alloy contained 84Ni-16Cr-7.5Al (at.%). (Adapted from H. Gleiter and E. Hornbogen, Acta Met., Vol 13, p 577, 1965)

decreases the mismatch. (Remember that most of the commercial superalloys will have Cr contents in the range of 20%.) These results are different from those shown in Fig. 5-18, based on the ternary Ni-Cr-Al alloys, where the mismatch vanishes at about 20% Cr. However, this just reflects the different behavior of the ternary and the quaternary alloys.

The effect of Al, Ti and Nb on the lattice mismatch in Ni-Cr-Al, Ni-Cr-Al-Ti and Ni-Cr-Al-Nb alloys is shown in Fig. 5-40. In these alloys, Al and Nb have little effect, but Ti increases the mismatch considerably. The interaction of Ti and Al is sometimes examined in terms of the sum of these elements, or in terms of the Ti/Al ratio. Figure 5-41(a) shows that increasing the Ti compared with the Al causes the mismatch to increase markedly.

Table 5-4. Approximate Antiphase Boundary Energy of γ′ in Different Alloys

Alloy composition	APB energy, erg/cm²
Ni-12.7 to 14.0Al (at. %)	153
Ni-18.5Cr-7.5Al (at. %)	104
Ni-8.8Cr-6.2Al (at. %)	90
Fe-Cr-Ni-Al-Ti	
Ti/Al = 1	240
Ti/Al = 8	300
Ni-19Cr-14Co-7Mo-2Ti-2.3Al (at. %)	170-220
Ni-33Fe-16.7Cr-3.2Mo-1.6Al-1.1Ti (wt %)	270

Adapted from N. S. Stoloff, in The Superalloys, ed. C. T. Sims and W. C. Hagel, Wiley, New York, 1972.

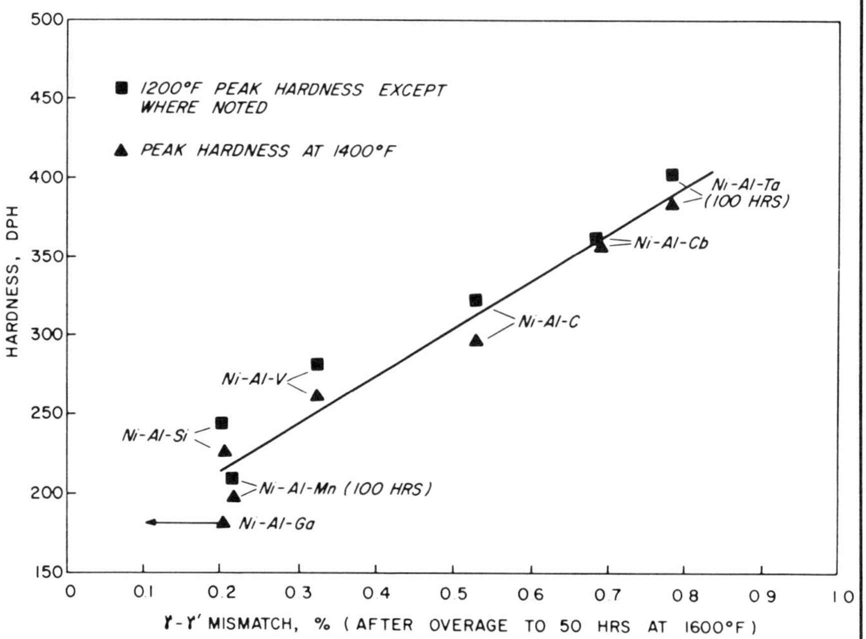

Fig. 5-39. Correlation of peak aged hardness with γ-γ′ lattice mismatch for several Ni-Al-X ternary alloys. The hardness was measured at 25 °C. (From R. F. Decker and J. R. Mihalisin, Trans. ASM, Vol 62, p 485, 1969)

However, as the alloy becomes more complicated this relation may not be valid, as shown in Fig. 5-41(b). The mismatch decreases as the sum of Ti and Al increases, although note here that the mismatch is low in either alloy.

The effect of Mo is shown in Fig. 5-42. This element lowers the mismatch, although note that in the highest Al alloy (Fig. 5-42b) the mismatch is increased by Mo (that is, it becomes more negative). Figure 5-43 shows the effect of W, Cu, Fe and Co. The behavior of W is similar to that of Mo, to be expected because of their similar atomic nature. The mismatch is decreased by Cu. The presence of Fe and Co has little effect; these

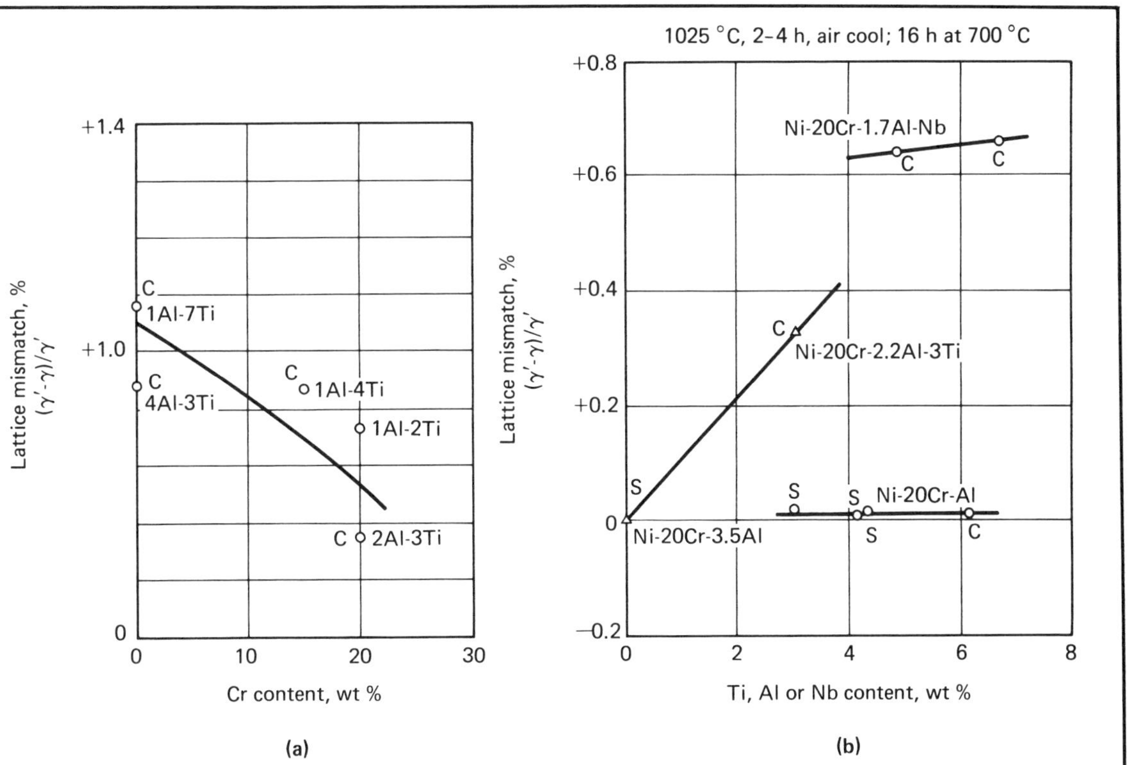

Fig. 5-40. (a) Approximate effect of Cr content on the γ'-γ lattice mismatch in Ni-Cr-Al-Ti alloys, and (b) effect of Al, Ti and Nb on the lattice mismatch in Ni-Cr-Al, Ni-Cr-Al-Ti and Ni-Cr-Al-Nb alloys. All compositions, wt %. (a) 1Al-7Ti and 4Al-3Ti, 500 h at 750 °C: adapted from B. A. Parker and D. R. F. West, J. Australian Inst. Metals, Vol 14, p 108, 1969; 1Al-2Ti, 1000 h at 760 °C: adapted from same source as Fig. 5-32; 1Al-2Ti, 50 h at 730 °C: adapted from same source as Fig. 5-33; 2Al-3Ti, >24 h at 750 °C: adapted from same source as Fig. 5-31. (b) adapted from same source as Fig. 5-31.

elements are similar to Ni and are probably substituting for it in the alloy, altering the character of the alloy very little. Particularly examined for its effect is Fe, because considerable saving in cost can be achieved if Fe can be substituted for Ni. Although Fe has little effect on the lattice mismatch, it must be added with caution because of other effects (to be discussed).

The effect of other elements on the lattice mismatch in Ni-base alloys can be inferred from measurements on Fe-base, austenitic stainless steel alloys in which the chemical composition is such that γ' can be precipitated from the face-centered cubic matrix. Figure 5-44 shows some of these data.

It is emphasized that whether a high or a low mismatch is desired depends upon the application. Generally, high mismatch means high strength, but at high temperature the γ' particles tend to coalesce more rapidly the higher the mismatch, and thus for creep applications low mismatch is desired (see Fig. 5-18).

The lattice mismatch affects the morphology (shape) of the precipitate. The shape the particle takes is one which minimizes the surface and strain energy per volume of precipitate. If the strain energy is negligible, and the

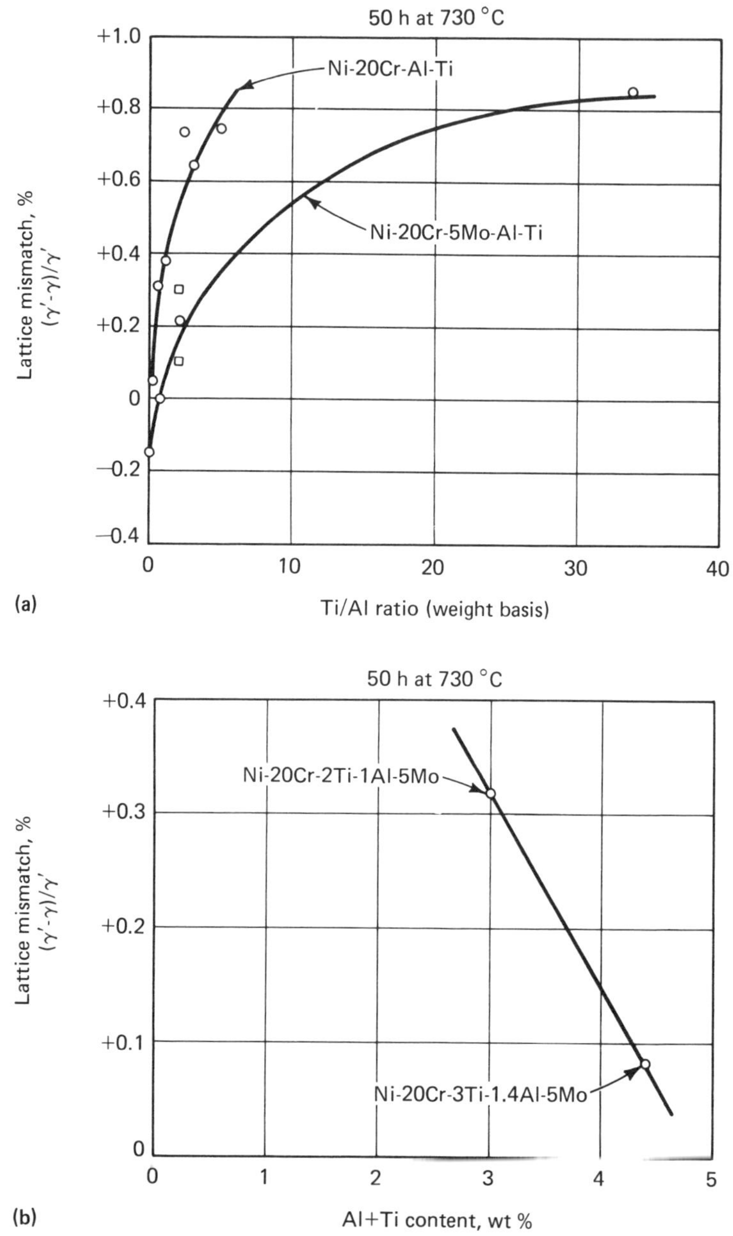

Fig. 5-41. Effect of Ti and Al on the γ'-γ lattice mismatch. For the alloys in (a), the (Al + Ti) was approximately constant at 3.5. Crosses are from data in (b). For the alloys in (b), the Ti/Al ratio was 2.2. All compositions, wt %. (a) adapted from data of G. N. Maniar, J. E. Bridge, H. M. James and G. B. Heydt, Met. Trans., Vol 1, p 31, 1970; (b) adapted from data from same source as Fig. 5-33.

surface energy is orientation independent (not necessarily true, but we will consider it to be so here), then the geometry which will minimize the surface and strain energy per volume of precipitate is a sphere. However, as the lattice mismatch increases, the strain energy is quite orientation dependent (anisotropic), and the geometry which minimizes the energy will not be

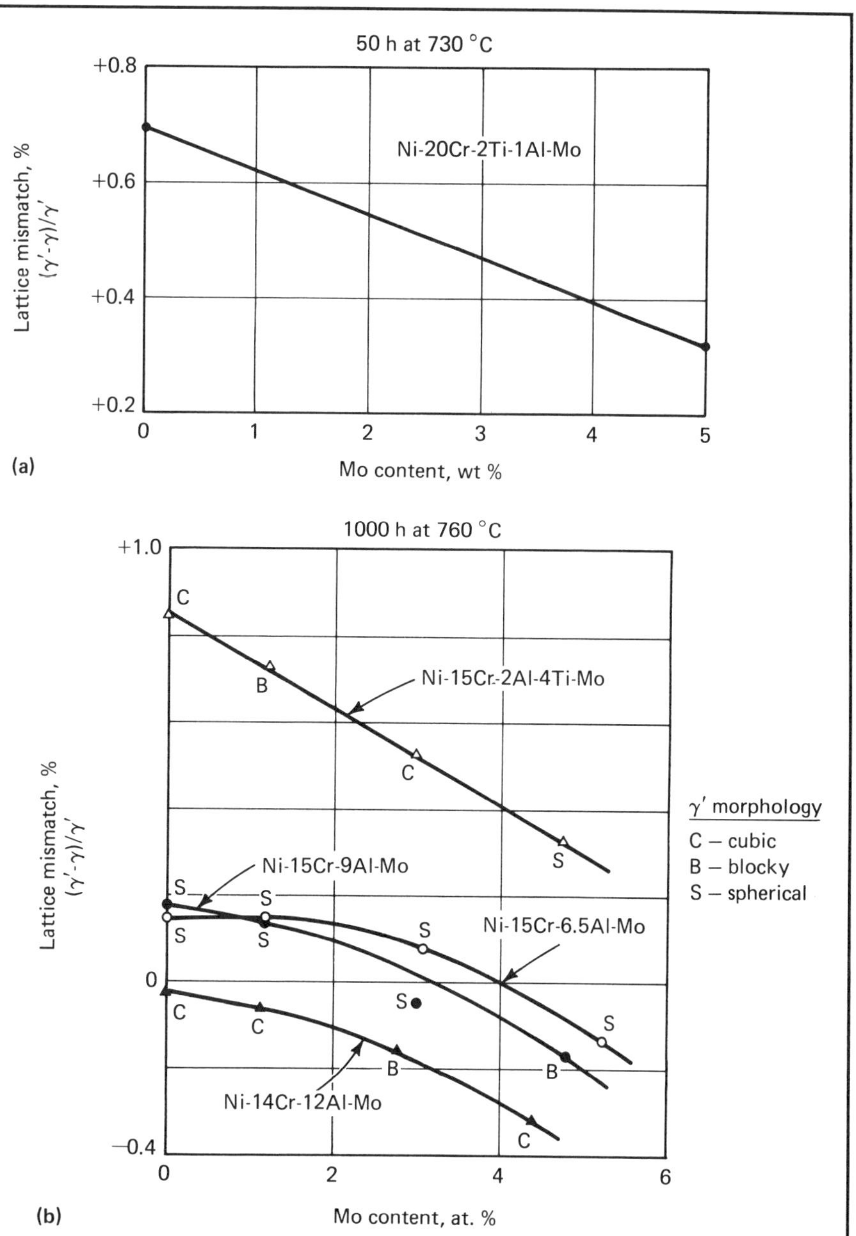

Fig. 5-42. Effect of Mo content on the γ'-γ lattice mismatch in various alloys. Compositions in (a) are in weight percent, in (b) in atomic percent. (a) adapted from same source as Fig. 5-33; (b) adapted from same source as Fig. 5-32.

spherical. Because of the unit-cell alignments of the γ and γ' (see Fig. 5-8) the cubic shape is preferred. (It is emphasized that the orientation relationship shown in Fig. 5-8 between γ and γ' is maintained regardless of the shape of the particle.) The microstructures in Fig. 5-45 illustrate the effect of lattice mismatch on the morphology, and some results are summarized in Fig. 5-46. Note that a cubic particle occurs if the mismatch is sufficiently high, around 0.6% or higher, regardless of the sign of the mismatch. In many

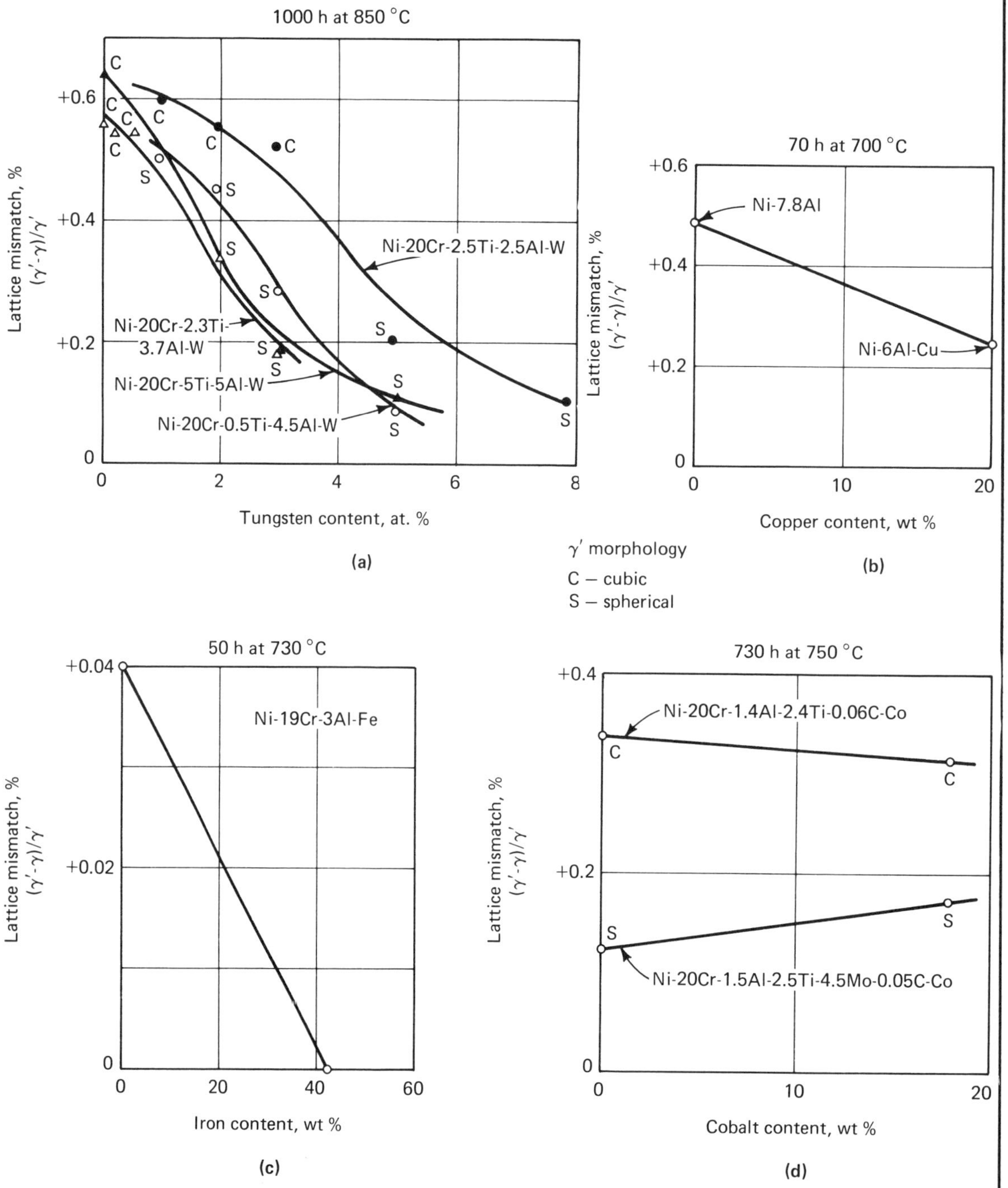

Fig. 5-43. Effect of W, Cu, Fe and Co on the γ′-γ lattice mismatch. (a) from data of A. Havalda, Trans. ASM, Vol 62, p 477, 1969; (b) from data of J. Manenc, Acta Met., Vol 7, p 124, 1959; (c) from data of G. N. Maniar, J. E. Bridge, H. M. James and G. B. Heydt, Met. Trans., Vol 1, p 31, 1970; (d) from data of W. Peter, H. Muller and E. Kohlhaas, Arch. Eisen., Vol 38, p 329, 1967.

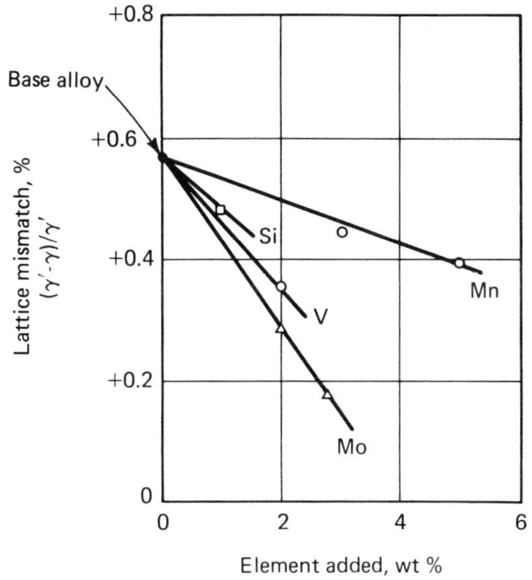

Fig. 5-44. Effect of several elements on the γ'-γ lattice mismatch in a Fe-base, austenitic FCC alloy containing 25% Ni, 15% Cr and 4% Ti. All alloys were aged for 200 h at 700 °C to precipitate γ'. (From data of F. G. Wilson, in Effect of Second-Phase Particles on the Mechanical Properties of Steel, The Iron and Steel Institute, London, 1971)

of the figures in this section, the morphology of the precipitate has been marked with a C for cubic, B for blocky and S for spherical.

6. Coarsening Rate of Gamma Prime

Generally, in alloys in which γ' precipitates, the equilibrium composition of the precipitate and matrix is reached in the early stages of precipitation. From this stage on, the γ' particles coarsen (ripen), but the equilibrium amount of γ' has been attained. Thus, as discussed in Section 5-2, the smaller particles dissolve and the larger ones grow. It has been found that the particle size varies with the aging time to the 1/3 power, and this relation is followed in most alloys, as illustrated in Fig. 5-47. (The growth relation is the same whether particles are cubic or spherical.)

For high-temperature applications, the coarsening should be retarded as much as possible, and in Section 5-2 it was pointed out that the proportionality constant relating particle size and aging time to the 1/3 power is related to the diffusion coefficient and to the γ'-γ surface energy. Although both of these are controlled by the composition of the alloy, there are not sufficient data to predict the effect that elements have on these, and thus on the coarsening rate. Available data* show that Ti and Al increase the coarsen-

*A. J. Ardell, R. B. Nicholson and J. D. Eshelby, Acta Met., Vol 34, p 1295, 1966; B. A. Baker and D. R. F. West, J. Australian Inst. Metals, Vol 14, p 102, 1969; W. I. Michell, Z. Metallkunde, Vol 55, p 614, 1964, and Vol 57, p 586, 1966.

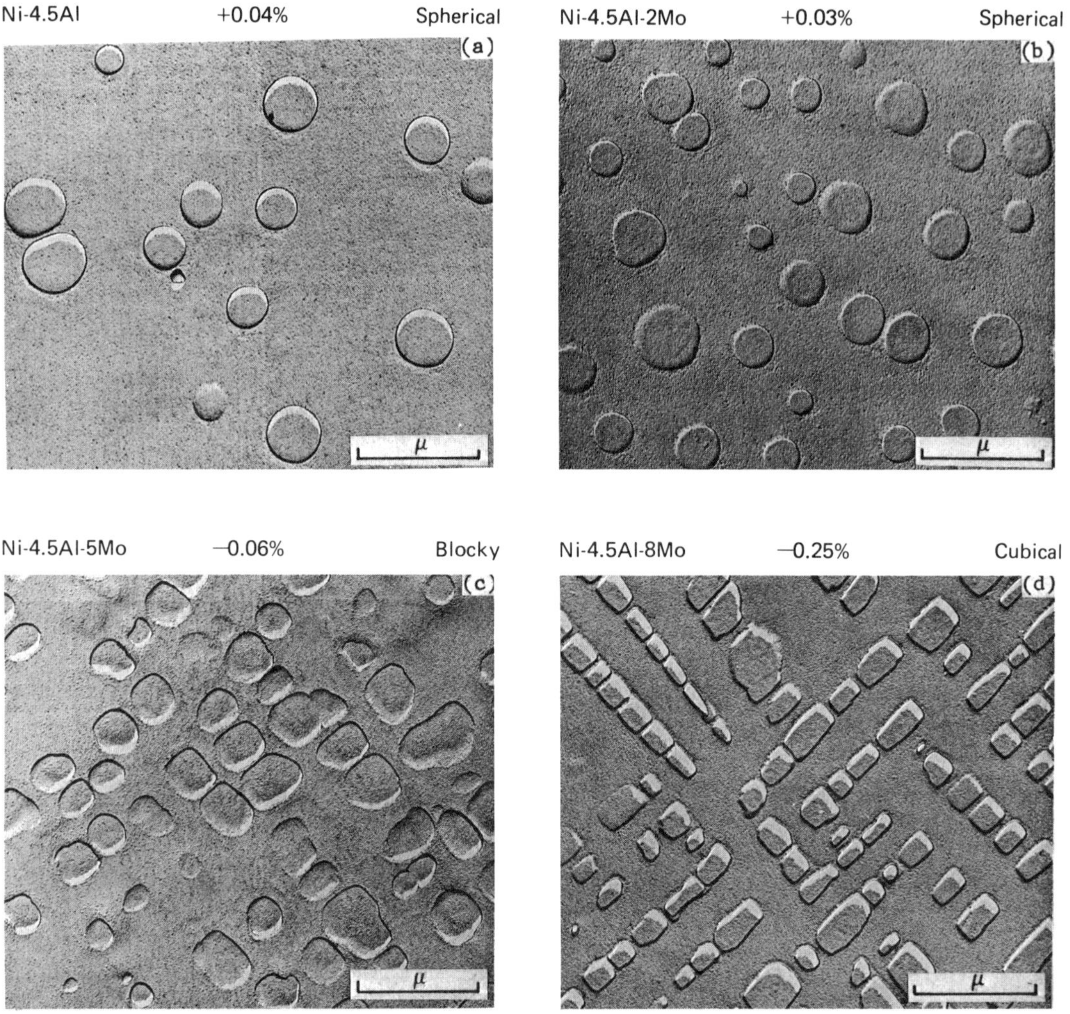

Fig. 5-45. Effect of Mo content on the γ' morphology in Ni-Al-Mo alloys. The alloys were aged for 112 h at 925 °C. The γ'-γ lattice mismatch values are shown, as well as the description of the γ' morphology. All compositions, wt %. (Adapted from V. Biss and D. L. Sponseller, Met. Trans., Vol 4, p 1956, 1973)

ing rate, and Mo (Fig. 5-47), Cr and Nb decrease it. The effect of Fe and Co is slight, and Zr and B have no effect. These data are the basis for the list in Table 5-3.

7. Oxidation and Corrosion Resistance

Evaluation of the effect of alloying additions on the corrosion resistance has been much more difficult to establish than for strength. The exception is Cr, where more Cr is nearly always beneficial. For the other elements, however, observed behavior in one environment will not necessarily be followed in

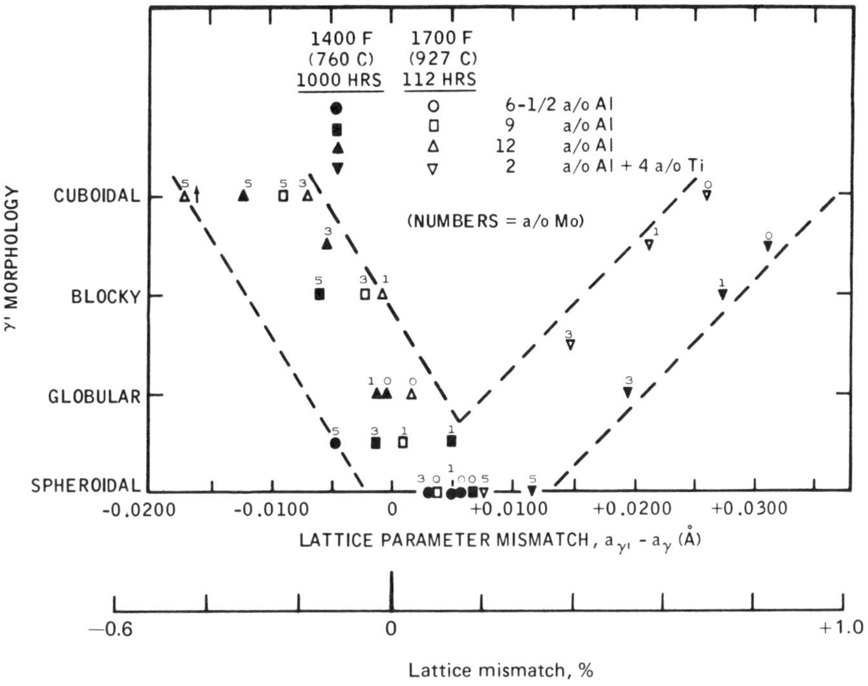

Fig. 5-46. Influence of lattice mismatch between γ and γ′ and the γ′ precipitate morphology for Ni-Al-Ti-Mo alloys. All compositions, at.%. (Adapted from same source as Fig. 5-32)

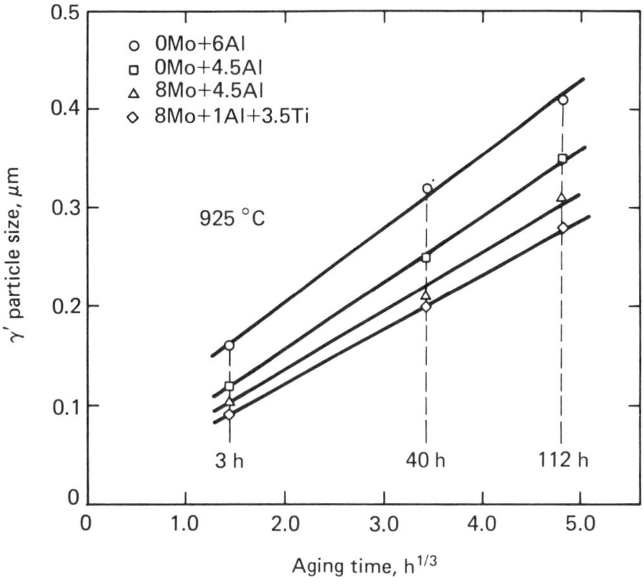

Fig. 5-47. γ′ particle size of Ni-Mo-Al and Ni-Mo-Al-Ti alloys as a function of aging time to the 1/3 power. All compositions, wt %. (Adapted from same source as Fig. 5-45)

another. This has made it quite difficult to predict the effect of chemical composition, and instead reliance is placed on detailed corrosion experiments.

We have already seen that a minimum Cr content for oxidation resistance is about 20%, and that the addition of Al to alloys containing Cr improves it further. Figure 5-48 shows the results of some static oxidation studies of several Ni-base alloys. The improvement in the oxidation resistance due to the Cr is apparent, as is the decreased oxidation resistance as other alloying additions are made and the alloys become more complex. These alloys and some others have been analyzed in terms of the effect of various elements on the oxidation resistance, and the results are illustrated in Fig. 5-49. Again, in these alloys Cr has its most potent effect around 20%. Note that increasing the sum of Mo + W improves the oxidation resistance slightly, whereas increasing Si lowers the resistance. The addition of Mn and Fe has little effect, and Nb, Cu and Ti decrease the oxidation resistance.

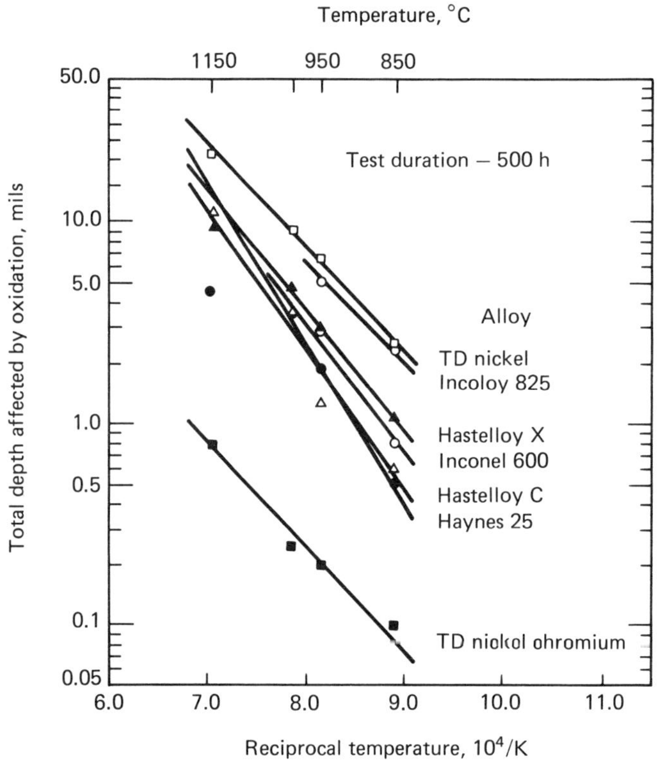

TD* nickel-chromium	78Ni-20Cr-2ThO$_2$
Inconel 600	75Ni-16Cr-7Fe-1Mn-1Si
Hastelloy C	57Ni-15Cr-5Fe-10Mo-4W-1Mn-1Si
Hastelloy X	49Ni-22Cr-18Fe-1Co-9Mo
Incoloy 825	42Ni-22Cr-29Fe-3Mo-1Mn-2Cu-1Ti
Haynes 25	10Ni-20Cr-3Fe-50Co-15W-2Mn-1Si
TD* nickel	98Ni-2ThO$_2$

*Thoria dispersed for strengthening.

Fig. 5-48. Oxidation resistance of several Ni-base alloys in static air. (Adapted from C. L. Angerman, Oxid. Metals, Vol 5, p 149, 1972)

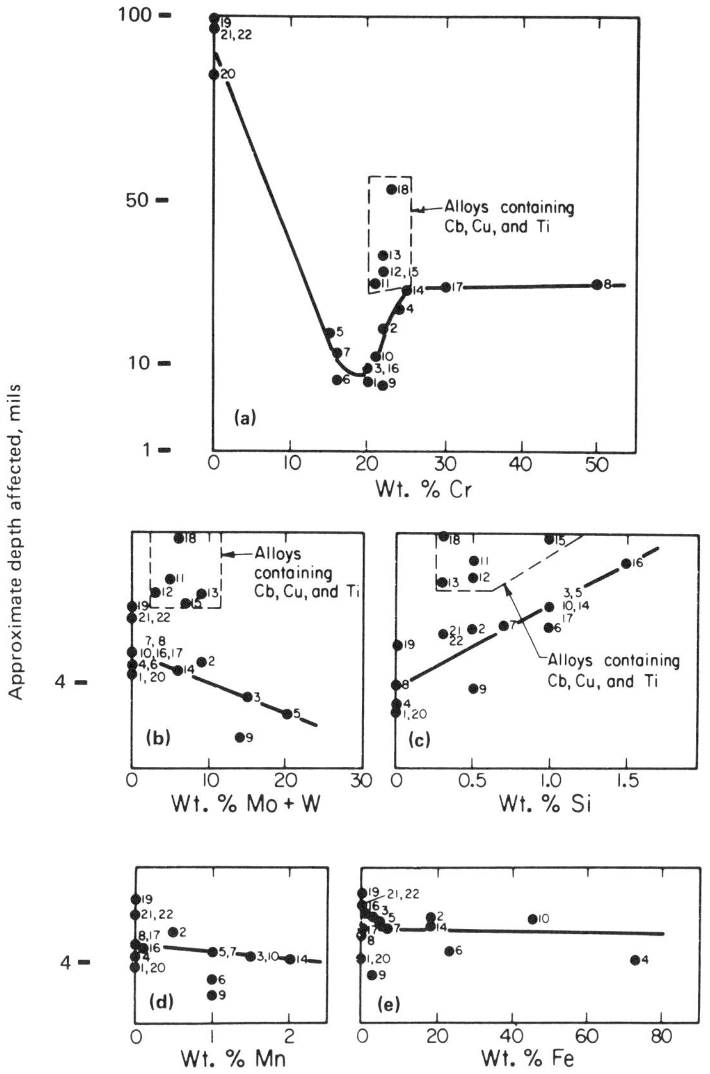

Fig. 5-49. Estimated depth of oxidation of several Ni-base alloys after exposure in static air at 1000 °C for 50,000 h. The approximate depth affected is shown. (From same source as Fig. 5-48)

We discussed in Section 5-3 the importance of cyclic oxidation resistance. Figure 5-50 compares the cyclic oxidation resistance of several alloys, including some superalloys. Generally, both the static and the cyclic oxidation resistance is less as the alloy deviates from the base 80% Ni–20% Cr composition. However, note that the addition of a small amount of yttrium allows the cyclic oxidation resistance to be retained. How Y brings about this improvement is not understood; it may be because its presence in the oxide scale makes the coefficient of expansion more nearly that of the base metal, lowering the thermal stresses.

Recall that Cr lowers the high-temperature strength, because it lowers the melting point of the alloy, and hence to improve the strength the Cr should be lowered. However, this conflicts with the need for Cr for oxidation

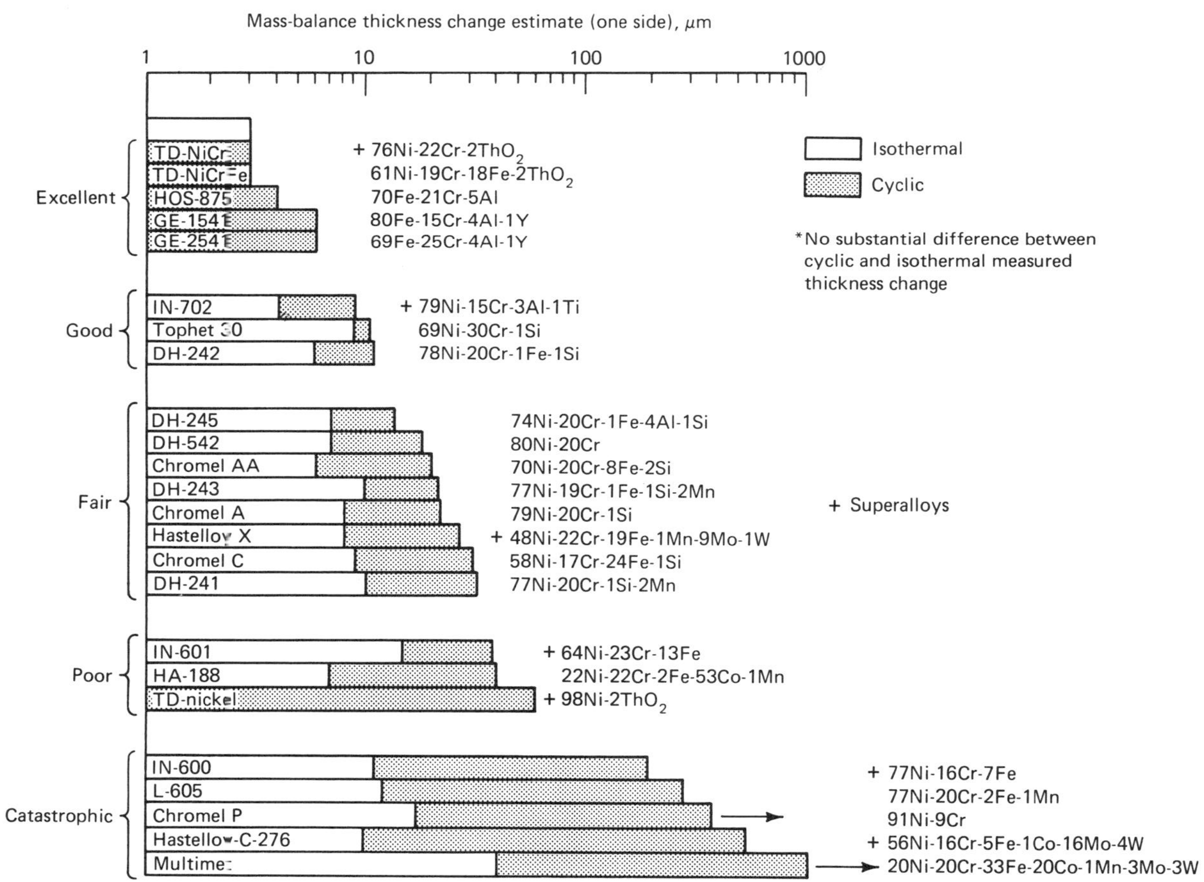

Fig. 5-50. Comparison of static (isothermal) and cyclic oxidation data for several Ni-base, Fe base and Co-base alloys. In static tests alloys were oxidized in still air at 1150 °C for 100 h; cyclic tests involved heating to 1150 °C, holding for 1 h, then cooling to 25 °C. Alloys considered Ni-base superalloys are marked with a +. (Adapted from C. A. Barrett and C. E. Lowell, Oxid. Metals, Vol 9, p 307, 1975)

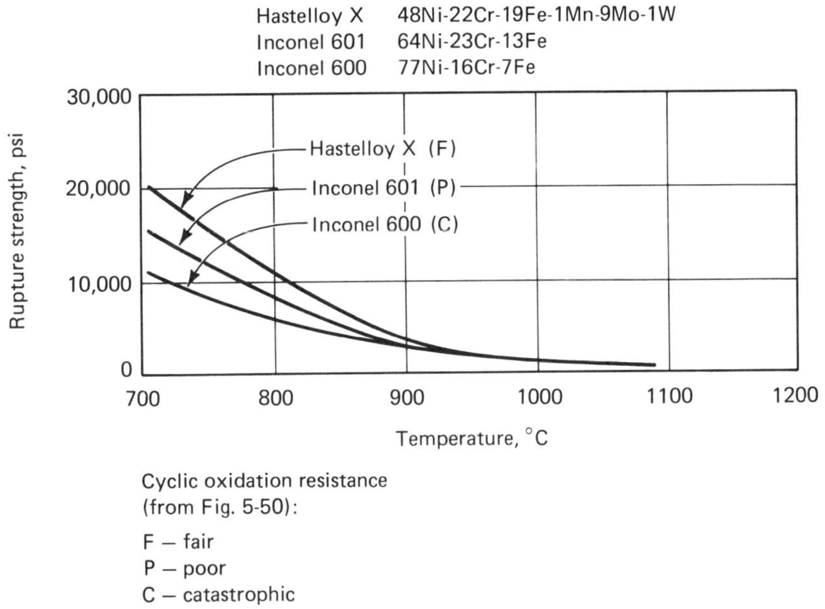

Fig. 5-51. Typical rupture strength values for three Ni-base superalloys. (Adapted from The Superalloys, ed. C. T. Sims and W. C. Hagel, Wiley, New York, 1972)

resistance. This is illustrated in Fig. 5-51, where rupture strength is shown for three of the superalloys shown in Fig. 5-50. Note that none of these alloys is strengthened by γ' precipitation, as they contain no Al or Ti. Instead, strengthening relies on solid solution hardening, due to Mo and W, and on carbides. The Mo and W markedly increase the strength of Hastelloy X, especially in the lower temperature range, and the higher Cr content retains the relatively good oxidation resistance. The low Cr ternary alloy has the lowest strength and the poorest oxidation resistance.

One approach to the problem of a balance between oxidation resistance and strength is to dispersion harden the matrix. This involves adding a stable, strong compound, thorium oxide (thoria) being commonly used, in a fine dispersion. It is emphasized that in these cases the dispersion is made by mechanically mixing the matrix alloy and the thoria, both in powder form, and then compacting and sintering the material to densify it. These particular alloys can be made by melting the matrix alloy and thoria, casting, then solution heat treating to dissolve the thoria followed by a controlled precipitation treatment. This processing requirement is at the present time a strong deterrent to the wide use of these dispersion-hardened alloys. However, Fig. 5-52 shows the marked improvement of thoria dispersion-hardened Ni and Ni-20Cr alloys over the three alloys of Fig. 5-51. Note that the TD-NiCr alloy is weaker than TD-Ni because of the lower melting point, but that the TD-NiCr alloy has by far the superior oxidation resistance.

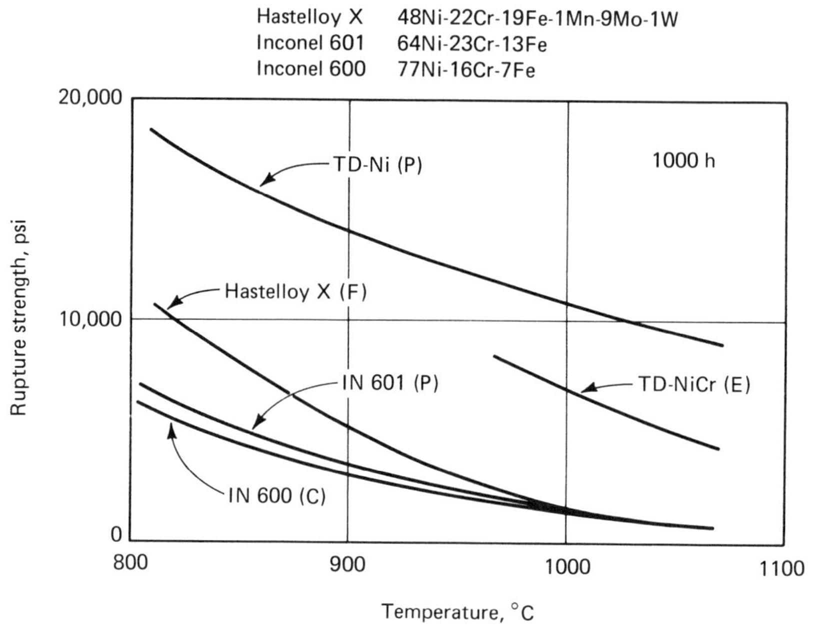

Cyclic oxidation resistance
(from Fig. 5-50):

E – excellent
F – fair
P – poor
C – catastrophic

Fig. 5-52. Typical rupture strength values for five Ni-base superalloys. (Adapted from same source as Fig. 5-51)

The clear advantage over the alloys shown in Fig. 5-52 of using a γ'-strengthened alloy is illustrated in Fig. 5-53. Note that the strength of the γ' alloys is markedly higher than that of Hastelloy X, and that the strength of 713C, IN-738 and MM-421 is about the same. Table 5-5 shows that for IN-738 and MM-421 the Al + Ti content is about the same, as well as the refractory metal content (Mo + W + Ta + Nb). Udimet 500 has the lowest refractory metal content and the lowest strength, whereas IN-500 has the highest Al + Ti and the highest strength. However, IN-100 has the least hot corrosion resistance (Fig. 5-54), mainly because of the low Cr content. Udimet 500, with the highest Cr content, has the best corrosion resistance but is the weakest. Of the alloys shown here, IN-738 would have the best combination of corrosion resistance and strength.

It is clear that it is reasonable to list Cr in Table 5-3 as increasing the corrosion resistance. However, no trend, independent of alloy and environment, is clear for the other common alloying additions, and therefore they are listed as variable.

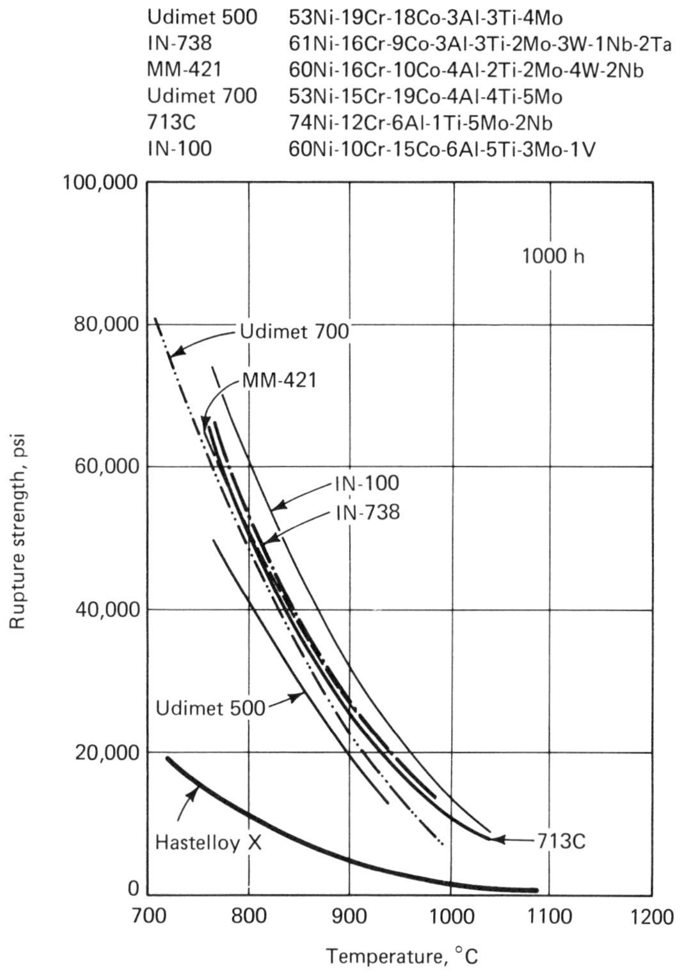

Fig. 5-53. Rupture strength as a function of temperature for six Ni-base superalloys. (Adapted from same source as Fig. 5-51)

Table 5-5. Comparison of the Effect of Chemical Composition on Hot Corrosion Resistance and Creep Strength of Six Ni-Base Superalloys (From Fig. 5-50 and 5-53.)

	Al+Ti	Mo+W +Ta+Nb	Alloy	Alloy	Cr	
Decreasing strength ↓	8	4	IN-100	IN-100	10	Increasing hot corrosion resistance ↓
	5	8	IN-738	713C	12	
	6	8	MM-421	Udimet 700	15	
	7	7	713C	MM-421	16	
	8	5	Udimet 700	IN-738	16	
	6	4	Udimet 500	Udimet 500	19	

Udimet 500	53Ni-19Cr-18Co-3Al-3Ti-4Mo
IN-738	61Ni-16Cr-9Co-3Al-3Ti-2Mo-3W-1Nb-2Ta
MM-421	60Ni-16Cr-10Co-4Al-2Ti-2Mo-4W-2Nb
Udimet 700	53Ni-15Cr-19Co-4Al-4Ti-5Mo
713C	74Ni-12Cr-6Al-1Ti-5Mo-2Nb
IN-100	60Ni-10Cr-15Co-6Al-5Ti-3Mo-1V

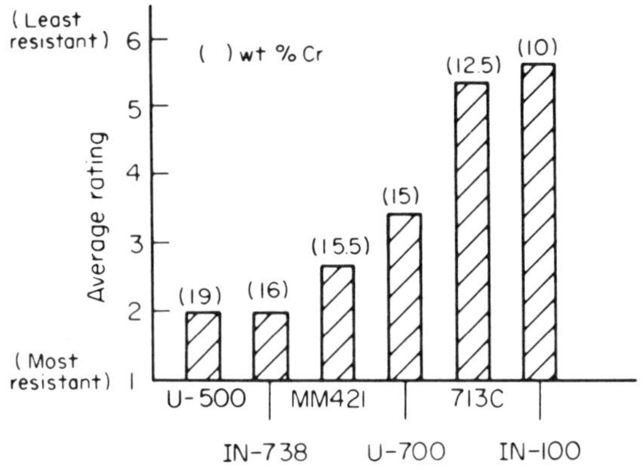

Fig. 5-54. Relative hot corrosion resistance of six Ni-base superalloys. (Adapted from R. W. Fawley, in The Superalloys, ed. C. T. Sims and W. C. Hagel, Wiley, New York, 1972)

5-6. Longtime Phase Stability

The discussion so far has emphasized the γ'-γ structures, but consideration must be given to the possibility of the presence of other phases (e.g., carbides), either in the original heat treated condition or formed during high-temperature use. These phases generally are carbides or intermetallic phases. Carbides are to be expected, as the alloys being considered have elements present which are strong carbide formers (e.g., Cr), and carbon is present (Appendix A) either because it is difficult to remove during refining or because it is added on purpose to allow carbide formation to improve creep properties. The possibility of the formation of intermetallic phases should be suspected from the experience of high-alloyed, Fe-base alloys (e.g., stainless steel), where the sigma phase (FeCr) can form. The carbide phases were not shown on the phase diagrams we have been examining because we were not considering the influence of carbon. The intermetallic phases may not form in the simple ternary and quaternary systems whose phase diagrams we examined, but also the alloys on which these diagrams are based may not have been heat treated for sufficient times for these phases to form.

Figure 5-55 shows the results of a study on 713C. The alloy was in the as-cast condition, and was held for 5000 h at temperatures from 760 to 1100 °C, and for 2000 h at 1150 °C. The M stands for a combination of metallic elements, such as Cr, Ni, Al, etc. The identification is based on chemically extracting the phases present by dissolving the matrix, then

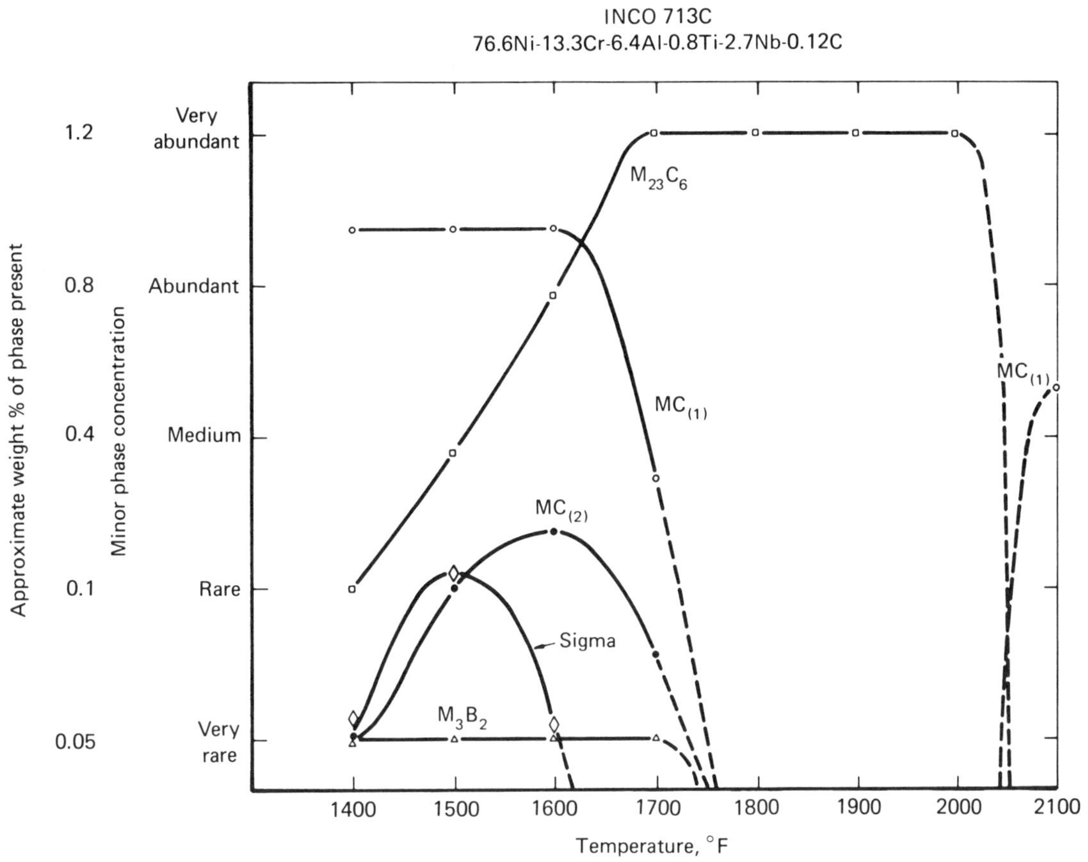

Fig. 5-55. Minor phase concentration in 713C after aging the as-cast alloy for 500 h from 760 to 1100 °C and 2000 h at 1150 °C. (Adapted from H. E. Collins, Trans. ASM, Vol 62, p 86, 1969)

analyzing an x-ray diffraction pattern. This allows establishment of the type of carbide, but not the exact chemical composition. It is important to realize that the level marked "very abundant" corresponds to about 1 wt % phase. Note that four different carbides can form, depending upon the temperature range. (MC_1 and MC_2 are carbides of different lattice parameters but the same crystal structure.) In the as-cast condition MC_1 and $M_{23}C_6$ are both present, along with the boride M_3B_2. Upon aging in the low temperature range, MC_2 and the intermetallic phase sigma (σ) appear.

Actually, curves as in Fig. 5-55 are related to time-temperature-transformation (TTT) diagrams, such as shown in Fig. 5-56 specifically for the formation of σ. In this alloy, at a temperature above about 925 °C (1700 °F) the σ is completely dissolved. At low temperature [e.g., 705 °C (1300 °F)] the atom mobility is too low for the rate of precipitation to be significant. At high temperatures, approaching the solubility limit, the energy difference between the supersaturated alloy and the alloy with the sigma precipitated is too low for the rate of precipitation to be significant. Thus, the rate is a maximum at an intermediate temperature range, giving a minimum time for the start of sigma formation.

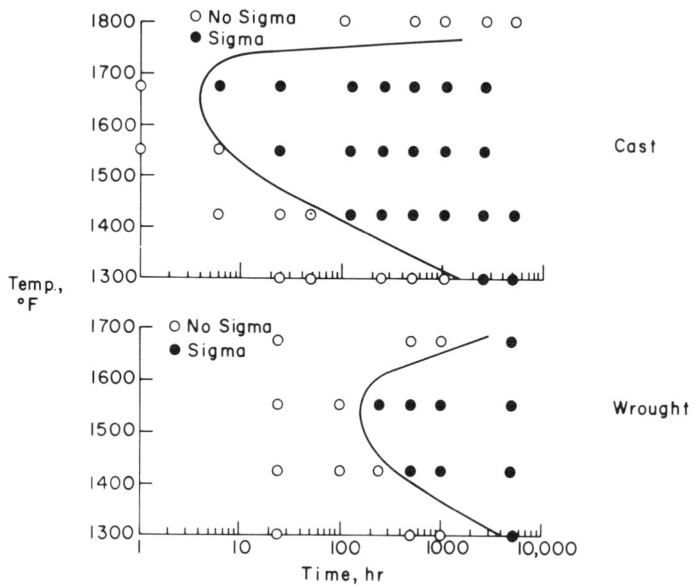

Fig. 5-56. Time-temperature-transformation diagram for the start of sigma formation in IN-100. (From C. T. Sims, in The Superalloys, ed. C. T. Sims and W. C. Hagel, Wiley, New York, 1972)

The formation of the carbides has an important effect on properties, especially creep properties, but the effect is very variable, depending upon the alloy and how it is processed. This makes it difficult even to lay down guidelines to follow in the heat treatment of these superalloys, but instead requires careful consideration of each alloy. In many of these alloys, carbides are present in the as-cast condition and the solution-annealed condition which are not stable, but disappear upon aging. This is evident in Fig. 5-57. The MC forms upon solidification and is stable at high temperatures, but it is not stable at low temperatures and thus upon aging begins to disappear. $M_{23}C_6$ is present in the lower temperature range but not at high temperatures [above about 875 °C (1600 °F)]. A common reaction is

$$MC + \gamma \rightarrow M_{23}C_6 + \gamma',$$

which may be depicted as

$$(Ti,Mo)C + (Ni,Cr,Al,Ti) \rightarrow (Cr,Mo)_{23}C_6 + Ni_3(Al,Ti).$$

The actual composition of MC and $M_{23}C_6$ formed in Udimet 700 is shown in Table 5-6. Note that basically MC is a carbide containing Ti and Mo, whereas $M_{23}C_6$ is mainly Cr and Mo. Thus, when the $M_{23}C_6$ carbide forms, the Cr content of the γ is lowered, increasing the percentage of Al and Ti to favor the formation of γ'. Figure 5-58 shows typical microstructures of a Udimet 700 alloy, with these features shown. Both types of carbides form

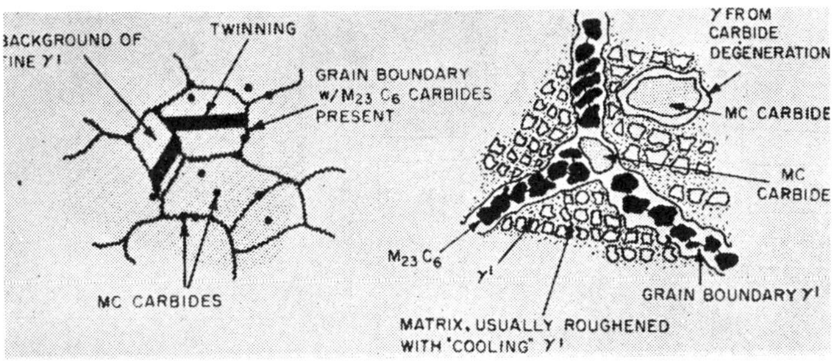

500X (at 110%) 10,000X (at 110%)

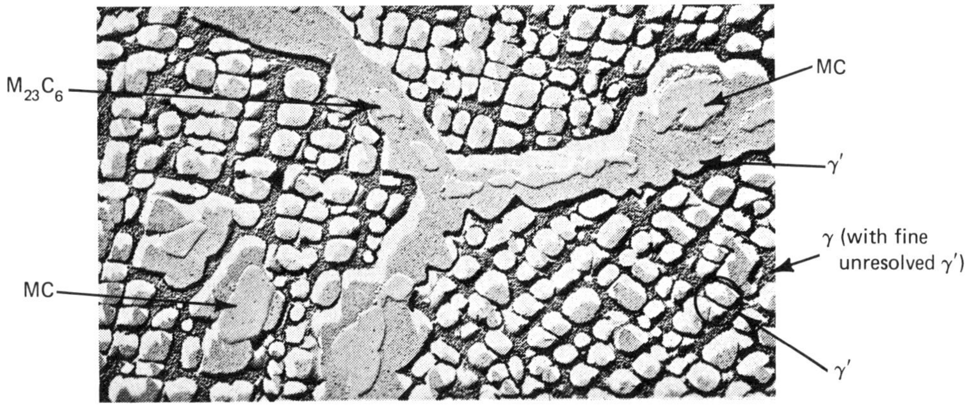

10,000X (at 110%)

Fig. 5-57. Illustration of the microstructure in Udimet 700 after aging under a creep stress of 36,500 psi at 845 °C (1550 °F) for 1387 h. In (a) the structures are depicted schematically; (b) is an actual micrograph. Note that the carbides are visible in optical microscopy, but to see the γ' requires electron microscopy. (Adapted from C. T. Sims, J. Metals, Vol 18, p 1119, 1966)

Table 5-6. Approximate Chemical Composition of MC and $M_{23}C_6$ Carbides in Udimet 700

Carbide	Composition, at. % C	Cr	Mo	Ni	Ti
MC	49	1	10	—	40
$M_{23}C_6$	21	68	7	2	2

From data of W. J. Boesch and H. B. Canada, J. Metals, Vol 20(4), p 46, 1968.

at the grain boundaries, but the MC is in the process of decomposing, forming the γ'. Thus, in this alloy the grain boundaries become coated with γ' and carbides. Note that at high temperature the γ-γ' structure changes. Also note that the sigma phase forms as long needles.

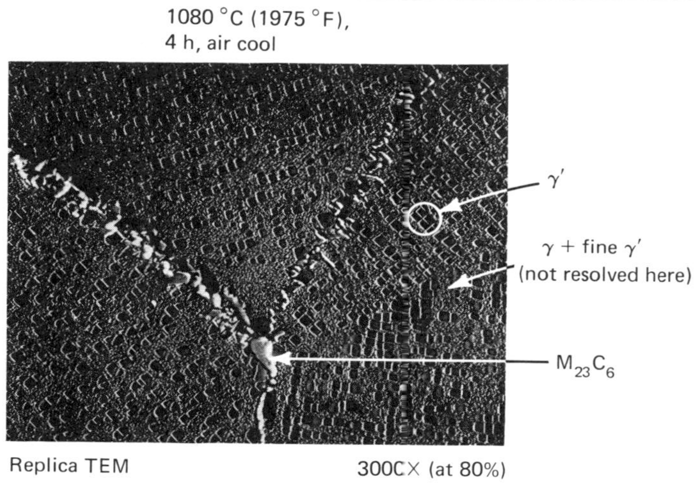

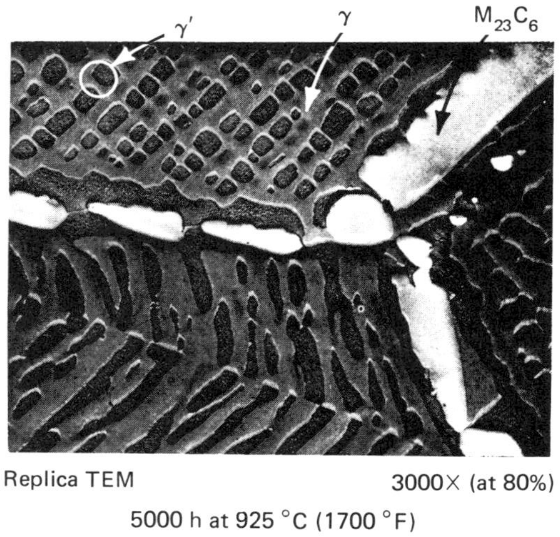

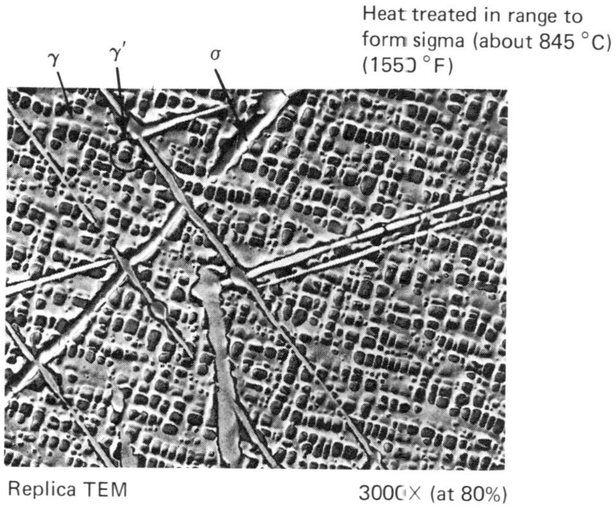

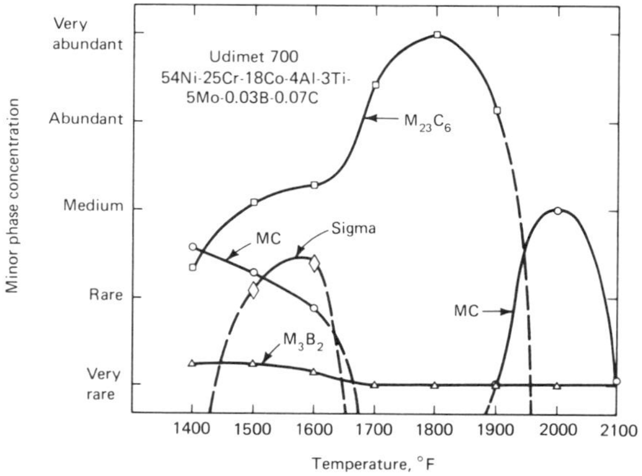

Fig. 5-58. Typical microstructures of a Ni-base superalloy, Udimet 700, and approximate relation to phase analysis for similar heat treatments. (Phase analysis adapted from same source as Fig. 5-54; microstructures adapted from W.J. Boesch and H.B. Canada, J. Metals, Vol 20(4), p 46, 47 and 50, 1968)

The effect of sigma formation on creep behavior of IN-100 alloy is shown in Fig. 5-59. Three alloys of slightly differing Co and Ti content show the same creep curve at 985 °C (1800 °F), and note that this is above the temperature range for the formation of σ. However, at 815 °C (1500 °F) the alloy high in Ti and Co has a significantly shorter stress-rupture life. The other two alloys do not form sigma, and retain their strength. Note that the presence of sigma does not affect the yield strength at 25 °C, but it does

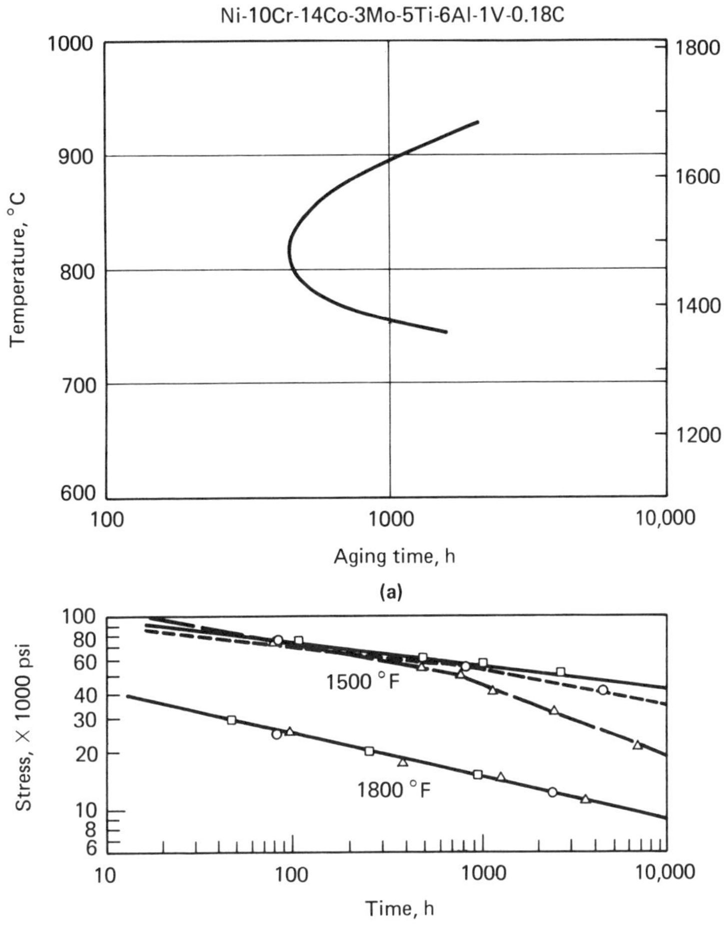

Room temperature tensile properties of specimens taken from stressed rupture bars

Code	$\bar{N}_V$	Time at 1500 °F	0.2% YS	UTS	El., %	R.A., %
5.3 Ti, 15 Co	2.70	6900 h	99,400	99,400	0.2	Nil
4.7 Ti, 15 Co	2.46	4355 h	109,000	136,000	3.0	3.5
4.7 Ti, 10 Co	2.23	2655 h	96,800	118,000	4.5	10.0

(b)

Fig. 5-59. (a) Time-temperature-transformation diagram for the start of sigma formation in IN-100 alloy. (b) Stress rupture data for three IN-100 alloys. (Adapted from J.R. Mihalisin, C.G. Bieber and R.T. Grant, Trans. Met. Soc. AIME, Vol 242, p 2399, 1968)

lower significantly the elongation at fracture. The sigma in such alloys precipitates in the form of long needles (e.g., Fig. 5-58). The sigma is brittle, and the long needles can easily propagate cracks once they are formed, which causes the low tensile ductility at 25 °C. Sigma contains a high refractory metal content (e.g., Mo, W), which comes from the γ-γ' matrix in which it precipitates. This causes weakening of the matrix, and hence reduced creep strength.

Sigma is a phase belonging to a class of phases which have complex crystal structures and are hard. These structures are characterized by close-packed layers of atoms separated by relatively large interatomic distances because of layers of the larger atoms. Since this is considered to be a characteristic topology, these phases are called topologically close-packed (TCP). In this category are also the phases designated μ and Laves.

The importance of avoiding the formation of the TCP phases has generated intense activity in attempts to predict, from the chemical composition of an alloy, whether this phase will be present at equilibrium. Based on some solid-state physics principles and some empirical observations, a method has evolved which is quite successful, and is widely used in alloy design. The approach is related to an effect recognized many years ago, namely that certain intermetallic compounds have the same electron/atom ratio; such compounds are called electron compounds. The electron to atom ratio (e/a) is calculated by using the number of valence electrons and the number of atoms in the compounds. For example, Cu has a valence of 1, Zn 2 and Ga 3. Thus in the compound CuZn, the number of atoms is 2, and the number of valence electrons is $(1 + 2) = 3$, or $e/a = 3/2$. In the compound Cu_3Ga, the ratio is $(3)(1) + (1)(3)/4 = 3/2$. Here the Cu valence is 1, but there are three Cu atoms in the compound; the Ga valence is 3 and there is one atom. It was also observed in several systems that if the e/a ratio is calculated at the boundary of the solvent, a common value is found. For example, in the Cu-Zn system, the solubility is about 37 at.% Zn. (The solubility, of course, is temperature dependent, so that the calculations are approximate.) Thus, $e/a = [(0.37)(2) + (0.63)(1)]/1 = 1.37$. For the Cu-rich solubility of the Cu-Ga system, a solubility of about 20 at.% Ga is found. Then $e/a = [(0.20)(3) + (0.80)(1)]/1 = 1.40$. In several Cu-base alloys, this value seems to predict the solubility of the solute in Cu.

The determination of the valence of elements such as Fe, Ni, Co, Mo, W, etc. (those of interest in Ni-base alloys), is not so straightforward. It is clear that the valence electrons play a role in determining phase stability. Usually, the bonding strength is related to the number of valence electrons, or to the number of orbital vacancies or "holes" available to fill a shell. For example, Cr normally is considered to have one electron in the outer shell (which in this case can hold two), and five in the next shell (which can hold ten). From magnetic measurements, it has been concluded that electron orbital sharing leads to 4.66 holes. A similar approach leads to the values tabulated in Table 5-7, which are then used to calculate an electron hole number, $\overline{N}_v$, for a given alloy. The chemical composition used is not that of the alloy, but that of the γ matrix. Thus, the alloy composition is corrected

Table 5-7. Possible Electron Hole Numbers ($\overline{N}_v$) for Several Elements

Group →	VIA	VIIA	VIIIA		
First long period	Cr (4.66)	Mn (3.66)	Fe 2.22	Co 1.71	Ni 0.61
Second long period	Mo (4.66)	Tc (3.66)	Ru (2.66)	Rh (1.66)	Pd (0.66)
Third long period	W (4.66)	Re (3.66)	Os (2.66)	Ir (1.66)	Pt (0.66)

From C. T. Sims, in The Superalloys, ed. C. T. Sims and W. C. Hagel, Wiley, New York, 1972.

for the presence of carbides and γ', and the resulting composition is used. If the phases present, and their chemical composition and amount, are known, then this correction can be made. This information is not always known, but approximations have been developed which are used.

An example of a calculation is given in Table 5-8. It is assumed that all the boron precipitates as M_3B_2 with the composition shown, and that all of the carbon is tied up as carbides, one-half as MC and the other as $M_{23}C_6$ of the chemical composition shown. Then all of the Al, Nb and Ti, and 1/10 of the Cr, combine with Ni to form Ni_3(Al, Ti, Nb). The results give an electron hole number of 2.72.

These calculations are called PHACOMP (for phase computation), and are widely used in alloy design. Correlations have been made with $\overline{N}_v$ and the presence or absence of sigma in alloys which have undergone longtime heat treatment. Roughly, a value of about 2.5 or less is desired to avoid sigma formation (Fig. 5-60). This type of calculation can be used to predict behavior of different alloys, but it also is useful in alloys in which $\overline{N}_v$ is close to the critical value, such as Udimet 700. Table 5-9 shows calculated values for several heats of this class of alloy, and this information can be used to assist in application of components made from these heats. This can be especially important if the alloy is used in the cast condition. In this case the cored structure (see Chapter 2) contains regions in which the chemical composition deviates considerably from the nominal composition. Local regions may be quite susceptible to sigma formation, whereas the average composition may not predict its formation. Thus the sensitivity of the calculation of $\overline{N}_v$ to the chemical composition must be examined.

5-7. Directional Solidification

As will be shown, a structure with few or no grain boundaries normal to the major stress direction allows significant improvement in creep properties over an equiaxed, polycrystalline structure. One of the major materials

Table 5-8. Calculation of the Electron Hole (Vacancy) Number, $\overline{N}_v$, for a Given Alloy

	Element							
	Al	Ti	Cr	Mo	Co	Ni	B	C
Alloy weight percent	4.43	3.49	15.3	5.0	18.6	52.8	0.015	0.13
Alloy atomic percent	9.03	4.01	16.2	3.0	17.4	49.6	0.076	0.60

Boride Precipitation

$(Mo_{0.5}Ti_{0.15}Cr_{0.25}Ni_{0.10})_3B_2$

Ni = Ni − Ni in boride = 49.6 − (0.30 × 0.076/2) = 49.58
Cr = Cr − Cr in boride = 16.2 − (0.75 × 0.076/2) = 16.17
Ti = Ti − Ti in boride = 4.01 − (0.45 × 0.076/2) = 3.99
Mo = Mo − Mo in boride = 3.0 − (1.50 × 0.076/2) = 2.94

Carbide Precipitation

$Cr_{21}Mo_2C_6$ and TiC form

C = available = 0.60
Ti = Ti − Ti in TiC = 3.99 − (0.60/2) = 3.69
Cr = Cr − Cr in $M_{23}C_6$ = 16.17 − (21 × 0.60/2)(1/6) = 15.12
Mo = Mo − Mo in $M_{23}C_6$ = 2.94 − (2 × 0.60/2)(1/6) = 2.84

Gamma Prime

Cr = Cr − Cr in γ' = 15.12 − 0.1(15.12) = 13.61
Ni = 49.58 − 3(3.69 + 9.03 + 1.51) = 49.58 − 3(14.23) = 6.89

The elemental amounts now remaining constitute the residual matrix. These amounts are then scaled to 100 to equal matrix composition atomic percent.

	Al	Ti	Cr	Mo	Co	Ni	B	C
Residual amount	–	–	13.61	2.84	17.40	6.89	–	–
Atomic percent	–	–	33.41	6.97	42.70	16.91	–	–
Weight percent	–	–	29.36	11.31	42.56	16.77	–	–

The atomic and weight compositions of the residual matrix are printed out, and the electron vacancy number is then calculated from the matrix atomic percent as follows:

$$\overline{N}_v = 4.66(Cr + Mo) + 1.71(Co) + 0.66(Ni) = 2.72$$

Adapted from L. C. Woodyatt, C. T. Sims and H. J. Beattie, Trans. Met. Soc. AIME, Vol 236, p 519, 1966.

applications advances has been the development of techniques to grow single-crystal and columnar grain structures through directional solidification, and the application of these techniques to the production of gas-turbine blades and vanes in superalloys. The principles involved are described in this section, and the data on one superalloy are used to illustrate the advantages of such structures.

In creep, the grain boundaries play an important role. Grain boundary sliding and void formation occur, and the boundary is the site for precipitation and hence the alteration of boundary structure. The problem is illustrated in Fig. 5-61. Note that voids have formed on the boundaries, being

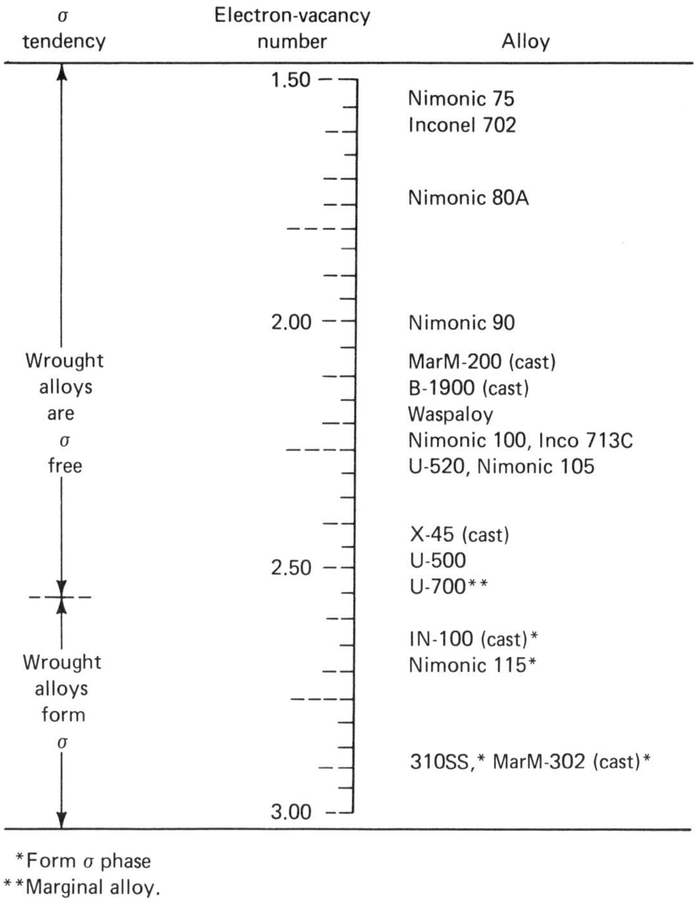

Fig. 5-60. Sigma-forming tendency versus electron hole (vacancy) number for several Ni-base superalloys. (From same source as Table 5-8)

especially prevalent on those boundaries normal to the stress axis. It would appear, then, that an improvement in creep performance would be obtained for larger grains. This is generally the case, if the structure remains relatively stable, and is illustrated in Fig. 5-62 for some face-centered cubic, Fe-base, single-phase structures.

On the basis of this type of information, an approach in designing creep resistance for applications where the stress is predominantly in one direction, such as in turbine blades and vanes, has been to develop a structure which has no grain boundaries normal to the stress direction. This has been accomplished by directional solidification, a solidification process whereby either a single crystal is formed, or at least a structure of columnar grains each having the length of the object, with the long axis of the columnar grains parallel to the stress axis. Both the single crystal and the columnar structure form an object (e.g., a turbine blade) which has no grain boundaries normal to the stress direction.

Table 5-9. Chemical Composition, Sigma Phase Occurrence and Electron Hole Number for Several Heats of Udimet 700 (From Same Source as Table 5-8)

Heat	Composition, wt %									
	C	Al	Ti	Cr	Mo	Co	Fe	B	Zr	Ni	$\overline{N}_v$
1	0.05	4.70	3.60	14.45	5.00	19.00	0.00	0.020	0.00	53.17	2.67(a)
2	0.13	4.43	3.49	15.30	5.00	18.60	0.13	0.015	0.02	52.92	2.62(a)
3	0.06	4.49	3.44	15.10	4.95	18.70	0.15	0.014	0.05	53.05	2.62(a)
4	0.06	4.45	3.45	15.20	4.95	18.70	0.00	0.031	0.00	53.16	2.61(a)
5	0.07	4.40	3.43	14.60	5.10	18.00	0.30	0.030	0.05	54.00	2.56(a)
6	0.05	4.30	3.31	15.20	5.00	18.40	0.10	0.030	0.05	53.60	2.55(a)
7	0.06	4.37	3.40	14.60	4.45	17.60	0.00	0.028	0.00	55.49	2.48
8	0.08	4.50	3.47	15.28	4.28	19.17	0.00	0.030	0.00	57.47	2.45
9	0.07	4.20	3.23	14.70	4.70	18.00	0.00	0.030	0.00	55.07	2.44
10	0.05	4.20	3.19	14.50	4.50	17.50	0.10	0.030	0.05	55.90	2.41
11	0.05	4.20	3.19	14.50	4.50	17.50	0.00	0.028	0.00	56.03	2.40
12	0.06	3.91	2.98	14.80	4.45	17.50	0.00	0.030	0.00	56.27	2.32

(a) σ-prone heats.

We will first examine some properties of single-crystal material, and then look at how blades and vanes are made in this form and in columnar form. The strength, of course, depends upon the orientation of the crystal relative to the stress axis. (See the discussion of the plastic deformation of single crystals in Chapter 1.) To illustrate the behavior of Ni-base superalloys, we take Mar-M200 as an example. This alloy contains nominally (wt %) 9Cr, 10Co, 5Al, 2Ti, 13W, 2Nb and 0.15C. The as-cast, single-crystal structure is cored (see Chapter 2), with about 60% γ' present which forms upon cooling following solidification. The γ' has the cube morphology described previously. Figure 5-63 shows some tensile stress-strain curves for the as-cast material. Note that for any orientation, the strength increases with temperature, as would be predicted from Fig. 5-13 because of the presence of γ'. At 25 °C, the strength is greatest for the [111] orientation, but at 760 °C the [100] orientation is as strong as the [111]. At 980 °C, the stress-strain behavior is independent of orientation. Heat treating the cast material for 4 h at 1230 °C, then aging for 64 h at 870 °C, increases the strength, as shown in Fig. 5-64.

The creep behavior of as-cast Mar-M200 crystals is shown in Fig. 5-65. Note that orientations near [100] generally have the best properties. The common orientation in single crystals formed by directional solidification is [100], so that this high-strength orientation can be utilized.

The elastic modulus of the single-crystal material is shown in Fig. 5-66. The modulus of crystals is known to be quite anisotropic, so that along the [100] direction it is markedly different from that of the average value characteristic of a randomly oriented, polycrystalline structure. (As will be discussed, a directionally solidified, columnar structure is such that each columnar grain has the [100] axis parallel, but rotated about this axis at

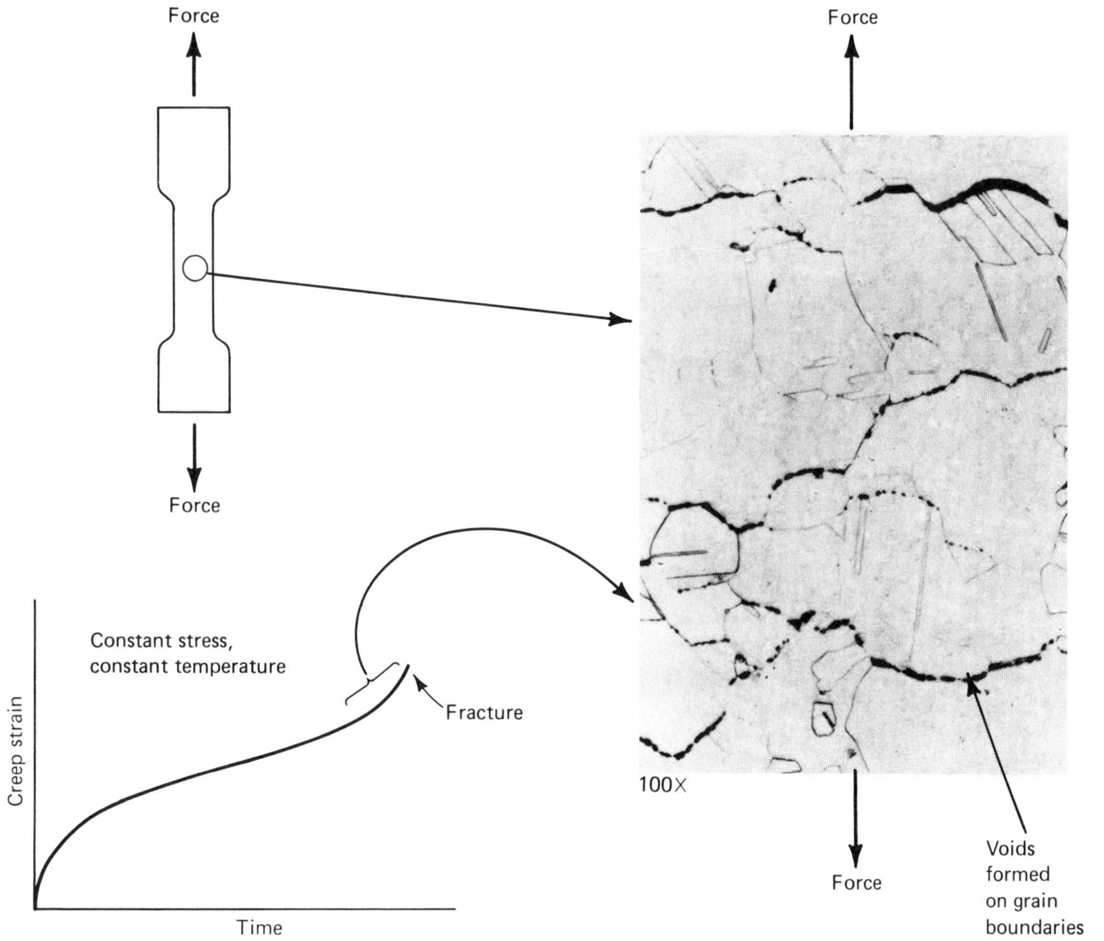

Fig. 5-61. Illustration of void formation on grain boundaries which are mainly normal to the stress axis. The microstructure was taken from a sample of copper creep tested at 385 °C under a stress of 2500 psi. The schematic sketches show the relation of the microstructure to the creep test specimen and to the final stage of creep in which the voids form. (Microstructure from F. L. VerSnyder, in High Temperature Materials, ed. R. F. Hehemann and G. M. Ault, Wiley, New York, 1959)

random. Thus, determination of the elastic modulus along the long axis of the columnar grained structure should give the same modulus as the single material, as shown in Fig. 5-66.) The lower modulus of the single-crystal material (and of the columnar material) leads to improved fatigue resistance, as the thermal stresses induced upon cooling and heating are less.

In the columnar structures, the long axis of the grains is parallel to the [100] direction, but each grain is randomly oriented about this direction. Thus, it would be expected that samples stressed parallel to the columnar grains would have better properties than those having an equiaxed structure.

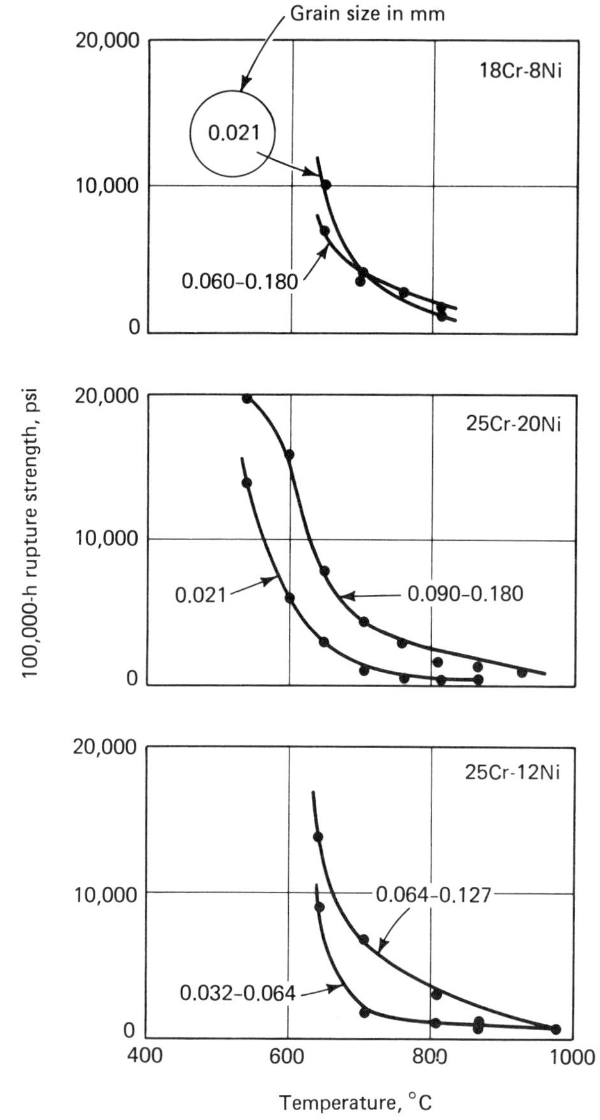

Fig. 5-62. Effect of grain size on creep rupture strength of three austenitic stainless steels. Chemical composition, wt %. (Adapted from C. L. Clark and J. W. Freeman, Trans. ASM, Vol 38, p 148, 1947)

Figure 5-67 shows the tensile strength and elongation at fracture to be superior, and Fig. 5-68 shows the superiority of the creep behavior. Note especially the increase in creep life. The creep properties are compared further in Fig. 5-69, which shows the stress-rupture properties. Note that if the columnar structure is tested with the stress normal to the long axis of the grains, so that many grains are under tensile loading, the stress-rupture behavior is identical to that of a polycrystalline, equiaxed structure.

The previous discussion has made it clear that the columnar and single-crystal structures have superior properties over the conventional poly-

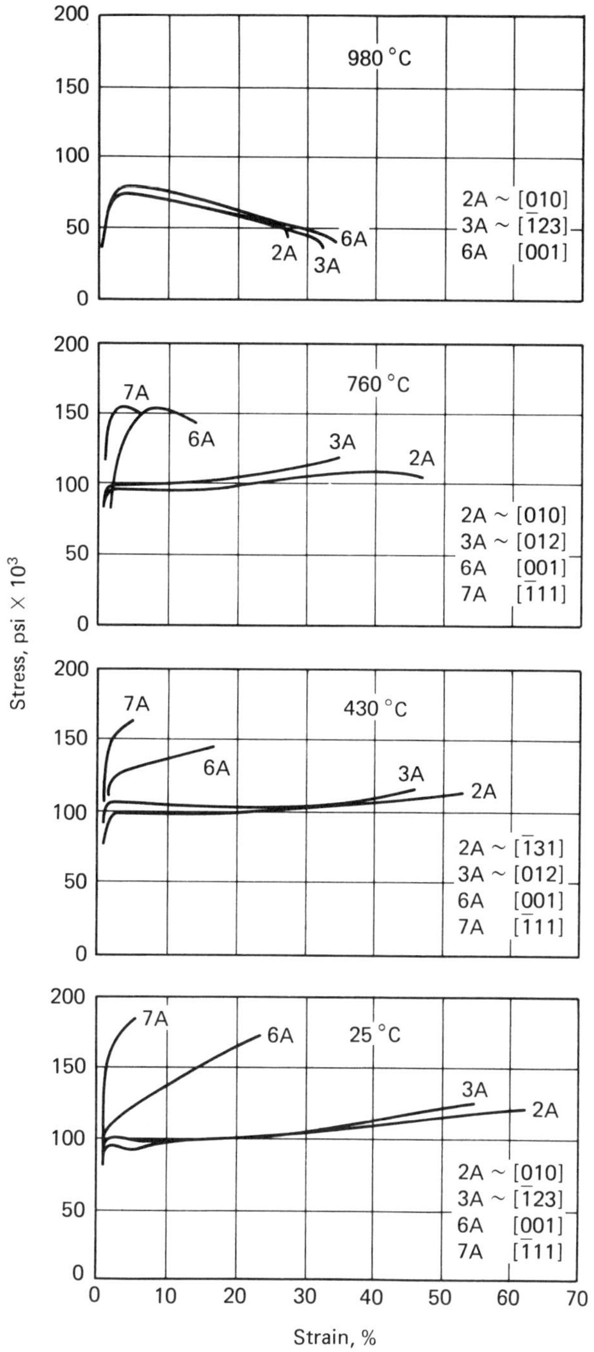

Fig. 5-63. Tensile stress-strain curves for single crystals of Mar-M200. The directions shown were parallel to the initial tensile axis in the tension test. (Adapted from B. H. Kear and B. J. Piearcey, Trans. Met. Soc. AIME, Vol 239, p 1210, 1967)

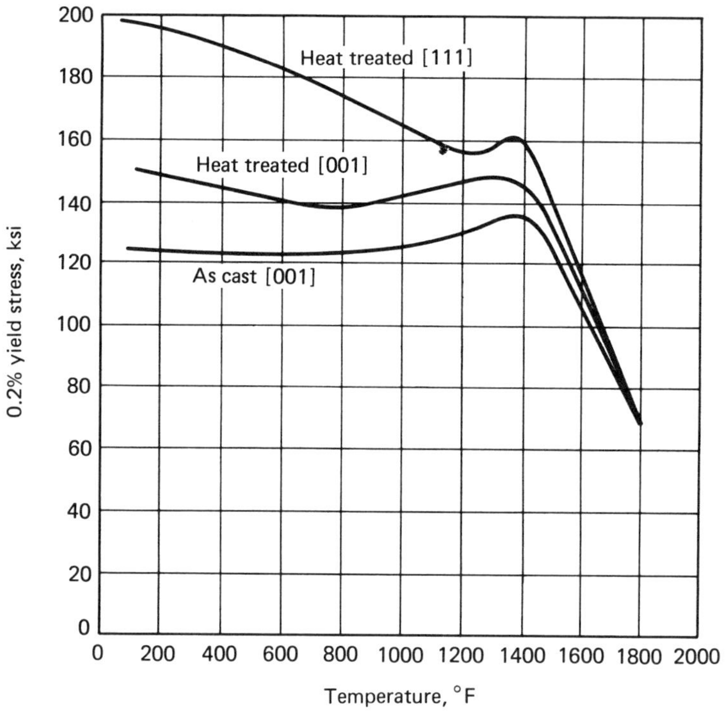

Fig. 5-64. Yield strength as a function of testing temperature for single crystals of Mar-M200. The directions shown were parallel to the tensile stress. (From B. H. Kear, S. M. Copley and F. L. VerSnyder, Proceedings of the International Conference on the Strength of Metals and Alloys, Japan Institute of Metals, 1968)

crystalline, equiaxed structures for applications such as turbine blades. We now examine how blades and vanes can be made by directional solidification.

The process of the formation of columnar grains was discussed in Chapter 2. In face-centered cubic structures, the direction of most rapid growth is ⟨100⟩. Thus, if molten metal comes into contact with a cold surface and many grains of various orientation nucleate, and if the temperature gradient is sufficiently large and unidirectional, then the grains will grow fastest along the gradient. However, as seen in Fig. 5-70, only those grains with the ⟨100⟩ axis parallel to the gradient will survive the early stages of crystal growth. What develops is a structure of columnar grains with parallel ⟨100⟩ axes, but each grain randomly oriented around its ⟨100⟩ axis. The success of this process depends upon a steep thermal gradient, usually obtained by pouring the molten alloy into a mold in which the bottom is a water-cooled copper plate.

The casting is achieved by using an investment casting process. Here a pattern has cast around it a ceramic oxide slurry which is then dried in place, and the pattern removed. This "shell" mold is then placed in a

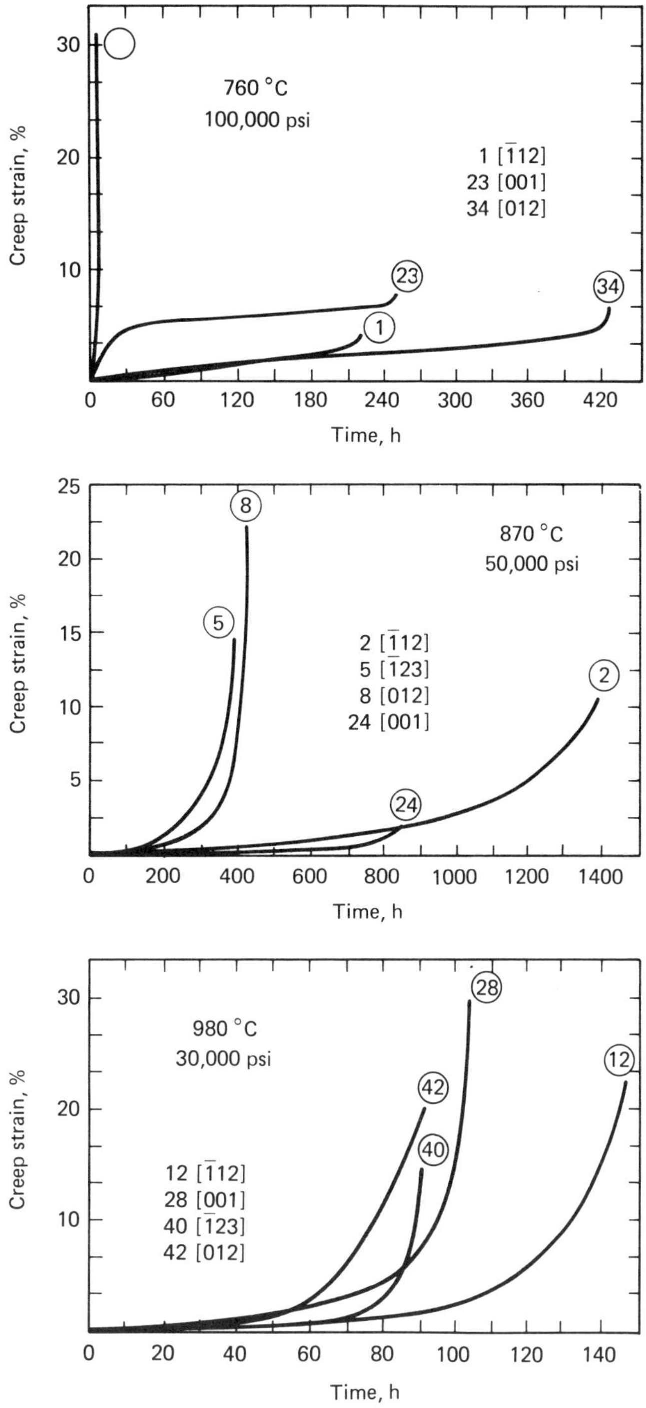

Fig. 5-65. Creep curves for single crystals of Mar-M200. The directions shown were parallel to the stress direction. (Adapted from same source, p 1213, as Fig. 5-63)

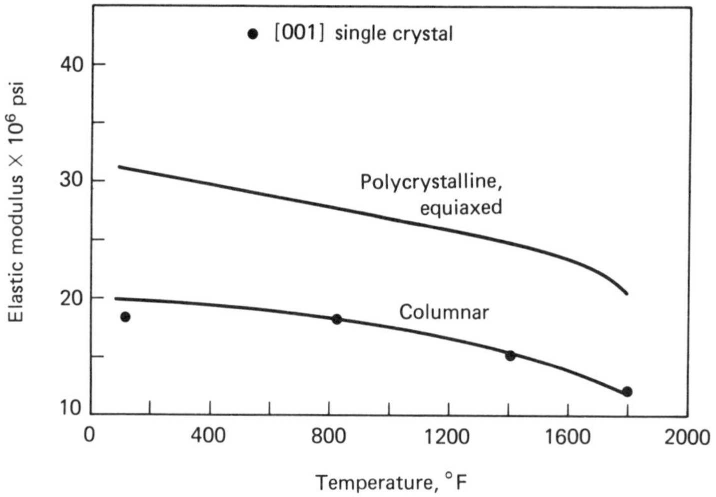

Fig. 5-66. Elastic modulus as a function of temperature for [001] oriented single-crystal, columnar and polycrystalline equiaxed structures of Mar-M200 alloy. (Adapted from B. J. Piearcey and F. L. VerSnyder, Pratt & Whitney Aircraft Report No. 66-007, 1966)

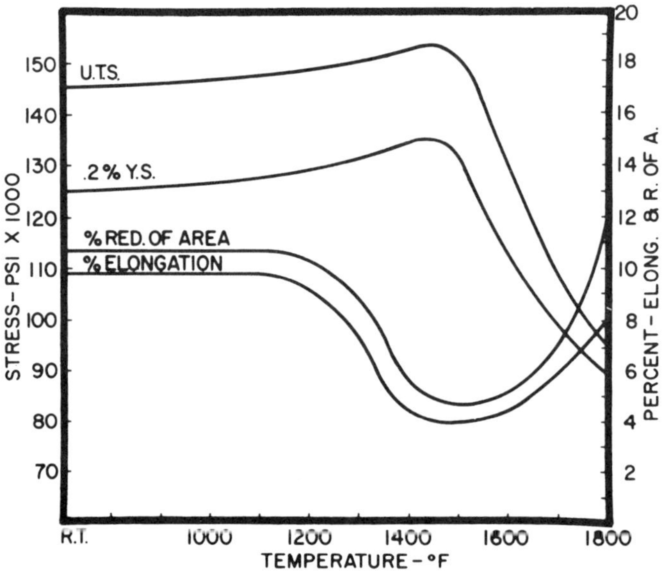

Fig. 5-67. 0.2% yield strength, tensile strength, % elongation and % reduction in area at fracture for columnar-structure Mar-M200 alloy. (From B. J. Piearcey and F. L. VerSnyder, Pratt & Whitney Aircraft Report No. 65-007, 1965)

container and the outside filled with insulating material. Around this is a heater, which is adjusted to establish the temperature gradient, and which then has the power slowly reduced to promote freezing. Thus, freezing proceeds from the bottom toward the top in a controlled fashion. Figure 5-71 shows such an arrangement, an investment shell mold and a casting containing four turbine blades.

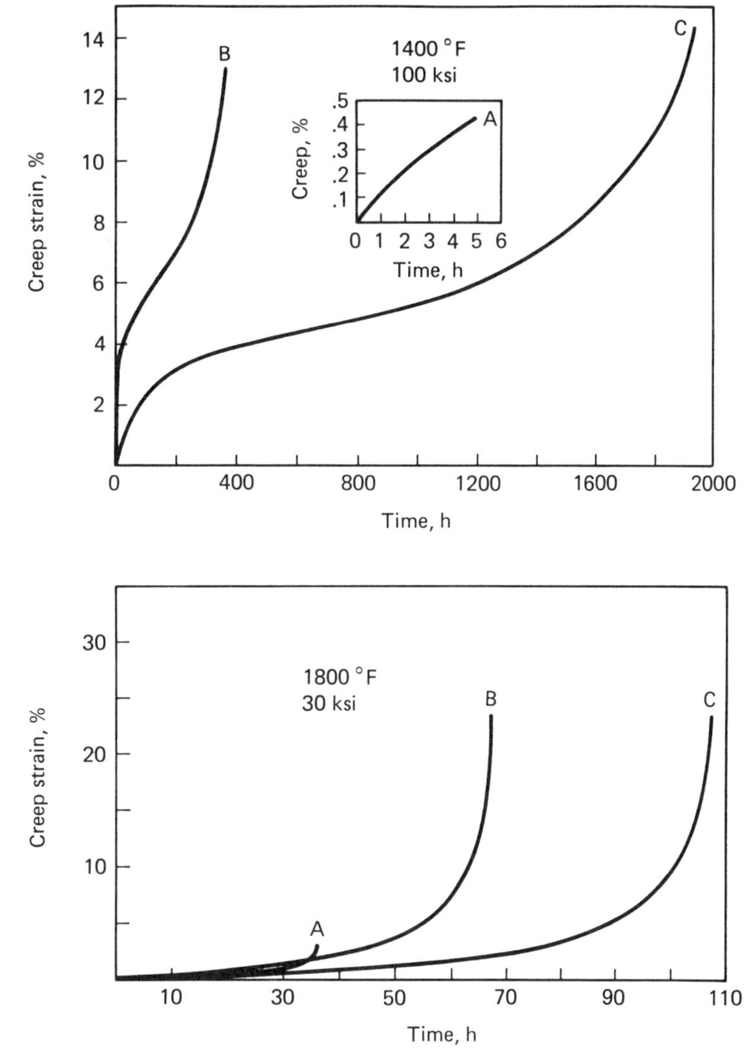

Fig. 5-68. Comparative creep curves of Mar-M200 in the polycrystalline equiaxed (A), [001] columnar (B) and [001] single-crystal (C) structures. (From same source as Fig. 5-64)

The columnar grain structure can be revealed by proper etching. Figure 5-72 shows a gas-turbine vane with the columnar grains delineated. If the structure is truly columnar, each grain should reach from the bottom to the top without interruption, which some of them clearly do.

To obtain single crystals, the mold is designed with a constrictor which allows only one crystal to grow into the cavity of the object to be made. This is illustrated in Fig. 5-73(a). The mold cavity for the object being cast is placed on top of the constrictor, as illustrated in Fig. 5-72(b). Figure 5-74 shows the type of investment shell assembly used, a shell mold, and a multiple constriction casting containing eight turbine blades.

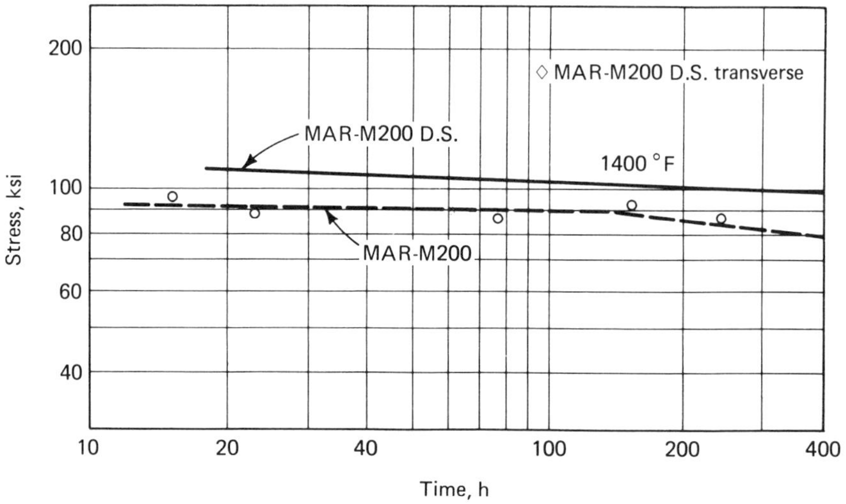

Fig. 5-69. Stress rupture data for Mar-M200 alloy, for the polycrystalline equiaxed structure (denoted Mar-M200) and the columnar structure (denoted Mar-M200 DS). Note that the creep properties are better for the columnar structure oriented with the stress axis parallel to the columnar grains (longitudinal); with the stress axis perpendicular to the columnar grains (transverse) the stress rupture data are the same as for the polycrystalline equiaxed structure. (From same source as Fig. 5-67)

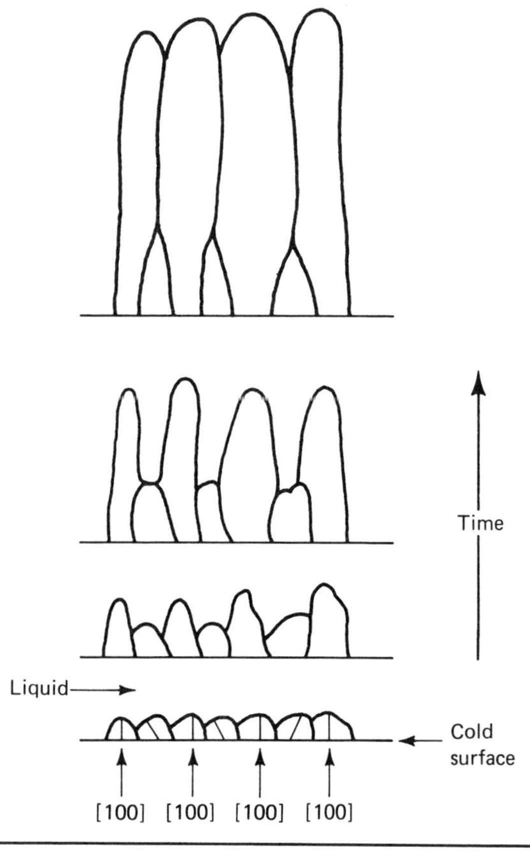

Fig. 5-70. Schematic illustration of the formation of columnar grains each having a [100] orientation. Each columnar grain is oriented randomly about the vertical axis. The [100] axis in each nucleated grain is shown by the line. Arrows indicate those with the [100] vertical.

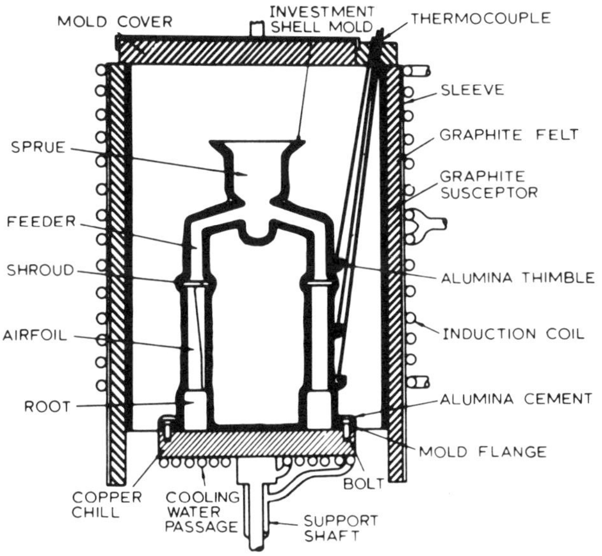

Induction-heated, investment shell mold assembly for casting directionally solidified gas-turbine blades

Four-blade silica investment shell mold

Four directionally solidified gas-turbine blades

Fig. 5-71. Illustration of a method of investment shell mold casting of directionally solidified gas-turbine blades. (Adapted from F. L. VerSnyder and M. E. Shank, Mater. Sci. Eng., Vol 6, p 213, 1970)

Figure 5-75 compares the microstructure of Mar-M200 in the conventionally cast condition (equiaxed polycrystalline structure) and in the directionally solidified condition (columnar structure). At low magnification, a typical cored dendritic structure is seen with MC carbides (not clearly resolved) and a γ-γ' eutectic constituent (too fine to resolve). Figure

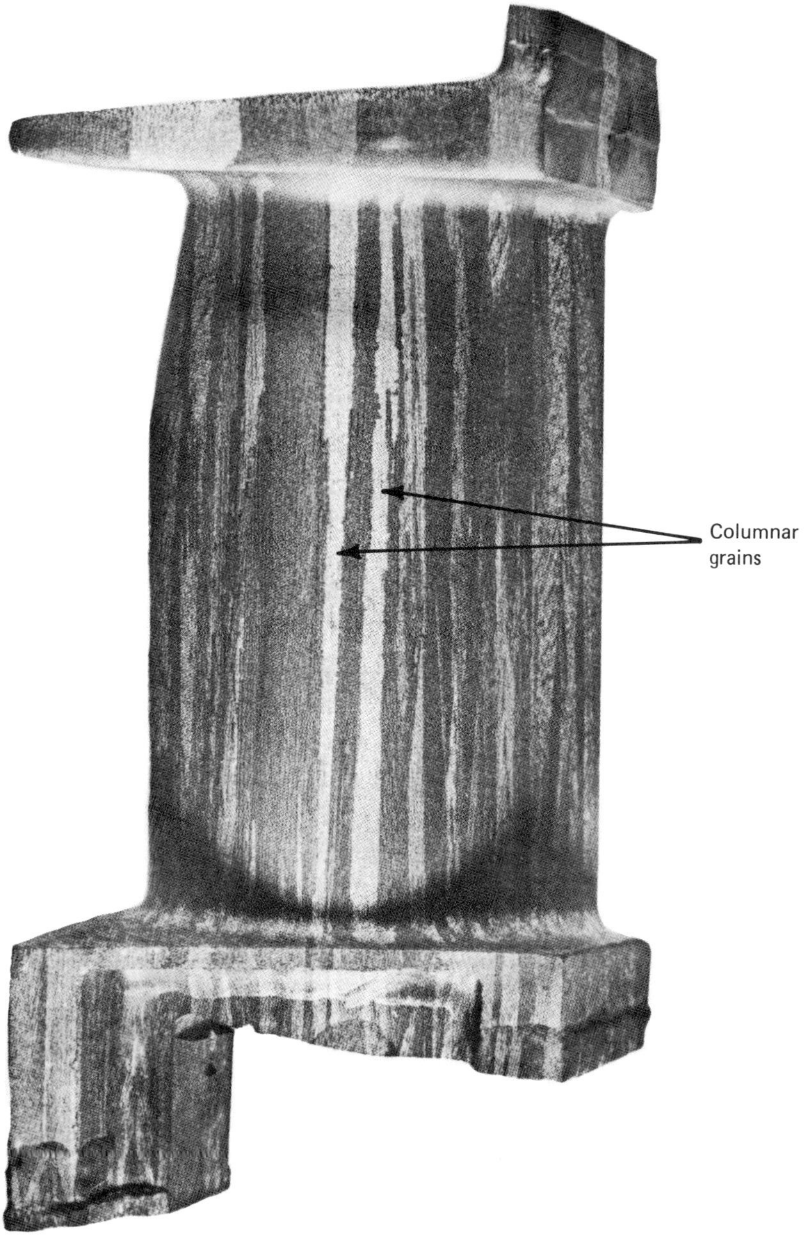

Fig. 5-72. Directionally solidified gas-turbine vane consisting of columnar grains, revealed by macroetching. (Adapted from same source as Fig. 5-67)

5-75 also shows the center of the dendrites at higher magnification, with blocky γ' in a matrix of very fine γ' in γ. The size and shape of the γ' are dependent upon the region of the dendrite observed, as the morphology is affected by the chemical composition, which varies across a dendrite because of coring.

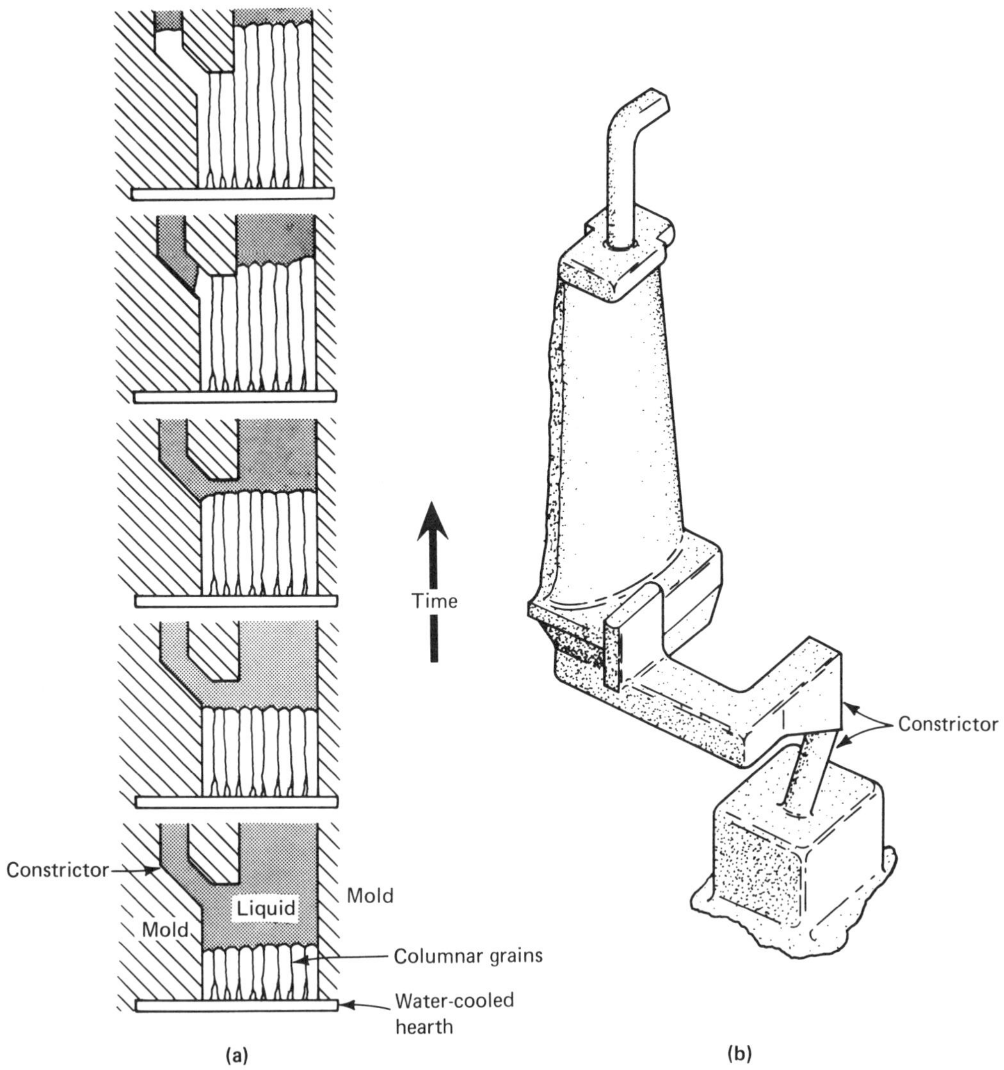

Fig. 5-73. (a) Schematic illustration of the solidification sequence which leads to a single crystal growing in the region of the constrictor. The mold cavity for the object being cast (e.g., a turbine blade) would be connected to the constrictor. (b) Constriction mold producing cast, single-crystal gas-turbine blades. (Adapted from same source as Fig. 5-71)

In the directional solidification to form a single crystal, the growing crystal solid: liquid interface covers the area of the cross section of the mold. The crystal form consists of several dendrites which are connected to form the single crystal. Each dendrite is cored (see Chapter 2), and upon etching a section through the cross section of such a casting, the dendrites are revealed. An example is shown in Fig. 5-76. The dendrite arms are parallel, which is the common alignment of the dendrites.

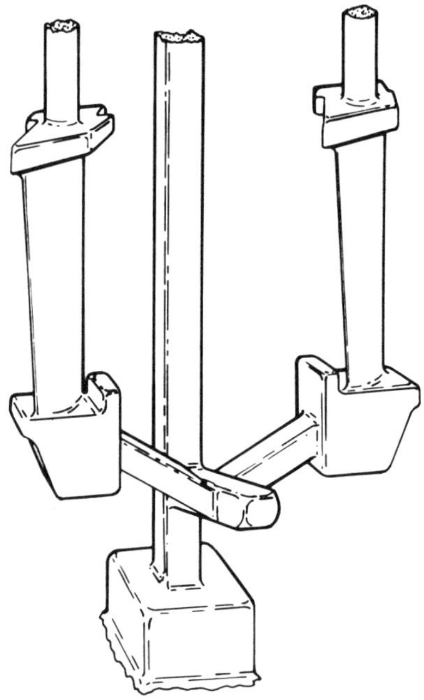

Constriction mold for bottom-filling of two blades

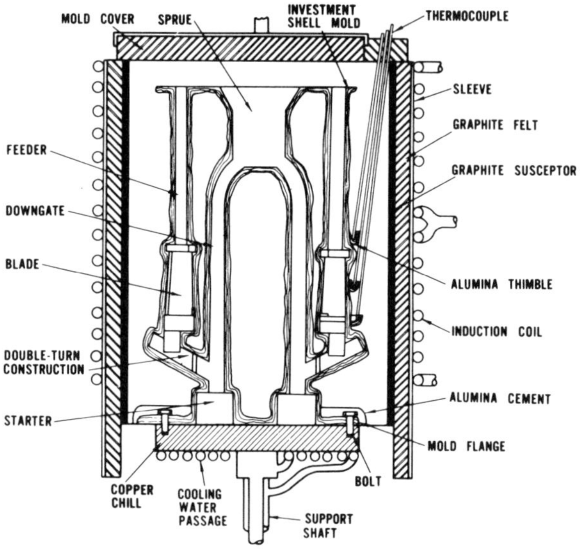

Induction heated investment shell mold assembly for casting alloy single-crystal gas-turbine blades

Multiple constriction, competitive growth investment shell mold

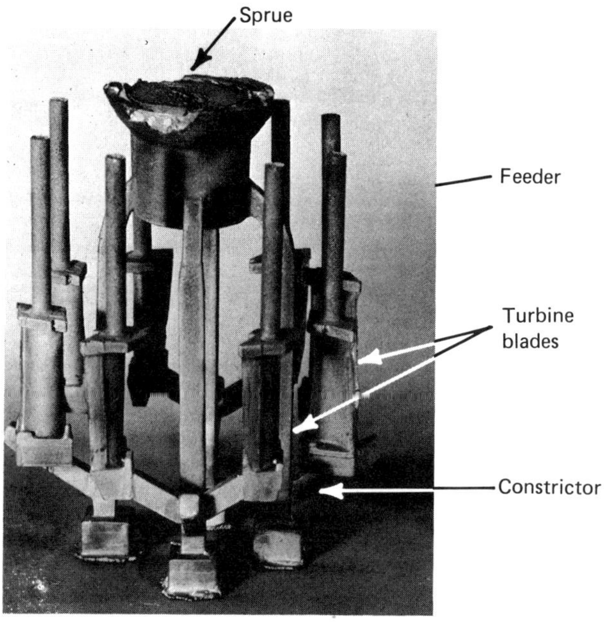

Multiple constriction, competitive growth investment casting of eight alloy single-crystal gas-turbine blades

Fig. 5-74. Illustration of use of multiple constriction investment casting technique to cast single-crystal gas-turbine blades. (Adapted from same source as Fig. 5-71)

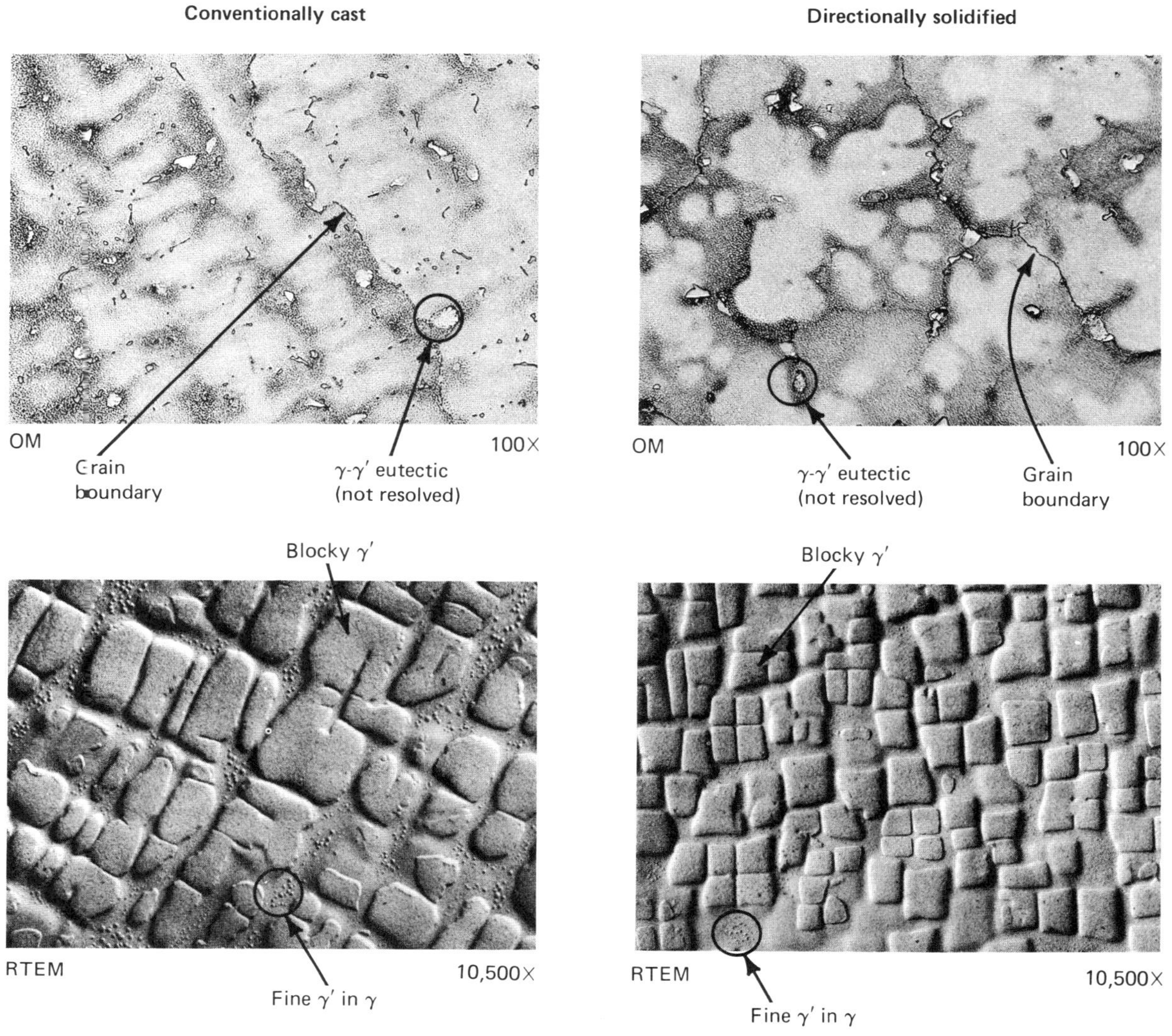

Fig. 5-75. Microstructures of conventionally cast (equiaxed polycrystalline structure) and directionally solidified (columnar structure) Mar-M200 alloy. (Adapted from B. J. Piearcey and B. E. Terkelsen, Trans. Met. Soc. AIME, Vol 239, p 1144 and 1145, 1967)

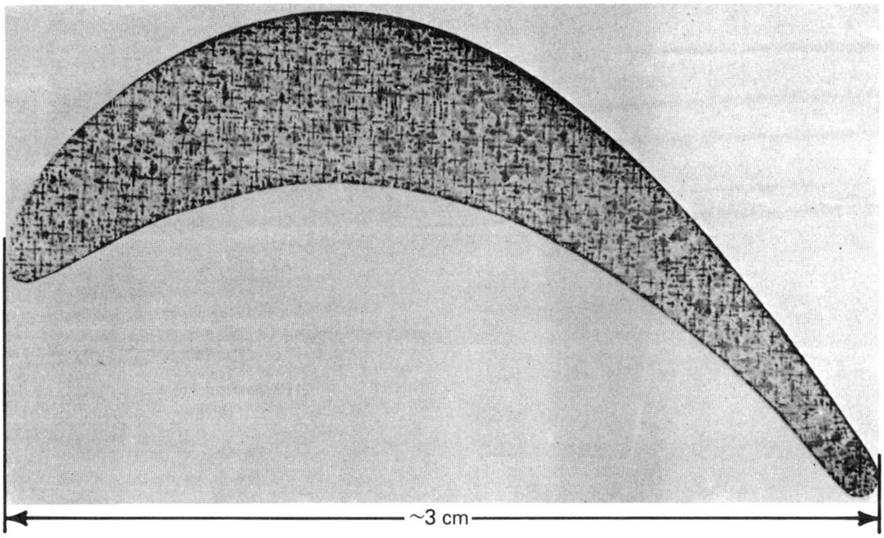

Fig. 5-76. Etched transverse section of a single crystal of Mar-M200 gas-turbine blade, showing the regular alignment of dendrites. (Adapted from same source as Fig. 5-71)

5-8. Microstructure of a Commercial Alloy

To illustrate the microstructure of a Ni-base alloy in which γ' forms, the alloy IN-100 will be examined. This is an alloy containing about 0.2% C, 10% Cr, 15% Co, 3% Mo, 5% Ti and 5% Al. The Ti and Al additions, of course, allow a considerable volume of γ' to form (Fig. 5-34 and 5-35), and this is enhanced by the Mo content (Fig. 5-32b and 5-33). The Mo, Al, Ti and Co contents favor a low γ-γ' lattice mismatch (Fig. 5-42b, 5-41, 5-43d), giving stability to the γ' structure for high-temperature applications. Also, the Mo and Co lower the coarsening rate of γ'.

The alloy IN-100 is designed to be used in the as-cast condition, or in the as-cast and then heat treated condition, but not wrought. The microstructure examined here is of the as-cast material. Figure 5-77 shows the structure at low magnification. It consists of an eutectic structure surrounded by a structure of a γ-γ' mixture. Also present are MC type carbides (mainly TiC) and $M_{23}C_6$ carbides (high in Cr). The origin of this as-cast microstructure is depicted schematically in Fig. 5-78. For this alloy, primary γ is first precipitated from the liquid, and when the eutectic temperature is reached, about 70% of the structure is primary γ, the rest being eutectic liquid. This liquid freezes to give a eutectic solid consisting of γ and γ'. Then, upon further cooling, the primary γ precipitates a considerable volume fraction (about 40%) of γ', having approximately the cuboidal shape. The microstructure is seen more clearly in Fig. 5-79. The exact morphology of the eutectic structure depends upon the cooling process (e.g., cooling rate)

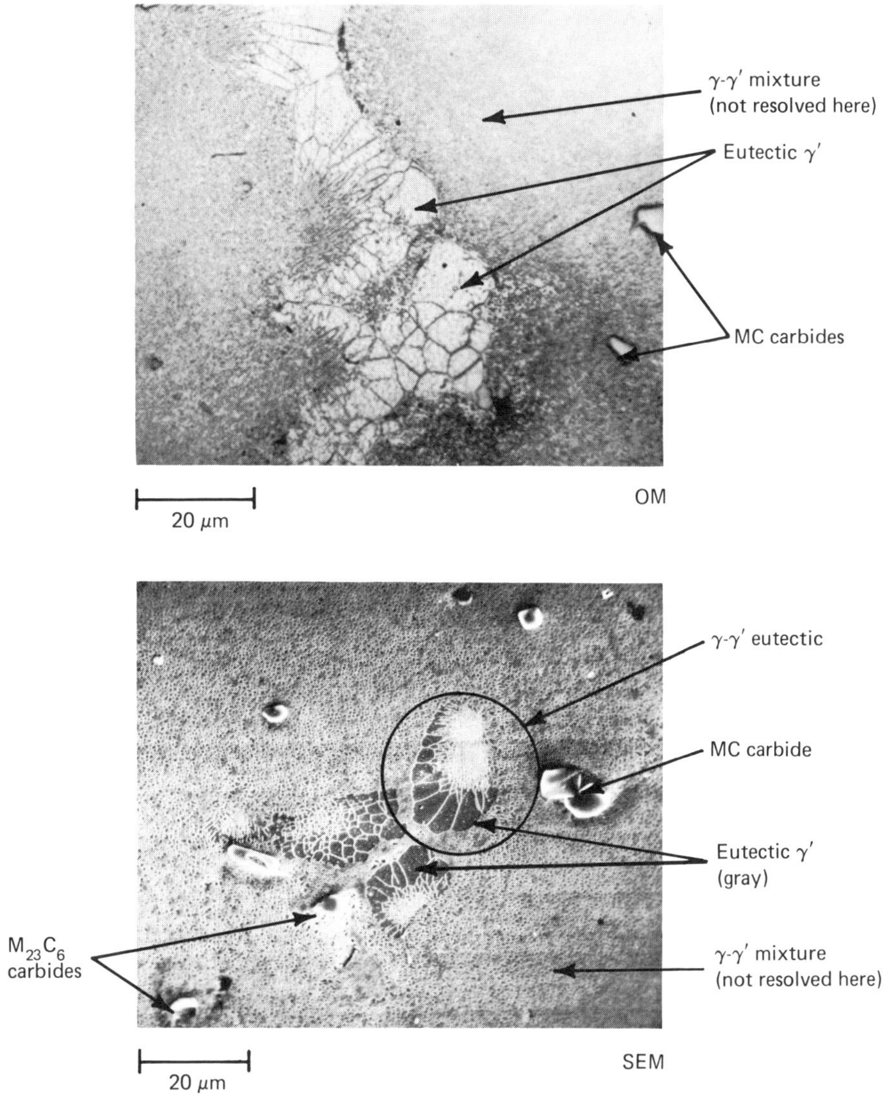

Fig. 5-77. Relatively low-magnification pictures of the microstructure of as-cast IN-100 superalloy.

during casting, which was responsible for the coarser eutectic structure that formed near the end of the eutectic reaction. Note in Fig. 5-79 the continuity of the eutectic γ and the γ in which the γ' precipitated. Note that in principle the eutectic γ should precipitate γ' upon cooling below the eutectic temperature, as did the primary γ. However, it was present as such a fine structure that only at the γ-γ' interface did some of the γ convert to γ' upon cooling, allowing the γ' to thicken, which cannot be easily detected in the microstructure.

Now we examine the microstructure of Udimet 500, an alloy which contains about 0.1% C, 19% Cr, 18% Co, 4% Mo, 3% Ti and 3% Al. For high-temperature oxidizing conditions, even an alloy with such a high Cr

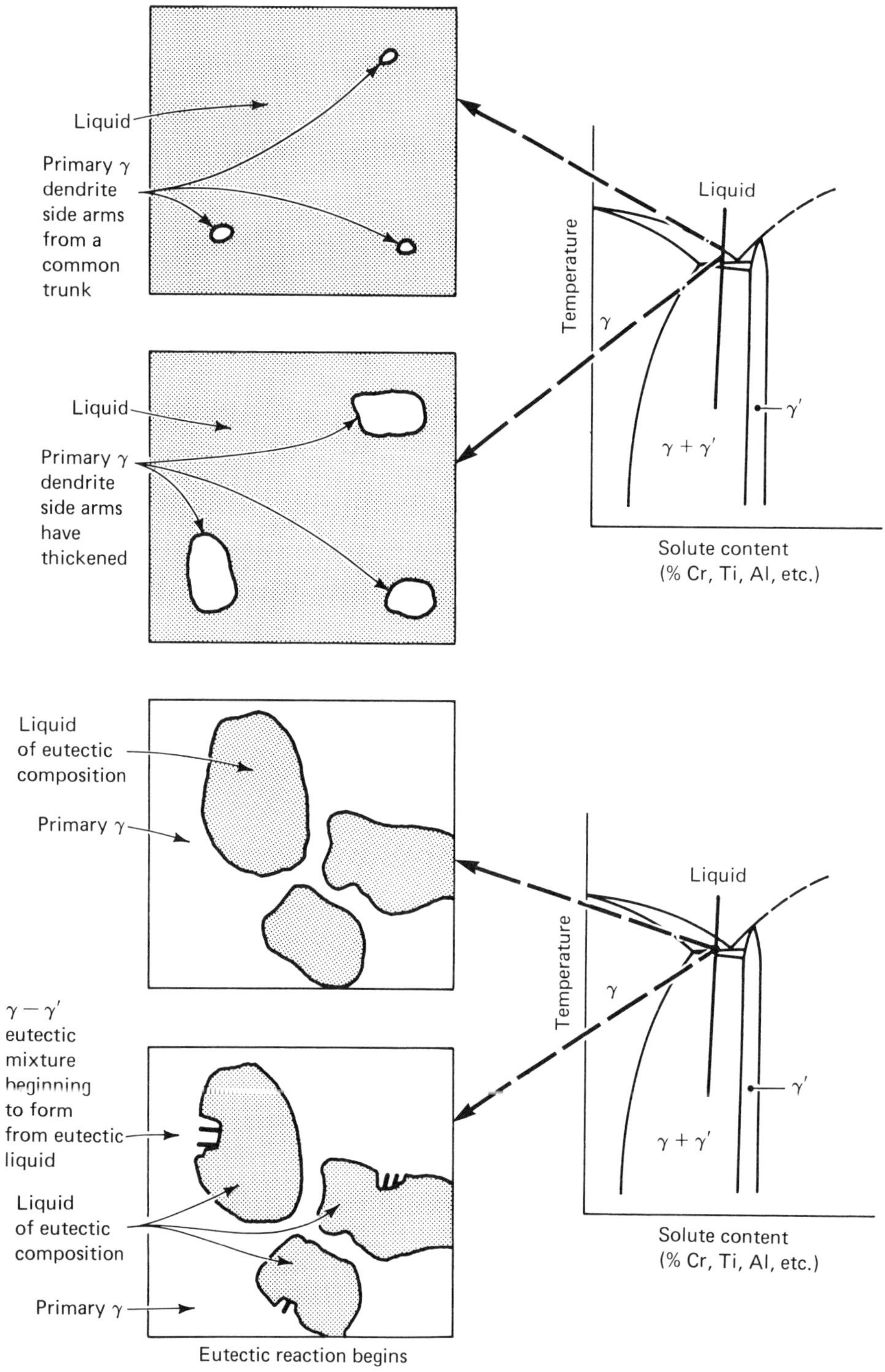

Fig. 5-78. Schematic illustration of the formation of the as-cast structure of IN-100. See text for explanation.

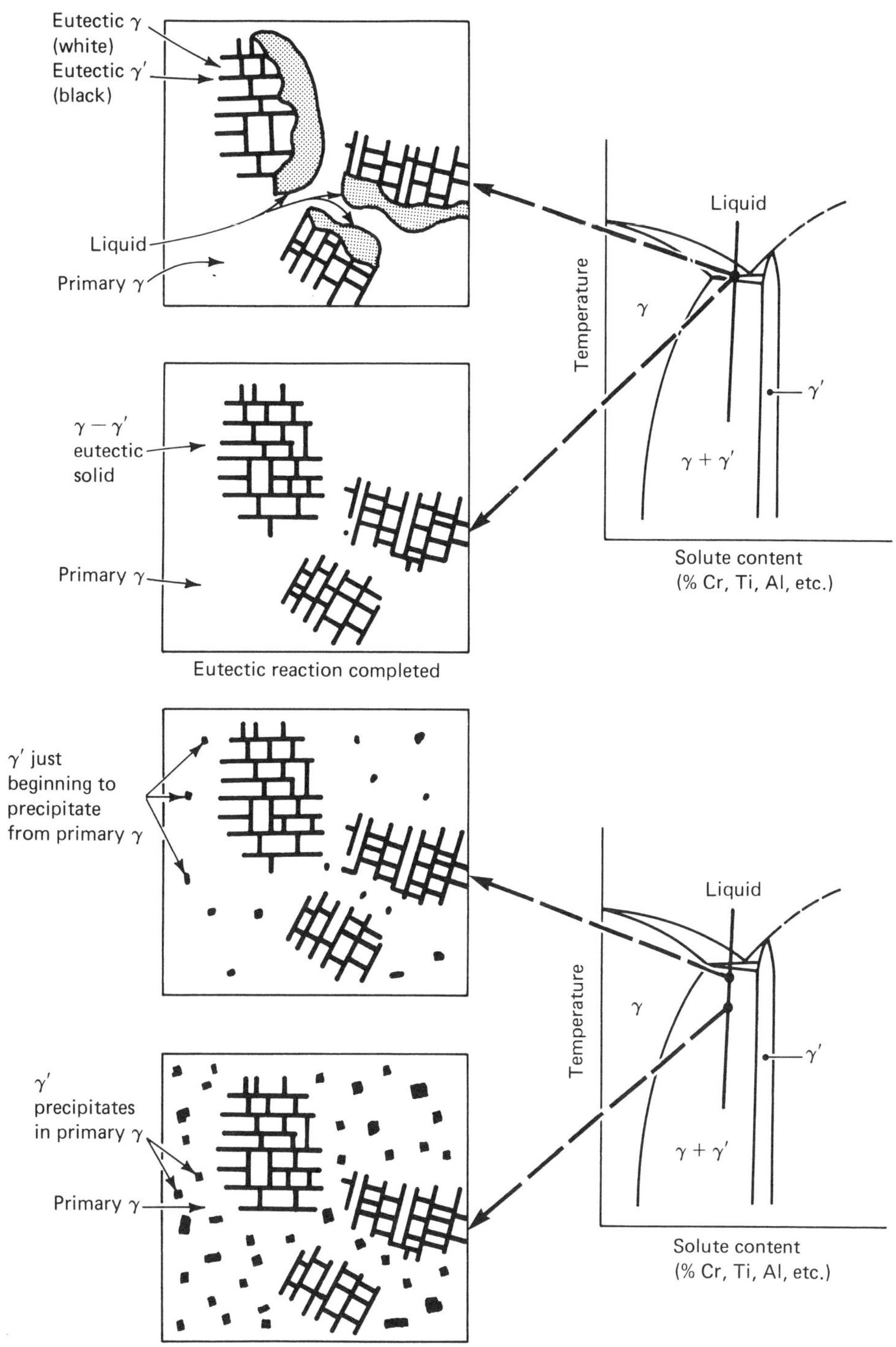
Eutectic γ
(white)
Eutectic γ′
(black)
Liquid
Primary γ
Liquid
Temperature
γ
γ′
γ + γ′
Solute content
(% Cr, Ti, Al, etc.)
γ − γ′
eutectic
solid
Primary γ
Eutectic reaction completed
γ′ just
beginning to
precipitate
from primary γ
Liquid
Temperature
γ
γ′
γ + γ′
Solute content
(% Cr, Ti, Al, etc.)
γ′
precipitates
in primary γ
Primary γ

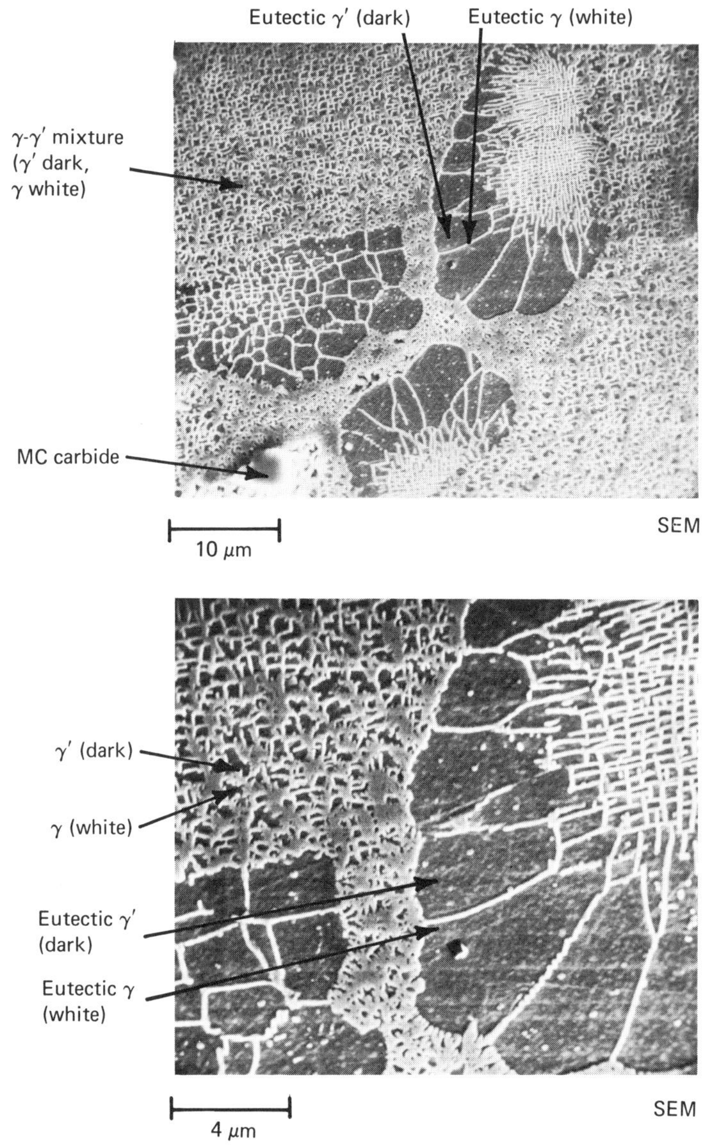

Fig. 5-79. Higher-magnification views showing distinctly the γ and γ' structure of the as-cast IN-100 alloy.

content may be coated to protect the surface. One method is to vapor deposit a layer of Al on the surface, then heat the alloy to allow the Al to react with the elements in the alloy to form an adherent, oxidation-resistant layer. The microstructure can be predicted somewhat by examination of the Ni-Al phase diagram (Fig. 5-7) at the temperature of the diffusion treatment. If the alloy is in the γ-γ' region on the phase diagram, and on the surface is initially pure Al, then upon heat treatment above 1200 °C, the Al rapidly diffuses into the alloy and reacts to form the NiAl phase. Thus, the microstructure should show a layer of NiAl on the surface, then some type of γ-γ' structure, depending on the specific composition of the alloy.

Figure 5-80 shows the microstructure of Udimet 500 after the diffusion anneal following Al coating by vapor deposition; the alloy was originally in the as-cast condition. On the surface is an Al-rich layer, then NiAl, then a carbide-rich zone, then the original alloy in which sufficient Al has diffused to form reaction products, then the unaffected structure.

Each region is shown in more detail in Fig. 5-80(b) through 5-80(e). The outer surface (Fig. 5-80c) consists of particles of Al_2O_3, which are the basis of the oxidation resistance of the coating. Its adherence to the alloy surface depends on the formation of the NiAl layer between the oxide and the alloy.

Immediately beneath this layer is the NiAl phase (which also contains in solution some Cr and Co), in which are found a few carbides and/or oxides. Then a carbide-rich layer is reached (Fig. 5-79d) in which the carbides are Cr-rich, being either $M_{23}C_6$ or M_7C_3 type carbides. Between this carbide-rich layer and the unaffected matrix is a reaction zone (Fig. 5-80d) in which $M_{23}C_6$ type carbides and the σ phase are found in a γ-γ' matrix. According to Table 5-9, increasing Al should enhance the σ-forming tendencies, so its presence in the reaction zone is not surprising. Although the diffusion time was not long (i.e., about 1 h), in some alloys σ can form in a short time (e.g., see Fig. 5-56).

Figure 5-80(e) shows the microstructure well removed from the diffusion region. It consists of $M_{23}C_6$ type carbides on the grain boundaries (along with some MC type carbides, not shown here), and the remainder of the structure is a mixture of γ and γ'. The γ' is in the form of fine cubes (about 0.2 μm in size) and makes up a considerable fraction (e.g., 40%) of the structure.

5-9. Conclusion

The evolution of four Ni-base superalloys, which illustrates some of the principles discussed, is shown in Table 5-10. The alloy IN-792 was developed to attain usable strength and hot corrosion resistance and yet minimize the amount of the expensive Cr. Beginning with the alloy 713C, introduced into application about 1955, the Cr was reduced and the Ti and Al were increased to increase the amount of γ', and hence the strength. This gave the alloy IN-100, with improved high-temperature creep strength (Fig. 5-81b) and hot corrosion resistance (Fig. 5-81c) over that of 713C. To improve the corrosion resistance further, the Cr content was increased. This would lower the strength, so that the Ti and Al content, as well as that of the refractory elements W, Ta and Nb, were increased. This increased the strength further (Fig. 5-81b) and the corrosion resistance (Fig. 5-81c). To reduce the cost of the alloy, the Cr content was reduced and the Co increased. This required increasing Ti, W and Ta to at least maintain the strength, and corrosion

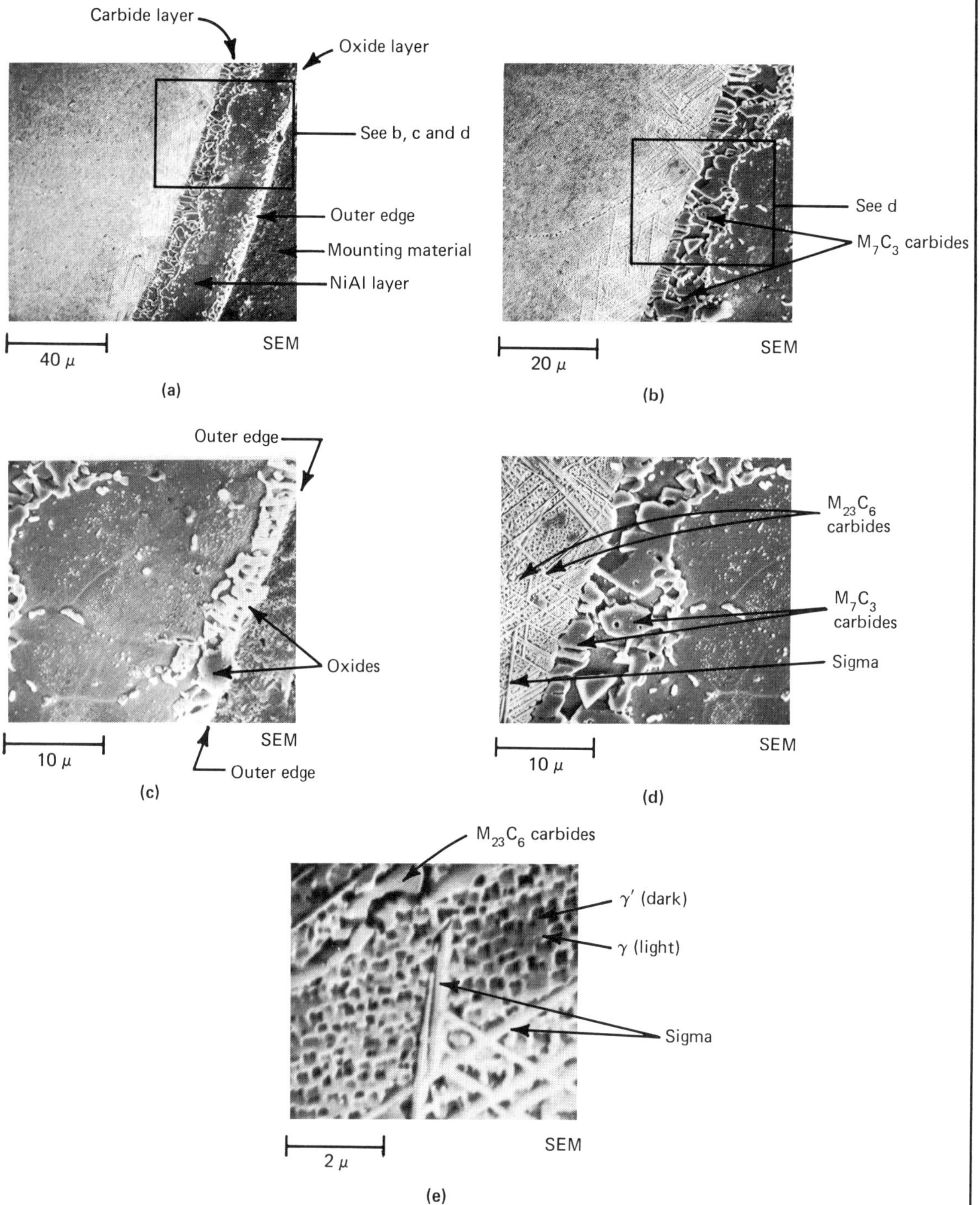

Fig. 5-80. Microstructure of coated region of Al-coated Udimet 500 alloy.

Table 5-10. Evolution of Four Ni-Base Superalloys

Approximate year of introduction	Alloy	Chemical composition, wt %: Cr	Co	Al	Ti	Mo	W	Nb	Ta	C
1955.....	713C	12.5	–	6.1	0.8	4.2	–	2.0	–	0.12
						↓ Reduce Cr, increase Ti+Al to increase amount of γ' and hence strength				
1962.....	IN-100	10.0	15.0	5.5	4.7	3.0	–	–	–	0.18
						↓ Increase Cr to increase hot corrosion resistance; to maintain strength, adjust Ti and Al, and add W, Ta and Nb				
1965.....	IN-738	16	8.5	3.4	3.4	1.8	2.6	0.9	1.8	0.17
						↓ Reduce Cr and increase Ti, W and Ta to increase strength; remove Nb to improve hot corrosion resistance				
1971.....	IN-792	12.4	9.0	3.0	4.5	2.0	4.0	—	3.9	0.21

Compiled partly from information after J. W. Schultz and W. R. Hulsizer, Metals Eng. Q., Vol 16(3), p 15, 1976.

studies showed that the Nb should be removed to assist in compensating for the corrosion loss due to the lower Cr content. This gave alloy IN-792, introduced in 1971. Note that this alloy has essentially the same creep strength as 713C, but markedly better hot corrosion resistance. Although the strength is not as good as IN-738, it has about the same corrosion resistance. Note (Fig. 5-81a) that for these alloys the approximate maximum temperature of application decreases with increased Cr content and increases with Al, Ti, Mo, W, Ta and Nb content. The steps taken in the evolution of these alloys to maintain and to increase the strength are consistent with the effects discussed in Section 5-5.

The evolution of the microstructures of the Ni-base superalloys is summarized in Fig. 5-82. The trend was to adjust the elements to increase the amount of γ', increasing the creep strength. This involved mainly increasing the Ti and Al content and decreasing the Cr content. Also, the effect of elements on the morphology of the carbides was studied, allowing the development of a distribution contributing to creep strength. In addition, methods of avoiding embrittling phases and morphologies were examined; especially critical was the development of methods to predict sigma for-

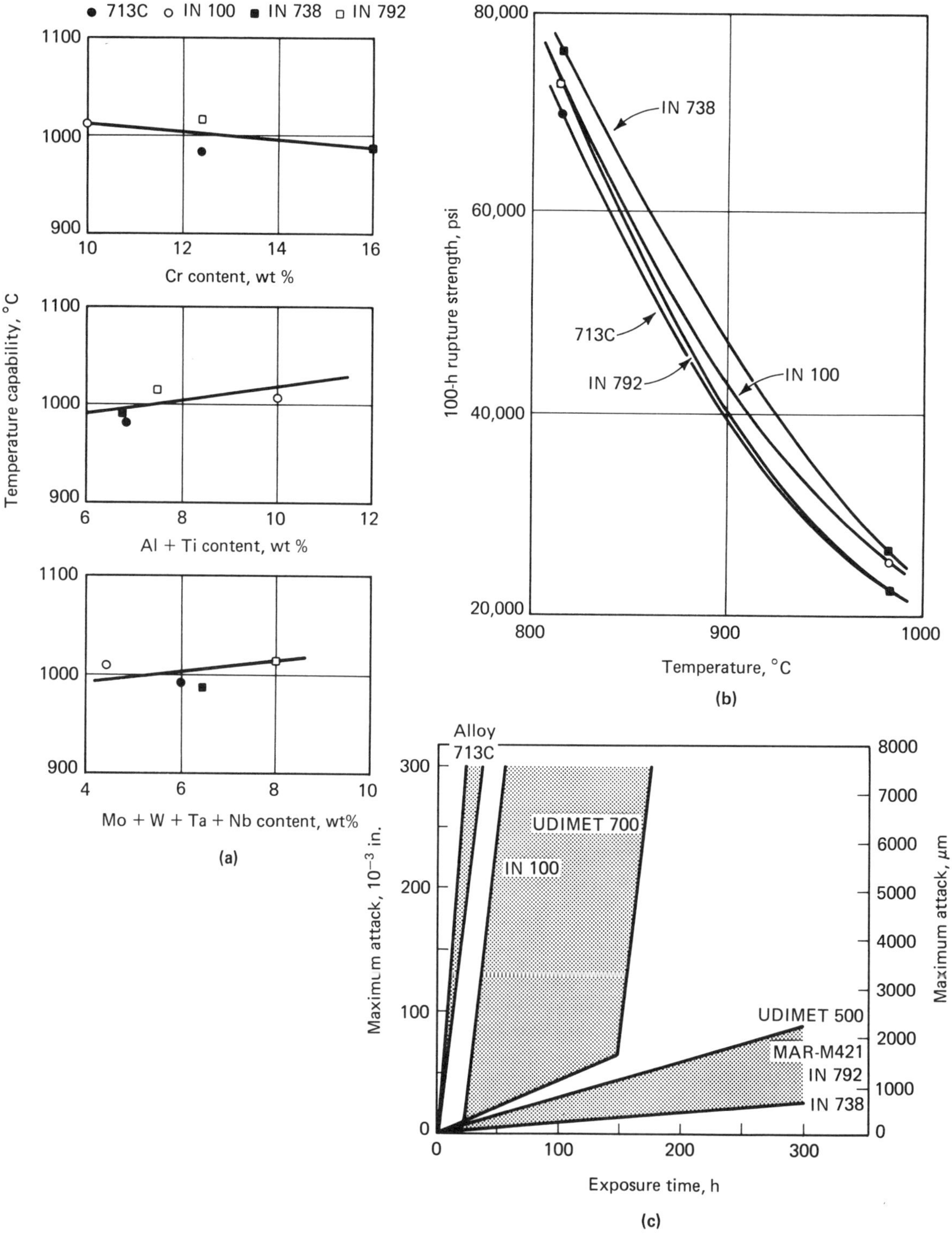

Fig. 5-81. Temperature capability, rupture strength and hot corrosion resistance for the superalloys 713C, IN-100, IN-738 and IN-792. (a) adapted from same source as Fig. 5-54; (b) adapted from Metal Progress 1978 Databook, American Society for Metals, Metals Park, Ohio, 1978; (c) from J. W. Schultz and R. W. Hulsizer, Metals Eng. Q., Vol 16(3), 1976.

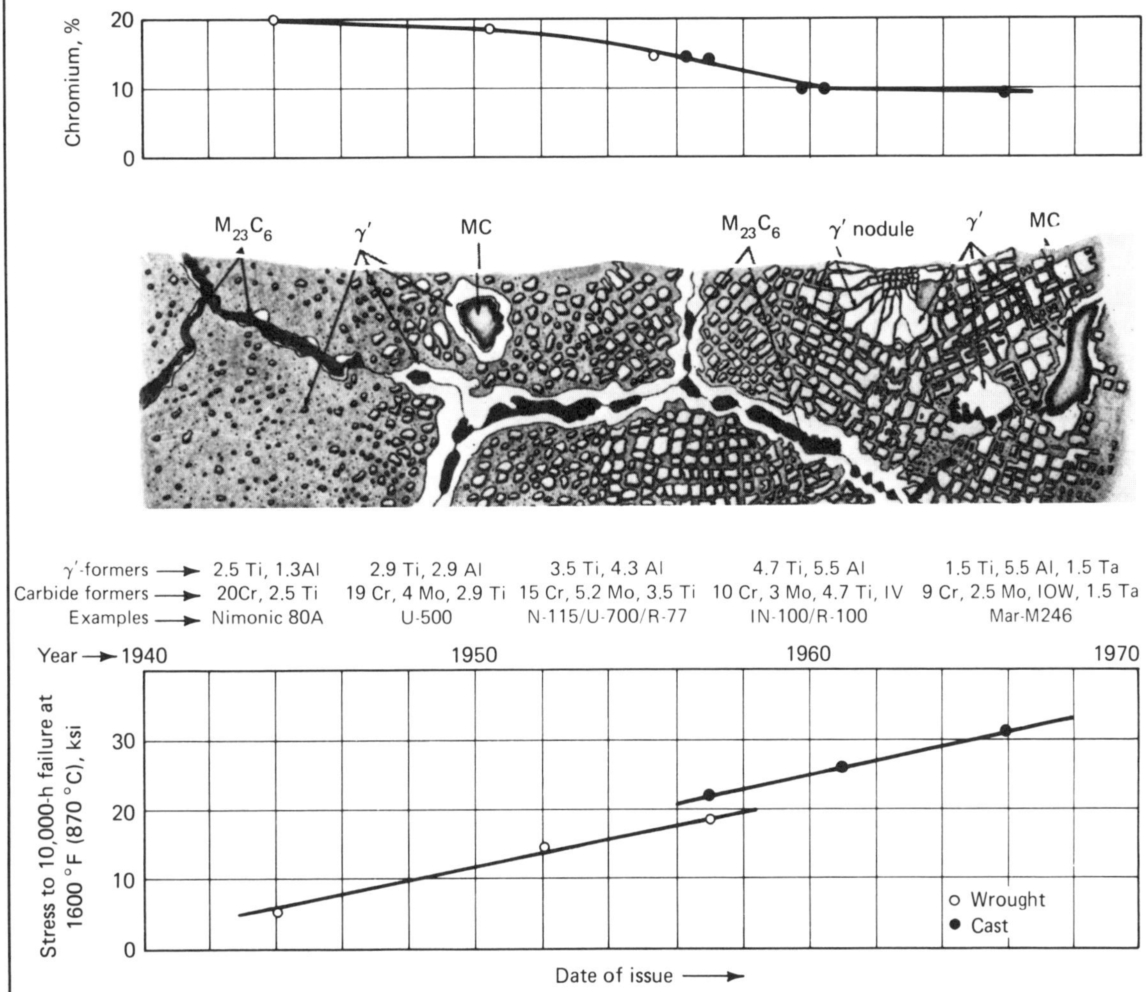

Fig. 5-82. Qualitative description of the evolution of the microstructure (schematic) and properties of Ni-base superalloys. (Adapted from same source as Fig. 5-56 and as reported by P. Elliott and A. F. Hampton, in Deposition and Corrosion in Gas Turbines, ed. A. B. Hart and A. J. B. Cutler, Wiley, New York, 1973)

mation. This evolution is summarized in Fig. 5-83. As of 1970, alloys such as Mar-M246 showed no sigma formation.

Looking back now, we can see the key role that Cr plays in these high-temperature alloys. It imparts an oxidation and corrosion resistance that cannot be developed by any other element. However, its presence generally causes weakening, and strengthening is attained by Al and Ti additions to develop the necessary γ' precipitate. The challenge, then, of alloy design is to find a proper chemical composition and structure to attain a usable balance between strength and corrosion resistance.

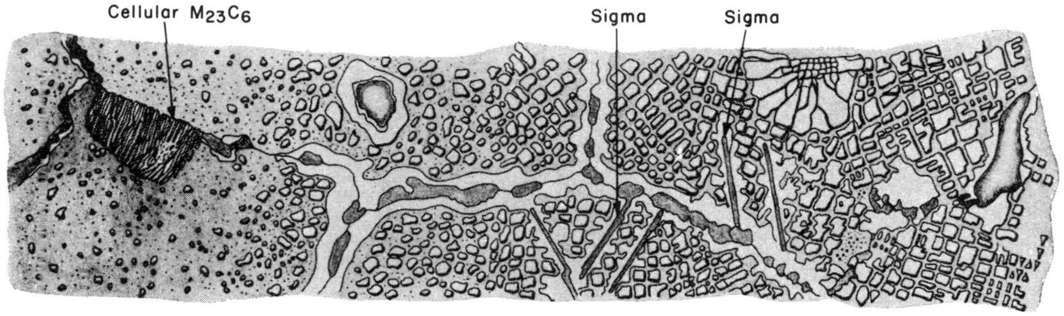

γ′-Formers →	2.5Ti, 1.3Al	2.9Ti, 2.9Al	3.5Ti, 4.3Al	4.7Ti, 5.5Al	1.5Ti, 5.5Al, 1.5Ta
Carbide Formers →	20Cr, 2.5Ti	19Cr, 4Mo, 2.9Ti	15Cr, 5.2Mo, 3.5Ti	10Cr, 3Mo, 4.7Ti, 1V	9Cr, 2.5Mo, 10W, 1.5Ta
Examples →	Nimonic 80A	U-500	N-115/U-700/R-77	IN-100/R-100	Mar-M246

Year → 1940		1950	1960	1970
Undesirable phase or structure	Cellular $M_{23}C_6$	None	Sigma phase	None yet
Characteristics	$M_{23}C_6$ grown in cells with γ interlayers	—	Hard TCP phase usually occurs as plates similar to μ or Laves	—
Effect on alloy	Shortened rupture life	—	Shortened rupture life; low-temperature brittleness	—
Cause	Poor control of microstructure	—	Poor control of alloy chemistry	—
Cure	Improve heat treatments	—	Control chemistry through "Phacomp"	—

Fig. 5-83. Common undesirable phases or structure in Ni-base superalloys; micrograph at 5000×. (From same source as Fig. 5-56)

References

A Contemporary View of Nickel-Base Superalloys, C. T. Sims, Journal of Metals, Vol. 18, p 1119 (1966).

Strengthening Mechanisms in Nickel-Base Superalloys, R. F. Decker, in Steel-Strengthening Mechanisms, Climax Molybdenum Co., Greenwich, Conn. (1969).

Microstructure of Nickel-Based Superalloys, G. P. Sabol and R. Stickler, physica status solidi, Vol 35, p 11 (1969).

The Development of Columnar Grain and Single Crystal High Temperature Materials Through Directional Solidification, F. L. VerSnyder and M. E. Shank, Materials Science and Engineering, Vol 6, p 213 (1970).

The Metallurgical Design of a Superalloy, R. G. Davies and T. L. Johnston, in Ordered Alloys — Structural Applications and Physical Metallurgy, ed: B. H. Kear, C. T. Sims, N. S. Stoloff and J. H. Westbrook, Claitor's Publishing Division, Baton Rouge (1970).

Nickel Alloys — The Heart of Gas Turbine Engines, C. T. Sims, ASME Reprint, 70-GT-24 (1970).

The Superalloys, ed. C. T. Sims and W. C. Hagel, Wiley, New York (1972).

The Nimonic Alloys and Other Nickel-Base High-Temperature Alloys, 2d Ed., ed. W. Betteridge and J. Heslop, Edward Arnold Ltd, London (1974).

Materials for Gas Turbines, R. J. Glenny, J. E. Northwood and A. Burwood-Smith, International Metallurgical Reviews, Vol 20, p 1 (1975).

Changes and Evolution of Aircraft Engine Materials, E. F. Bradley and M. J. Donachie, Metals Engineering Quarterly, Vol 15(2), p 55 (1975).

A Review of the Strengthening Mechanisms in Iron and Nickel Based Fe-Ni-Cr Alloys Used at High Temperature, G. Wallwork and J. Croll, Reviews on High-Temperature Materials, Vol 3(2), p 69 (1976).

Corrosion-Resistant Nickel-Base Alloys for Gas Turbines, J. W. Schultz and W. R. Hulsizer, Metals Engineering Quarterly, Vol 16(3), p 15 (Aug 1976).

The Physical Metallurgy of Superalloys and Titanium Alloys, C. Hammond and J. Nutting, Metal Science, Vol 11, p 474 (1977).

Microstructural Objectives for High-Temperature Alloys in Advanced Energy Systems, C. T. Sims, in MiCon 78: Optimization of Processing, Properties and Service Performance through Microstructural Control, ed. H. Abrams, G. N. Maniar, D. A. Nail and H. D. Solomon, American Society for Testing and Materials, Philadelphia, Pa. (1979).

Physical Metallurgy and Effects of Process Variables on the Microstructure of Wrought Superalloys, D. R. Muzyka, in MiCon 78: Optimization of Processing, Properties and Service Performance through Microstructural Control, ed. H. Abrams, G. N. Maniar, D. A. Nail and H. D. Solomon, American Society for Testing and Materials, Philadelphia, Pa. (1979).

CHAPTER 6

Cobalt-Base Alloys

The melting point of cobalt is 1495 °C and that of nickel is 1455 °C; this is an indication that the high-temperature properties of cobalt-base alloys might be comparable to those of nickel-base alloys. Cobalt is face-centered cubic (FCC) above 417 °C, and hexagonal close-packed (HCP) below this temperature, although the rate of formation of the HCP structure is low. Thus, at high temperature cobalt has the same FCC structure as nickel, and because of the similarity of the atomic structure of the two elements, it is reasonable to search for alloying additions such that the ordered γ' phase, so important in strengthening nickel-base alloys, can be formed. However, few cobalt-base systems form γ' and do not develop usable strengthening by this route. Figure 6-1 compares the creep properties of nickel-base and cobalt-base alloys, where it is seen that up to about 900 °C the nickel-base alloys are superior, as the cobalt-base alloys cannot develop the γ' precipitate. However, above about 900 °C the cobalt-base alloys are comparable in strength to nickel-base alloys, and some cobalt-base alloys may be superior. The advantage of cobalt-base alloys occurs because in the nickel-base alloys the strengthening γ' phase begins to dissolve at these high temperatures. Also, cobalt-base alloys are strengthened by the formation of carbides, which are more stable than in nickel-base alloys, where they tend to dissolve at high temperatures. Thus, in the high-temperature range, cobalt-base alloys may be used in preference to nickel-base alloys.

The number of cobalt-base alloys available is more limited than that of nickel-base alloys, as seen by comparing the tables in Appendix 3 and Appendix 4. Note that most of the cobalt-base alloys have chromium contents around 20-30%, and little or no aluminum or titanium. They also have a much higher carbon content than the nickel-base superalloys.

In this chapter, we will examine alloying additions and heat treatments required to develop corrosion resistance and strength in cobalt-base superalloys.

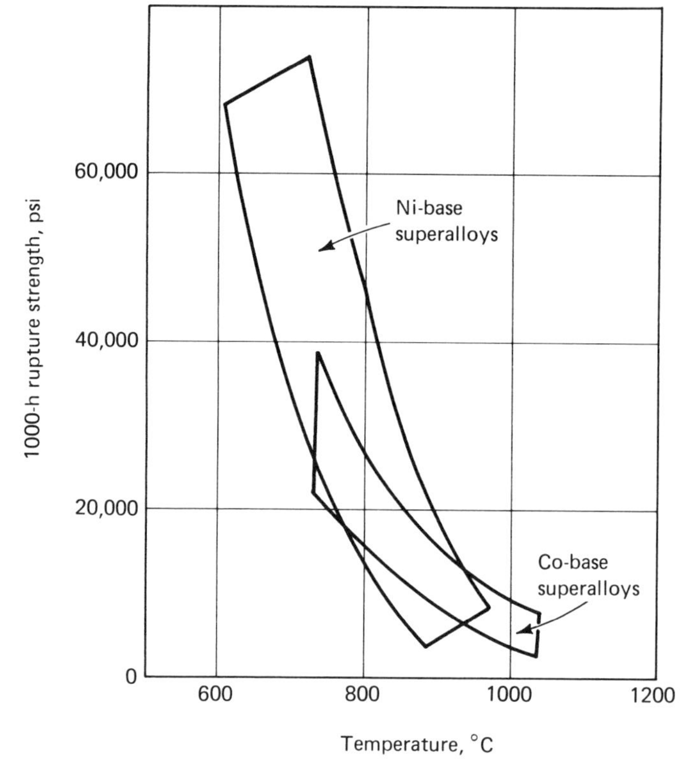

Fig. 6-1. Comparison of the 1000-h rupture strength for Ni-base and Co-base superalloys. (Adapted from Cobalt Monograph, Centre d'Information du Cobalt, Brussels, 1960)

6-1. Oxidation of Cobalt-Base Alloys

The most important element to add to cobalt to improve oxidation resistance is chromium, as in the case of nickel-base alloys. Figure 6-2 shows that the critical Cr content is between 20 and 30%, where the scaling constant for parabolic oxidation is reduced about four orders of magnitude from that of pure Co. (Co-30Cr alloy is the base for a widely used alloy for surgical implants and dental applications. This alloy is extremely corrosion resistant at 25 °C.) For most of our discussion, then, we will center attention on Co-Cr alloys, as the commercial Co-base superalloys all contain Cr, most of them in the range of 20-30% Cr (see Appendix 3).

The effect of elements other than Cr on the static and cyclic oxidization and hot corrosion resistance must be examined. The binary alloys of Co-5Si and Co-10Al show good oxidation resistance, with the Co-Al alloy better than Co-Cr alloys. However, the binary Co-Al alloys do not have sufficient high-temperature strength to consider this combination as a base for developing Co alloys. It has been found that in Co-20Cr alloys, tested from 900 to 1100 °C in air and in oxygen for up to 240 h, Fe and Ni have no effect,

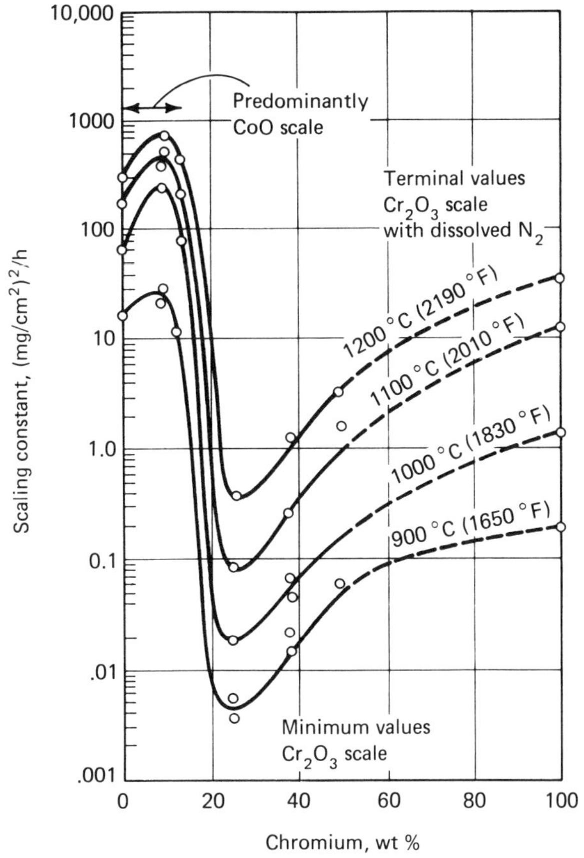

Fig. 6-2. Parabolic oxidation constant for Co and Co-Cr alloys oxidized in static air. (From C. A. Phalnikar, E. B. Evans and W. M. Baldwin, J. Electrochem. Soc., Vol 103, p 429, 1956, reprinted by permission of the publisher, The Electrochemical Society, Inc.)

but Mn increases the spalling tendency. In some situations, Mo, V and W can be detrimental. It is especially important to note that the hot corrosion resistance of Co-Cr–base alloys generally is not affected by the presence of sulfur in the hot gas as that of the Ni-base alloys is.

As in Ni-base alloys, Cr is the key element, and other elements, added for strengthening, must be examined carefully in terms of their effect on oxidation and hot corrosion resistance.

6-2. Strengthening of Cobalt-Base Alloys

Solid Solution Strengthening

As we did with Ni-base alloys, we can examine the relationship between the difference in atom size between Co and solute atoms, and the maximum solubility, to obtain an indication of the magnitude of the solid solution strengthening effect to be expected. This information is shown in Table 6-1, which shows that Cr, W, Ta, Nb and Mo look promising. It is best to examine the solubility in the ternary alloys of 30% Cr, and this shows that W and Mo are promising. (No data seem to be available for the effect of Ta and Nb.)

The effect of various solutes on the hardness is illustrated in Fig. 6-3 and 6-4, where it is seen that there is general agreement with the information in Table 6-1. Figure 6-5 shows the strong hardening effect of Mo, Ta and W on Co-Cr-X ternary alloys. It would be expected that Nb, being similar to these three elements, would have a similar effect.

Table 6-1. Approximate Atom Diameter Size Difference and Approximate Solubility in Co and Co-30Cr Alloys of Several Solutes

Solute	Approximate atom diameter size difference, % $(d_{Co} - d_M)/d_{Co}$	Approximate solubility in Co at 1000 °C, wt %	Approximate solubility in Co-30Cr alloys at 1200 °C, wt %
Chromium	−0.1	35	—
Nickel	+0.2	100	100
Iron	+0.6	100	30
Manganese	+10	76	—
Tungsten*	−10	25	15 (at 25°C)
Tantalum*	−15	12	—
Molybdenum*	−9	20	100
Niobium*	−15	3	—
Titanium	−17	8	—
Aluminum	−15	6	—
Carbon	+40	0.3	0.3

Note. Solutes marked with an asterisk (*) show the best combination of large size difference and high solubility and should be the most potent solid solution strengtheners.

Precipitation of Gamma Prime

Now we turn our attention to the possibility of precipitation hardening. From what we saw of the potent strengthening of Ni-base alloys upon precipitation of $Ni_3Al(\gamma')$, we would like to develop an alloy in which this type of structure precipitates, but based, of course, on Co_3Al or Co_3Ti. Figure 6-6 shows that neither the Co-Al nor the Co-Ti system will precipitate a γ'-type phase. In the Co-Al system, the precipitate from the Co-rich terminal solid solutions is based on CoAl (ζ phase), and in the Co-Ti it is based on Co_3Ti (γ). However, we know that Cr will be required for oxidation resistance, so that the Co-Cr-X ternary phase diagrams should be examined, but these have not been determined. The data in Fig. 6-7 do show that precipitation hardening occurs in Co-Cr-Ti alloys, and the precipitate is Co_3(Cr, Ti) of the γ' structure. Note that the strength is not quite as high as that for Ni-base alloys

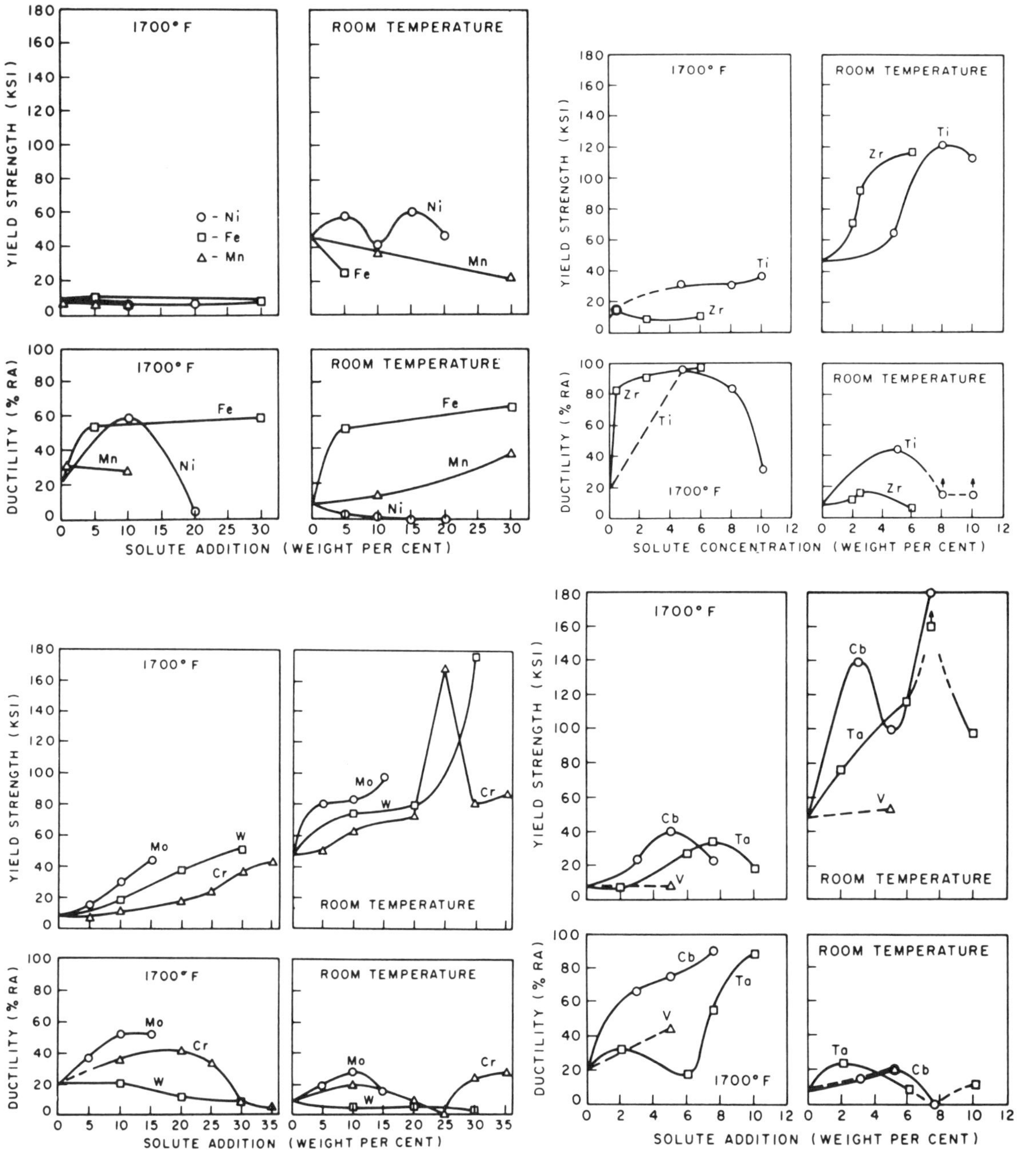

Fig. 6-3. Effect of addition of several solutes on the tensile mechanical properties of Co at 25 and 925 °C (1700 °F). (From J. J. Rausch, J. B. McAndrew and C. R. Simcoe, in High Temperature Materials II, ed. G. M. Ault, W. F. Barclay and H. P. Munger, Gordon and Breach, New York, 1963)

in which γ' precipitates. In this Co-base alloy, the γ' dissolves at a lower temperature than it does in Ni-base alloys. Note that the binary alloy Co-8.5Ti shows considerable strength, but the lack of Cr precludes its use in an oxidizing environment.

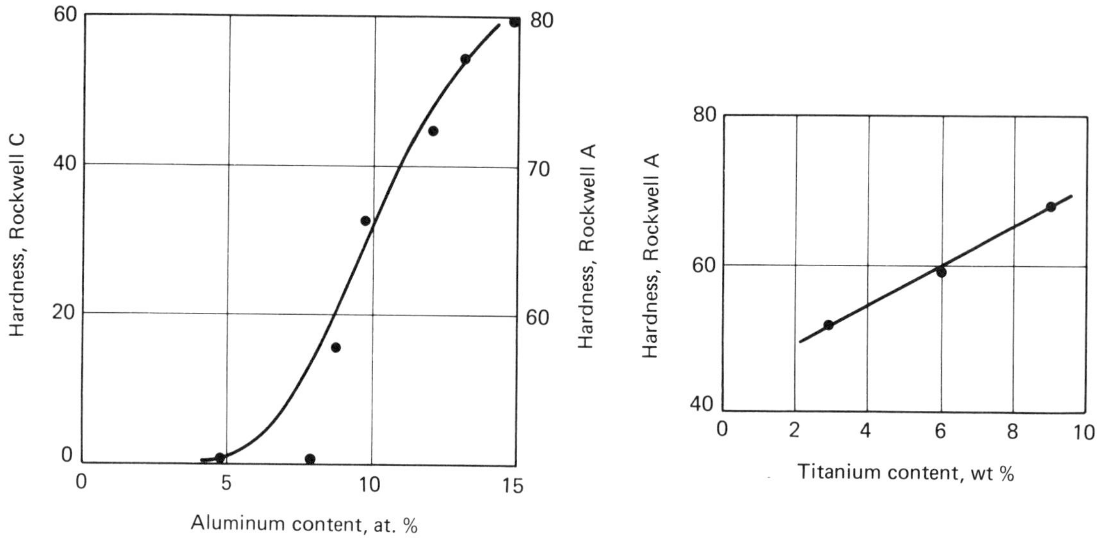

Fig. 6-4. Effect of the Al and Ti content of binary Co-base alloys on the hardness at 25 °C. (Adapted from G. A. Fritzlen, W. H. Faulkner, B. R. Barrett and R. W. Fountain, in Precipitation From Solid Solution, American Society for Metals, Metals Park, Ohio, 1958)

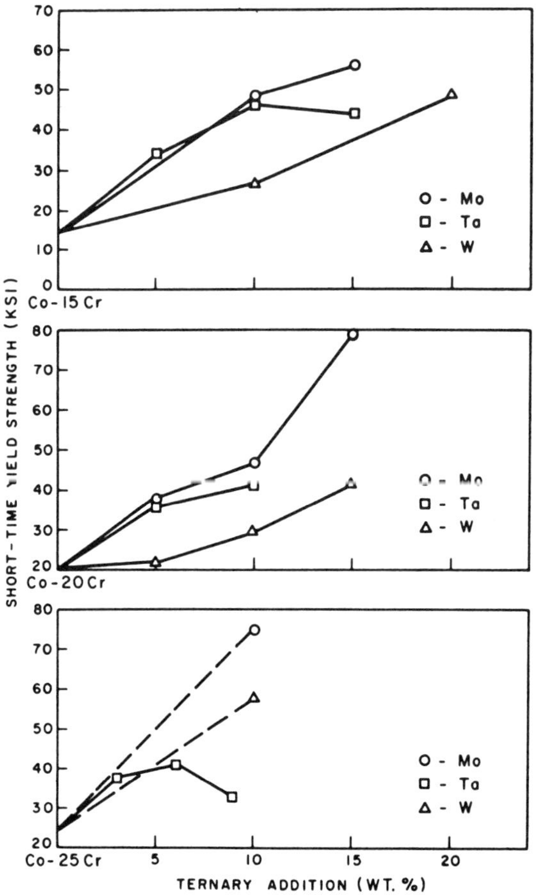

Fig. 6-5. Effect of Mo, Ta and W on the yield strength at 925 °C for ternary alloys of Co-Cr-X. (From same source as Fig. 6-3)

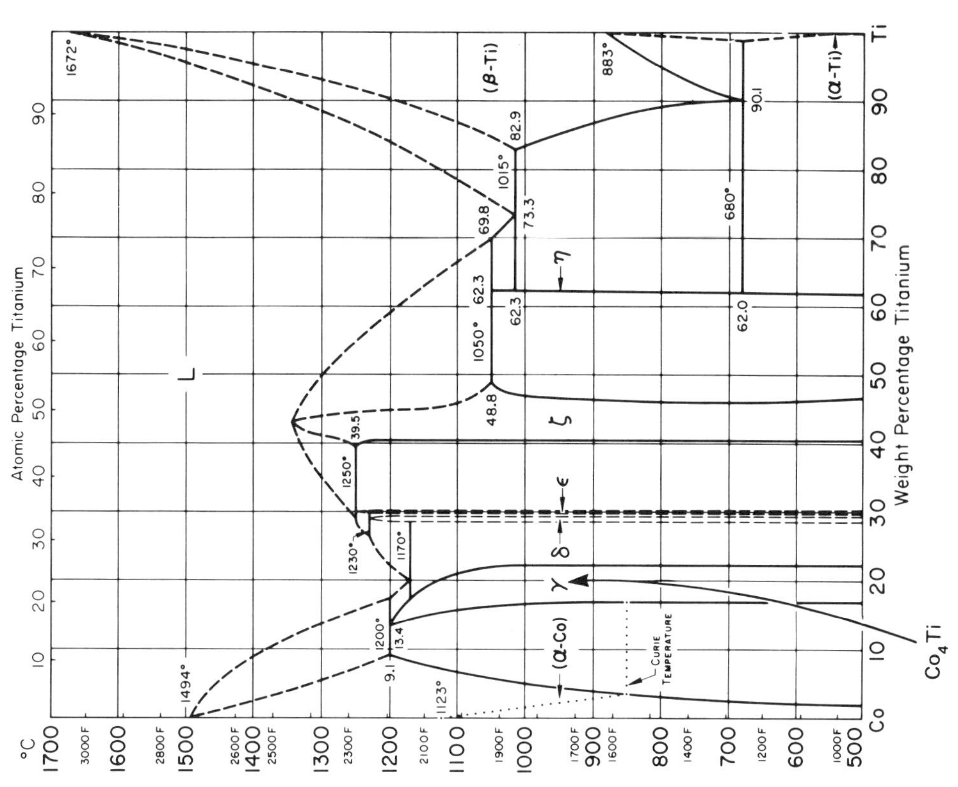

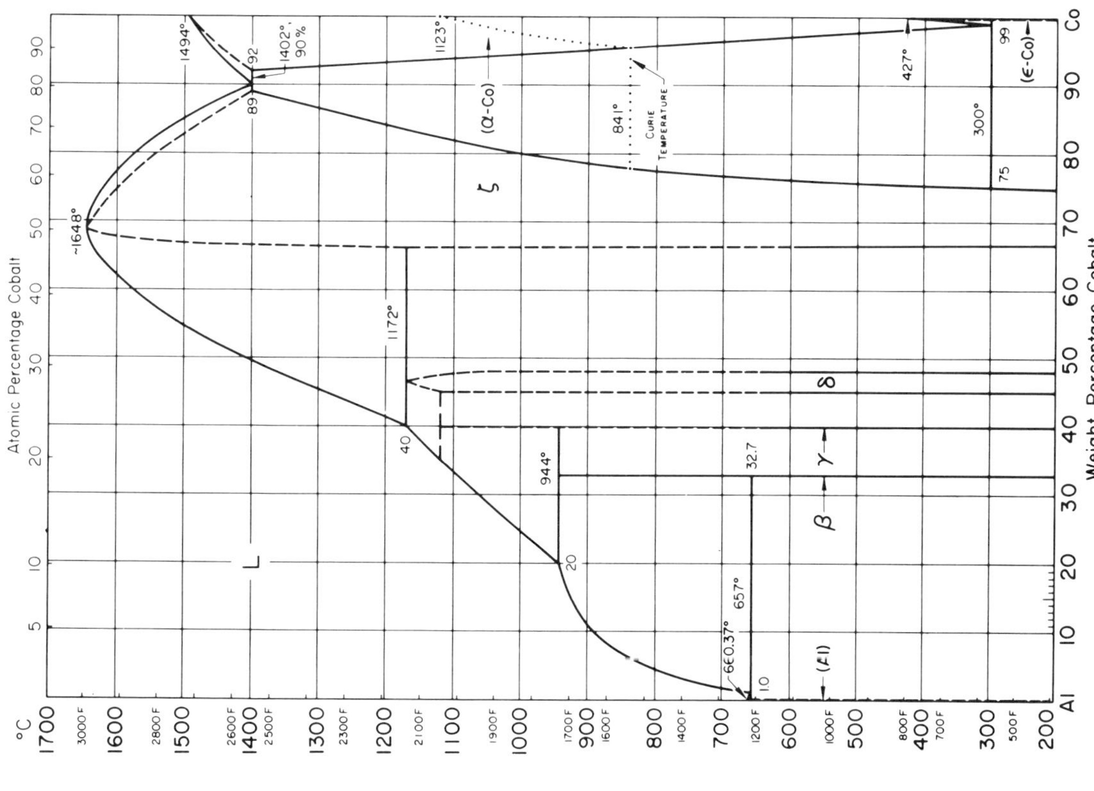

Fig. 6-6. The Co-Al and Co-Ti phase diagrams. (Adapted from Metals Handbook, 8th Ed., Vol 8, American Society for Metals, Metals Park, Ohio, 1973)

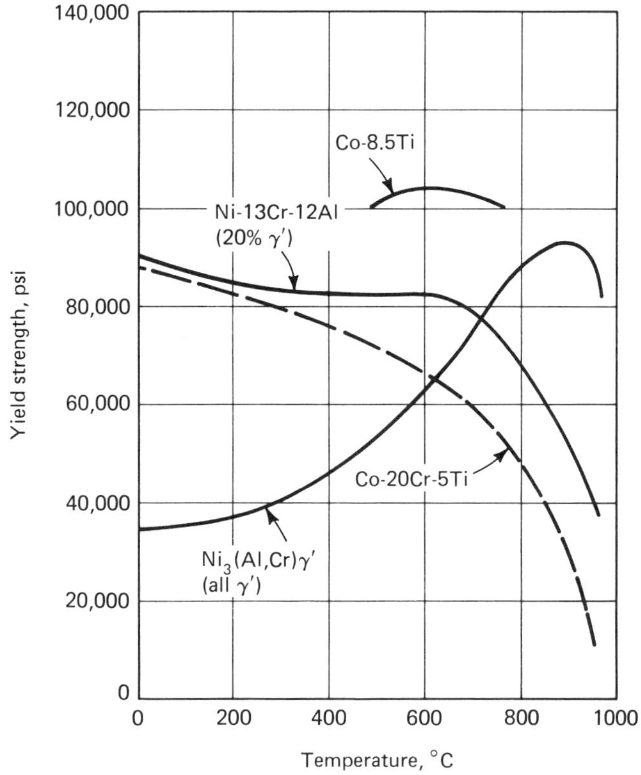

Fig. 6-7. Temperature dependence of the yield strength of some Co-base and Ni-base alloys containing γ'. In the Ni-base alloys the γ' is Ni_3 (Al, Cr); in the Co-base it is Co_3 (Cr, Ti). (Ni-base alloy curves from Fig. 5-16; Co-base alloy curves from P. Viatour, J. M. Drapier and D. Coutsouradis, Cobalt, No. 3, p 67, 1973)

Stacking Faults

Stacking faults were described in Chapter 1, where it was seen that in FCC alloys a dislocation can disassociate into two partial dislocations, between which exists a region in which the structure three atom layers thick is now hexagonal close-packed (HCP). The extent of the separation of the partial dislocations is a balance between decreased elastic energy of the dislocations and increased energy due to the presence of the nonequilibrium HCP region. The lower the stacking fault energy, the greater the separation. Stacking faults are important in strengthening the solid solution, as cross slip cannot occur until the stress attains a value sufficient to force the partial dislocations together. Thus, the higher the stacking fault energy, the stronger the alloy.

Pure cobalt exists in two crystal modifications, FCC above 427 °C, and HCP below this temperature. It is not unexpected, then, that the stacking fault energy is very low, so that extensive stacking faults are seen in the FCC phase in pure cobalt. The solubility of Cr in both the FCC and HCP phases is extensive (Fig. 6-8), so that stacking faults are expected to form easily in Co-rich Co-Cr alloys. In the range of 20-30% Cr the stable phases are the

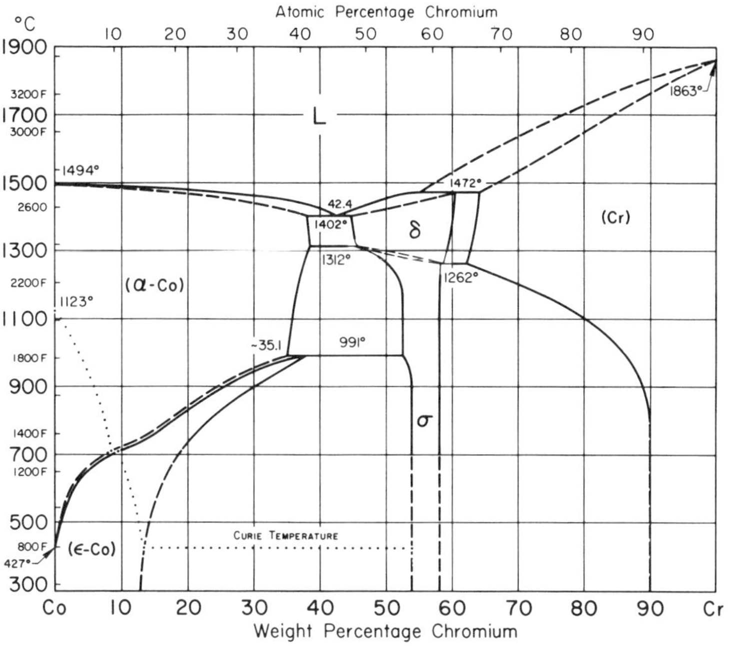

Fig. 6-8. The Co-Cr phase diagram. (From same source as Fig. 6-6)

HCP phase and σ. However, their formation from the high-temperature FCC phase is very slow, so that the common structure is still FCC, except upon extensive aging.

Alloying additions markedly affect the stacking fault energy. For example, FCC Ni stabilizes the FCC structure and increases the stacking fault energy, so that the dissociation is less. The effect of alloying additions can be gleaned by examination of their effect on the temperature of the HCP-to-FCC phase transformation in pure Co. This is summarized in Fig. 6-9. Note that Ni and Fe are very important because, although they do not have a very potent effect on the basis of each percentage added, their solubility in Co is quite high, so that the overall effect is significant. On the other hand, Nb has a potent effect, but the solubility is low so that little can be dissolved in the FCC matrix, and hence this effect cannot be utilized. The BCC elements, such as Mo and W, lower the stacking fault energy, favoring extensive separation of the partial dislocations and making the FCC matrix stronger but less ductile. Adding Ni or Fe counteracts this effect, stabilizing the FCC structure and decreasing the separation between the partial dislocations.

Stacking faults also play a key role in carbide precipitation. It has been found that in these Co-base alloys (and in some austenitic stainless steels) the stacking faults serve as nucleating sites for the precipitation of carbides. Thus upon aging, a fine dispersion of carbides can be developed which contributes greatly to tensile and creep strength.

Figure 6-10 is a microstructure showing stacking faults typical of those in Co-base alloys, with carbides precipitated on them. Note the fineness of the carbides.

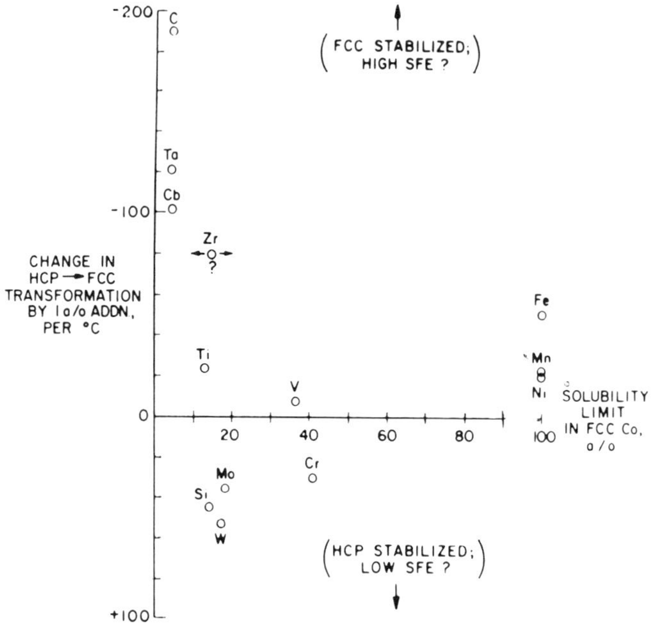

Fig. 6-9. Effect of alloying additions on the temperature of the HCP-to-FCC transformation in Co, shown in relation to the solubility in Co. (From C.T. Sims, J. Metals, Vol 21, p 35, 1969)

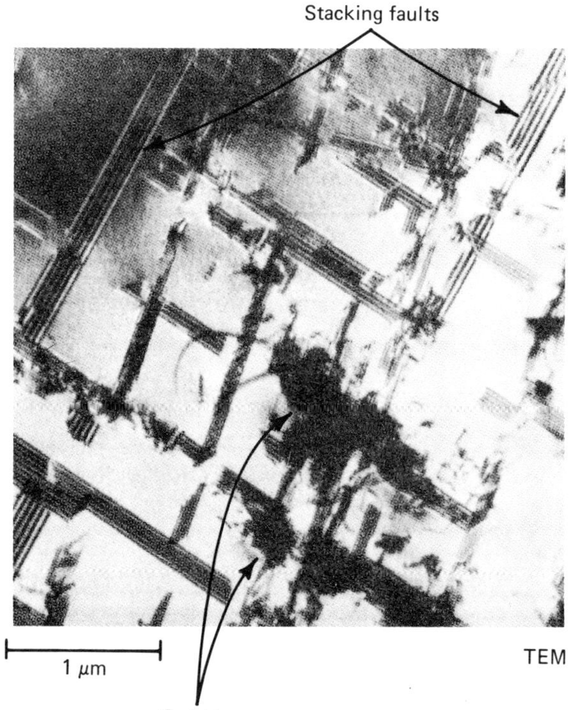

Fig. 6-10. Transmission electron micrograph of a Co-base alloy showing stacking faults and carbides precipitated on them. (Adapted from C.T. Sims, in The Superalloys, ed. C.T. Sims and W.C. Hagel, Wiley, New York, 1972)

Carbide Formation

In Co-base alloys, a number of different carbides can form, depending sensitively upon the chemical composition of the alloy. Alloying elements are adjusted to favor the carbide distributions and stability that develop properties best, especially high-temperature creep properties. Figure 6-11 illustrates the effect of chemical composition on the type of carbides formed. Because the Cr content in these alloys is usually high (e.g., greater than 20%), M_7C_3 and M_3C_2 are seldom found, or if present, will decompose upon aging. Note that the Cr content favors the formation of $M_{23}C_6$ type carbides, and they are the most common. The exact chemical composition of the $M_{23}C_6$ carbide depends upon the specific alloy. In one alloy (MM-509) this carbide was found to be $[Cr_{.77}Co_{.27}W_{.03}(Ni+Ta)_{.03}]_{23}C_6$, which can be represented approximately by the formula $[Cr_{18}Co_4(W,Ta)_1]C_6$. Note in Fig. 6-11 that the addition of Zr, Ti, Nb and Ta favor the formation of M_7C_3 and MC (such as TaC) type carbides. Additions of Mo and W favor the formation of M_6C type carbides. These effects are generally seen in examination of the binary phase diagrams shown in Fig. 6-12. Note that Co forms no carbides, but examination of the Co-Cr-C phase diagram (Fig. 6-13) shows that $M_{23}C_6$ will be the stable carbide for carbon and Cr contents usually encountered in Co-base alloys (e.g., 0.2-1.0% C, 10-30% Cr).

Figure 6-14 illustrates the effect of heat treatment on the carbide distribution in Mar-M 509, an alloy containing both MC and $M_{23}C_6$ carbides in the as-cast condition. Upon solution heat treating at 1275 °C for 4 h, the network carbides are dissolved, with the principal carbide now being M_6C but with some $M_{23}C_6$ present. The M_6C stability is assured by the presence

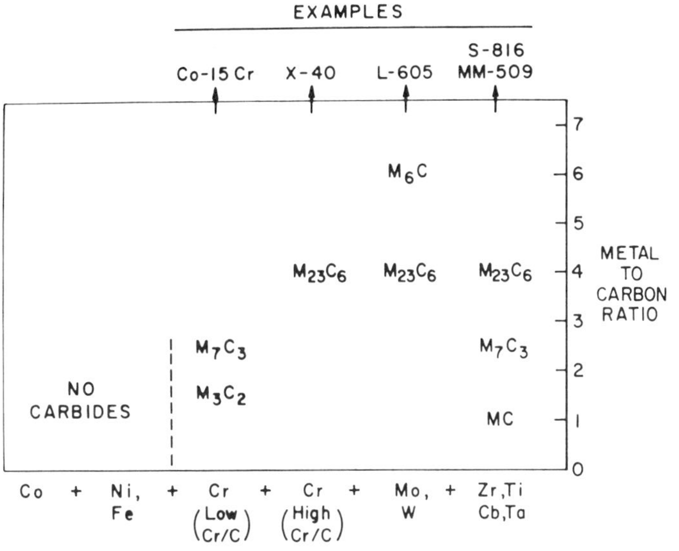

Fig. 6-11. Effect of chemical composition on the type of carbides commonly found in Co-base alloys for carbon contents from 0.1 to 0.6%. (From same source, p 30, as Fig. 6-9)

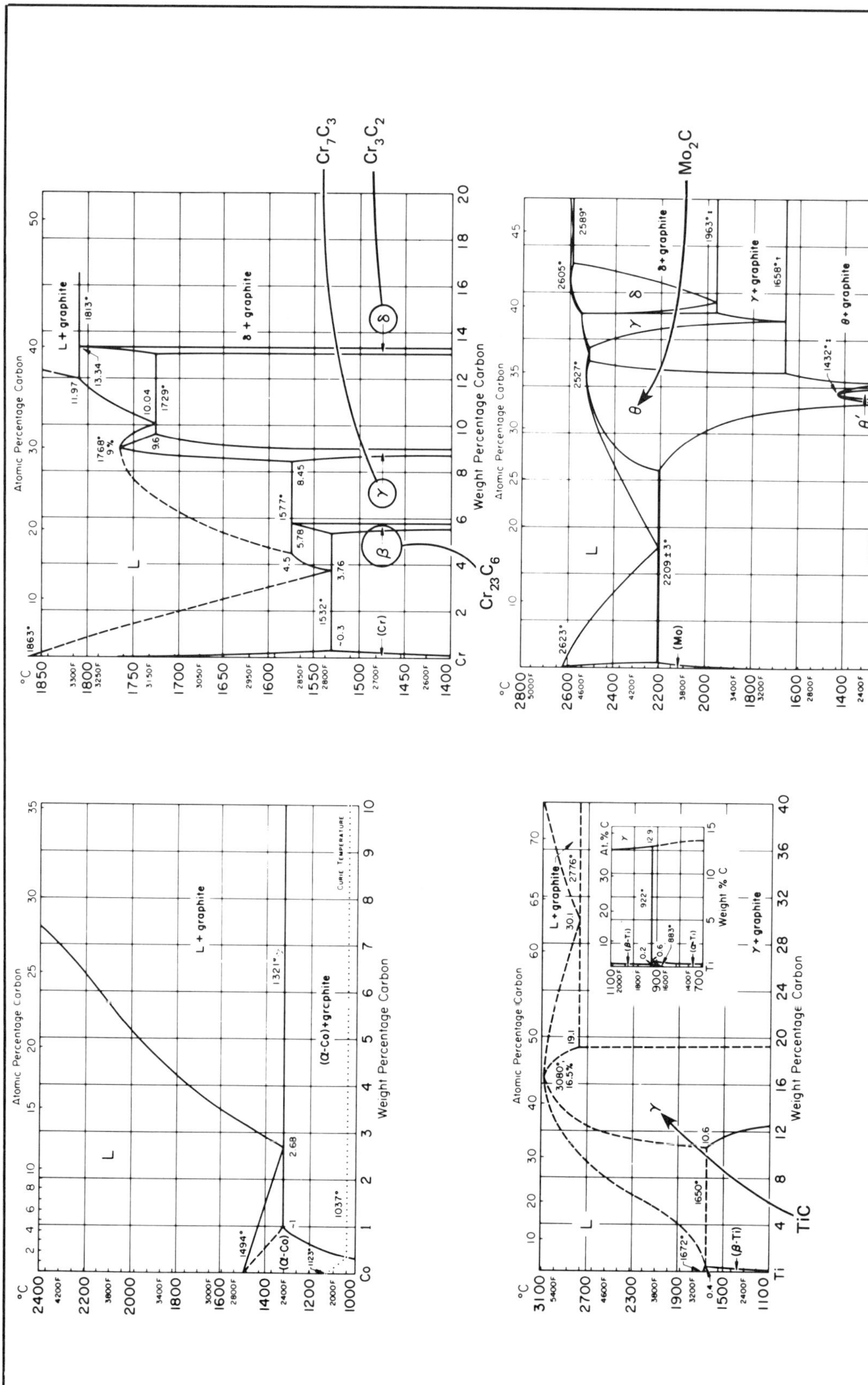

Fig. 6-12. The Co-C, Cr-C, Ti-C and Mo-C phase diagrams. (Adapted from same source as Fig. 6-6)

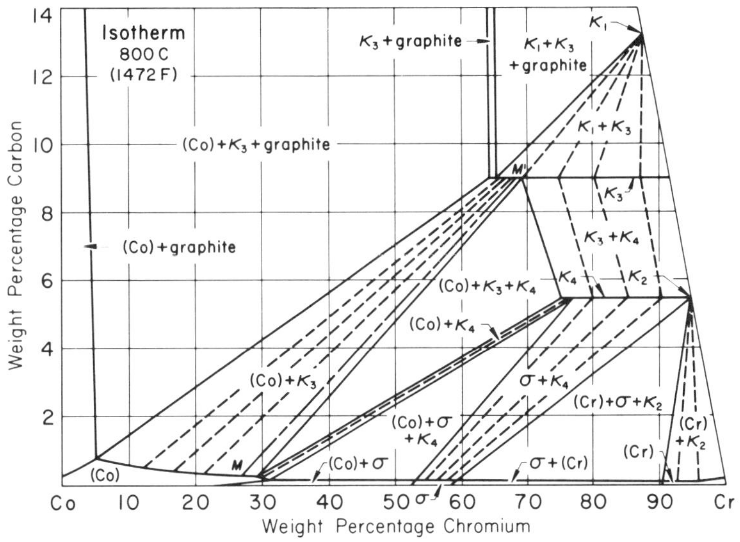

Fig. 6-13. The Co-Cr-C phase diagram at 800 °C. K_1 is Cr_3C_2; K_2 is $Cr_{23}C_6$; K_3 is $(Cr,Co)_7C_3$; K_4 is $(Cr,Co)_{23}C_6$. (Adapted from same source as Fig. 6-6)

of W in this alloy. Aging for 24 h at 927 °C produces a profusion of $M_{23}C_6$ carbides. Some are present as plates parallel to the {111} planes of the FCC matrix, and others are present as finely dispersed particles. In addition, a very fine dispersion of MC carbides is present, much too fine to be resolved in Fig. 6-14. The Ti, Zr and Ta content of this alloy assures the formation of the MC carbides. This fine and complex dispersion of carbides is what gives this alloy a high creep strength. However, if the aged alloy is placed in service at 1095 °C, after 723 h the $M_{23}C_6$ carbides have coarsened appreciably and visibly. The fine MC (mainly WC) carbides precipitated during aging have reacted with some of the Cr in solution to form more $M_{23}C_6$ carbides and allow the W to re-enter the FCC matrix.

The carbide distribution desired is one which has a fineness to give strength but which retains usable ductility by avoiding too many carbides (this generally is controlled by limiting the carbon content) and avoiding any type of continuous or semicontinuous film of carbides. Development of a usable carbide distribution for a given alloy is a matter of trying heat treatments and examining the results. There is no usable theoretical foundation which can be used to do this.

6-3. Formation of Intermetallic Phases

As in Ni-base alloys, consideration must be given to the possibility of the formation of the topologically close-packed (TCP) phases (see Chapter 5) such as sigma and Laves. In Ni-base alloys, their formation causes em-

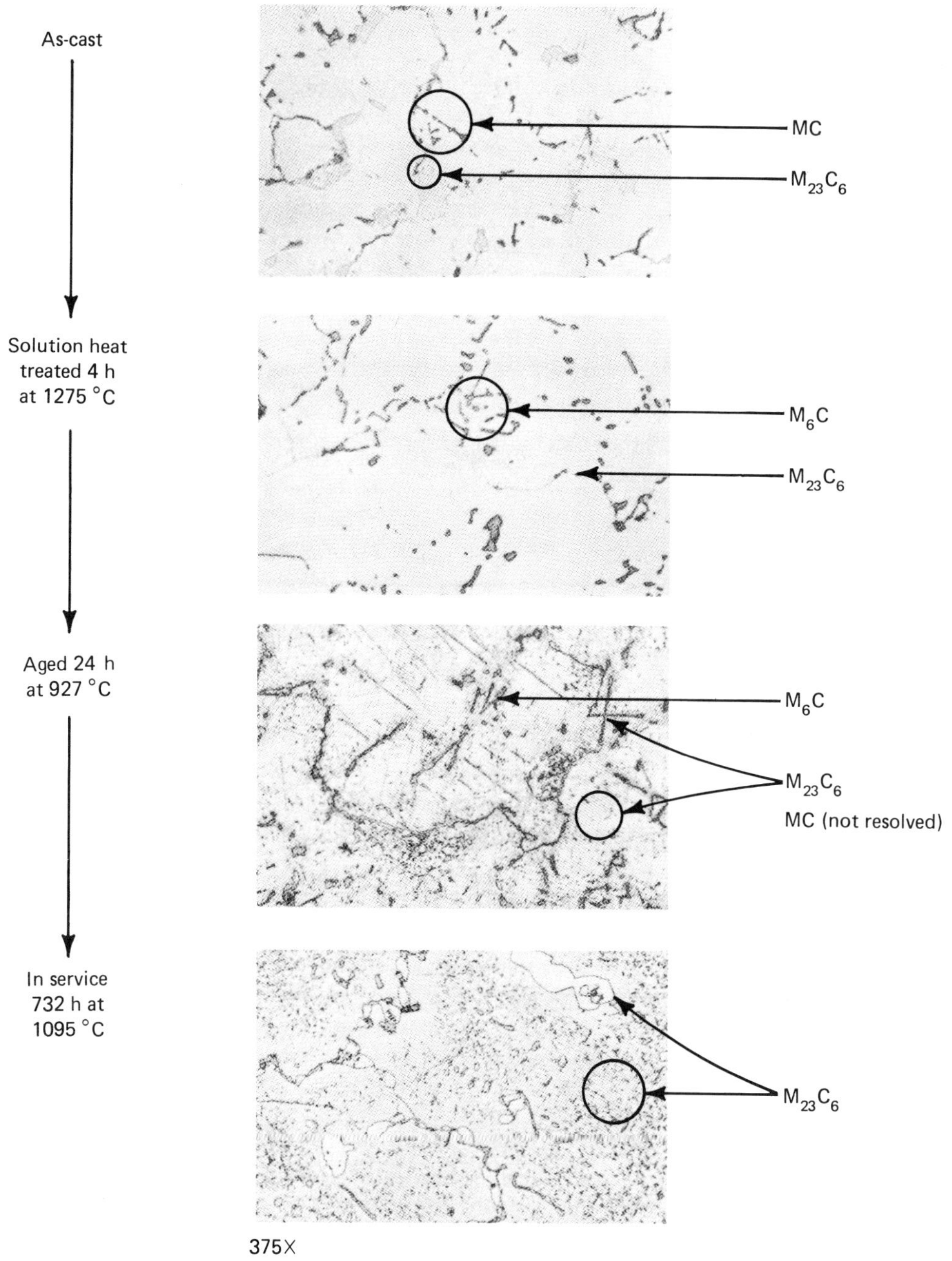

Fig. 6-14. Microstructures of Mar-M509 for different conditions. This alloy contains 10% Ni, 23.5% Cr, 7% W, 3.5% Ta, 0.2% Ti, 0.5% Zr and 0.6% C. (Adapted from same source as Fig. 6-10)

brittlement and loss of creep strength, and methods to predict their formation were described which are used in designing the alloys. The same approach is taken in predicting the presence of these phases in Co-base alloys.

The presence of phases such as sigma is predicted from the appropriate phase diagrams, such as shown in Fig. 6-15 for the Co-Cr-Ni and Co-Cr-Mo

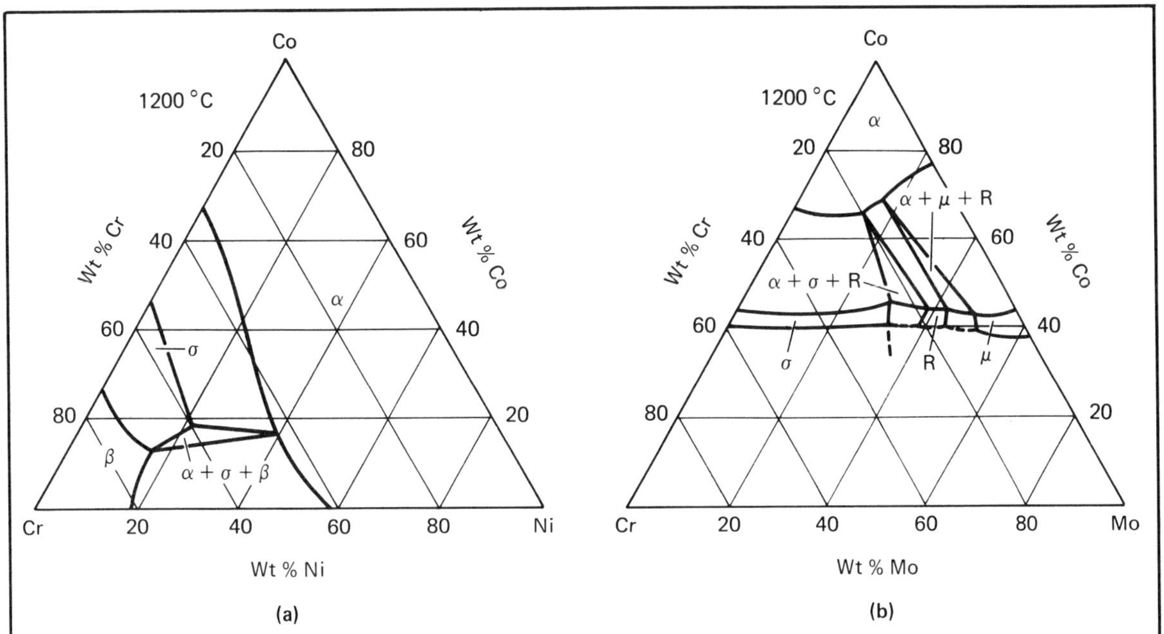

Fig. 6-15. The Co-Cr-Ni and Co-Cr-Mo ternary phase diagrams at 1200 °C. (Adapted from S. Rideout, W. D. Manly, E. L. Kamen, B. S. Lement and P. A. Beck, Trans. AIME, Vol 191, p 873 and 874, 1951)

systems. Using 30% Cr as a base, it clearly is seen that at 1200 °C the gamma phase is stable for any amount of Co or Ni. However, this is not true for the Co-Cr-Mo system (at 30% Cr) for high Mo content, and the solubility would be more restricted at lower temperatures. Thus, Mo increases the stability of phases such as σ at lower temperatures, compared to the influence of Ni. Figure 6-16 depicts the general phase fields for five quaternary systems. Note that the TCP phases are stable in all of these systems for certain compositions.

A useful method of displaying the composition range of phase stability is to use polar phase diagrams (Fig. 6-17). Instead of using a triangular isotherm, the section is shown as a segment of a circle. Adjacent to the segment is shown that of another ternary system with one of the elements common to the other ternary. Note in Fig. 6-17(a) that the Cr-Mn-Fe system shows a σ region across the middle, moving from the binary at about Mn–20% Cr to the Fe-Cr binary at 50% Cr. This region extends on into the Cr-Fe-Co system, shown adjacent to the Cr-Mn-Fe section. Figure 6-17(a) shows the influence of Fe, Mn, V and Ni on σ formation. (Also note the range of the HCP ε phase, although its rate of formation may be relatively slow.) A diagram for a constant W content at 10% is shown in Fig. 6-17(b). Point A falls near a typical commercial composition of 55Co-25Cr-10Ni-10W. (This is in weight percent, whereas Fig. 6-17 is in atomic percent.) Note that the presence of this amount of W does not shift the phase boundaries appreciably, but higher W contents can place the alloy critically close to the phase boundary. The effect of temperature on these boundaries is, of course, critical.

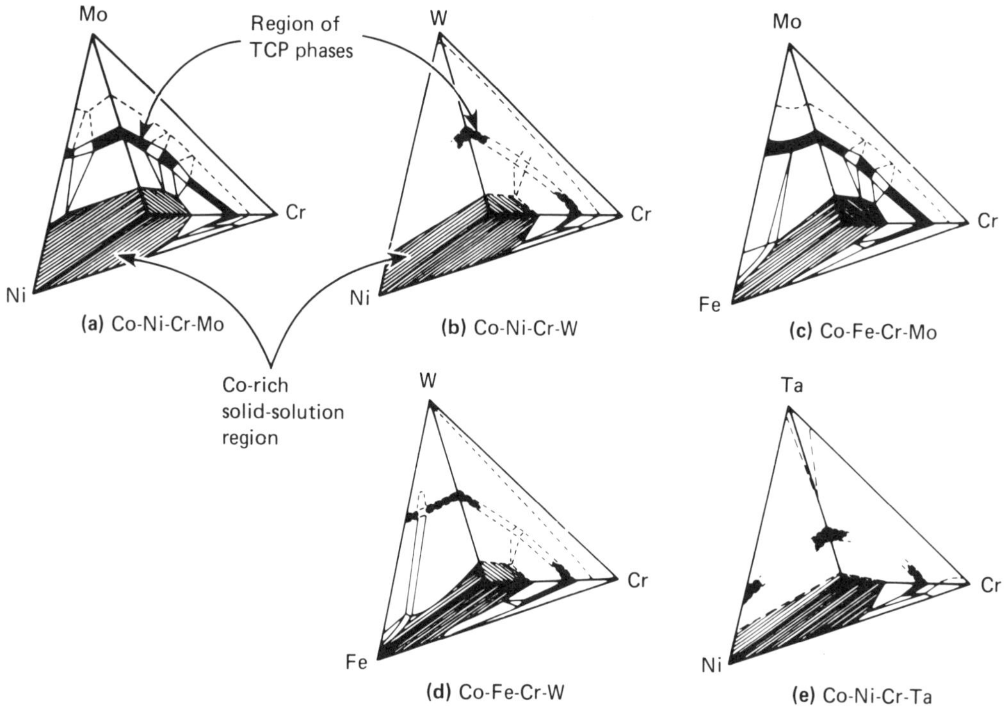

Fig. 6-16. Approximate quaternary phase diagrams at 1200 °C. In these figures, the apex of the figures is Co. The Co-rich solid solution is the large shaded region in each diagram. The smaller shaded regions are the TCP phases (e.g., sigma). (Adapted from same source, p 29, as Fig. 6-9)

The importance of control of the formation of the TCP phases is illustrated in Fig. 6-18, where it is shown that the formation of the Laves phase induces embrittlement in the alloy. The amount of the Laves phase formed upon aging at 870 °C clearly correlates with the ductility.

The occurrence of TCP phases is predicted by methods similar to the one described for Ni-base alloys in Chapter 5. The alloy composition is first corrected for the presence of carbides, and then the remaining composition is used to calculate the electron hole number, $\overline{N}_v$. Experience (somewhat more limited than with Ni-base alloys) indicates that the critical value of $\overline{N}_v$ is about 2.70, above which TCP phases will be prone to form. Some results are shown in Fig. 6-19; for a comparison with the results for Ni-base alloys, see Fig. 5-60.

6-4. The Microstructure of a Commercial Cobalt-Base Alloy

To illustrate the complexity of the carbide morphology in Co-base alloys, the microstructure typical of the as-cast alloy X-45 is described. This alloy contains about 26% Cr, 11% Ni, 8% W, 1% Mn, 1% Si and 0.25% C.

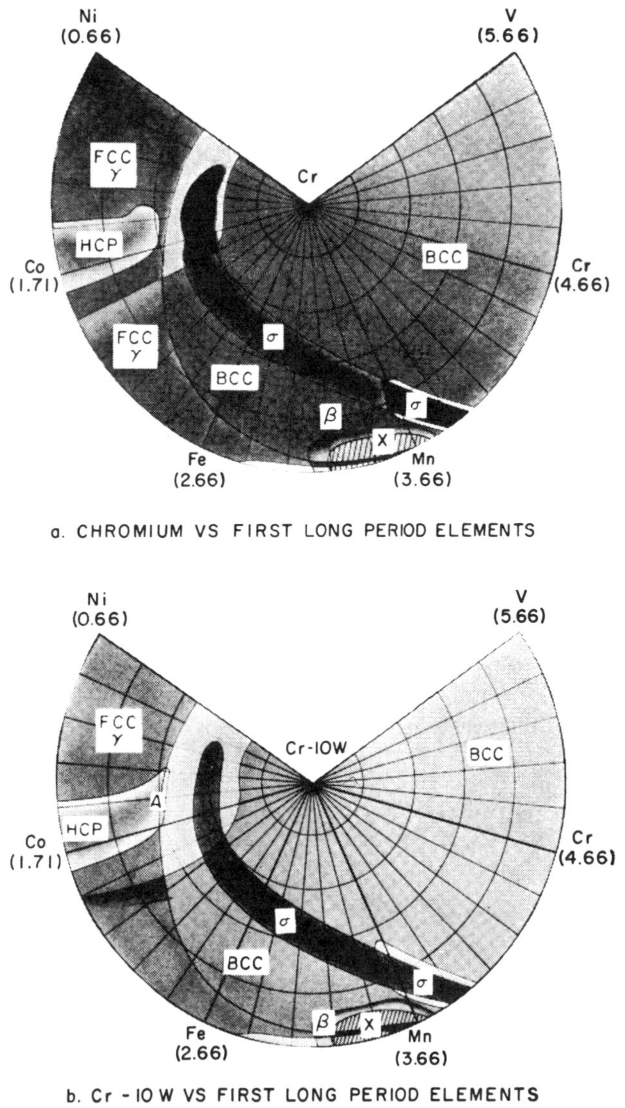

Fig. 6-17. Polar phase diagrams for Cr and Cr-10W alloys. (From same source as Fig. 6-9)

Figure 6-20 shows optical micrographs of the as-cast structure. The non-uniformity which is apparent is caused by coring (see Chapter 2), which affects the carbide morphology, and hence the appearance of the microstructure. Although not all of the steps in the solidification and cooling of this alloy are well understood, the process is probably something like this: When the liquid is cooled, the FCC matrix (γ) precipitates in the liquid (see Fig. 6-21). As cooling progresses, these crystals increase in size, but the central regions are richer in Co and poorer in carbon and other elements than the outer regions which form just before the eutectic temperature is reached. The eutectic reaction involves the formation of a two-phase structure of the FCC Co-rich phase and $M_{23}C_6$ type carbides.

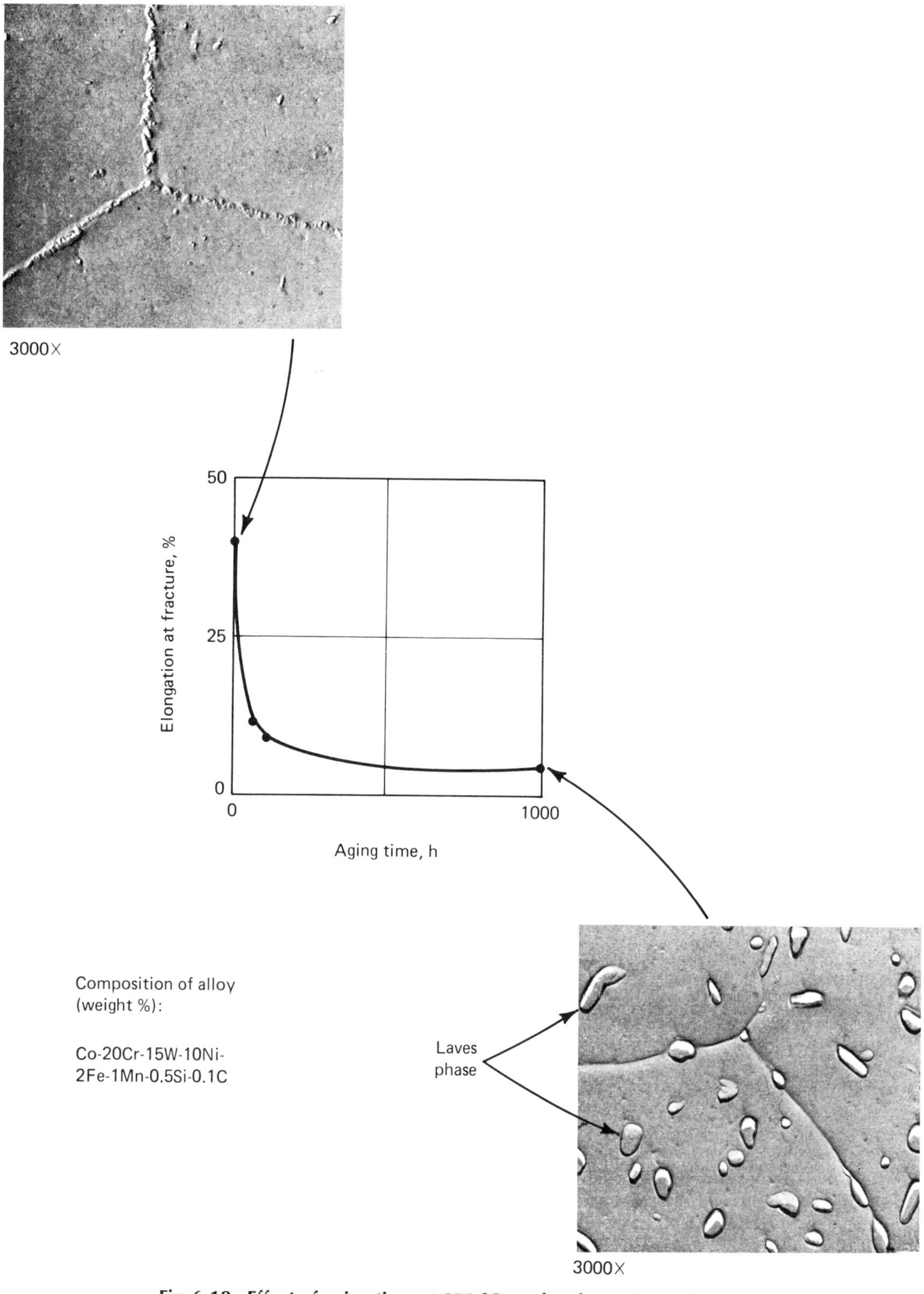

Fig. 6-18. Effect of aging time at 870 °C on the elongation at fracture in a tension test at 25 °C and on the microstructure, showing the precipitation of a Laves phase in a Co-base alloy. (Adapted from S. T. Wlodek, Trans. ASM, Vol 56, p 290, 1963)

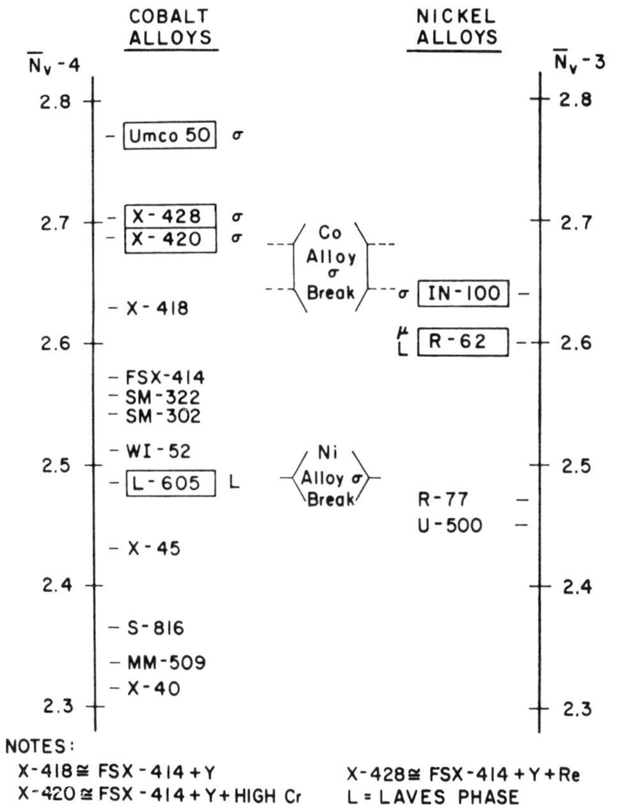

Fig. 6-19. The tendency of Co-base alloys to form TCP phases (sigma and Laves) as a function of electron hole number. The alloys in which sigma or Laves has been observed are shown boxed. (Adapted from same source, p 39, as Fig. 6-9)

Additional carbides of the $M_{23}C_6$ type are precipitated from the primary γ phase by solid-state reaction, either by very slow cooling following solidification or by reheating to allow precipitation (Fig. 6-21). The carbide precipitation is more probable in those regions of primary γ which are richest in both carbon and the strong carbide formers, especially Cr and W. Because the primary γ crystals are cored, those regions in which additional $M_{23}C_6$ precipitation is favored are the areas which were the last to freeze, and which are in the immediate vicinity of the eutectic regions. This is why in Fig. 6-20 there are regions which are free of carbide precipitates, and the carbides are concentrated around the eutectic regions. Also, the precipitation of the carbides from the primary γ leaves the γ adjacent to the carbides depleted in carbon and elements such as Cr and W, and hence these areas etch differently and appear dark in Fig. 6-20. This depleted zone is especially obvious around the long rods of $M_{23}C_6$. Note that the rods in Fig. 6-20 consist of four sets of parallel rods, and have probably formed parallel to the {111} planes of the matrix. Since there are no apparent grain boundaries in the primary γ in Fig. 6-20, and these four groups of carbide rods are distributed over the entire area shown, it is apparent that the primary γ in the figure is a common crystal.

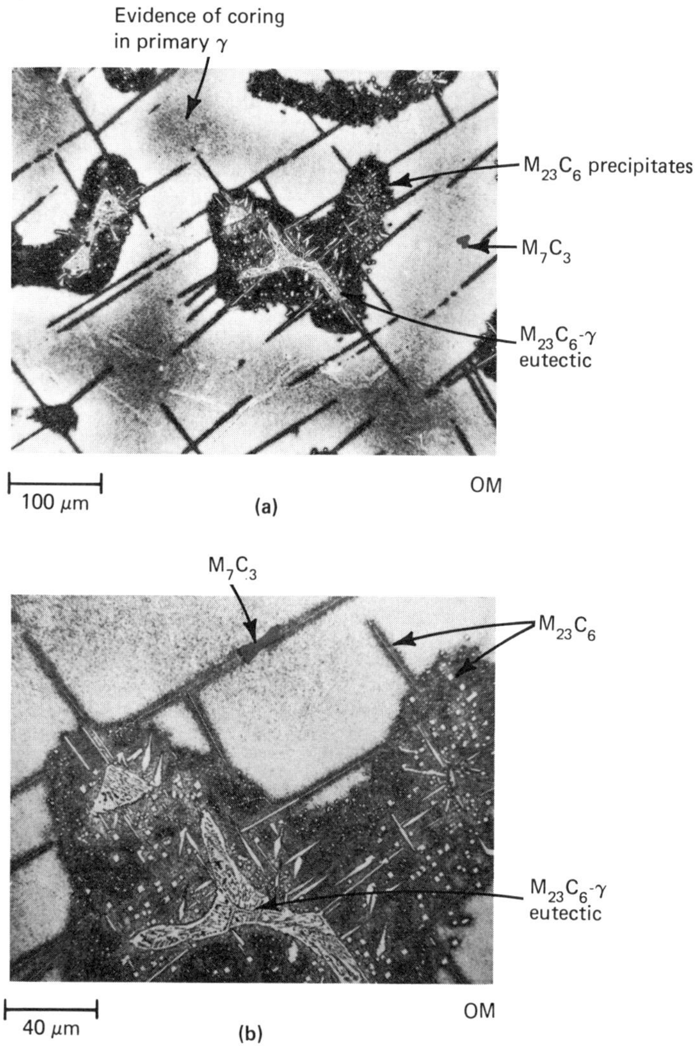

Fig. 6-20. Microstructure of alloy X-45 in the as-cast condition.

The fineness of the carbide cubes is shown clearly in Fig. 6-22. At this magnification it is seen that the depleted γ regions around the carbide precipitates are pitted, and hence in optical microscopy they appear dark (Fig. 6-20) as the pits scatter the light.

In this alloy, M_7C_3-type carbides are occasionally found (Fig. 6-20a), although they react with the γ to form $M_{23}C_6$ carbides upon sufficiently long aging.

Figure 6-23 shows an area in which the eutectic reaction in the liquid has apparently been initiated by the presence of a MnS inclusion.

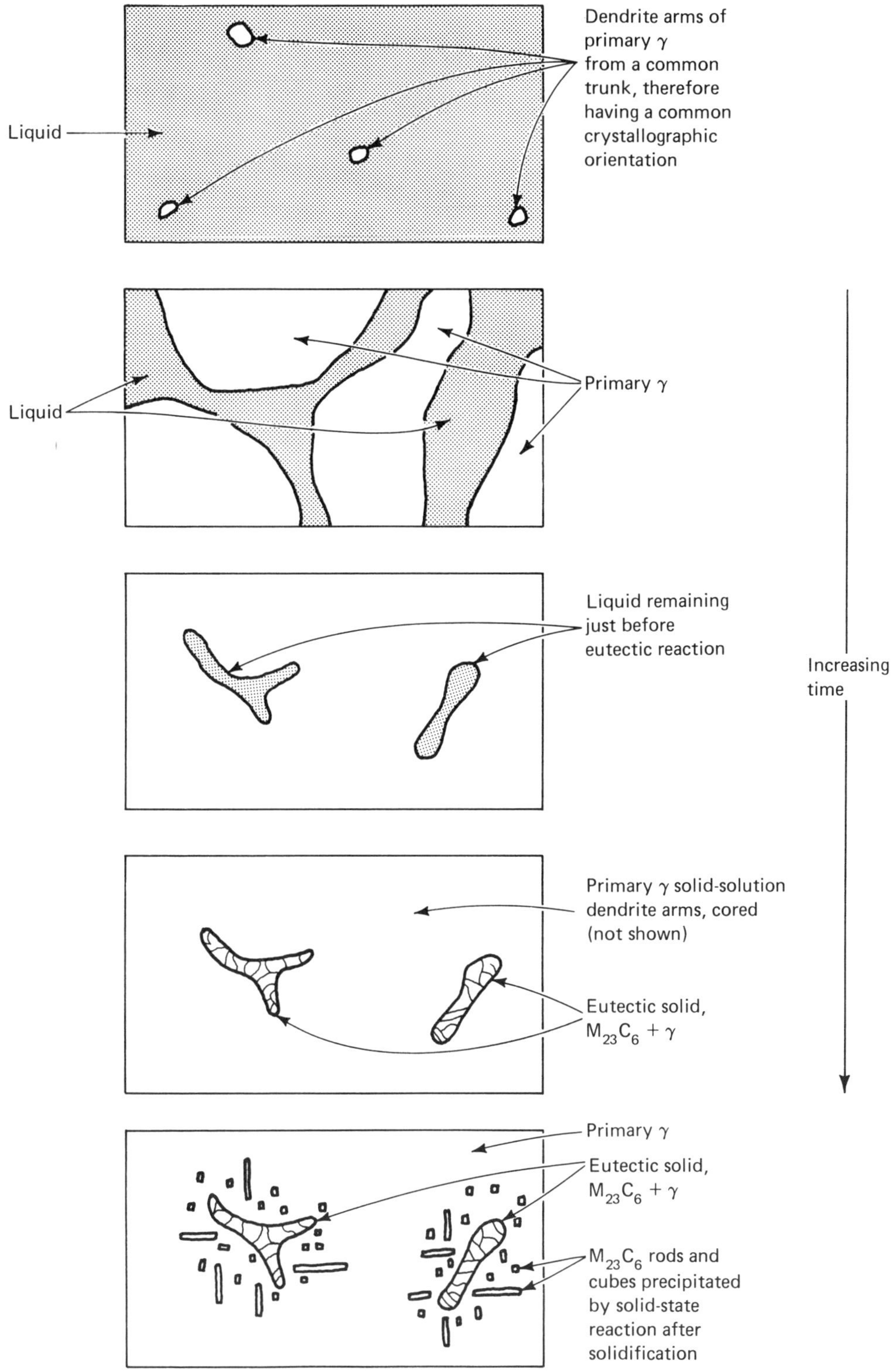

Fig. 6-21. Schematic illustration of the solidification and cooling sequence leading to the microstructure in Fig. 6-20.

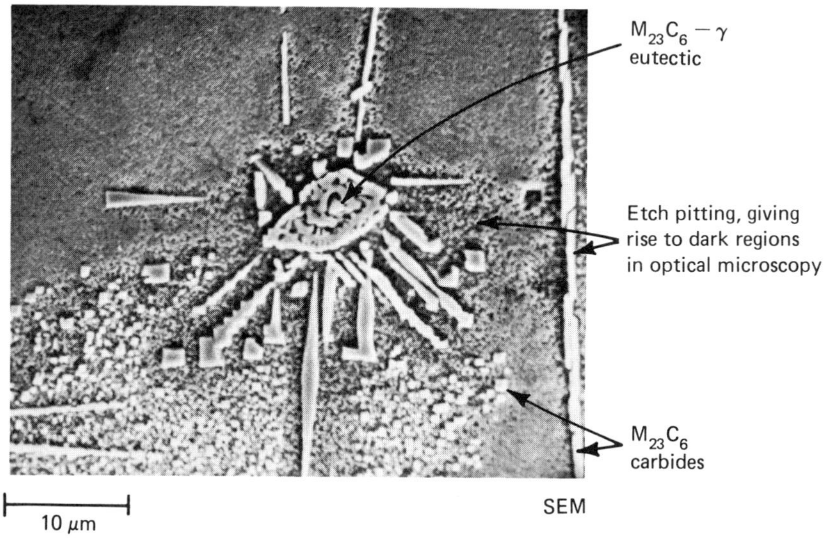

Fig. 6-22. Typical region showing the eutectic structure and the carbides precipitated in the primary γ. Note the etch pitting of the regions adjacent to the precipitated carbides, which causes these areas to appear dark in optical microscopy (see Fig. 6-20).

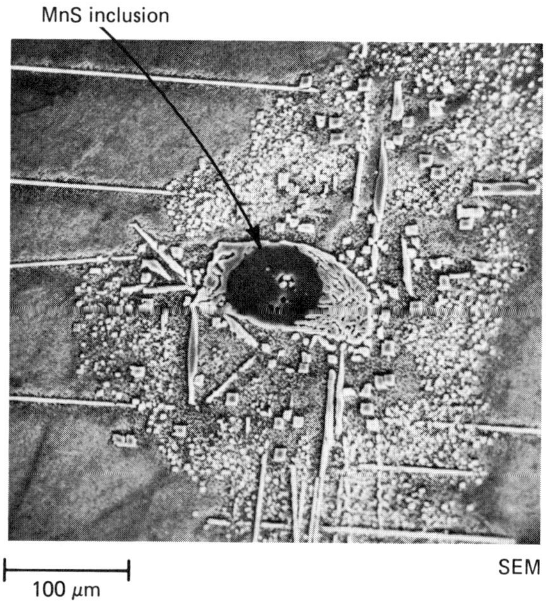

Fig. 6-23. Area showing the formation of the eutectic structure around a MnS inclusion.

6-5. Conclusion

In spite of the similarity of Ni and Co, a usable Co-base alloy in which a precipitate having the characteristics of γ' in Ni-base alloys has not yet been developed. Thus, Co-base alloys in general are not as suitable as Ni-base alloys for high-temperature application until the temperature range is attained at which in Ni-base alloys γ' is dissolved. Then Co-base alloys become competitive. Developing in these alloys sufficient high-temperature strength relies on solid solution strengthening and on the formation of a properly dispersed carbide structure. This latter effect is achieved by the addition of sufficient carbon (e.g., 0.3%) and carbide formers (e.g., Ti, Zr), and by the choice of heat treatment. The function of alloying element in Co-base alloys is summarized in Table 6-2.

Co-base alloys are used in both the wrought and the cast condition, and research is in progress to develop a directional solidification method similar to that described in Chapter 5 for Ni-base alloys.

Table 6-2. Summary of the Function of Alloying Elements in Co-Base Alloys (From Same Source as Fig. 6-9)

Elements	Nickel	Chromium	Tungsten	Titanium, zirconium, columbium, tantalum	Carbon
Principal function....	Austenite stabilization	Surface stabilization, carbide formation	Solid solution strengthening	Carbide formation	Carbide formation
Problems(a)	—	Forms TCP phases	Forms TCP phases	Harms oxidation resistance	Decreases ductility
Examples, wt %:					
X-40	10	25	7.5	—	0.45
MM-509 ..	10	24	7.0	3.5Ta, 0.5Zr, 0.2Ti	0.60
L-605.....	10	20	15	—	0.10
HS-188 ...	22	22	14	—	0.08

(a) When added in excess.

References

A Contemporary View of Cobalt-Base Alloys, C. T. Sims, Journal of Metals, Vol 21, p 27 (1969).

Cobalt-Base Alloys, C. T. Sims, in The Superalloys, ed. C. T. Sims and W. C. Hagel, Wiley, New York (1972).

Cobalt Monograph, Centre d'Information du Cobalt, Brussels (1960).

Materials for Gas Turbines, R. J. E. Glenny, J. E. Northwood and A. Burwood-Smith, International Metallurgical Reviews, Vol 20, p 1 (1975).

The Physical Metallurgy of Superalloys and Titanium Alloys, C. Hammond and J. Nutting, Metal Science, Vol 11, p 474 (1977).

The Physical Metallurgy of Cobalt-Base Superalloys, H. J. Wagner and A. M. Hall, DMIC Report 171, Defense Metals Information Center, Battelle Memorial Institute, Columbus, Ohio (1962).

CHAPTER 7

Magnesium-Base Alloys

The extremely low density of magnesium makes it attractive for consideration for engineering applications; even compared to aluminum (2.70 gm/cm^3 density), magnesium (1.74 gm/cm^3 density) is considerably lighter. This property makes it important to compare different metals and alloys with magnesium on a strength-to-weight ratio. For example, if stiffness is desired in a component, a large size may be required, and this favors magnesium because of the favorable strength/weight ratio. This effect is illustrated in Fig. 7-1.

Table 7-1 compares the yield strength and the yield strength divided by the specific gravity of a magnesium alloy, an aluminum alloy, a titanium alloy, an austenitic stainless steel and a low-alloy, high-strength steel. Note that the Mg alloy has the lowest yield strength in this group but has a strength/weight ratio about the same as that of the Al and the Ti alloy and quite superior to that of the two steels.

There are a variety of Mg-base alloys available (see table in Appendix 5), offering the engineer and designer a variety of properties. The specific elements may be present to induce solid solution strengthening, to improve creep properties, to allow precipitation hardening or to improve castability. In this chapter, a few alloys are examined to illustrate the reasons for their specific chemical composition and heat treatments.

One problem with Mg-base alloys is their low ductility. These alloys are hexagonal close-packed (or at least the matrix is), and thus have only three slip systems. This is also true of Zn, but the c/a ratio is such that Zn can mechanically twin in tension, and hence bring into play new slip systems. This leads to great ductility (e.g., 50%) in Zn alloys. However, as pointed out in Chapter 1, the c/a ratio of Mg is such that it can mechanically twin only in compression. Thus, in a tension test of a polycrystalline sample, twinning cannot be used to activate new slip systems, and Mg-base alloys

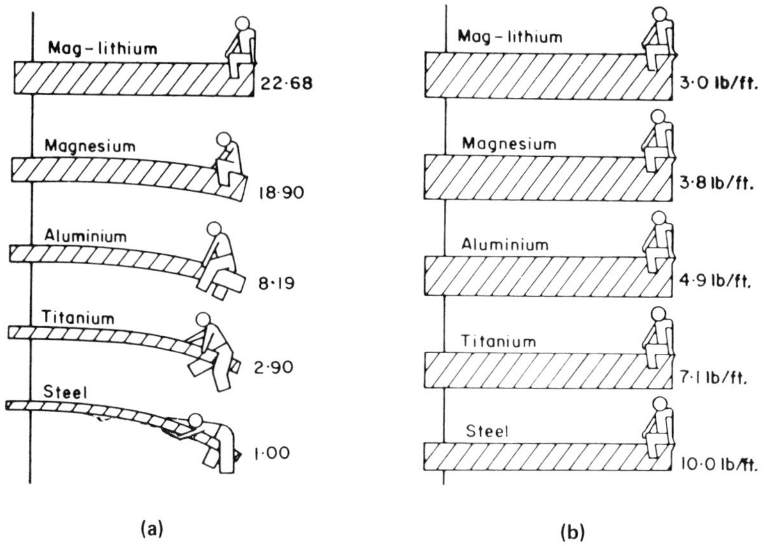

Fig. 7-1. Approximate comparison of (a) beams of equal weight and (b) beams of equal stiffness in various materials. (From E. F. Emley, Pergamon Press, New York, 1966)

Table 7-1. Comparison of Strength/Weight Ratio of an Mg Alloy and Other Alloys

Alloy	Specific gravity	Yield strength, psi	Yield strength divided by specific gravity
Magnesium alloy Type AZ80X Extruded HTA	1.8	40,000	22,000
Aluminum alloy 7075 T6	2.8	72,000	25,500
Titanium alloy Full hard	4.5	110,000	24,500
Stainless steel Type 302 30% cold worked	7.9	140,000	17,500
Low-alloy, high-strength steel	7.9	55,000	7,000

From Physical Metallurgy for Engineers, by D. S. Clark and W. R. Varney. Copyright ©1952, Van Nostrand Reinhold Company, New York. Used with the permission of Brooks/Cole Publishing Company.

invariably have low (e.g., less than 10%) ductility in tension. However, some plastic forming operations (e.g., rolling, extrusion) have a compression component, and hence Mg-base alloys can be fabricated by such cold forming operations.

7-1. Precipitation Hardening in Magnesium-Base Binary Alloys

From the viewpoint of strength, alloying additions would be sought which would increase the strength of Mg when present as solutes; then additional precipitation hardening would be examined in alloys in which the solubility limit is exceeded. Figure 7-2 shows the effect of some solutes on the critical resolved shear stress (see Chapter 1) for slip on the basal plane of the HCP structure. These alloys had their solute concentration restricted to the single-phase region, so that only solid solution strengthening is being examined.

Further data on the effect of solutes on solid solution strengthening are given in Table 7-2. Here, the percentage increase in the yield strength, or in the hardness, for each precentage increase in solute is listed for several elements. Also listed is the difference in the atom size of the solute and Mg.

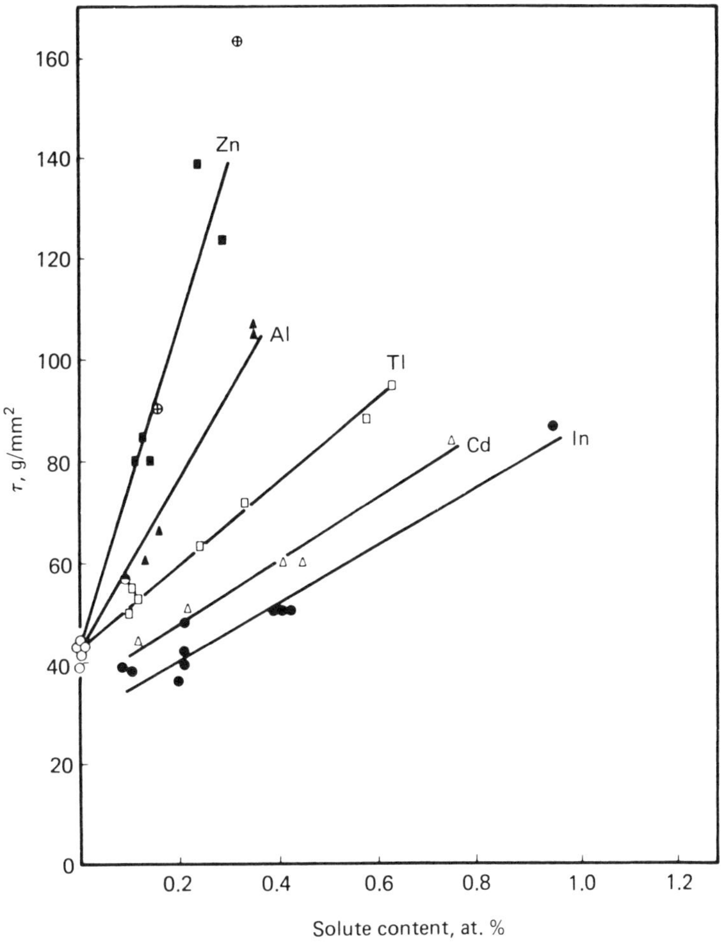

Fig. 7-2. Effect of solute content in Mg-base solid solutions on the critical resolved shear stress (at 25 °C) for slip on the basal plane of the HCP structure. (Adapted from E. D. Levine, W. F. Sheely and R. R. Nash, Trans. AIME, Vol 215, p 523, 1959)

Table 7-2. Approximate Solid Solution Strengthening Effect on Mg of Several Solutes

Solute	Approximate atom diameter size difference, % $(d_{Mg} - d_M)/d_{Mg}$	Approximate solubility in Mg at 200 °C, wt %	Percentage increase for each wt % of solute: Yield strength	Hardness
Aluminum....	+10	3	25	8
Zinc	+16	2	45	7
Silver	+9	5	23	7
Calcium ..	−24	1	110	—
Cerium ...	−14	2	148	—
Thorium ..	−13	1	212	—
Lithium...	+5	5	—	3
Cadmium .	+7	50	10	1
Bismuth ..	+2	1	—	5
Indium ...	+2	43	—	1
Tin	+5	0	26	3
Copper ...	+20	2	35	—

Yield strength data adapted from J. C. McDonald, Trans. AIME, Vol 143, p 179, 1941, and Vol 137, p 430, 1940; hardness data adapted from D. Hardie and R. N. Parkins, J. Inst. Metals, Vol 85, p 449, 1956-57.

As was seen for Ni- and Co-base alloys, there is a general correlation between the size difference and the solid solution strengthening effect. Note in Table 7-2 that Ca, Ce and Th are especially potent strengtheners. However, their rather large atom size, relative to Mg, precludes wide solubility, and thus this effect is limited.

Thus it is seen that the elements in Table 7-2 will cause some solid solution strengthening, especially for the first small additions, but the solubility of most of these limits this method as a means of greatly strengthening Mg. Therefore, we turn to an examination of the possibility of precipitation hardening as a means of further strengthening.

Figures 7-3, 7-4 and 7-5 show three Mg binary alloys which can be precipitation hardened, and the strengthening is not insignificant. The hardness can be increased by about 20% (Fig. 7-3 and 7-4) to 70% (Fig. 7-5). In some of these Mg alloys, the precipitation hardening is caused by the formation of solute-rich zones and transition phases, as was described in Chapter 4 for Al-base alloys. For example, Fig. 7-6 shows the fine transition precipitate of the MgZn type which forms near the maximum in the hardness curve (see Fig. 7-5). Upon overaging, this transition phase changes to the equilibrium MgZn γ phase (see Fig. 7-5).

Not all Mg-base systems in which precipitation occurs harden, however. For example, Fig. 7-7 shows that in both Mg-Pb and Mg-Sn alloys the Mg-rich solid solution can dissolve a considerable amount of solute at high temperature, and considerable precipitation will occur upon aging at lower temperature. However, in neither of these systems does significant hard-

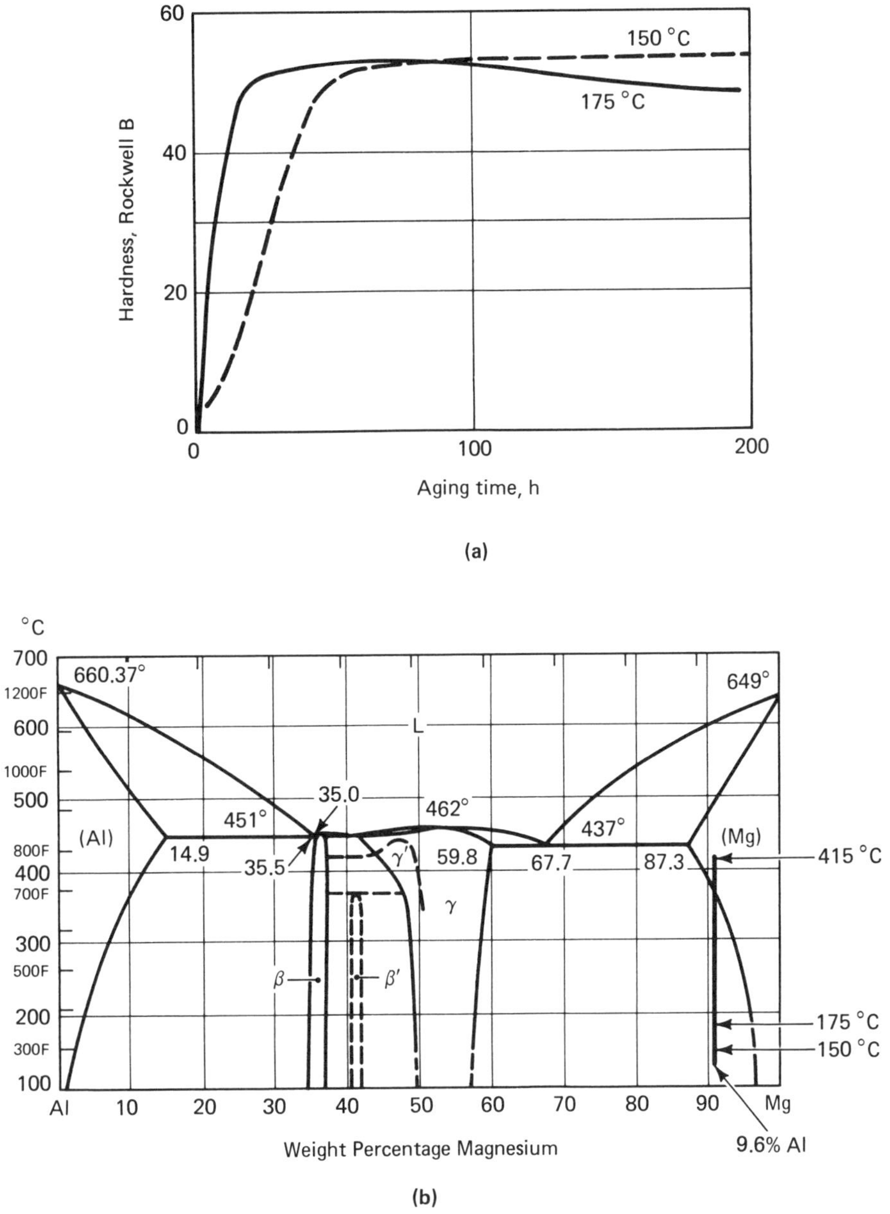

Fig. 7-3. Age hardening of a Mg-9.6Al alloy. The samples were solution treated at 415 °C for 24 h, then water quenched before aging. (a) adapted from A. M. Talbot and J. T. Norton, Trans. AIME, Vol 122, p 301, 1936; (b) adapted from Metals Handbook, 8th Ed., Vol 8, American Society for Metals, Metals Park, Ohio, 1973.

ening accompany the precipitation. Although a fine dispersion of the precipitate forms, lattice coherency is lost early in the formation of the precipitate, and significant hardening cannot occur. Thus Pb and Sn are not found in common Mg alloys (Appendix 5).

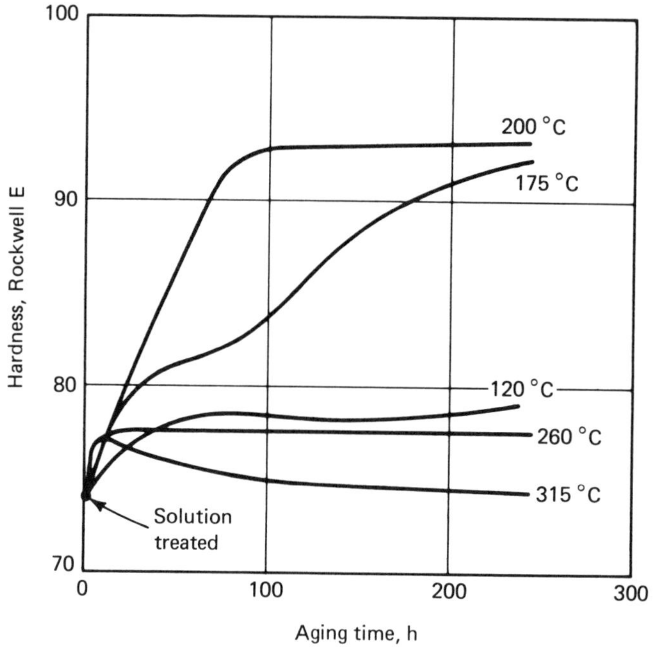

(a)

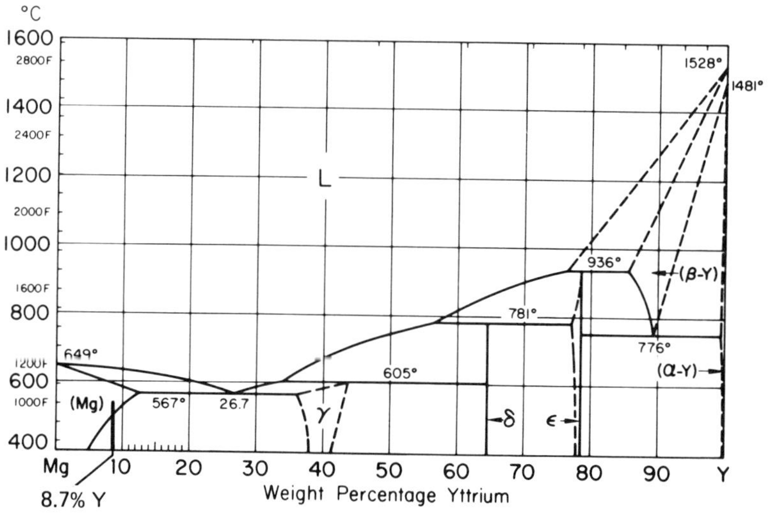

(b)

Fig. 7-4. Age hardening of a Mg-8.7Y alloy. The samples were solution treated at 510 °C for 1 h, then water quenched before aging. (a) adapted from D. Mizer and B.C. Peters, Met. Trans., Vol 3, p 3262, 1972; (b) adapted from same source as Fig. 7-3b)

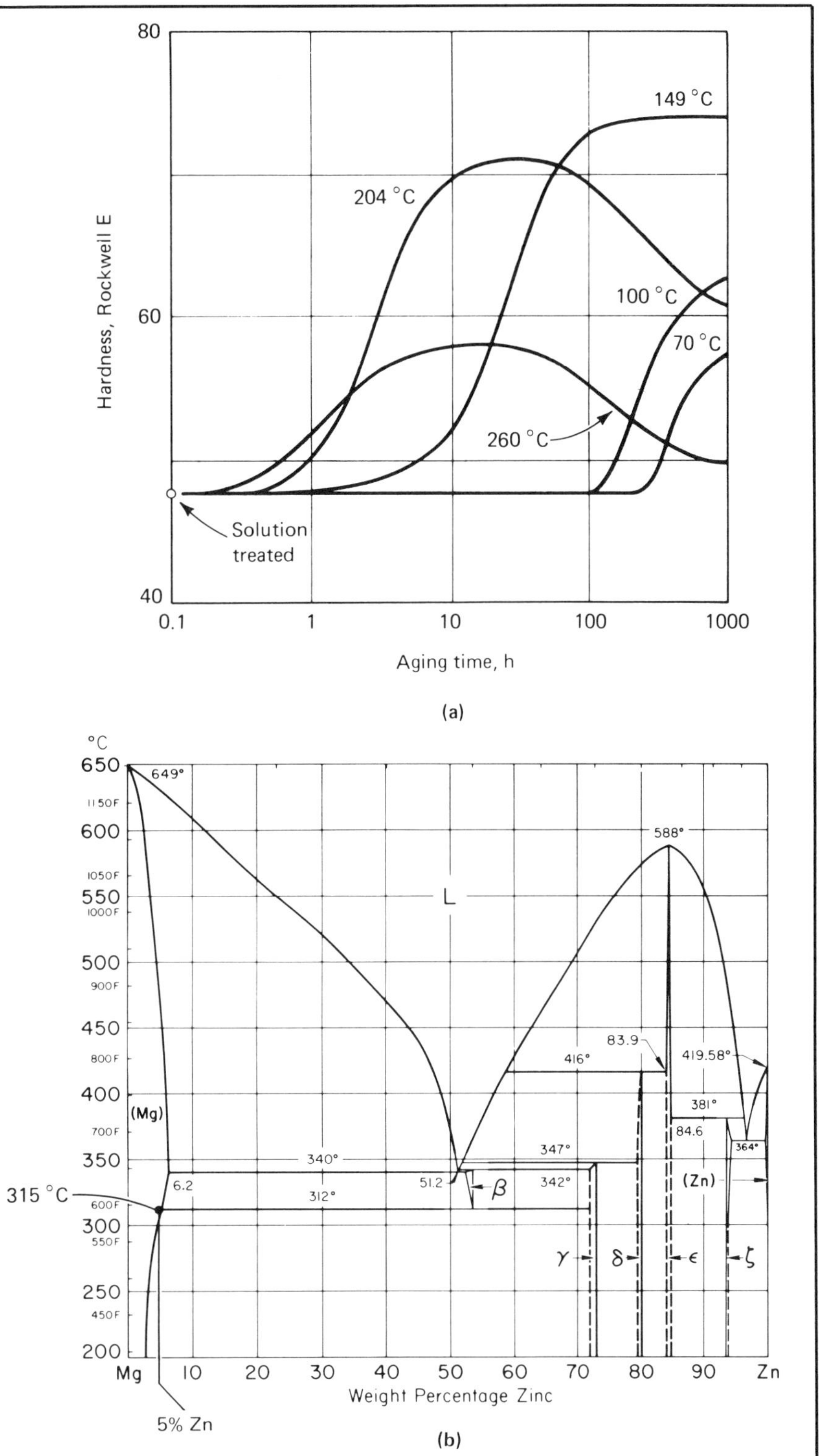

Fig. 7-5. Age hardening of a Mg-5Zn alloy. The samples were solution treated at 315 °C for 1 h, then water quenched before aging. (a) adapted from J. B. Clark, Acta Met., Vol 13, p 1281, 1965; (b) adapted from same source as Fig. 7-3b)

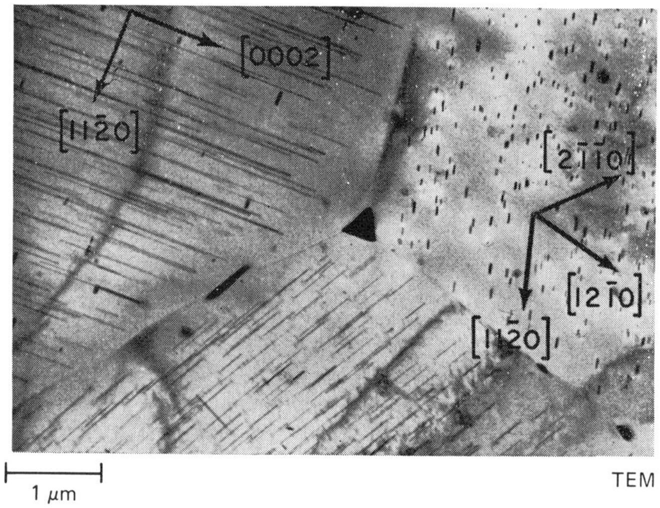

Fig. 7-6. Transmission electron micrograph showing the fine MgZn transition precipitate which forms upon aging a Mg-5Zn alloy for 16 h at 204 °C. Note from Fig. 7-5 that this microstructure corresponds to the maximum in the hardness curve. (From same source, p 1282, as Fig. 7-5a)

7-2. Magnesium-Aluminum-Zinc Alloys

We are now in a position to examine some of the commercial alloys, and we begin with AZ92A, an alloy which is basically Mg-9Al-2Zn. The Mg-Al-Zn ternary phase diagram is shown in Fig. 7-8. The liquidus, solidus and solvus temperatures are shown in Fig. 7-8(d). Table 7-3 gives the recommended heat treatments for AZ92A; note the agreement with the values predicted from the phase diagrams.

In the Mg-9Al-2Zn alloy, the solid phase in equilibrium with the Mg-rich solid solution is $Mg_{17}Al_{12}$ (γ in Fig. 7-3), the same phase as in the binary alloy. This phase forms by an eutectic reaction, and even if the particular alloy upon slow cooling passes from the liquid into the Mg-rich solid solution region without occurrence of the eutectic reaction, the eutectic reaction may occur as a result of coring if the alloy is cooled sufficiently rapidly from the liquid. (This was discussed in detail in Chapter 3.) A typical as-cast microstructure for the Mg-9Al-2Zn alloy is shown in Fig. 7-9, where massive γ is present from this eutectic reaction. (The darker regions are solid-state reaction products which formed upon cooling in the lower temperature region. This is discussed in a following section.) If an alloy with this phase present is reheated to the recommended solution treatment temperature of 408 °C (Table 7-3), then cooled rapidly to 25 °C, the microstructure in Fig. 7-10 is typical of that formed. The grain boundaries are slightly

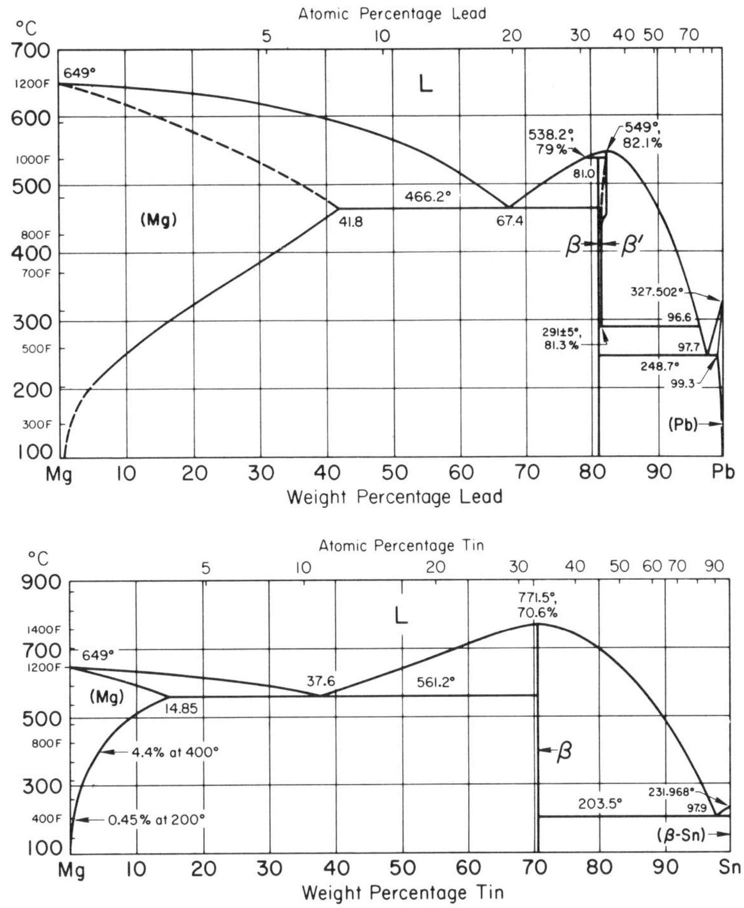

Fig. 7-7. The Mg-Pb and Mg-Sn phase diagrams. (From same source as Fig. 7-3b)

wavy because of the formation of an unresolved, discontinuous precipitate even upon the rapid cooling. The darker regions are Mn and Mg_2Si particles which did not dissolve at the solution temperature.

Now, if the Zn content is sufficiently high, Fig. 7-8 shows that the three-phase region will be entered, with the ternary intermetallic phase ϕ being present. This compound has a liquidus temperature of about 363 °C (see Fig. 7-8a), and thus if this compound formed upon casting (or upon cooling in the solid-state region), then upon solution treating at 408 °C (Table 7-3) these local regions may melt before equilibrium is attained. These liquid zones lead to porosity (by mechanisms that are not clear) in heat treated material, and an example is shown in Fig. 7-11. Such an occurrence is referred to as burning or incipient melting. To avoid this problem, the alloy is heated slowly, or in stages, to the solution-treating temperature.

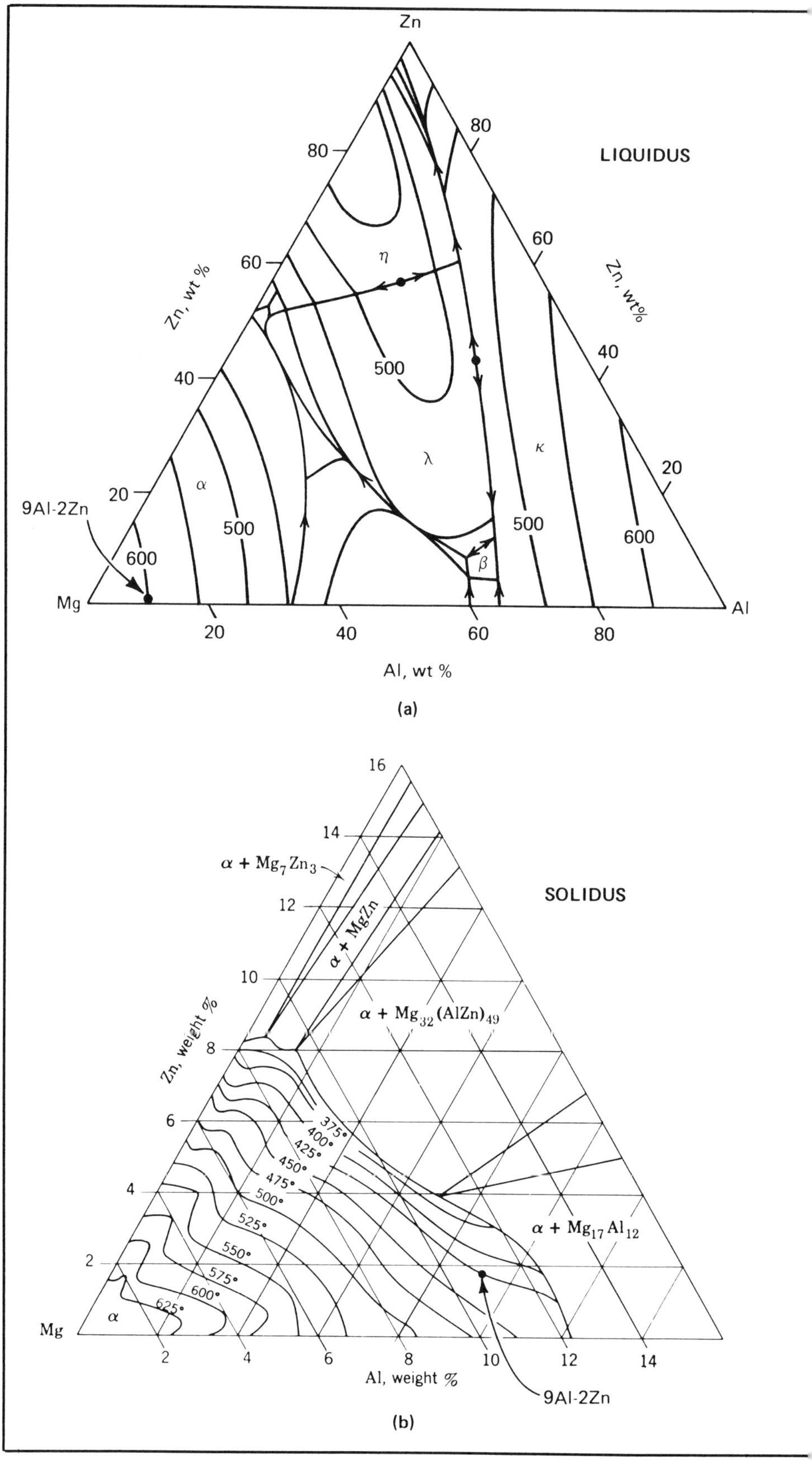
Zn
LIQUIDUS
80
60
40
20
Zn, wt %
Zn, wt%
η
λ
κ
α
β
500
600
9Al-2Zn
Mg
Al
Al, wt %
(a)
16
14
12
10
8
6
4
2
SOLIDUS
α + Mg_7Zn_3
α + MgZn
α + $Mg_{32}(AlZn)_{49}$
α + $Mg_{17}Al_{12}$
Zn, weight %
375°
400°
425°
450°
475°
500°
525°
550°
575°
600°
625°
α
Mg
Al, weight %
9Al-2Zn
(b)

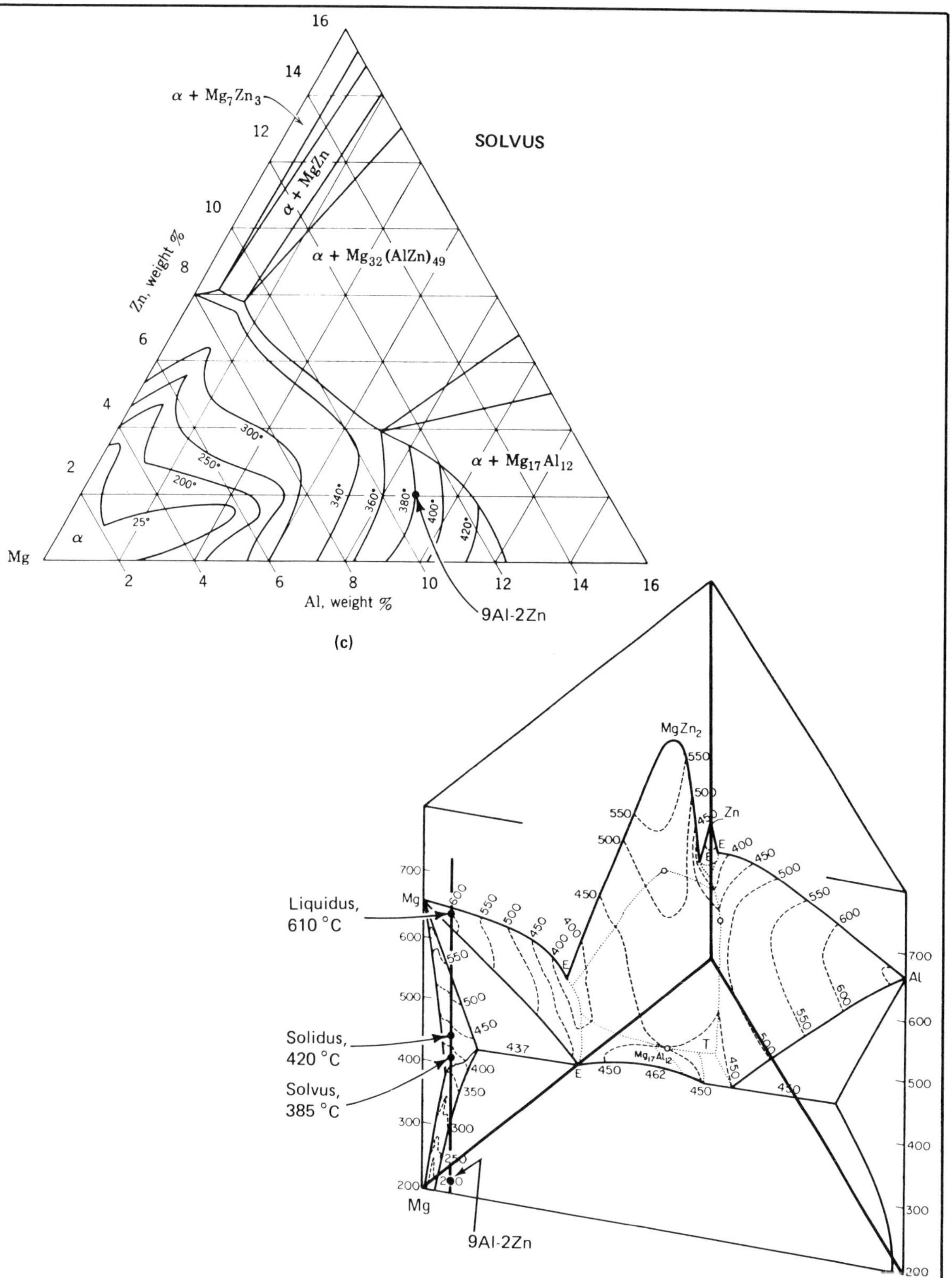

Fig. 7-8. Liquidus, solidus and solvus surfaces for the Mg-Al-Zn system, showing the location of the Mg-9Al-2Zn alloy. The liquidus, solidus and solvus temperatures for this alloy are shown in (d). (Adapted from same source as Fig. 7-1)

Table 7-3. Liquidus, Solidus and Solvus Temperatures and Recommended Heat Treatments for Mg Alloy AZ92A (Nominal Composition, 9% Al, 2% Zn)

Liquidus temperature: 593 °C
Solidus temperature: 443 °C
Incipient melting temperature: 410 °C
Solution heat treatment:
408 ± 6 °C for 16 to 24 h, cooled by strong air blast
Precipitation heat treatment:
Solution heat treatment, then 220 ± 5 °C for 5 h
or:
Solution heat treatment, then 260 ± 5 °C for 4 h
or:
Solution heat treatment, then 205 °C minimum for not less than 6 h

From Metals Handbook, 8th Ed., Vol 2, American Society for Metals, Metals Park, Ohio, 1964.

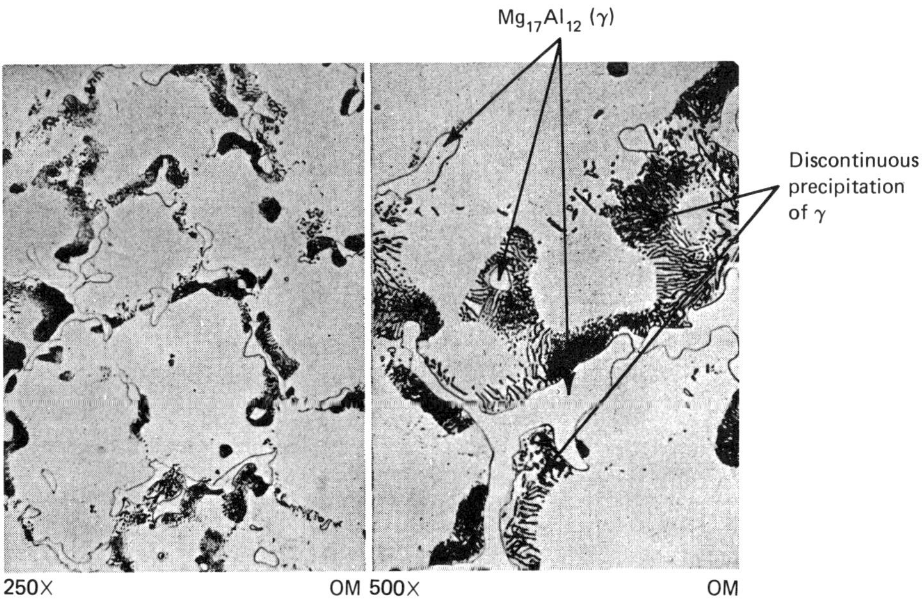

Fig. 7-9. Microstructure of a Mg-9Al-2Zn alloy in the as-cast condition. The $Mg_{11}Al_{12}(\gamma)$ phase is part of a divorced eutectic, formed perhaps as a result of coring which occurred during rapid cooling. Slow cooling after solidification permitted discontinuous precipitation of γ. (Adapted from Structure and Properties of Alloys, 2d Ed., by R. M. Brick and A. Phillips. Copyright © 1949, McGraw-Hill, New York. Used with the permission of McGraw-Hill Book Company)

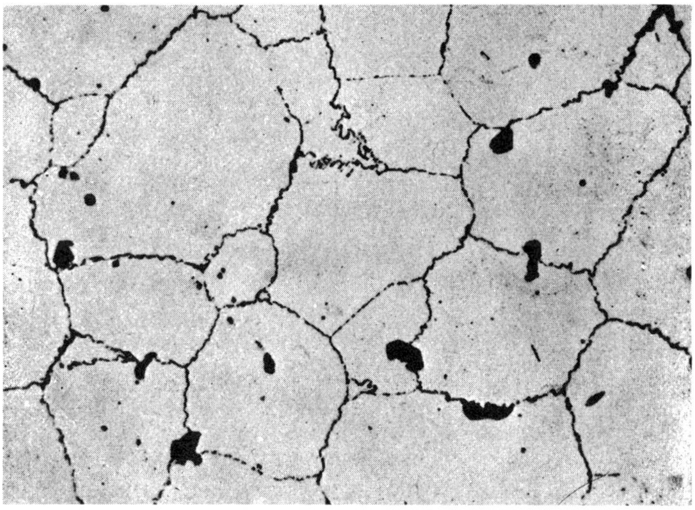

Fig. 7-10. Microstructure of a Mg-9Al-2Zn alloy which has been solution treated. The uneven grain boundaries are caused by a small amount of unresolved, discontinuous precipitate which formed even upon rapid cooling from the solution temperature. The dark regions are Mn or Mg_2Si particles which did not dissolve. (From J. B. Hess and P. F. George, Trans. ASM, Vol 31, p 438, 1943)

Figure 7-12 shows that this commercial alloy precipitation hardens, and the recommended aging temperatures and times listed in Table 7-3 clearly are based on responses similar to those shown in Fig. 7-12. The influence of Zn additions on the precipitation-hardening response of Mg-Al-base alloys is shown in Fig. 7-13. It is seen that Zn does increase the yield strength of a base Mg-9Al alloy about 15% at 2% Zn, but the tensile strength and the percentage elongation are decreased. In the precipitation-hardened condition, the yield strength is increased, the tensile strength decreased, and the elongation unaffected (but very low for any Zn content).

The aging results depend upon the amount of the potential precipitate which is dissolved. The rate of solution in a Mg-9Al-2Zn commercial alloy is illustrated in Fig. 7-14, where it is seen that the strength after aging remains low if the solution treatment time is too short. Note that these data indicate that this alloy should be solution treated about 20 h for optimum effect, and this is reflected in the recommended solution treatment in Table 7-3.

The addition of Zn above about 2% decreases the elongation at fracture, especially in the solution-treated condition (Fig. 7-13). This makes the alloy more susceptible to cracking from thermal stresses upon quenching from the solution temperature. Figure 7-15 is the isotherm at 335 °C, which shows that the equilibrium precipitate in the Mg-9Al-2Zn composition is the same phase as in the binary Mg-Al alloy. The shape of the α-γ boundary shows

Table 7-4. Galvanic Series of Some Commercial Metals and Alloys in Seawater

↑		Platinum
		Gold
Noble or cathodic		Graphite
		Titanium
		Silver
	[	Chlorimet 3 (62 Ni, 18 Cr, 18 Mo)
		Hastelloy C (62 Ni, 17 Cr, 15 Mo)
	[	18-8 Mo stainless steel (passive)
		18-8 stainless steel (passive)
		Chromium stainless steel 11-30% Cr (passive)
	[	Inconel (passive) (80 Ni, 13 Cr, 7 Fe)
		Nickel (passive)
		Silver solder
	[	Monel (70 Ni, 30 Cu)
		Cupronickels (60-90 Cu, 40-10 Ni)
		Bronzes (Cu-Sn)
		Copper
		Brasses (Cu-Zn)
	[	Chlorimet 2 (66 Ni, 32 Mo, 1 Fe)
		Hastelloy B (60 Ni, 30 Mo, 6 Fe, 1 Mn)
	[	Inconel (active)
		Nickel (active)
		Tin
		Lead
		Lead-tin solders
	[	18-8 Mo stainless steel (active)
		18-8 stainless steel (active)
		Ni-Resist (high-Ni cast iron)
		Chromium stainless steel, 13% Cr (active)
	[	Cast iron
		Steel or iron
		2024 aluminum (4.5 Cu, 1.5 Mg, 0.6 Mn)
Active or anodic		Cadmium
		Commercially pure aluminum (1100)
		Zinc
↓		Magnesium and magnesium alloys

From Corrosion Engineering, 2d Ed., by M. G. Fontana and N. D. Greene. Copyright ©1978, McGraw-Hill, New York. Used with the permission of the McGraw-Hill Book Company.

that the addition of Zn will allow more γ precipitate to form, at a fixed Al content of 9% (see Fig. 7-16), and this is probably why the ternary alloy is stronger in the precipitation-hardened condition than the binary alloy. However, if the Zn content exceeds about 4% a three-phase region is entered involving the ternary intermetallic phase ϕ (Fig. 7-16). According to Figure 7-8(a), this compound involves an eutectic reaction at a eutectic temperature of about 360 °C. Thus, for example, if this compound formed upon casting, then local melting would occur upon reheating to the solution temperature range (e.g., in the range 400 °C). (Recall that such an eutectic structure can form upon rapid cooling from the liquid, even if it is not an equilibrium structure. See Chapter 3.) These local liquid regions which form

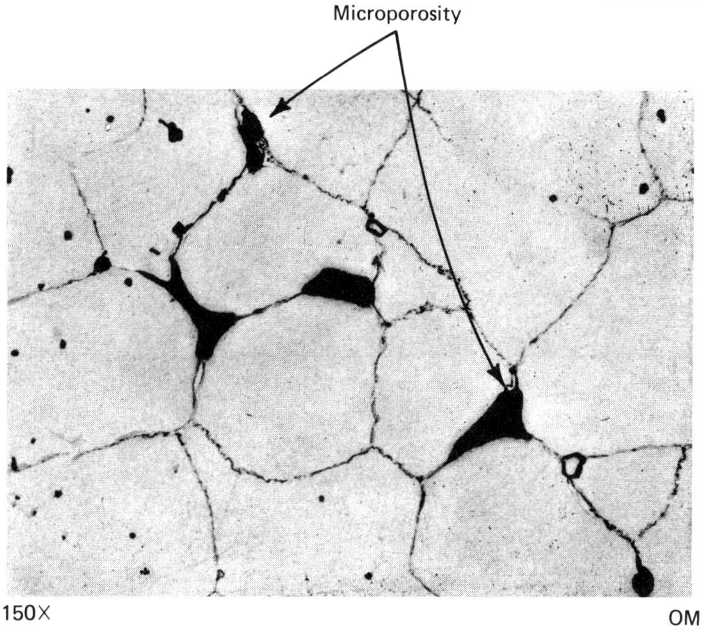

Fig. 7-11. Microstructure of a Mg-9Al-2Zn alloy showing microporosity caused by localized melting. (Adapted from same source, p 440, as Fig. 7-10)

will eventually disappear at the solution temperature, or refreeze upon cooling, leaving behind porosity. This problem can be minimized by heating slowly to the solution temperature, or preheating, to allow time for the compound to disappear by solid-state reaction.

In the Mg-9Al-2Zn commercial alloy, the precipitation occurs both continuously and discontinuously, in a fashion similar to that for the binary Mg-Al alloy described in the previous section. The amount of discontinuous precipitate affects the mechanical properties. Also, the cooling rate from the solution temperature affects the precipitation upon cooling, and the amount of discontinuous precipitation upon subsequent aging. These effects are illustrated in Fig. 7-17. Note that higher cooling rates allow the aged material to be stronger and somewhat more ductile, and that this more favorable condition corresponds to having a microstructure with less discontinuous precipitate. Figure 7-18 shows microstructures illustrating both types of precipitate structures.

7-3. Corrosion Resistance of Magnesium-Base Alloys

A problem that requires careful attention in Mg alloys is the sensitivity of the corrosion resistance to minor amounts of certain elements (Fig. 7-19). The reasons for this lie in the phase relations. For example, Fig. 7-20 shows that

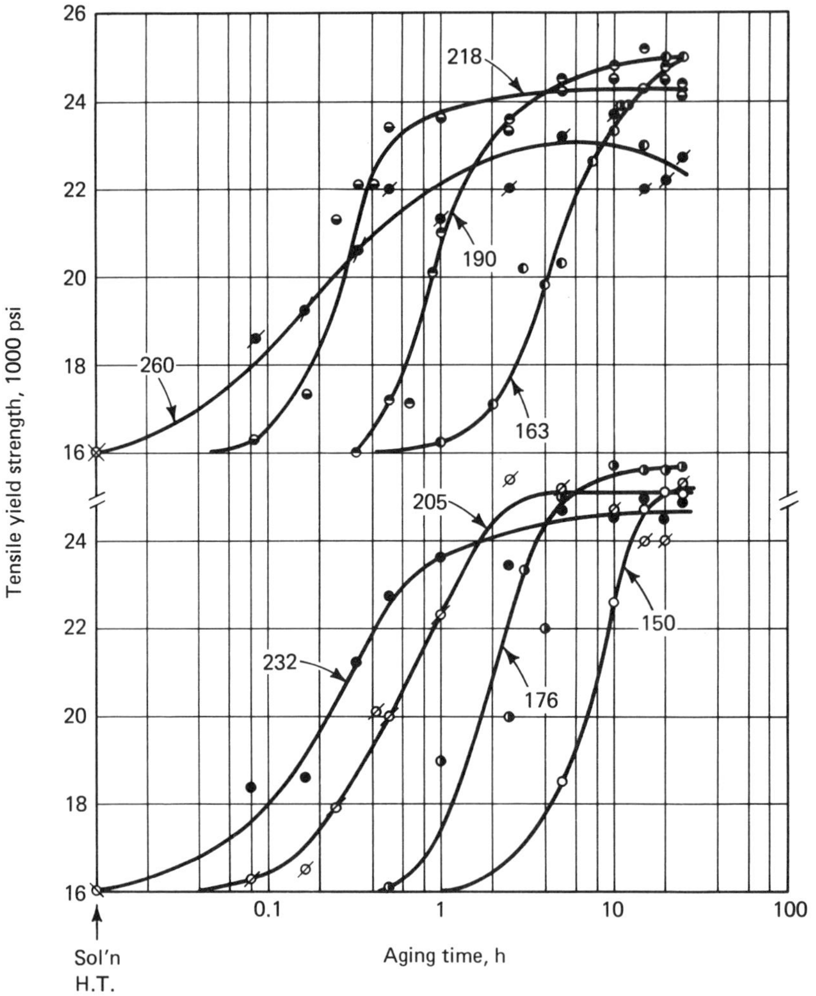

Fig. 7-12. Effect of aging temperature and time on yield strength for Mg alloy AZ92A, which contains nominally 9Al-2Zn-0.2Sn. The alloy was solution heat treated by heating from 260 °C to 415 °C in 2 h, then holding for 24 h at 416 °C, followed by air cooling to 25 °C. (Adapted from T. E. Leontis and C. E. Nelson, Trans. AIME, Vol 191, p 120, 1951)

the solubility of Fe in Mg is quite low and that the precipitate is pure iron, not an intermetallic phase. On the galvanic series Mg and Fe are widely separated (Table 7-4), and if these two metals are coupled in an electrolyte, the Mg will dissolve at the expense of the Fe. (Indeed, this is a method of protecting steel in environments such as seawater, using sacrificial Mg anodes which dissolve and leave the steel intact.) Thus, if the Fe content exceeds the solubility limit (which is quite low) in Mg or Mg alloys, the microstructure will consist of a Mg matrix with a dispersion of Fe particles. These two constituents are electrically coupled, of course, and hence the Mg will be readily dissolved in aqueous solutions. Similar behavior is observed with Ni, Co and Cu, and in all of these cases the solubility is quite low, although it is not the pure element but a compound that precipitates.

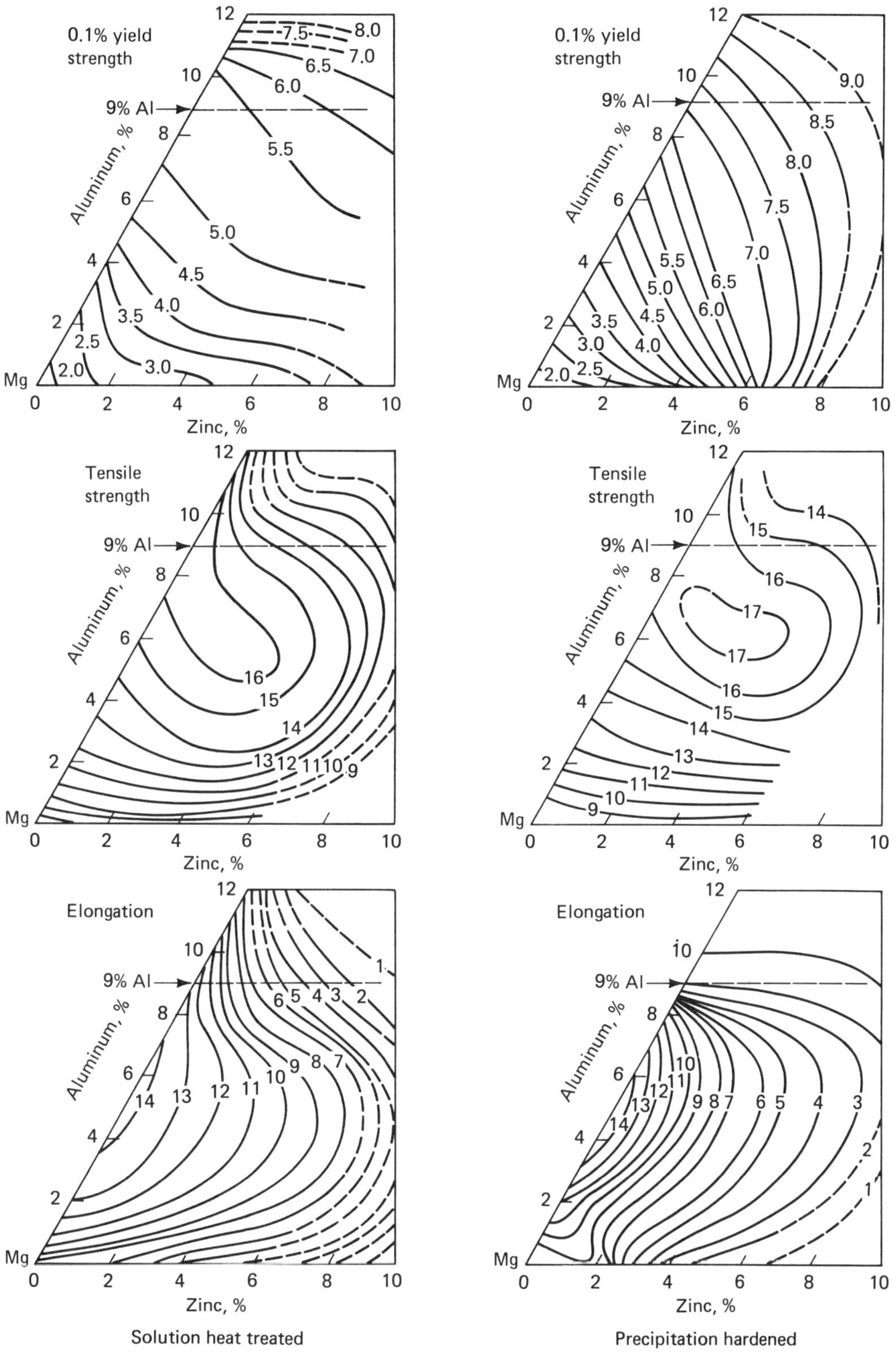

Fig. 7-13. Tensile mechanical properties for solution-treated (left) and precipitation-hardened (right) Mg-Al-Zn alloys. The dashed line is for a fixed Al content of 9%. The alloys were solution heat treated for 30 h at 15 °C below the liquidus. Precipitation-hardened alloys were then held at 200 °C for 15 h. (The units of strength are 10,000 psi.) (Adapted from F. A. Fox, J. Inst. Metals, Vol 71, p 415, 1945)

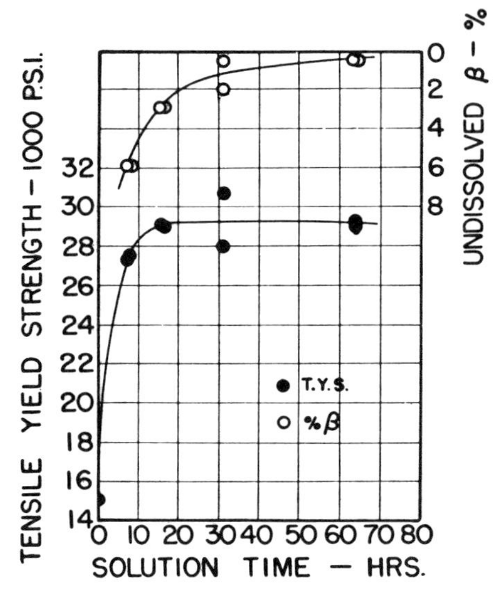

Fig. 7-14. Effect of the solution time at 412 °C on the amount of unsolved precipitate (β) present, and on the yield strength after aging at 190 °C. The samples were quenched from 412 °C into water at 95 °C. (From A.E. Flanigan, I.I. Cornet, R. Hultgren, J.T. Lapsley and J.E. Dorn, Trans. AIME, Vol 175, p 551, 1948)

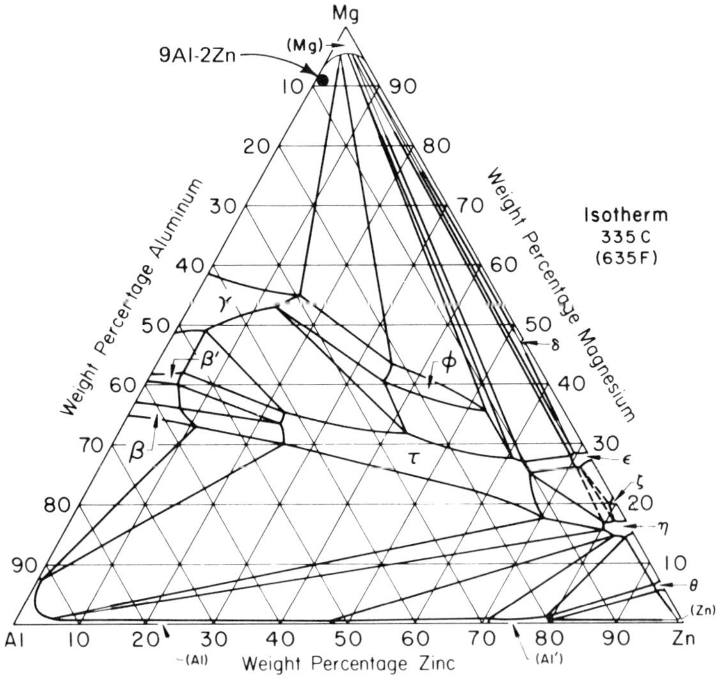

Fig. 7-15. Isothermal section at 335 °C for the Mg-Al-Zn system. (Adapted from same source as Fig. 7-3b)

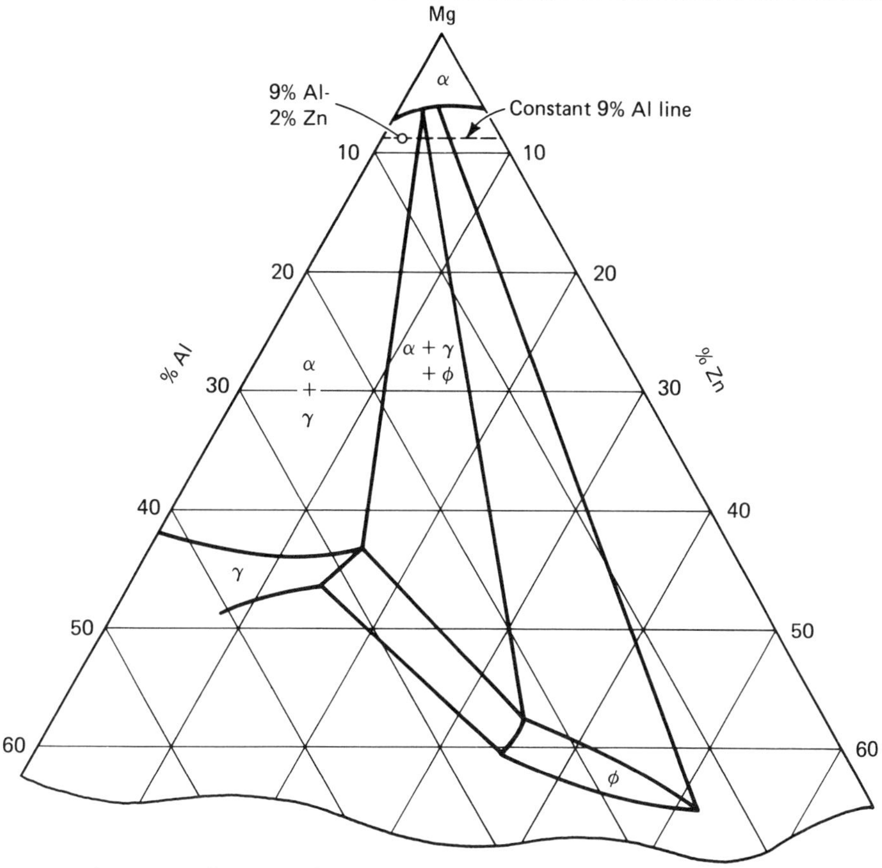

Fig. 7-16. The Mg-rich region of the Mg-Al-Zn phase diagram at 335 °C, illustrating that if the Zn content is too high, the φ phase can form.

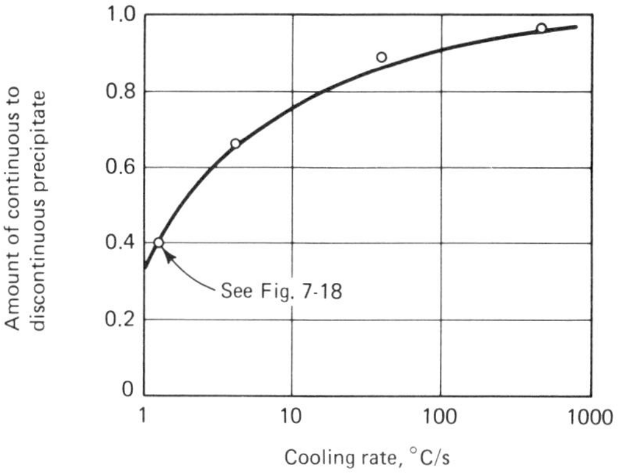

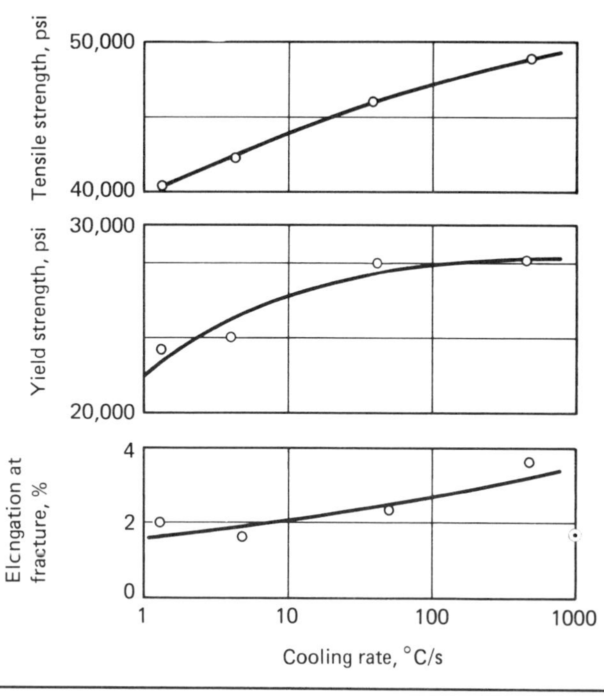

Fig. 7-17. Effect of cooling rate from the solution heat treating temperature on the amount of continuous to discontinuous precipitate and on the mechanical properties of Mg alloy AZ92A (9Al-2Zn-0.2Sn). After cooling, the alloys were aged to 18 h at 177 °C, then tested. (Adapted from R. S. Busk and R. E. Anderson, Trans. AIME, Vol 161, p 278, 1945)

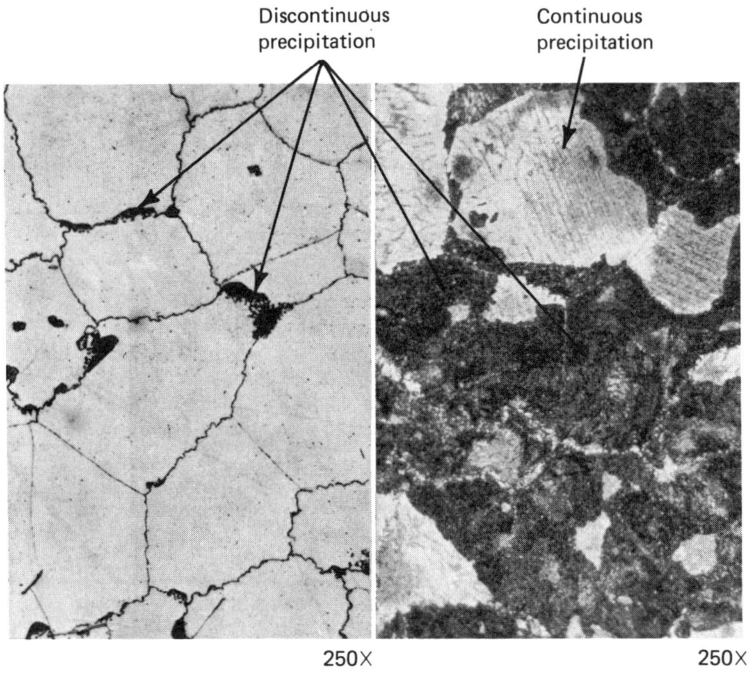

Fig. 7-18. Microstructure of a Mg-9Al-2Zn commercial alloy which was air cooled from 410 °C, then aged for 18 h at 177 °C. (Adapted from R. S. Busk, in Precipitation From Solid Solution, American Society for Metals, Metals Park, Ohio, 1958)

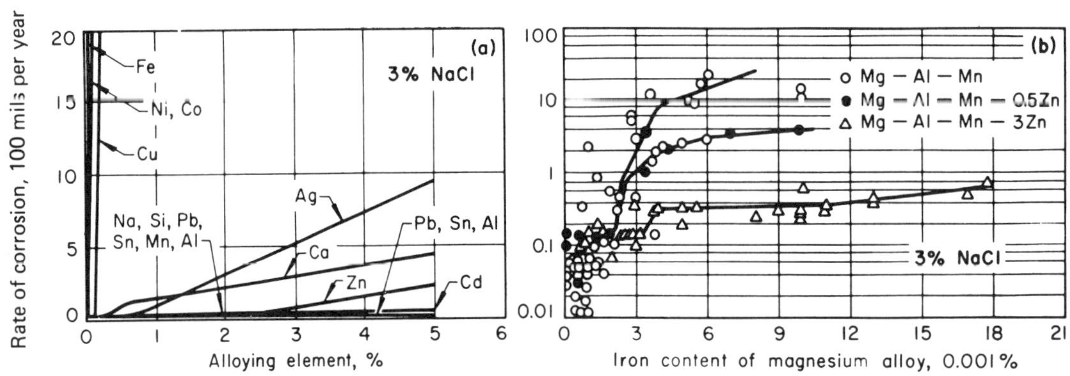

Fig. 7-19. Effect of elements on the corrosion resistance of Mg alloys. Note that Mg is especially sensitive to quite small levels of Fe, Ni, Co and Cu. (From Metals Handbook, 8th Ed., Vol 1, American Society for Metals, Metals Park, Ohio, 1961)

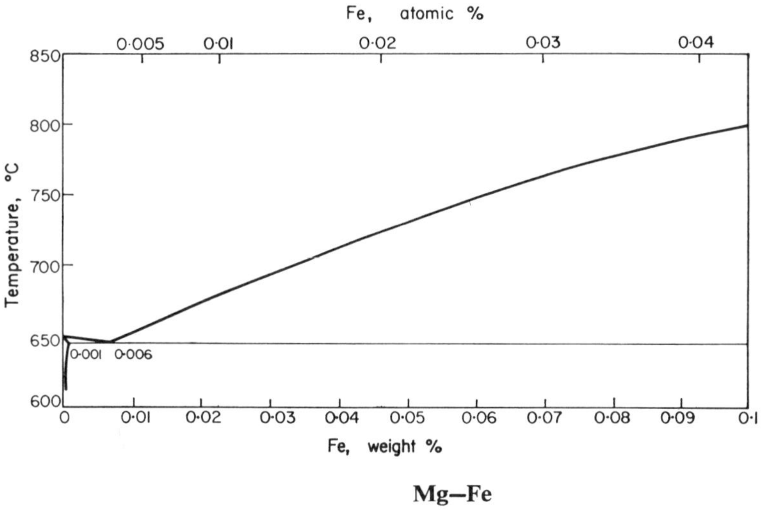

Fig. 7-20. The Mg-Fe phase diagram, showing the extremely low solubility of Fe in solid Mg. (From same source as Fig. 7-1)

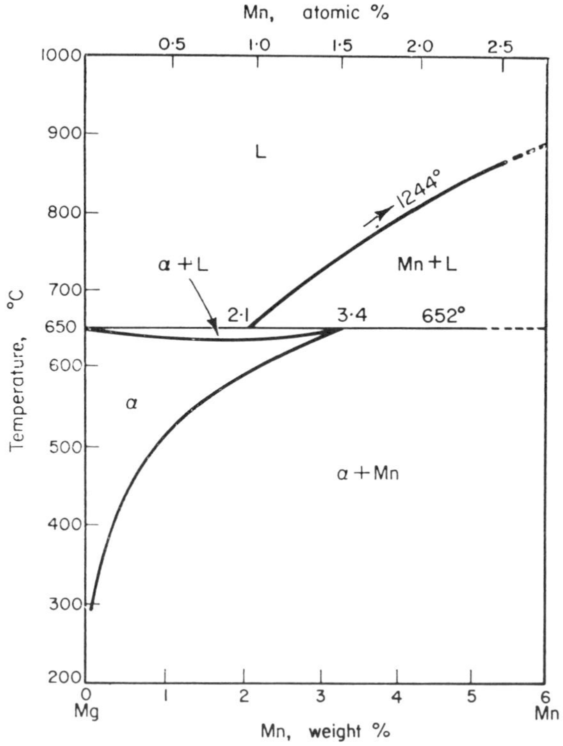

Fig. 7-21. The Mg-Mn phase diagram, showing that in the solid state pure Mn precipitates. (From same source as Fig. 7-1)

The removal of these elements is achieved by treatment of the liquid Mg alloy, additions being made to the melt to precipitate the undesirable elements. This is the origin of the Mn in Mg-Al-Zn alloys (see table, Appendix 5). The Mn is added to the melt, and after several hours in the molten state the Fe is precipitated. Sufficient Mn is added to guarantee this effect, and some Mn is invariably left in the liquid alloy and then precipitates as pure Mn upon heat treating in the solid state. The reason pure Mn forms is shown by the phase diagram in Fig. 7-21, and this is the origin of the Mn particles in Fig. 7-10. For Mg-Zn alloys, the precipitation of these undesirable elements is achieved by a similar method using Zr, which is the reason for this element in these alloys.

References

Magnesium Alloys, R. S. Busk, in Precipitation From Solid Solution, American Society for Metals, Metals Park, Ohio (1958).

The Physical Metallurgy of Magnesium and Its Alloys, G. V. Raynor, Pergamon Press, New York (1959).

Magnesium and Its Alloys, C. S. Roberts, John Wiley & Sons, New York (1960).

Metallic Materials in Engineering, C. H. Samans, Macmillan Co., New York (1963).

Principles of Magnesium Technology, E. F. Emley, Pergamon Press, New York (1966).

Designing Alloys for Die Casting, G. S. Foerster, Metals Engineering Quarterly, Vol 13(1), p 154 (Feb 1973).

Production, Properties and Industrial Use of Magnesium and Its Alloys, P. A. Fisher, International Metals Reviews, Vol 23, No. 6, p 269 (1978).

CHAPTER 8

Copper-Base Alloys

Copper and copper-base alloys have a wide range of properties that account for their extensive use as engineering materials. Their high electrical and thermal conductivity, ease of formability and excellent corrosion resistance under certain conditions are three characteristics that make them attractive.

The most common use of copper is in applications requiring high electrical conductivity, and the reason that this element (or its alloys) is preferred is illustrated in Table 8-1. (Although silver has a slight advantage over copper in this respect, it is considerably more expensive.) To attain high conductivity requires the use of copper in the "pure" form, and therefore in this chapter we will discuss commercially pure copper first. Because the electrical conductivity of copper is very impurity sensitive, considerable attention is paid to the effects of minor amounts of other elements present. The brasses, copper-zinc alloys which are especially noted for their formability, are then discussed. The copper-tin and copper-aluminum alloys are described, and finally the age-hardening copper-beryllium alloys. Although there are many other copper-base alloys of importance, the group chosen for close examination serves well to illustrate the variety of copper-base alloys available and the application to these commercial alloys of the principles of physical metallurgy reviewed in the first three chapters.

A comment about the use of the terms brass and bronze is in order here. Generally, brass means copper-zinc alloys. However, bronze is used to describe many different copper-base alloys, and an attempt is sometimes made to distinguish them (e.g., aluminum bronze, a Cu-Al alloy). In any case, it is difficult to employ the two terms, brass and bronze, without being very careful to modify them when referring to specific alloys. Therefore, in this chapter, these terms are not employed extensively.

Table 8-1. Electrical and Thermal Conductivity of Copper and Several Other Commercially Pure Metals at 20 °C

Metal	Relative electrical conductivity (Copper = 100)	Relative thermal conductivity (Copper = 100)	Metal	Relative electrical conductivity (Copper = 100)	Relative thermal conductivity (Copper = 100)
Silver	106	108	Cobalt	18	17
Copper	100	100	Iron	17	17
Gold	72	76	Steel	13-17	13-17
Aluminum	62	56	Platinum	16	18
Magnesium	39	41	Tin	15	17
Zinc	29	29	Lead	8	9
Nickel	25	15	Antimony	4.5	5
Cadmium	23	24			

From J. S. Smart, Jr., in Copper, ed. A. Butts. Copyright © 1954, Van Nostrand Reinhold Company, New York. Used with the permission of Brooks/Cole Publishing Company.

8-1. Commercially Pure Copper

There are several grades of commercially pure copper, the variety reflecting mainly the need for high electrical conductivity but with other properties. For example, solid solution elements will strengthen pure copper but lower the electrical conductivity; thus, some alloys have a chemical composition which strives for a balance between these two effects. In this section, only three commercially pure copper alloys, which are widely used mainly for their high electrical conductivity, will be examined in some detail: deoxidized copper, oxygen-free copper and tough-pitch copper.

Any disturbance of the periodicity of the copper lattice will scatter electrons and hence decrease the electrical conductivity. Solute elements in the copper lattice, with their different atomic size, cause locally elastically strained regions and hence decrease conductivity. This is the major effect of impurities. However, if the second element combines with the copper to form a second phase (e.g., Cu_2O), then the presence of this second phase usually reduces the conductivity mainly because there is less volume of the copper to carry the current. This effect is not as important as when the second element is in solid solution.

Figure 8-1 shows the effect of several solutes on the conductivity of copper. Note that Ag, Pb, Zn and Cd reduce the conductivity only slightly, whereas P has a potent effect. Note also that these curves cover the range up to only 0.1 wt%, showing that the impurity levels of importance are quite low. Now, in the manufacture of pure copper from the ore, these elements are present from the ore itself, and most are reduced to acceptable levels, usually by selective oxidation. That is, most of these elements have a greater affinity for oxygen than for copper, and hence in liquid copper will react with oxygen more readily than with copper, forming oxides which are transferred to a slag layer which is immiscible with the liquid copper. Once these elements are reduced to low values, then oxygen will be present in the liquid

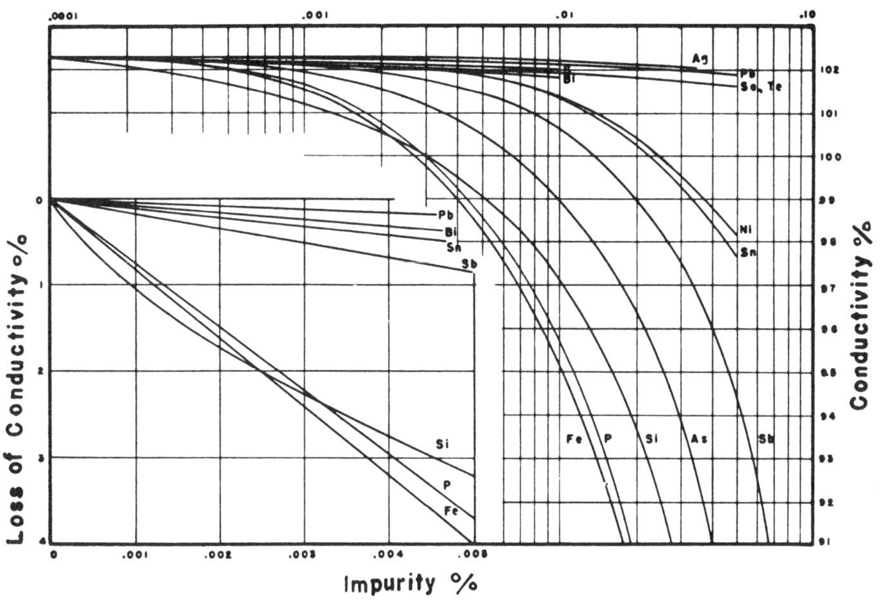

Fig. 8-1. Effect of solute content on the electrical conductivity of oxygen-free copper. The curves give the conductivity of the alloys in terms of the percentage of that of pure copper. (From same source as Table 8-1)

copper. Upon solidification, this oxygen has an effect on the conductivity, and hence it must be contended with.

The copper-oxygen phase diagram is shown in Fig. 8-2. In the range of interest, the alloys undergo an eutectic reaction, forming upon cooling the copper-rich terminal solid solution (α) and the oxide Cu_2O. Note that in the solid state the solubility of oxygen is quite low, so that invariably most of the oxygen is in the oxide. The eutectic composition is 0.39% oxygen.

In the as-cast condition, the structure, if the oxygen content is below 0.39%, will consist of primary α crystals showing coring (see Chapter 2) and eutectic solid. Figure 8-3 shows typical microstructures. After extensive plastic deformation and annealing, the eutectic structure loses its identity, and the oxides may appear as elongated inclusions, as shown in Fig. 8-4. The effect of these oxide inclusions on mechanical properties is slight, as shown in Fig. 8-5. Here the effects of cold working and of subsequent annealing on tensile mechanical properties and on hardness show little difference between tough-pitch copper with about 0.05% oxygen and oxygen-free, high-conductivity copper (OFHC) with essentially no oxygen. (These two grades of copper will be discussed shortly.) Figure 8-6 shows that increasing the amount of Cu_2O does lower the impact energy, although the material remains relatively tough. However, where extensive plastic deformation is required in fabrication, a copper containing lower oxygen may have to be used.

The grades of commercially pure copper available are related to the refining processes used. The most common practice is to take concentrated copper sulfides and pass oxygen (usually as air) through the molten material,

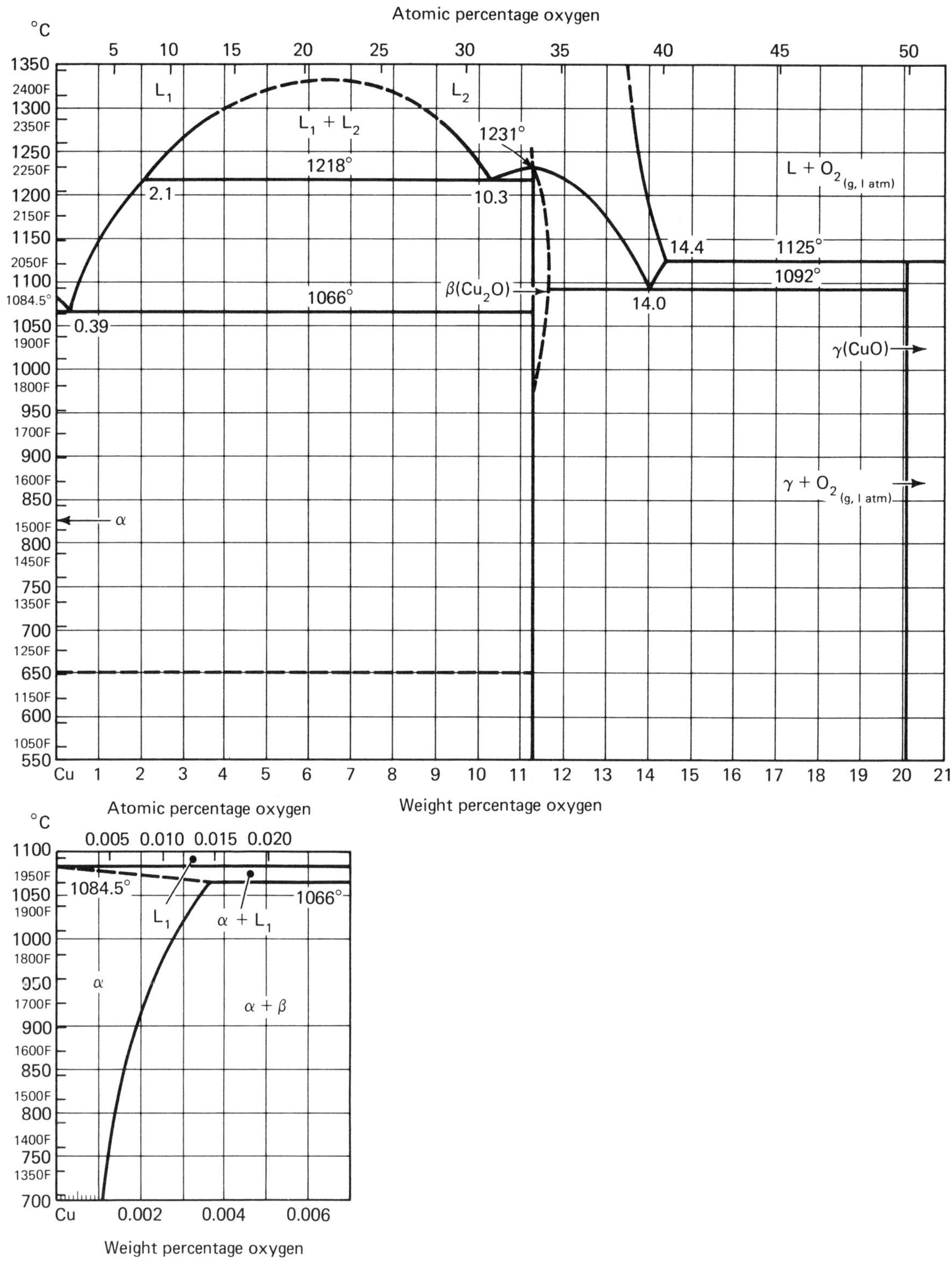

Fig. 8-2. The Cu-O phase diagram. (Adapted from Metals Handbook, 8th Ed., Vol 8, American Society for Metals, Metals Park, Ohio, 1973)

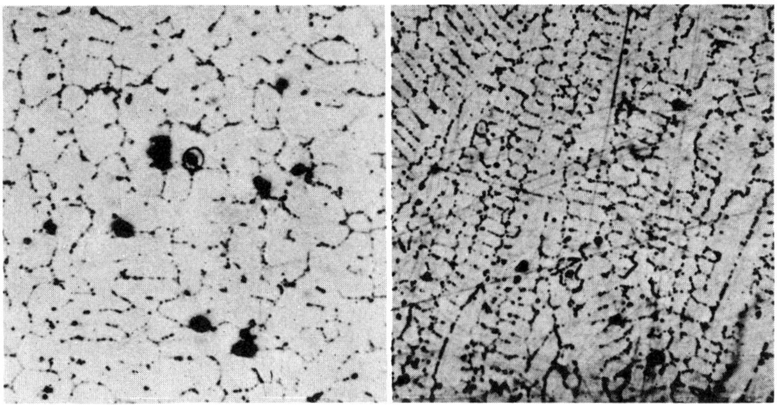

Fig. 8-3. Microstructures typical of as-cast Cu-O alloys. The light areas are the copper dendrite cells, outlined by the dark Cu-Cu_2O eutectic network. The larger black regions are shrinkage voids. (From Metals Handbook, 8th Ed., Vol 7, American Society for Metals, Metals Park, Ohio, 1972)

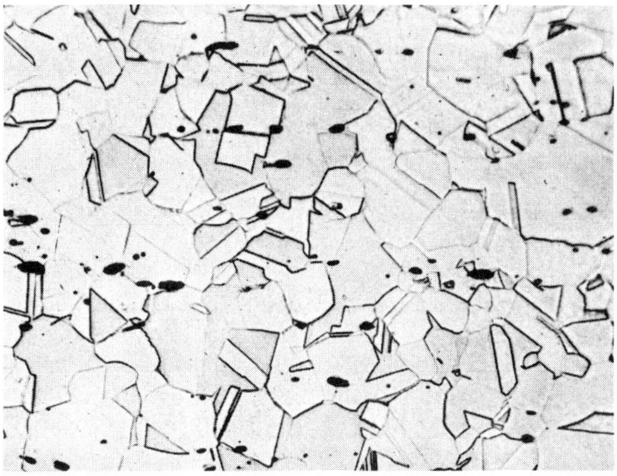

Fig. 8-4. Microstructure typical of wrought and then annealed Cu-O alloys. This is electrolytic tough-pitch Cu which has been hot rolled. The dark particles are Cu_2O inclusions. (Adapted from same source as Fig. 8-3)

effecting separation of an impure copper product known as *blister copper*. This product contains sufficient impurities that it must be further refined to obtain usable industrial copper. One method is to melt the blister copper and further oxidize the molten copper to remove most impurities to an immiscible liquid slag or to vaporize them. This method is usable because of the greater affinity of the impurities for the oxygen dissolved in the liquid copper than for the copper itself. To effect removal of these elements to a sufficiently low

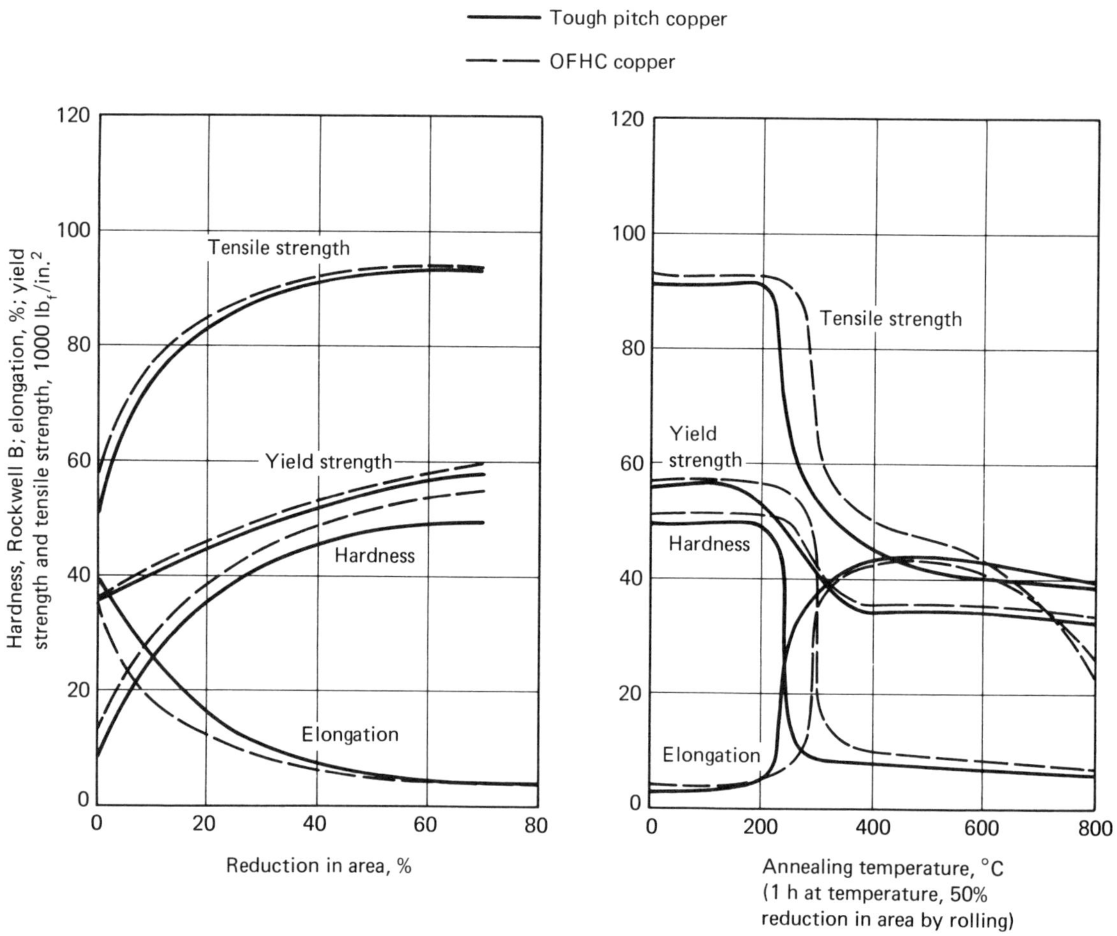

Fig. 8-5. Comparison of the effect of cold working (by rolling at 25 °C) and subsequent annealing on the tensile mechanical properties and hardness of tough-pitch copper (0.05% O) and oxygen-free, high-conductivity (OFHC) copper. Note that the two materials are affected in essentially the same way. (Adapted from R. A. Wilkins and E. S. Bunn, Copper and Copper Base Alloys, McGraw-Hill, New York, 1943)

value, however, requires using excess oxygen, so after these impurities are lowered to an acceptable level, oxygen remains dissolved in the liquid copper and may also be present as Cu_2O if the solubility of oxygen is exceeded (Fig. 8-2). Thus, the oxygen must then be lowered to an acceptable level. This is accomplished by the use of reducing gases, which convert the dissolved oxygen to gaseous oxides and reduce the Cu_2O back to copper. If only pure copper remains, shrinkage during solidification leaves a depression on the surface of a test ingot. However, if some oxygen is present, Cu_2O forms during solidification, along with the low oxygen content copper (α) (see the phase diagram, Fig. 8-2). The oxide has a density of about 6 gm/cm^3 compared to 9 for pure copper, and hence when the copper-oxygen liquid freezes, less contraction occurs than if the liquid were pure copper. That is, the formation of the less dense oxide compensates

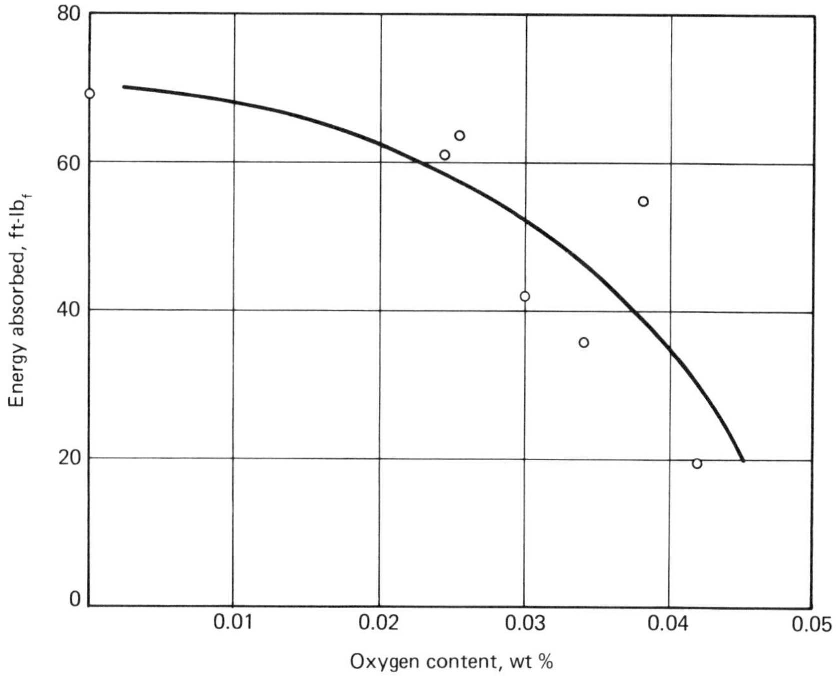

Fig. 8-6. Average Charpy impact energy at 20 °C for electrolytic tough-pitch copper of different oxygen contents. The datum at zero oxygen is for oxygen-free copper. (Adapted from W. R. Opie, P. W. Taubenblat and Y. T. Hsu, in Copper and Its Alloys, The Institute of Metals, London, 1970)

partially for the volume decrease when the solid copper forms from the liquid. In addition, some dissolved hydrogen is released and also compensates for contraction. If the test ingot shows little shrinkage of the surface upon freezing, then the copper is called *tough-pitch*. The oxygen content is about 0.05%, nearly all present in the solidified copper as Cu_2O (about 0.5% Cu_2O present). Note from the phase diagram that 0.05% oxygen will give upon solidification primary α (almost pure copper) and a small amount of eutectic solid consisting of α and Cu_2O. This is shown in the microstructures in Fig. 8-3.

If oxygen-bearing copper is heated above approximately 400 °C in a hydrogen-containing atmosphere (or in some cases other reducing gases), the hydrogen dissolves readily in the copper and then diffuses rapidly. Upon encountering the Cu_2O particles, the reaction

$$\underset{\text{(dissolved in copper lattice)}}{2\underline{H}} + Cu_2O = 2Cu + \underset{\text{(gas)}}{H_2O}$$

occurs. The H_2O molecules formed combine readily into gas (steam) pockets, as these molecules are insoluble in the copper. This leaves a porous structure, and the strength is reduced drastically. Thus, in manufacturing copper tubing and pipe, if annealing is to be in hydrogen to reduce oxidation, tough-pitch copper cannot be used. Figure 8-7 is a microstructure showing void formation due to heating in hydrogen.

There are two important reasons why tough-pitch copper with the Cu_2O particles is undesirable: such copper cannot be used in reducing gases at elevated temperatures, and fabrication, especially by cold working, may be difficult. One method of reducing the oxygen content is to add to the molten copper an element which has a greater affinity for oxygen than for copper. One element commonly used is phosphorus. In the molten state, it will readily react with essentially all of the dissolved oxygen to form phosphorus oxide, which will pass into the immiscible slag. It is difficult, however, to control the reaction so that the amount of phosphorus added is just sufficient to combine with the oxygen but not leave any unreacted phosphorus dissolved in the copper. A practical compromise must be made, and this leads to the presence of some excess phosphorus, which then after solidification shows up dissolved in the solid copper lattice. There are different grades of phosphorus-deoxidized copper, depending upon the amount of residual phosphorus. *Deoxidized low-phosphorus copper (DPC)* contains from 0.01 to 0.04% P. According to Fig. 8-1, this should lead to a significant reduction in the electrical conductivity, and this is substantiated by the values in Table 8-2. The phosphorus-deoxidized copper has an electrical resistivity about 15% higher than that of tough-pitch copper (15% lower conductivity). Thus, this copper is more widely used for piping and tubing than for electrical applications.

If high electrical conductivity and a low concentration of Cu_2O are desired, it is common to use *oxygen-free electronic copper* (OFHC), previously referred to as oxygen-free, high-conductivity (OFHC) copper. This copper is made by very carefully controlling deoxidation during refining, so

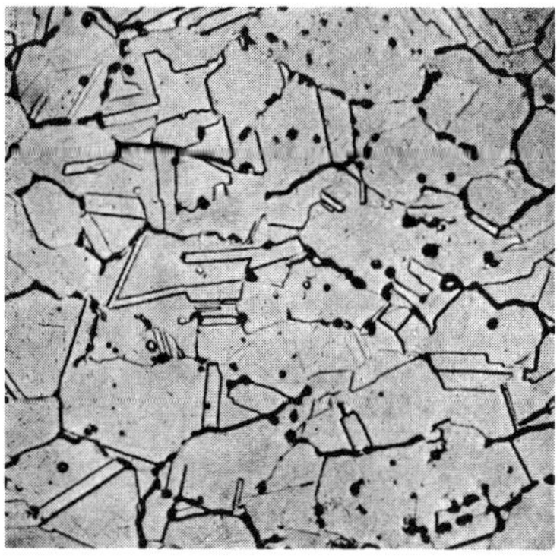

300×

Fig. 8-7. Microstructure of a wrought tough-pitch copper sample which has been heated in hydrogen (2 min at 850 °C) to form water vapor pockets, which impair the mechanical properties. (From C. H. Samans, Metallic Materials in Engineering, Macmillan Co., New York, 1963)

Table 8-2. Approximate Oxygen Content and Properties of Three Commercially Pure Coppers and a Silver-Bearing Copper

	Copper content, wt %	Oxygen content, wt %	Others, wt %	Electrical resistivity at 20 °C, m-Ω/mm²	Thermal conductivity at 20 °C (W/m °C)	Hardness, Rockwell F
Tough-pitch copper.....	99.0 min (Cu + Ag)	0.04-0.05	—	58.6	226	40
Deoxidized low-phosphorus copper.....	99.90	0.01	0.004-0.012 P	49.3	196	40
Oxygen-free electronic copper.....	99.99 min	0.001 max	—	58.6	226	40
Silver-bearing.....	99.90	—	0.03-0.05 Ag (10-15 oz per ton)	58.0	226	40

Adapted from several sources, including A. Davidson, ed., Handbook of Precision Engineering, Vol 2: Materials, McGraw-Hill, New York, 1971.

that both the residual oxygen content and the deoxidizer content are low. Table 8-2 shows that this copper has an electrical resistivity about the same as that of tough-pitch copper, but a much lower oxygen (and hence Cu_2O) content. Thus, this copper can be used where extensive cold working is involved and can be heat treated in reducing gases, whereas, as described above, tough-pitch copper cannot.

Note that all three commercially pure coppers in Table 8-2 have the same hardness. Thus the presence of Cu_2O and of solute atoms in the lattice has negligible effect on the strength properties, but they have a potent effect on the electrical resistivity and also on the thermal conductivity. Also, recall that tough-pitch copper is desirable because the solidification shrinkage is small, so that the ingot solidifies with a rather flat top. Ingots cast from oxygen-free or from deoxidized copper will have on top a large "pipe," or shrinkage cavity, and this portion will have to be removed before hot working the ingot, contributing to scrap loss.

These commercially pure coppers were chosen to illustrate the principles involved in the uses of the pure coppers. There are many reasons that specifications call for control of specific elements and that alloying additions are made to copper. We now look at two examples.

We just saw that one advantage of tough-pitch copper is that it solidifies with a flat top on the ingot, so that the entire ingot can be hot worked. A chemical-composition specification on electrolytic tough-pitch copper is that lead should not be present in amounts greater than 0.005% if the copper is to be hot worked. The reason for this can be seen by examining the copper-lead phase diagram, Fig. 8-8. The melting points of copper and lead differ

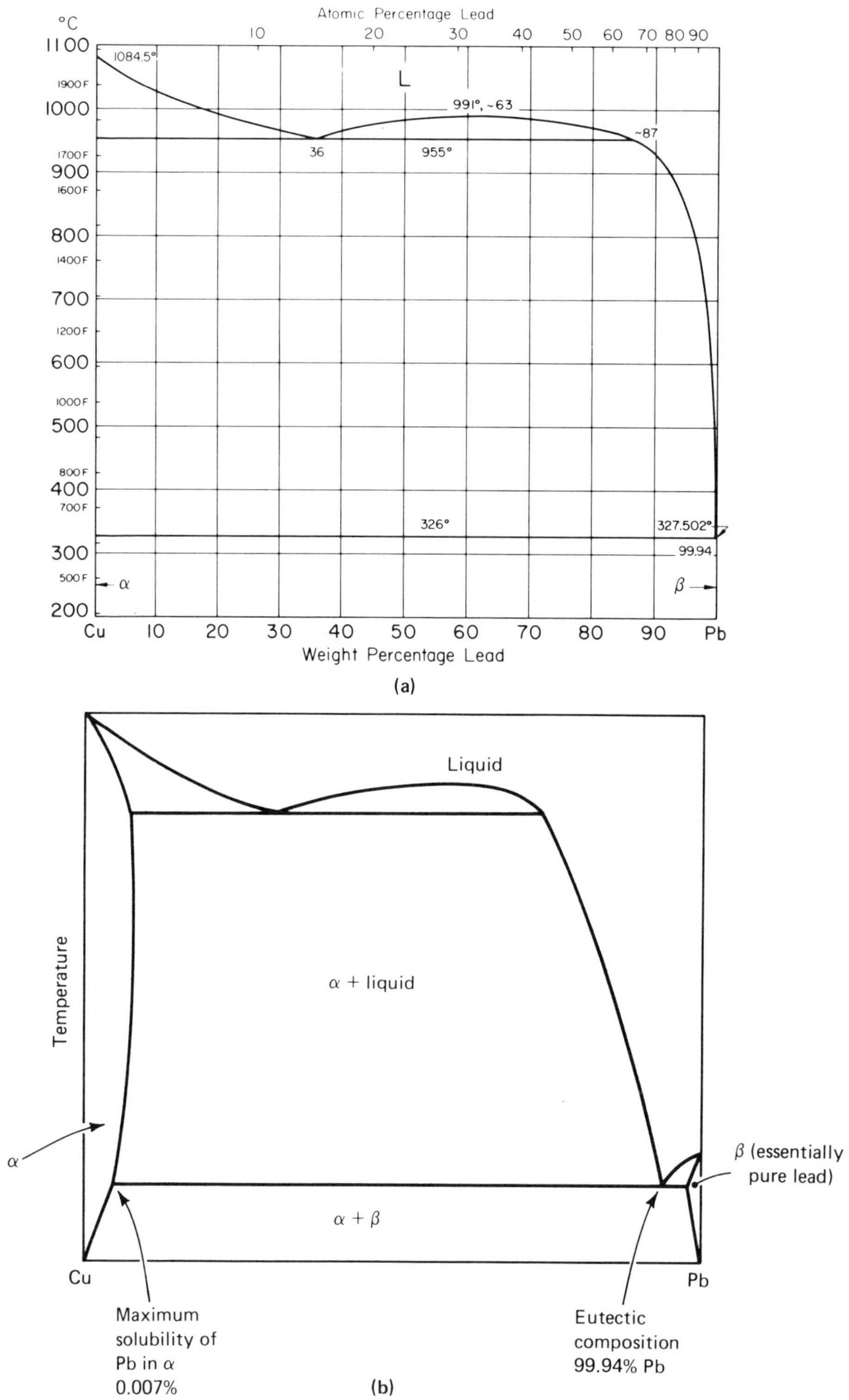

Fig. 8-8. The Cu-Pb phase diagram. The schematic diagram in (b) shows more clearly the location of the eutectic at almost pure lead and the low solubility of lead in copper. (a) adapted from same source as Fig. 8-2.

greatly, and there is an eutectic at 99.94% Pb. From 326 to 955 °C, a two-phase region exists with lead-rich liquid in equilibrium with the copper-rich, terminal solid solution. The solubility of lead in solid copper is extremely small, being about 0.007%. Thus, if the lead content of the copper exceeds this value, upon heating the copper to the hot working range (500-900 °C), the alloy is in the two-phase region. The liquid wets the α grain boundaries and forms a film on them, and upon hot working the alloy disintegrates. (This effect is called *hot shortness.*) For this reason, the chemical specification of lead in tough-pitch copper is set at 0.005% maximum.

The silver-bearing copper is an interesting example of the reason for the addition of a specific element. In many electrical applications, strength is required and is obtained by cold working. However, copper has a rather low recrystallization temperature (e.g., 140 °C), and it is desirable to alloy it with an element which will raise the recrystallization temperature (Chapter 1) and yet retain usable electrical conductivity. Figure 8-1 shows that the element which reduces the electrical conductivity least is silver; the addition of about 0.05% reduces the conductivity only about 1%. However, in most solid solutions the addition of only a very small amount of solute reduces the atom mobility and hence increases the recrystallization temperature. Figure 8-9 shows the effect of several solutes on the recrystallization temperature of copper, and it is seen that the addition of 0.05% silver increases this temperature from 140 to about 340 °C. Although some of the other elements have a comparable effect, they very markedly lower the electrical conductivity (Fig. 8-1). Thus, silver is the most obvious element to add to retard softening of cold worked copper and yet retain excellent electrical conductivity. Table 8-2 compares properties of a silver-bearing copper with those of the three commercial coppers previously discussed. Note the high electrical and thermal conductivity.

8-2. Copper-Zinc Alloys

Copper-zinc solid solution alloys are probably the most widely used copper-base alloys. They retain the good corrosion resistance and formability of copper but are considerably stronger. Zinc is hexagonal close-packed, so the solubility in copper cannot be complete. However, copper is face-centered cubic, a close-packed structure, and the atom size difference is only about 4%, so extensive solubility is expected. The phase diagram is shown in Fig. 8-10, where it is seen that the maximum solubility of zinc in copper is about 38%, and at 20 °C it is about 35%. (The dashed lines indicate that equilibrium is slow to attain, so that the solubility boundary at about 200 °C is valid at 20 °C.) Note that four intermetallic compounds form in this sytem.

As in previous chapters, it is useful to examine the size difference between copper and solute atoms to predict the extent of solid solution strengthening. Recall that a large size difference should favor strengthening,

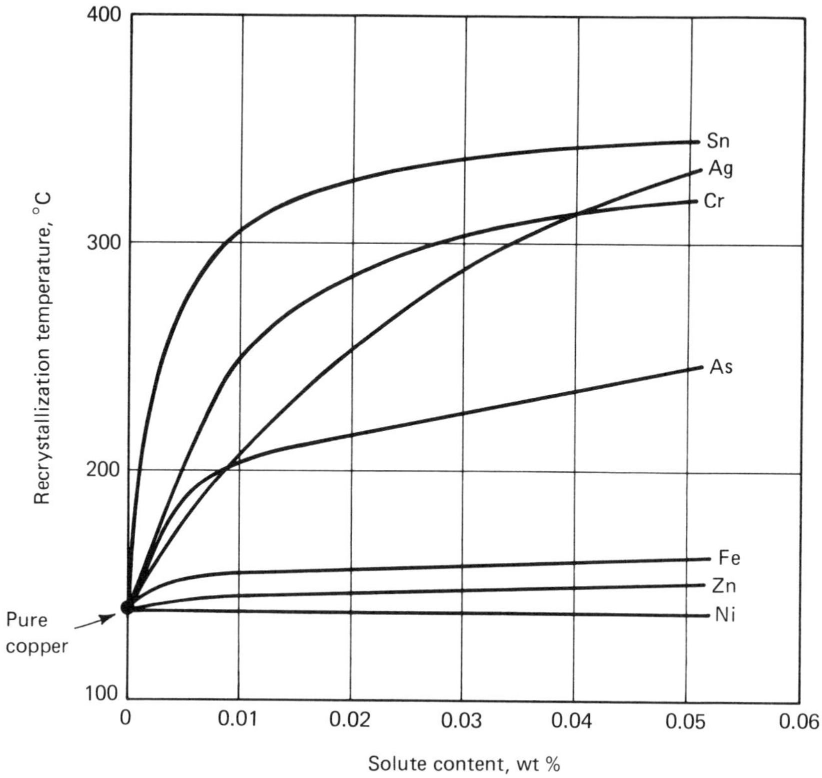

Fig. 8-9. Effect of solute content on the recrystallization temperature of pure copper. These data are for oxygen-free copper, but quite similar curves are obtained for tough-pitch copper. The samples were annealed for 30 min at 600 °C, then cold rolled at 20 °C to a reduction in thickness of 75%. Note that pure copper has a recrystallization temperature of 140 °C. (Adapted from Copper, ed. A. Butts. Copyright © 1954, Van Nostrand Reinhold Company, New York. Used with the permission of Brooks/Cole Publishing Company)

but this usually causes limited solubility. Thus the more usable solutes for solid solution strengthening are those for which the size difference is modest and the solubility large. Table 8-3 lists the atom size difference for several solutes, along with the approximate maximum solubility. Note that zinc has the best combination of size difference and solubility (except for gold, which is expensive).

The data in Table 8-3 are consistent with the data in Fig. 8-11, which show the effect of several solutes on the critical resolved shear stress (see Chapter 1). Those elements which show strong strengthening of copper generally have low solubility. For example, the solubility of indium is only about 1 at.%. (Note in Table 8-3 that tin is an attractive possibility, and Fig. 8-11 shows that it has a significant strengthening effect; the Cu-Sn alloys will be discussed in the next section.)

The effect of zinc is not as potent as that of several of the other solutes, but its high solubility allows the strengthening to be significant at contents in the 30% range (see Fig. 8-12). The solid solution strengthening effect of

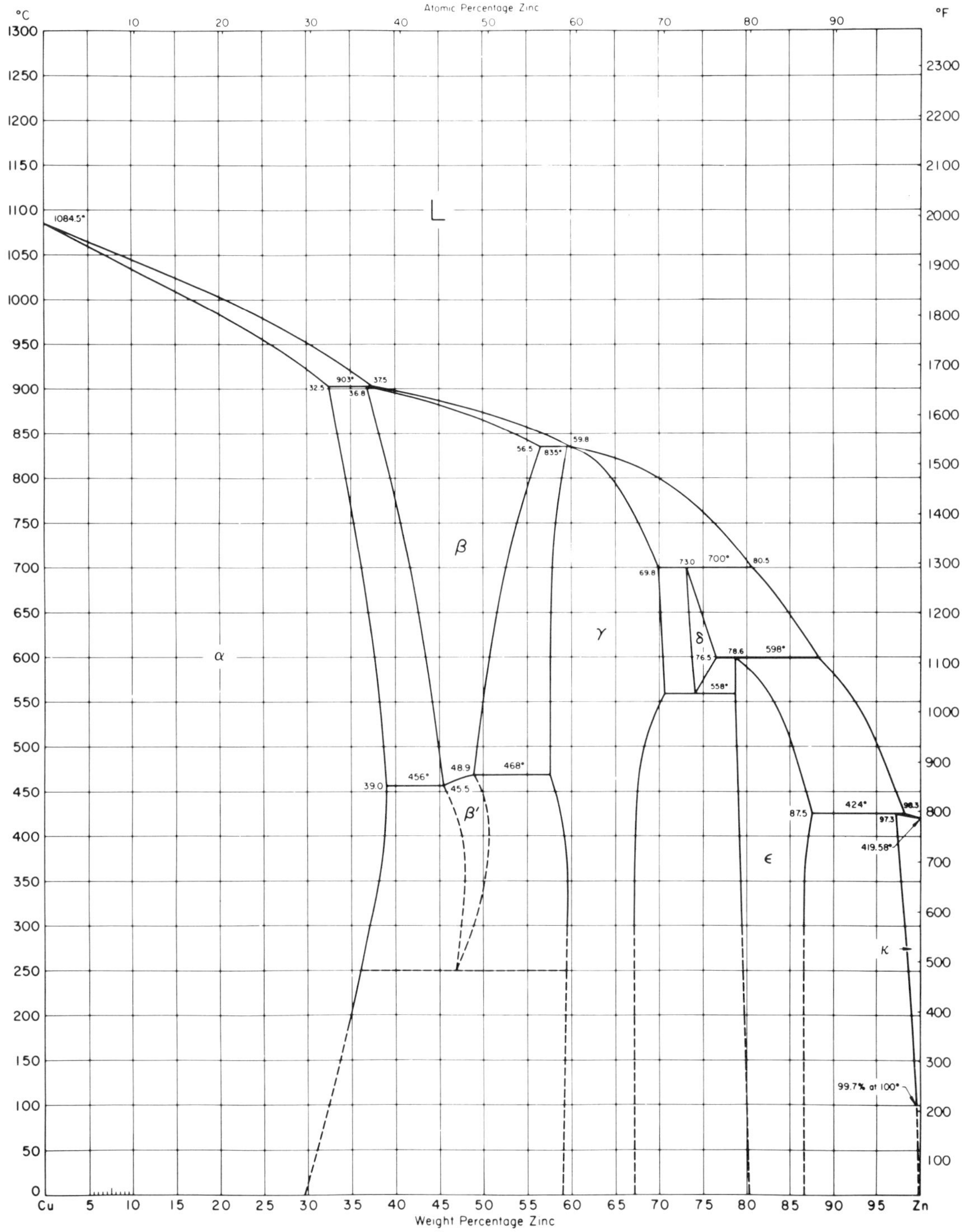

Fig. 8-10. The Cu-Zn phase diagram. (Adapted from same source as Fig. 8-2)

zinc shown by its effect on the critical resolved shear stress was seen previously in its effect on the hardness of polycrystalline Cu-Zn alloys (Fig. 1-23).

Table 8-3. Approximate Atom Diameter Size Difference and Approximate Solubility in Cu of Several Solutes

Solute	Approximate atom diameter size difference, % $(d_{Cu} - d_M)/d_{Cu}$	Approximate solubility in Cu, wt %	Solute	Approximate atom diameter size difference, % $(d_{Cu} - d_M)/d_{Cu}$	Approximate solubility in Cu, wt %
Oxygen	+113	0	Silver	−11	8
Beryllium	+ 14	2	Cadmium	−14	3
Aluminum	− 11	9	Tin	−15	15
Silicon	+ 9	5	Antimony	−12	11
Phosphorus	+ 70	2	Gold	−11	100
Nickel	+ 2	100	Bismuth	−18	0
Zinc	− 4	39	Lead	−27	0
Arsenic	+ 2	8			

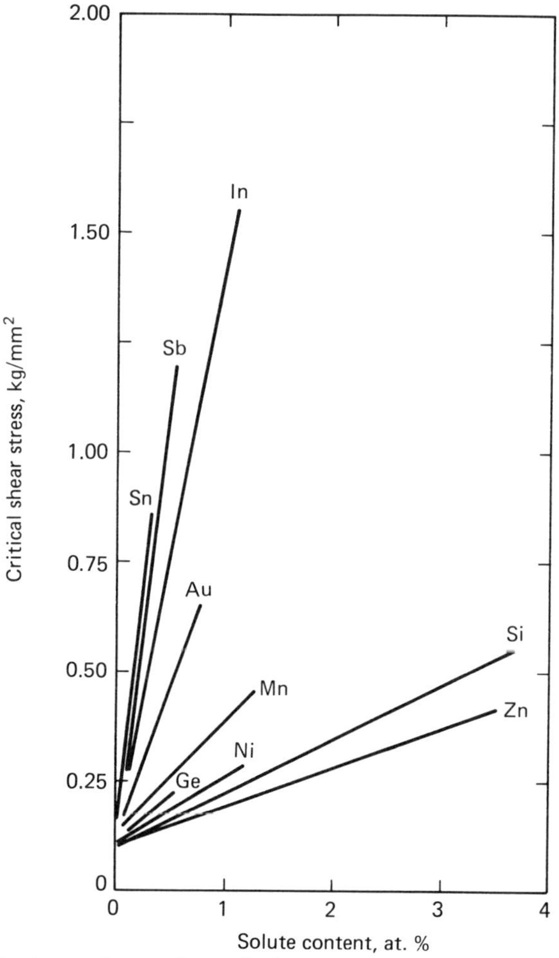

Fig. 8-11. Approximate effect of solute content on the critical resolved shear stress at 20 °C of dilute copper-base solid solutions. (Adapted from W. K. Honeycombe, The Plastic Deformation of Crystals, Edward Arnold Ltd, London, 1968; V. Goler and G. Sachs, Z. Physik, Vol 55, p 575, 1929; and D. McLean, Mechanical Properties of Metals, Wiley, New York, 1962)

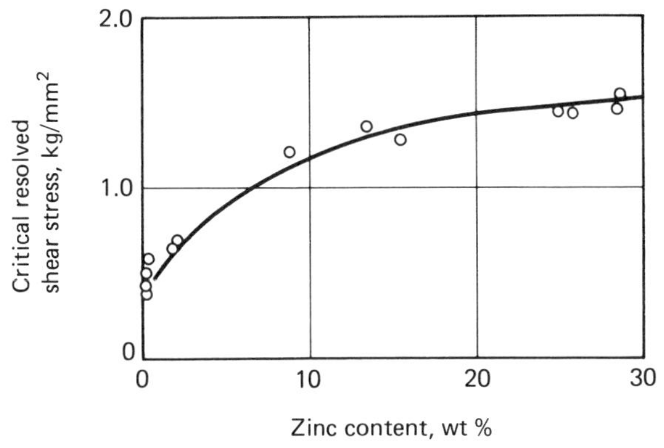

Fig. 8-12. Effect of Zn content on the critical resolved shear stress at 20 °C of Cu-Zn solid solution alloys. (Adapted from V. Goler and G. Sachs, Z. Physik, Vol 55, p 575, 1929)

One of the main uses of copper is in fabricating intricate shapes by deformation (such as deep drawing). The copper-zinc solid solution alloys have the advantage that they retain this deformation ease yet have higher strength than copper. Figure 8-13 shows the tensile strength and elongation at fracture (at 20 °C) of commercially pure copper and several commercial copper-zinc alloys as a function of the amount of cold rolling. Note that even in the initial, annealed condition the copper-zinc alloys containing 20 and 30% zinc show a greater elongation than pure copper and the alloys with less zinc.

Although the copper-zinc alloys are stronger than copper, they work harder at a lower rate (see Fig. 8-14), and necking occurs at a higher strain. Thus upon forcing these alloys into a die, the metal can be deformed extensively before the danger of necking appears, leading to an undesirable local dimensional change and to fracture. Note also in Fig. 8-13 that the 20 and 30% Zn alloys retain their superiority as they are cold worked, although all of the alloys show less ductility with increasing cold deformation.

The curves in Fig. 8-13 are typical of these alloys. The exact curve is sensitive to the beginning grain size. From the viewpoint of fabrication by deformation, the sensitivity of the elongation to the grain size, as illustrated by Fig. 8-15, is quite important. If the grain size becomes approximately the same as, or exceeds, the thickness of the tensile specimen (or the sheet in the case of the data in Fig. 8-15), then there are few or no grain boundaries to inhibit slip and contribute to work hardening, and as the grain size increases the metal shows less elongation at fracture. Thus control of grain size *before* cold deformation is quite important, and the prior annealing processes (see Chapter 1) must be carefully regulated to yield a fine grain size.

Now let us see how these mechanical properties are related to the ease of fabrication by cold deformation operations. To illustrate the relationship, we will consider deep drawing. The response of the metal to drawing depends upon a number of factors, such as loading rate, lubrication and die radius. The description given here is simplified and is designed to illustrate

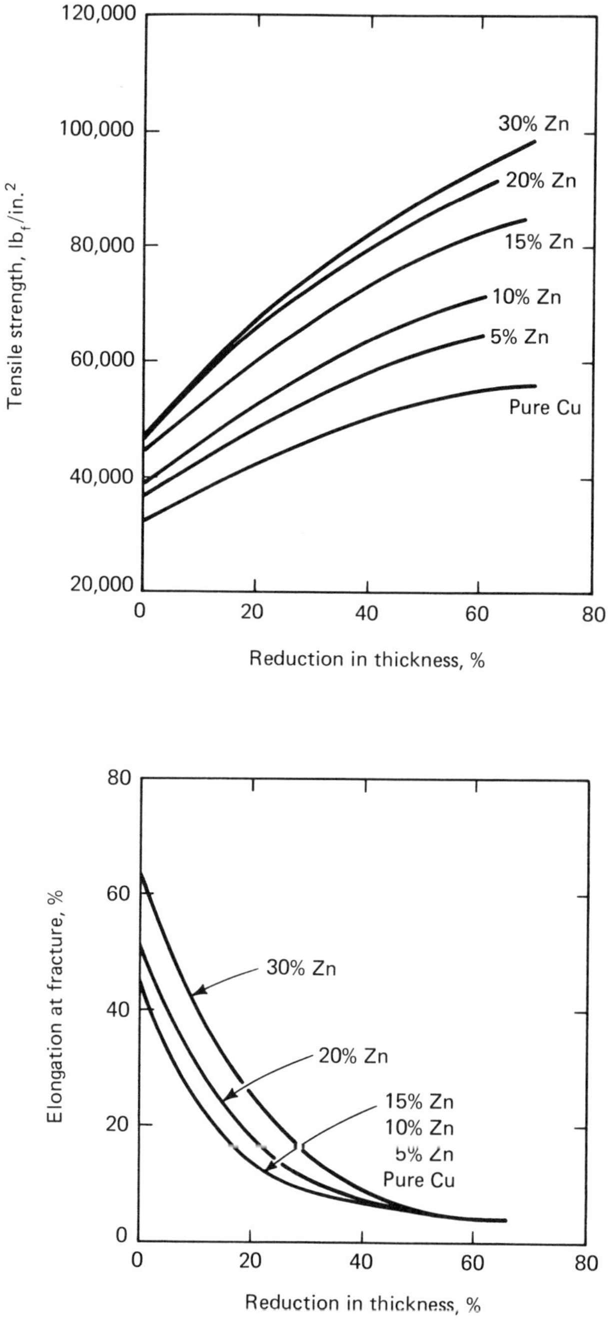

Pure copper . . ETP copper	15% Zn . . . red brass
5% Zn . . . . . . gilding	20% Zn . . . low brass
10% Zn . . . . . commercial bronze	30% Zn . . . cartridge brass

Fig. 8-13. Effect of cold rolling at 20 °C on the tensile strength and elongation at fracture of 0.040-in.-thick sheet of Cu and Cu-Zn single-phase commercial alloys. (Adapted from Metals Handbook, 8th Ed., Vol 1, American Society for Metals, Metals Park, Ohio, 1969)

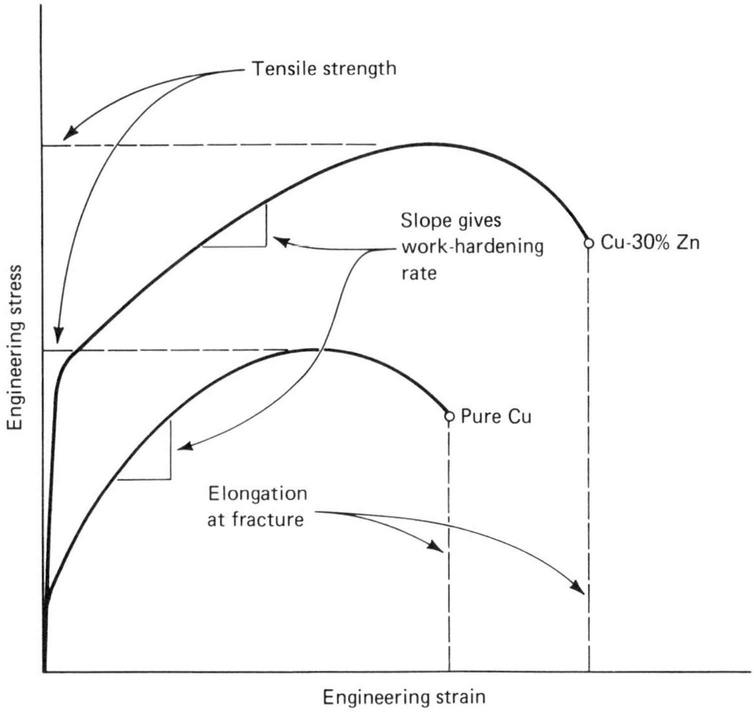

Fig. 8-14. Schematic engineering stress–engineering strain curves for pure Cu and a Cu-30Zn solid solution alloy, showing that the alloy has a higher tensile strength and a lower work hardening rate, necks at a higher strain and has a greater fracture strain (elongation at fracture).

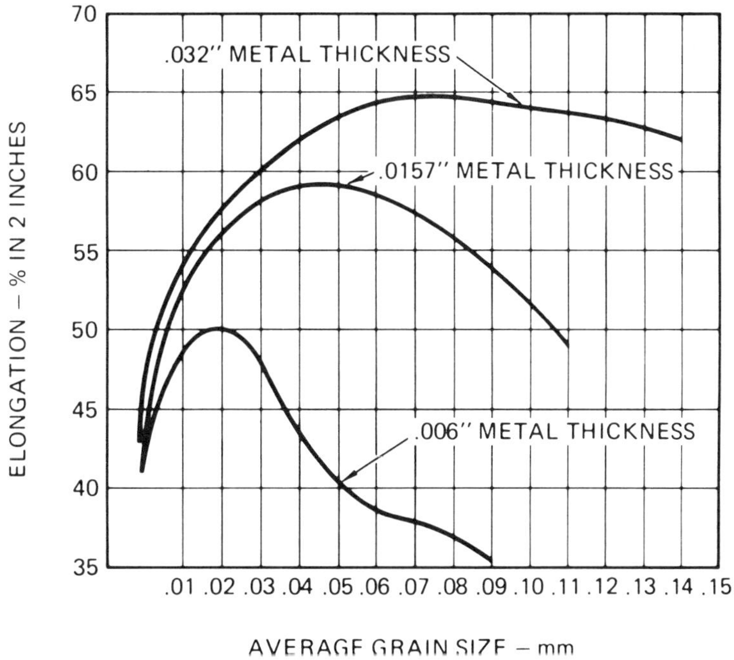

Fig. 8-15. Influence of initial grain size and sheet thickness on the elongation at fracture for the commercial alloy cartridge brass (Cu-30Zn). (From Understanding Copper Alloys, ed. J. H. Mendenhall, Olin Corporation, East Alton, Ill., 1977)

only the general aspects of the process. One type of deep drawing is depicted schematically in Fig. 8-16(a). Here a circular blank is drawn into a cup. The blank is forced into a die by a punch, and in the case shown the clearance between the die and punch is the same as the thickness of the blank. During cold working, density changes only slightly so that as the punch forces the metal into the opening, the tube elongates (compare step 3 with step 4). As the metal is drawn into the die, the thickness does not change and plastic strain occurs only tangentially and longitudinally (plain strain). A continued force must be exerted to cause the metal to flow, and the material must have sufficient strength to withstand the longitudinal force and the ability to plastically strain without necking to prevent thinning of the wall of the tube.

As the diameter of the circular blank increases, necessary to form a longer tube, the force required to form the tube is greater than for a smaller-diameter blank. This is because a greater volume of metal is forced into the same die opening. A measure of drawability of a metal is the maximum ratio of the diameter of the blank to that of the punch (*drawing ratio*) allowable without fracture or wrinkling. Table 8-4 shows the drawing ratio for pure copper and for a Cu-30Zn alloy (cartridge brass) for various amounts of prior work. Note that both materials retain their deformation ability quite well as the degree of prior cold work increases.

In the drawing operation in Fig. 8-16 it is obvious that a wrinkling effect will occur. To avoid this, it is common to use a plate on the blank to maintain the thickness constant and to force the blank to move horizontally (for the configuration in Fig. 8-16) into the die. Also, the clearance between the die and the punch can be less than the blank thickness, so that the tube height is increased. All of these factors complicate the flow analysis and clearly point out that a material with high ductility and high strength is required.

We now examine the higher-zinc alloys in which the β phase is present (Fig. 8-10). These alloys have rather complex phase transformations, which give rise to a variety of microstructures and hence properties. Their usefulness relies on excellent hot workability and excellent machinability. To illustrate these alloys, we will discuss the 40% zinc alloy in some detail. This alloy is essentially the commercial alloy Muntz metal.

Table 8-4. Drawability(a) of Pure Cu and Cu-30 Zn alloy.

	Approximate % reduction in thickness by rolling at 20 °C (temper designation)	Maximum drawing ratio using hemispherical punch
High-conductivity copper	0 (0)	2.20
	11 (1/4 H)	2.17
	21 (1/2 H)	2.12
	37 (H)	1.72
70/30 brass	0 (0)	2.17
	11 (1/4 H)	2.12
	21 (1/2 H)	2.07
	37 (H)	1.80

(a) Maximum ratio of the diameter of a circular blank to that of the punch; see Fig. 8-16. Adapted from C. J. Smithells, Metals Reference Book, Vol 2, Butterworths, London, 1962.

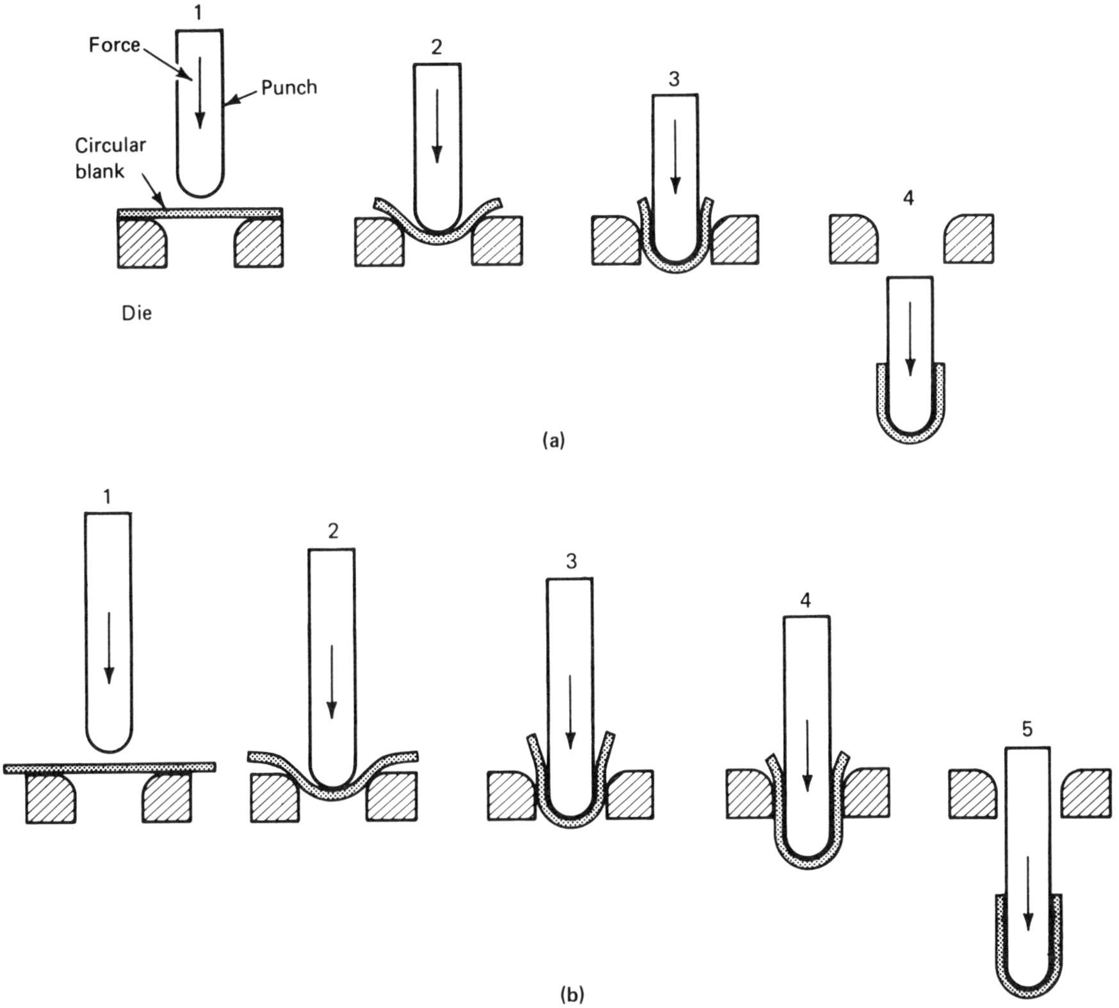

Fig. 8-16. Schematic illustration of a drawing operation. If the ratio of the diameter of the circular blank to that of the punch increases (b), the force to draw the tube increases, and the wall of the tube must withstand more force. This ratio of diameters is called the drawing ratio, and the maximum ratio which will withstand fracture is a measure of drawability. Some data are given in Table 8-4.

The β phase is referred to as an intermetallic compound, meaning a phase based on an approximate stoichiometry of the elements. Thus, note in Fig. 8-10 that the β phase is approximately centered at CuZn. However, the chemical composition of intermetallic compounds can usually be varied from the stoichiometric ratio and yet retain the crystal structure of the compound. For the β phase in the Cu-Zn system, at 800 °C it is stable from about 39 to 55% Zn. This range decreases as the temperature is lowered, being stable only from 45 to 49% Zn at 500 °C.

The β phase is body-centered cubic, and, in spite of the approximate stoichiometry, at temperatures above about 470 °C the Cu and Zn atoms are located at random on the lattice sites. There is a binding energy between the

Cu and the Zn atoms, and hence a tendency for the atoms to take on preferential positions relative to each other in the lattice. However, at high temperatures the thermal energy, in the form of lattice vibrations, is sufficient to keep the atoms dispersed randomly on the lattice sites. As the temperature decreases, the vibrations are reduced, and a temperature is reached (called the critical temperature) below which this preferential siting occurs. The phase is then said to be ordered (long-range ordered), or to have formed a *superlattice*. This is the β' phase in Fig. 8-10. The crystal structure is shown in Fig. 8-17.

The critical temperature for the ordering process is dependent upon the composition of the β phase, and this is the line shown in Fig. 8-10 separating the β and the β' (the ordered structure) region. The lines extending into the $\alpha + \beta$ and into the $\beta + \gamma$ regions give the critical temperature for the β phase in equilibrium with either α or γ.

Figure 8-10 shows that the β' phase undergoes an eutectoid reaction at 250 °C, forming α and γ. However, note that the phase boundaries are shown dashed, indicating considerable experimental uncertainty in their location. Also, the eutectoid reaction requires considerable time to initiate. Therefore the phase boundaries shown in the phase diagram in Fig. 8-18 can be used. Here, the β' phase is considered stable to 20 °C.

The process by which the ordered structure forms from the disordered lattice can be quite complex and in some alloys is not well understood. However, a simplified description of a possible process is useful in pointing out some factors which might affect mechanical properties. In Fig. 8-19(a) is depicted a (110) plane for a 50 at.% Zn-Cu alloy. The atoms are placed randomly in the schematic drawing. The picture in (a) would change with time, as the lattice thermal vibrations would allow atoms to move from one lattice site to another. This would occur mainly by vacancy movement, but the vacancies are not shown in the figure and they are neglected for simplicity in the following description.

The description in Fig. 8-19(a) typifies that of the alloy above the critical temperature. However, as the temperature is lowered below the critical temperature, the lattice vibrations no longer are sufficiently strong to

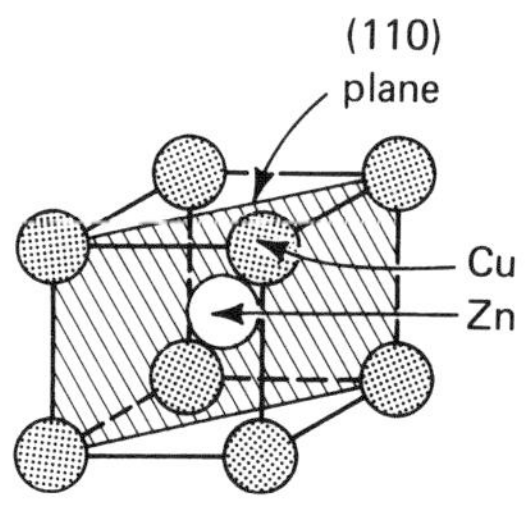

Fig. 8-17. The ordered (superlattice) β' phase for the exact composition of equal numbers of atoms of Cu and Zn (about 49 wt % Zn). The unit cell could just as well have been drawn with a Cu atom in the center and Zn atoms on the corners.

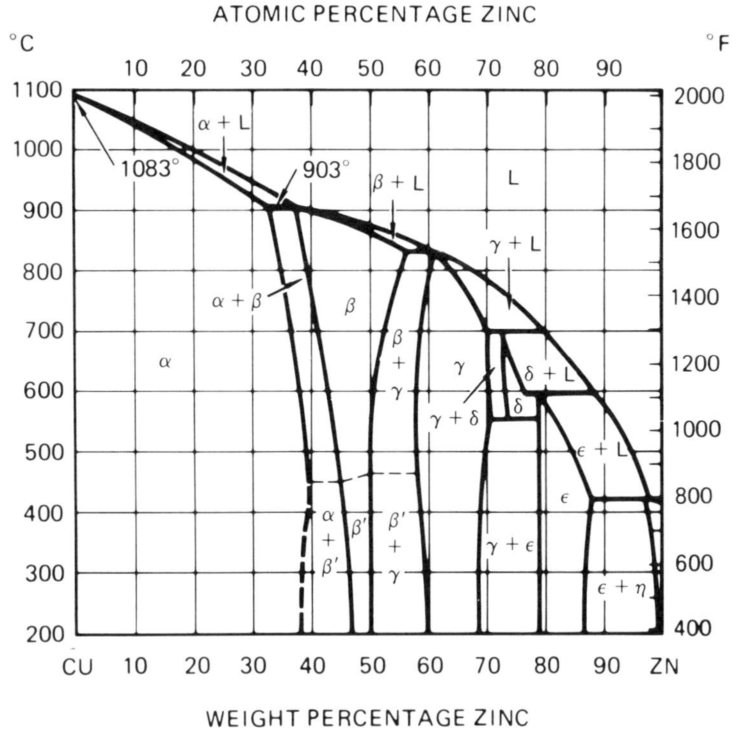

Fig. 8-18. The Cu-Zn phase diagram in which the β' phase is stable below the critical temperature. This diagram will be used in discussing the heat treatment of alloys containing the β' phase.

overcome the attraction between the Cu and the Zn atoms, and if the atoms become arranged as they are in Fig. 8-17 the configuration will be stable. Now, in spite of the fact that the arrangement in Fig. 8-19(a) was achieved by placing the closed and open circles randomly on the array, small regions can be found which have the ordered structure. When the alloy is cooled below the critical temperature, we will assume that these regions act as nuclei for the ordering process. In Fig. 8-19(b) are shown two perfectly ordered regions in the random arrangement. We will assume that only these two act as nuclei. For these to grow, atoms at the interface must relocate to extend the arrangement. The required atom exchanges, shown by the arrows in Fig. 8-19(b), lead to the perfectly ordered regions increasing their size as shown in Fig. 8-19(c). These small ordered regions are called *domains*.

For the domains to grow further requires the atoms in the disordered material right at the domain-disordered lattice (or β'-β) interface to continue to rearrange in order to propagate the correct arrangement. If the atoms are moved as indicated in Fig. 19(d), the two domains eventually contact. However, note that the particular choice of the original domain nucleus leads to a mismatch at the interface of the two domains. Such an interface is called an antiphase *domain boundary*.

The perfection of the order within a domain is referred to as the *degree of order*. For example, each of the domains depicted in Fig. 8-19(d) is

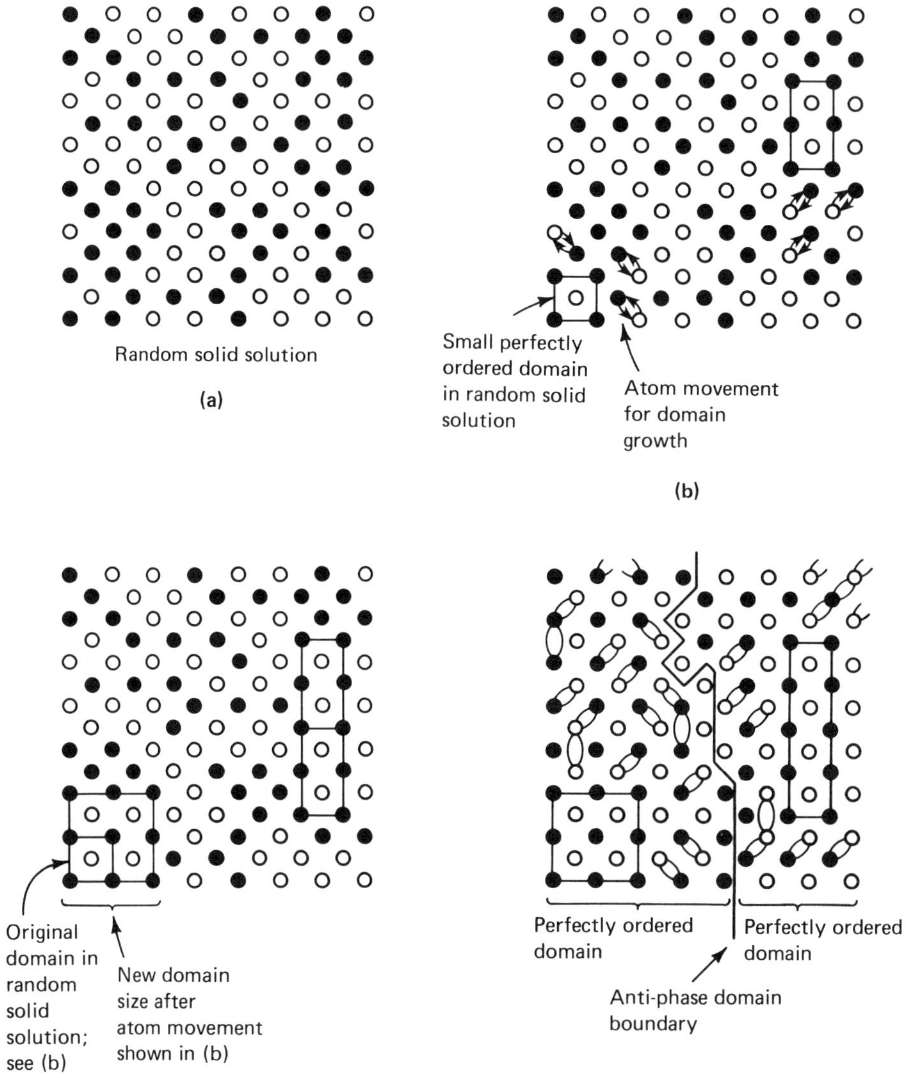

Fig. 8-19. Schematic illustration of a possible mechanism of the formation of two ordered domains. The plane shown is a (110) type (see Fig. 8-17).

perfectly ordered (with degree of order $S = 1$). If a few Cu and Zn atoms swapped positions, the degree of order would be less than perfect (S between 1 and 0). If the arrangement were random, $S = 0$. In the β' structure, the equilibrium degree of order is essentially unity below about 250 °C, and approaches zero as the critical temperature is approached. However, the actual degree of order in an alloy may be influenced by the heat treatment.

Two characteristics can be established here. One is that in the Cu-Zn β phase, even rapid cooling from β cannot suppress the ordering. This is reasonable from the description in Fig. 8-19, as it is expected that many nuclei exist and that only a few atom exchanges will bring the domains into contact, with the alloy locally completely ordered. The other characteristic is that a fine domain size should be established, with many antiphase boundaries. This should contribute to strengthening. (See the discussion in Chapter 5 of the effect of the ordered structure on properties.)

We now examine some mechanical properties of the β' phase. Figure 8-20 shows that the strength of the ordered β' can be increased by aging in the intermediate temperature range (200 to 500 °C) below the critical temperature. This reflects the influence of the heat treatment on the degree of order and on the domain structure. Figure 8-21 shows the influence of cooling rate from the β region on the hardness. Very slow cooling should induce the equilibrium amount of order and allow development of a relatively coarse domain size. Increased cooling rate from the β region should develop a fine domain size and perhaps allow less than the equilibrium amount of order to develop. This fine domain size is probably responsible for the increased hardness as the cooling rate increases. Again, it is necessary to remember that even very rapid cooling cannot suppress the ordering; it just develops a very fine domain size.

The data in Fig. 8-22 show that above the critical temperature the ductility and toughness of the β-composition alloys increase markedly.

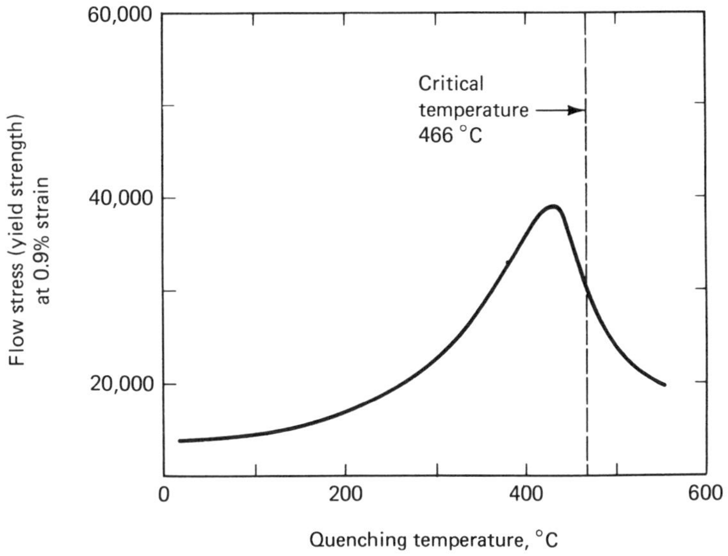

Fig. 8-20. Yield strength as a function of quenching temperature for β brass (Cu-49.5Zn). The alloy was originally slowly cooled from 500 °C to 25 °C (giving a completely ordered structure), reheated for 15 min at temperature, then water quenched. In this condition the flow stress was measured at 25 °C. From 20 to about 200 °C, the alloy is essentially in the condition obtained by slow cooling from 500 °C, which is why the strength is about the same. However, from 200 to about 500 °C, the degree of order depends on the temperature and hence affects the strength measured after quenching. (Adapted from H. Green and N. Brown, Trans. AIME, Vol 197, p 1240, 1953)

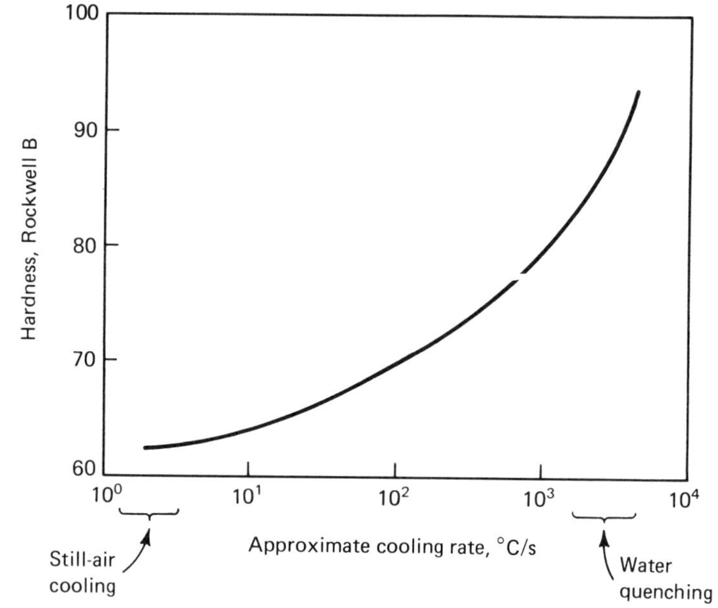

Fig. 8-21. Effect of cooling rate from 500 °C on the hardness at 25 °C of β brass (Cu-47Zn). The alloy was held for 15 min at 500 °C, then cooled in different media. The cooling rates shown are estimated, ranging from water quenching to cooling in still air. (Adapted from N. Brown, Acta Met., Vol 7, p 210, 1959)

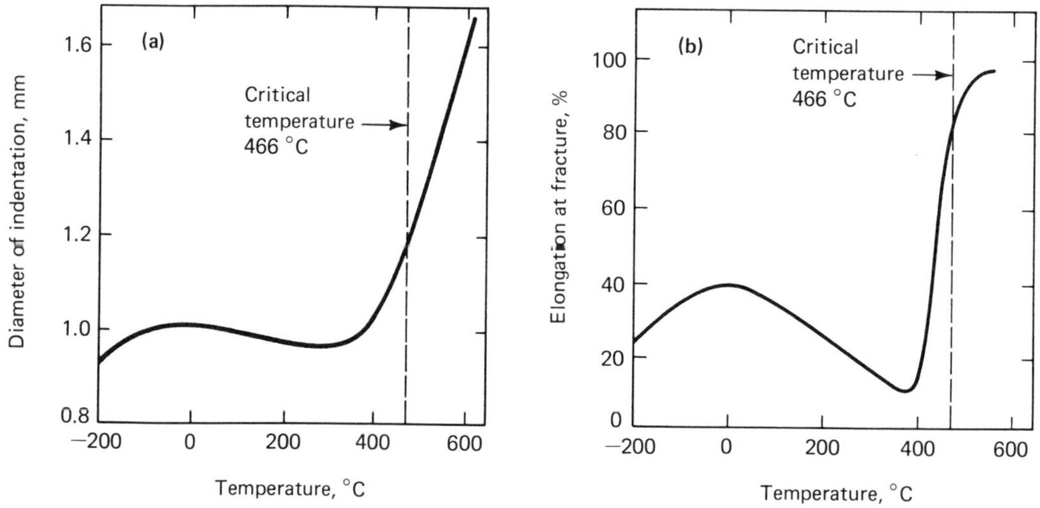

Fig. 8-22. Increased ductility of β brass (approximately Cu-49Zn) above the critical temperature. (a) Plot of the diameter of the impression of a steel ball dropped from a constant height, giving the impact hardness (adapted from C. S. Barrett, Trans. AIME, Vol 200, p 1003, 1954). (b) Plot of the elongation at fracture in a tension test (adapted from E. C. W. Perryman, Trans. AIME, Vol 203, p 693, 1955).

Thus, the high-temperature, disordered β phase deforms much easier than the ordered β' structure. One important characteristic of copper-zinc alloys containing β and α is that they show better hot workability (but not cold workability) than alloys consisting of only the α phase.

With this brief description of the β and β' phases, we return to a consideration of a Cu-40Zn alloy. The β' alloy is not suitable for commercial use, as it is brittle. However, alloys in which the β' phase coexists with the ductile α phase are useful. The Cu-40Zn alloy can be heat treated at high temperature so that it is all β. The structure developed at lower temperatures depends upon the heat treatment, as this controls the precipitation and formation of the α phase. If the alloy is cooled slowly from 800 °C, the phase diagram (Fig. 8-18) shows that at 20 °C the alloy should consist of approximately equal amounts of α and β'. Figure 8-23 shows a typical microstructure. The α phase can be identified by the presence of annealing twins, characteristic of the face-centered cubic structure (see Chapter 1).

A curve in Fig. 8-24 shows that the amount of β' influences the hardness. The alloy was cooled slowly from 700 °C, where it was mostly β, to 20 °C, then reheated to temperature for 30 min, followed by rapid cooling. Upon heating at 800 °C, the structure is all β, and upon cooling rapidly little α forms. However, the β orders to β', giving a hardness around 90. Reheating for 30 min in the lower temperature range, 20 to 500 °C, was not sufficient to significantly affect the originally slowly cooled structure, and the hardness remains constant. In this temperature range, the structure is

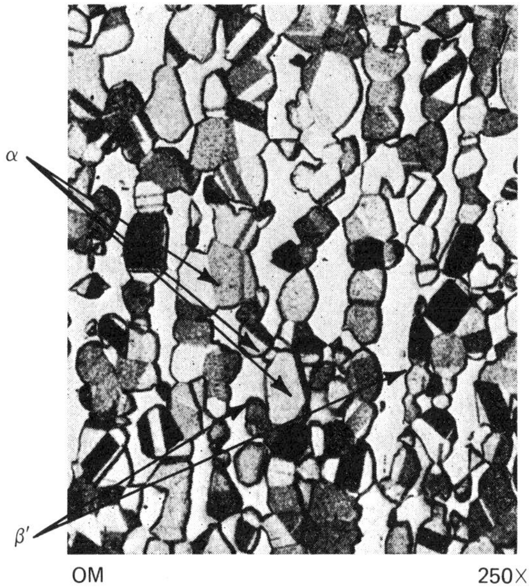

Fig. 8-23. Typical microstructure of annealed Muntz metal (Cu 40Zn). The clear, white regions are the β', and the dark and gray regions showing annealing twins are α. (Adapted from D. K. Crampton, Metal Progress, Vol 46, p 276, 1944)

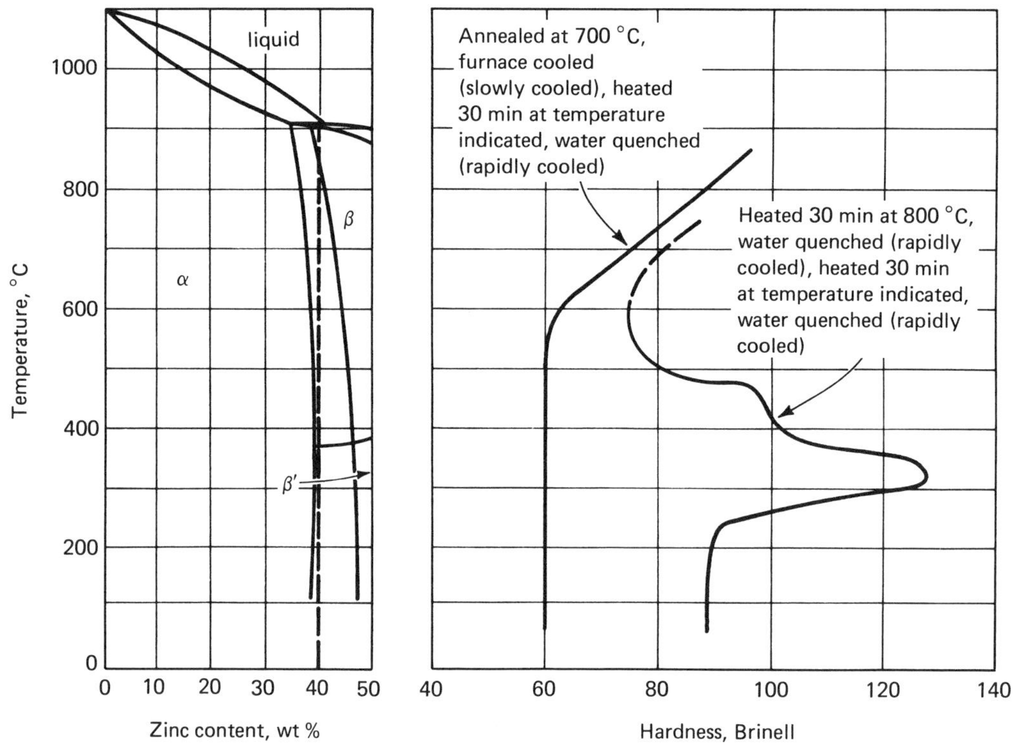

Fig. 8-24. Influence of heat treatment on the hardness at 20 °C of a Cu-40Zn alloy. (Adapted from T. Matsuda, J. Inst. Metals, Vol 39, p 67, 1928)

approximately equal amounts of α and β'. However, as the temperature increases from 500 °C, 30 minutes is sufficient time to allow the equilibrium amounts and α and β to form. Thus, as the temperature increases, increasing amounts of β and decreasing amounts of α are present at temperature, giving increasing amounts of β' upon cooling rapidly to 20 °C, and hence a rise in hardness.

If the Cu-40Zn alloy is cooled rapidly to 20 °C after sufficient holding (e.g., 30 mins) above about 750 °C, a structure of essentially all β' is obtained. Often some α is observed to have formed in the β grain boundaries, and the morphology will vary somewhat depending upon the exact cooling rate. Usually, the α is present as "needles" emanating from the boundaries, with a clear crystallographic relation between the α and the β' in which it has formed. Figure 8-25 shows two examples.

Upon reheating β' in the intermediate temperature range, the morphology of the α formed will vary depending upon the exact heat treatment. Also, recall that reheating will influence the change in the ordered structure. Both changes affect properties, and the hardness can be increased considerably by judicious treatment. In Figure 8-24 are shown hardness data for a Cu-40Zn alloy after reheating for 30 min following initial treatment of quenching from 800 °C (giving a structure similar to those in Fig. 8-25).

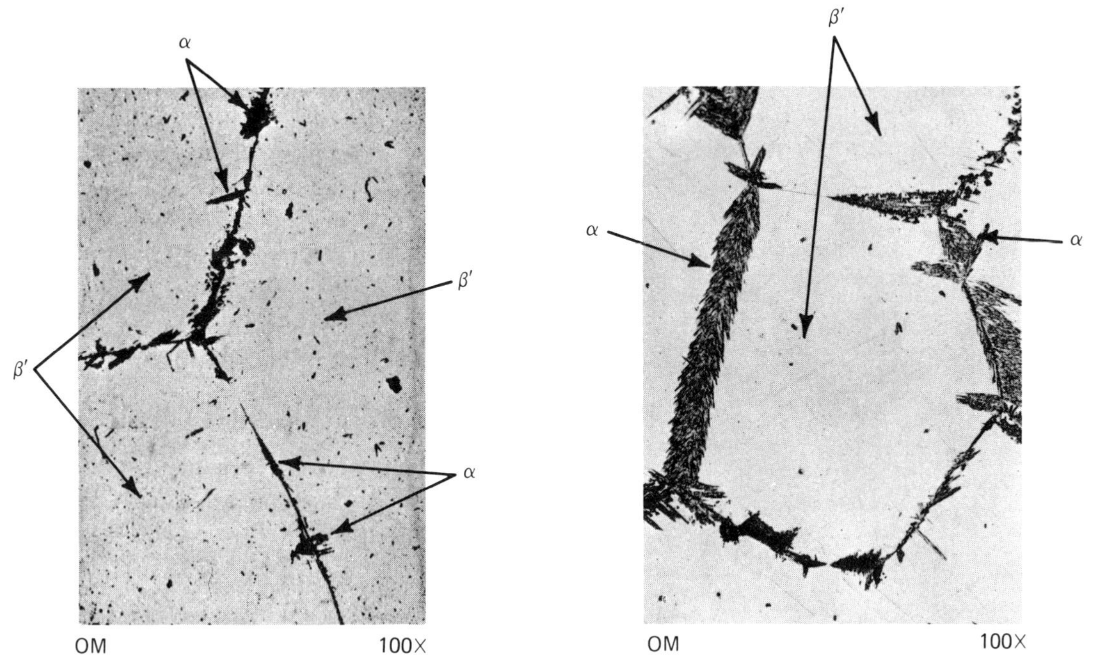

(a) (b)

Fig. 8-25. Microstructures typical of Cu-40Zn alloys cooled rapidly from the β region to 20 °C. Even rapid cooling has not prevented some α from forming. (a) Cu-40Zn alloy, quenched into ice water from 825 °C (adapted from Engineering Physical Metallurgy, by R. H. Heyer, copyright © 1939, Van Nostrand Reinhold Company; used with the permission of Brooks/Cole Publishing Company). (b) Quenched Muntz metal (adapted from same source as Fig. 8-3).

Supposedly the maximum hardness obtained by treatment around 300 °C is caused by the formation of a fine α precipitate and some changes in the ordered β' phase. The type of microstructures obtained for such heat treatments is illustrated in Fig. 8-26 for a Cu-42Zn alloy. In this alloy the zinc content is sufficiently high to completely suppress any α formation upon rapid cooling from β, giving at 25 °C only β' (Fig. 8-26a). Reheating for 30 min at 400 °C gives a fine α precipitate on the β' grain boundaries, and a fine intercrystalline precipitate of α (Fig. 8-26b). Reheating for 30 min at a higher temperature, 600 °C, gives a coarser α structure, shown in Fig. 8-26(c).

If the rate of cooling from the α region is quite slow (several hours to 20 °C), then α nucleates at a high temperature where the nucleation rate is low, and the α crystals grow relatively large as few crystals nucleate. This gives a rather coarse structure, typified by Fig. 8-27(a). As the cooling rate increases, the nucleation rate increases, but the individual α crystals do not have time to grow large before the temperature is too low for significant growth to continue. This gives a finer structure (see Fig. 8-27b) and will increase strength. Eventually, the cooling rate becomes sufficient to suppress

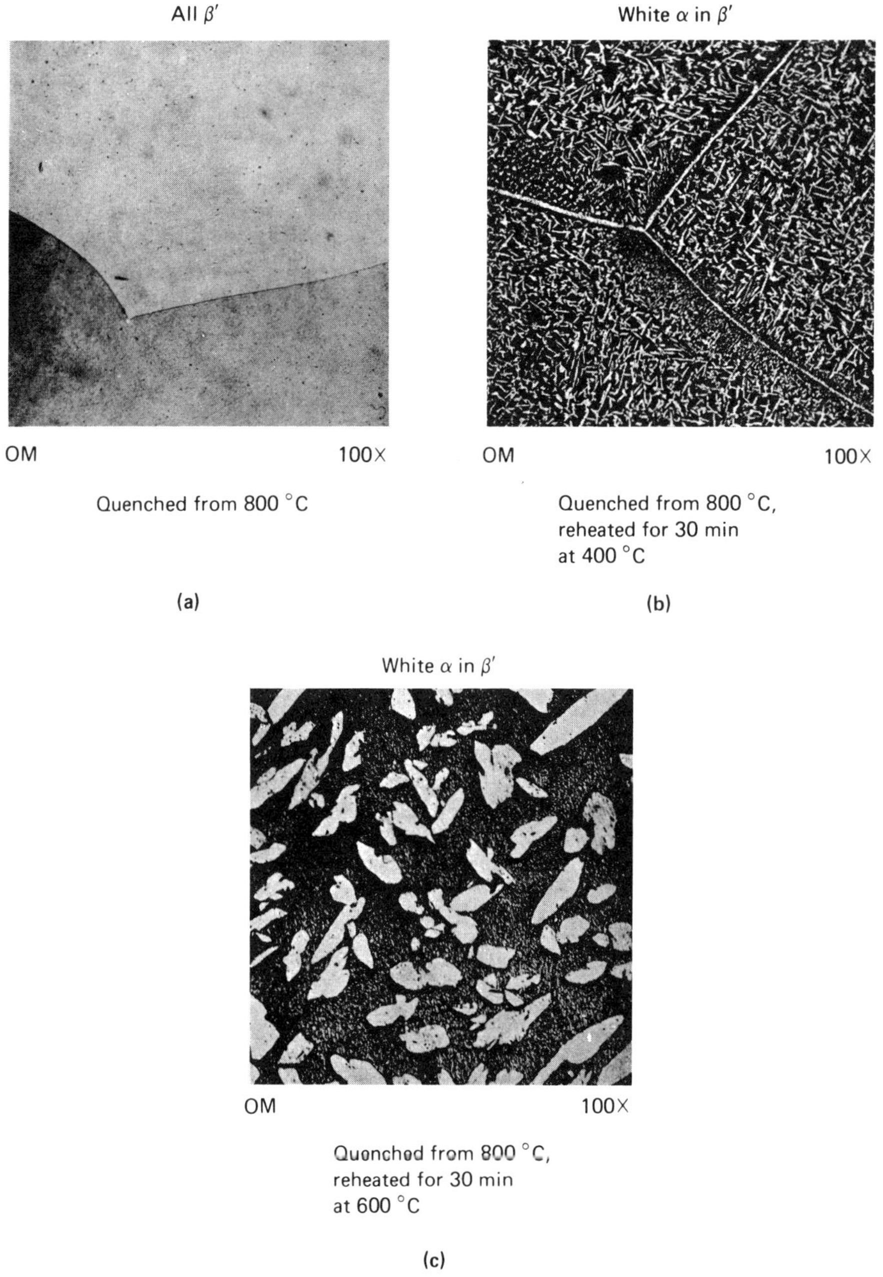

Fig. 8-26. Microstructures of a Cu-42Zn alloy quenched from the β region, then reheated to develope an α precipitate structure. The higher reheating temperature gives a coarser structure and hence a softer material. (Adapted from same source as Fig. 8-7)

the formation of α altogether, giving a structure entirely of highly unstable β' at 20 °C. As was shown in Fig. 8-25, however, it is difficult in the Cu-40Zn alloy to completely suppress some α formation even upon rapid cooling. Figure 8-28 illustrates the influence of cooling rate from β on the hardness.

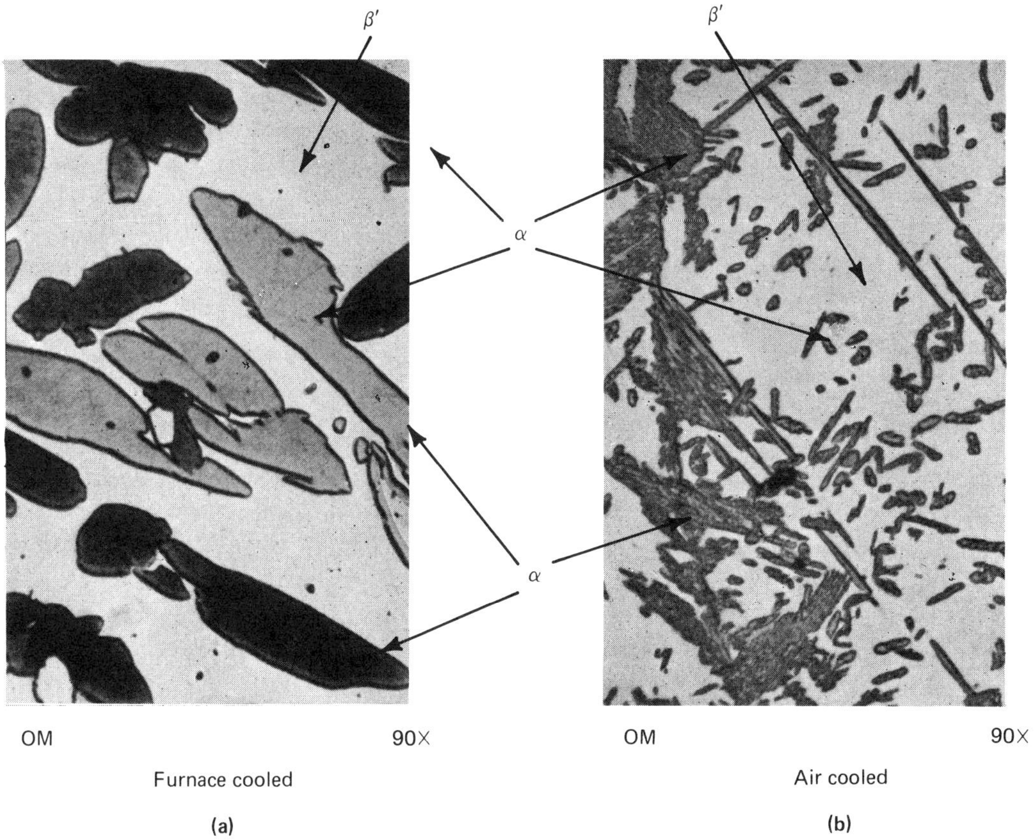

Fig. 8-27. Microstructure of a Cu-43Zn alloy after cooling from 700 °C, the β region. (Adapted from R. F. Mehl and G. T. Marzke, Trans. AIME, Vol 93, p 123, 1931)

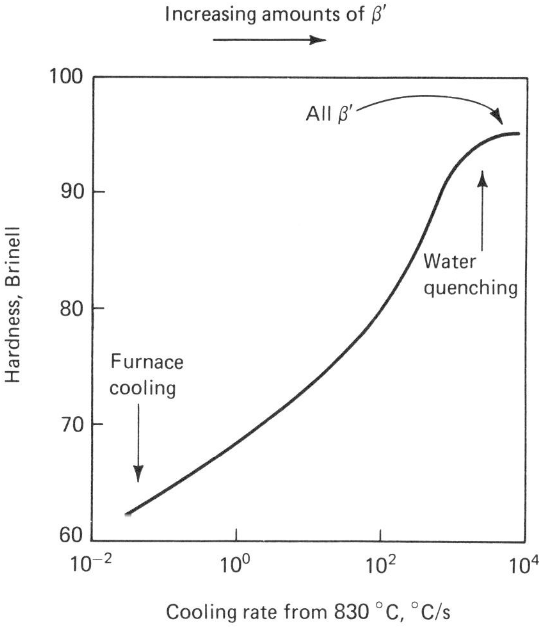

Fig. 8-28. Effect of cooling rate from the β region on the hardness of a Cu-40Zn alloy. (Adapted from same source as Fig. 8-24)

Table 8-5. Mechanical Properties Typical of Cartridge Brass (Cu-30Zn) and Muntz Metal (Cu-40Zn)

	Range of properties available		
	Tensile strength, $lb_f/in.^2$	Yield strength, $lb_f/in.^2$	Elongation, %
Cartridge brass	44,000-130,000	11,000-65,000	63-3
Muntz metal	57,000-74,000	21,000-55,000	52-10

		Approximate properties for specific treatments		
		Tensile strength, $lb_f/in.^2$	Elongation, %	Hardness, Rockwell B
Cartridge brass	Annealed	50,000-60,000	65-50	10-50
	Annealed + 40% cold work	80,000-90,000	8-5	84-90
	Annealed + 70% cold work	95,000-105,000	4	92-95
Muntz metal	Annealed	53,000-57,000	55-47	30-38
	Annealed + 40% cold work	80,000-90,000	10-5	85-90
	Annealed + 70% cold work	101,000-103,000	6-4	93

Adapted from Metal Progress Databook 1975, American Society for Metals, Metals Park, Ohio, 1975, and R. A. Wilkins and E. S. Bunn, Copper and Copper Base Alloys, McGraw-Hill, New York, 1943.

In Table 8-5 are shown the range of properties usually available for the commercial alloy Muntz metal, which contains 40% Zn, and for commercial cartridge brass, which contains 30% Zn. In the 30% Zn alloy the range is obtained by choice of the amount of cold work and of the annealing treatment. In the 40% Zn alloy, the range is achieved by cold working and by annealing. However, in this alloy the annealing and cooling rate from the high temperature affect the α structure and the ordered β' structure, which affect the hardness.

Again, alloys which consist of all β' are too brittle for industrial use, but the lower-zinc alloys with structures of $\alpha + \beta'$ are usable. However, the data in Table 8-5 show that the Cu-40Zn alloy is not preferred because of strength. Instead, recall that the β phase has excellent hot workability, and thus the 40% Zn alloy is frequently chosen to fabricate objects by hot working. Further, it is found that the 40% Zn alloy has better machinability than the 30% Zn alloy, although the latter is considered to have good machinability. This characteristic of the 40% Zn alloy is believed to be due to the presence of the brittle β', allowing removal of material with less energy consumption and the development of a fine surface finish.

8-3. Copper-Tin Alloys

Examination of Table 8-3 reveals that tin is a solute that has a large size difference and reasonably high solubility, so that the copper-tin solid solutions should have quite usable strength. Indeed, the Cu-Sn solid solutions are considerably stronger than the Cu-Zn alloys in dilute solutions (Fig. 8-11), and Fig. 8-29 shows that a Cu-8Sn alloy is harder than a Cu-30Zn alloy. However, note in Fig. 8-29 that the elongation decreases with increasing tin content. Although the Cu-8Sn alloy is quite ductile, it does not have nearly the workability of a Cu-30Zn solid solution alloy.

The copper-tin phase diagram in Fig. 8-30 shows that the solubility of tin decreases markedly with decreasing temperature. However, below about 300 °C the rate of precipitation of ε is low, so that alloys up to about 10% Sn will be single-phase α after proper homogenization and cooling to 25 °C. It also appears that these alloys would be precipitation hardenable, since δ has a high hardness. But the precipitation process is quite slow, so that the time for precipitation is too long to make such a treatment commercially feasible. The data in Table 8-6 show that aging a Cu-10Sn alloy for 5 h produces no significant difference in tensile properties and hardness from the as-cast condition.

The wide temperature range between the liquidus and solidus makes these alloys very susceptible to coring (see Chapter 2). Also, cast alloys with

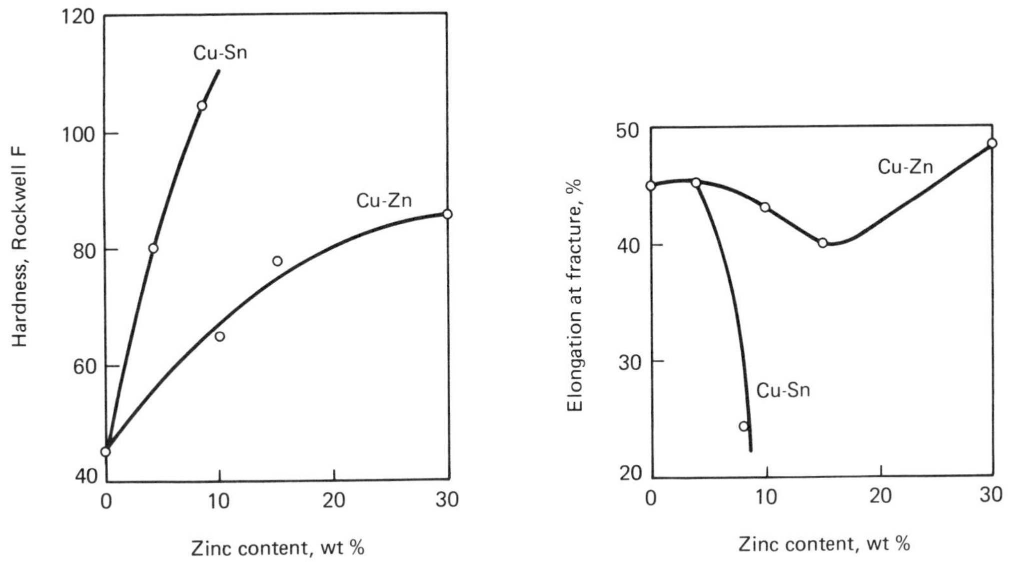

Fig. 8-29. Comparison of the effect of Zn and Sn on the hardness of the binary solid solutions. Sn has a greater solid solution strengthening effect than Zn but the elongation decreases, so that Cu-Sn alloys are not as readily fabricated as Cu-Zn alloys. (Data adapted from same source as Fig. 8-5)

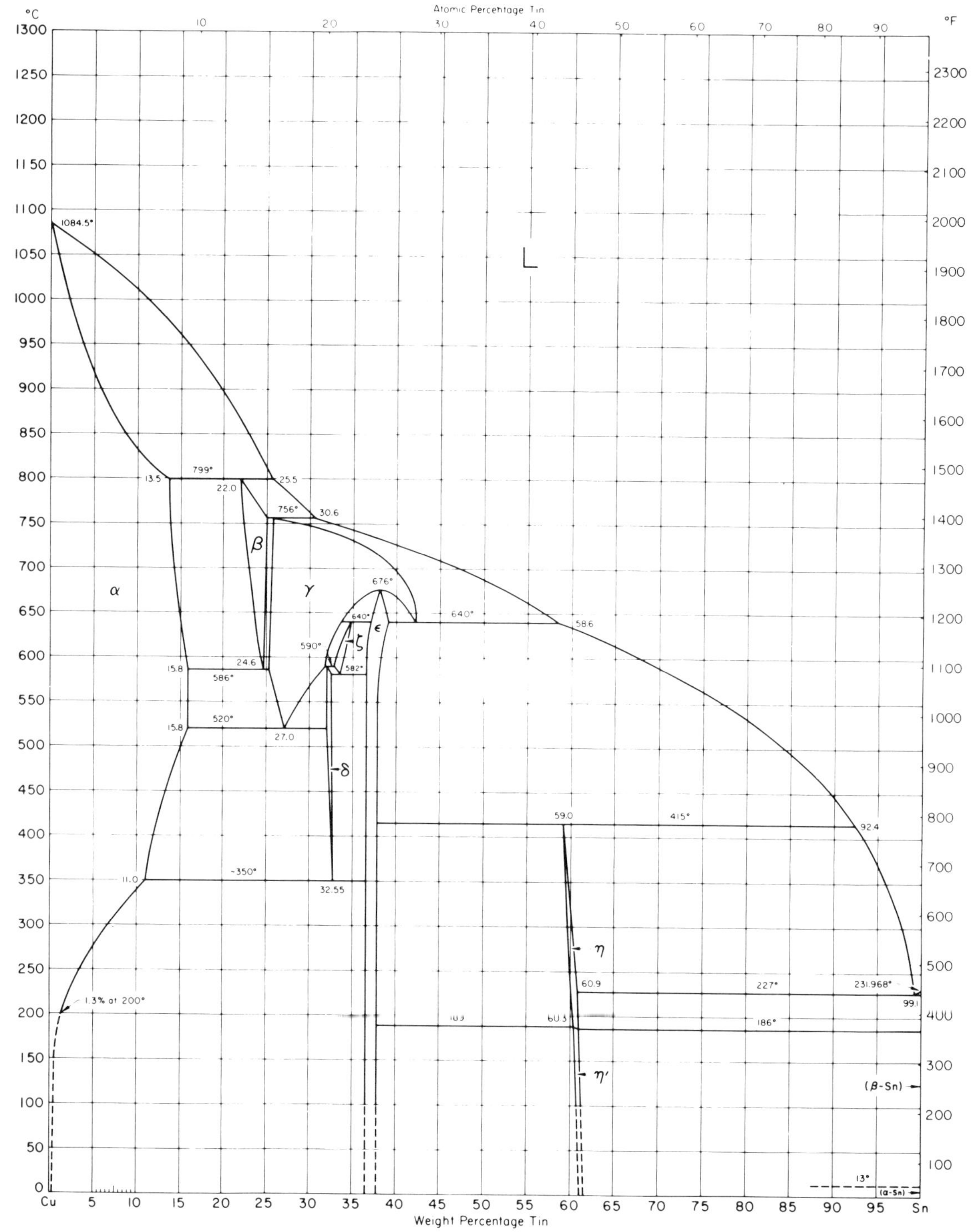

Fig. 8-30. The Cu-Sn phase diagram. (Adapted from same source as Fig. 8-2)

tin contents as low as 8% Sn frequently contain δ, a result of the fact that coring allows the outside of the α dendrites to attain about 13.5% Sn. This composition of α will react with the liquid to form some β by a peritectic

Table 8-6. Effect of Aging on the Strength of a Cu-10Sn Alloy

Property	As-cast	Cast, then annealed for 10 h at 760 °C, water quenched, aged 5 h at 315 °C
Yield strength, $lb_f/in.^2$	21,000	20,000
Tensile strength, $lb_f/in.^2$	44,000	43,000
Elongation, %	25	25
Hardness, Brinell	62	60

Adapted from E. M. Wise and J. T. Eash, Trans. AIME, Vol 111, p 218, 1934.

reaction. This β then decomposes by an eutectoid reaction to α and γ; the γ then decomposes to form α and δ. This sequence can be quite complex and dependent upon the cooling rate. Figure 8-31 shows an as-cast microstructure of a Cu-Sn alloy.

The Cu-Sn alloys containing up to 10% Sn (and higher) are used for bearing applications, which require high strength to support heavy loads and wear resistance. Apparently, the hard δ phase is important in the wear resistance, and the solid solution strengthening allows the development of strength. These alloys are frequently used for bearing applications in the as-cast condition, with no subsequent heat treatment, so the initial properties are dependent upon the development of δ during the solidification process. If the alloy is to be used in the as-cast condition, then the properties can only be controlled by the casting process. The main factor is the primary α grain

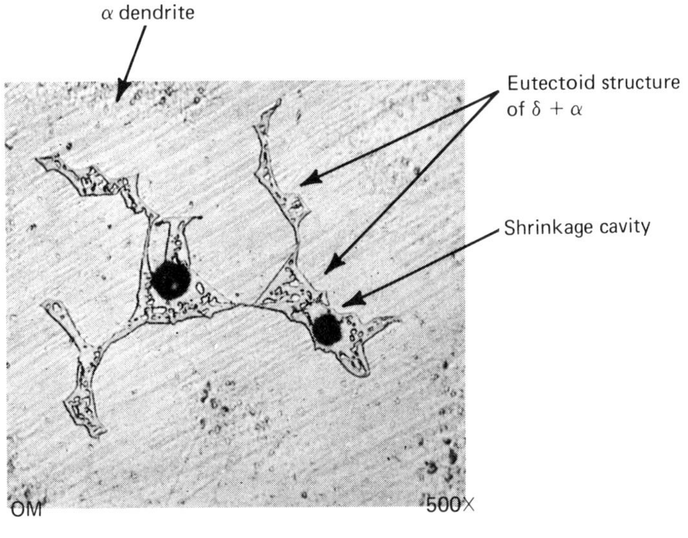

Fig. 8-31. Microstructure of a Cu-10Sn alloy in the as-cast condition showing the presence of the δ phase. (Adapted from T. F. Pearson and W. A. Baker, J. Inst. Metals, Vol 67, p 231, 1941)

size, and the δ-containing structure. The primary α grain size can be reduced by increasing the nucleation rate from the liquid, either by inoculation or by increasing the cooling rate (see Chapter 2). The effect of the cooling rate is shown in Fig. 8-32, where the faster cooling has formed considerably finer α dendrites.

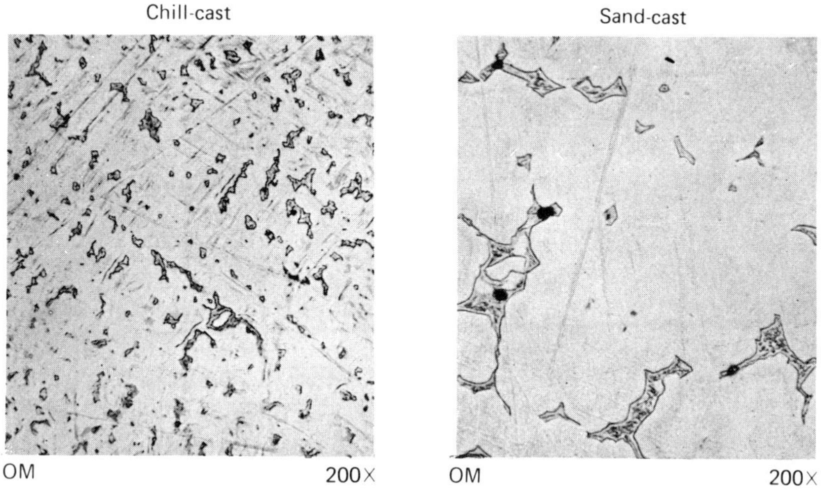

Fig. 8-32. Microstructure of Cu-10Sn alloys cooled at two different rates, showing that faster cooling (chill casting) has formed considerably finer α dendrites. (Adapted from same source as Fig. 8-31)

8-4. Copper-Zinc-Tin Alloys

The solid solution strengthening effect of zinc on copper should be enhanced by substituting some tin for zinc. The ternary phase diagram (Fig. 8-33) shows that the combined solutes have extensive solubility in α. For example, an alloy containing 10% Zn will dissolve up to about 9% Sn at 500 °C; however, the solubility at 300 °C is only 4%. For a 30% Zn alloy, the solubility at 300 °C is only about 1%. Although the solubility of tin in α in the lower-zinc alloys is sufficient to make the addition of tin attractive, the problem of the formation of the brittle δ phase during casting limits the amount of tin to values less than 2%. However, even low tin contents measurably strengthen the Cu-Zn binary alloys, as shown by the data in Table 8-7.

The addition of Zn to Cu-Sn alloys improves the soundness of the castings by affecting the solidification process. This has led to a group of alloys popular for making castings which must be pressure tight, such as for high-pressure water valves.

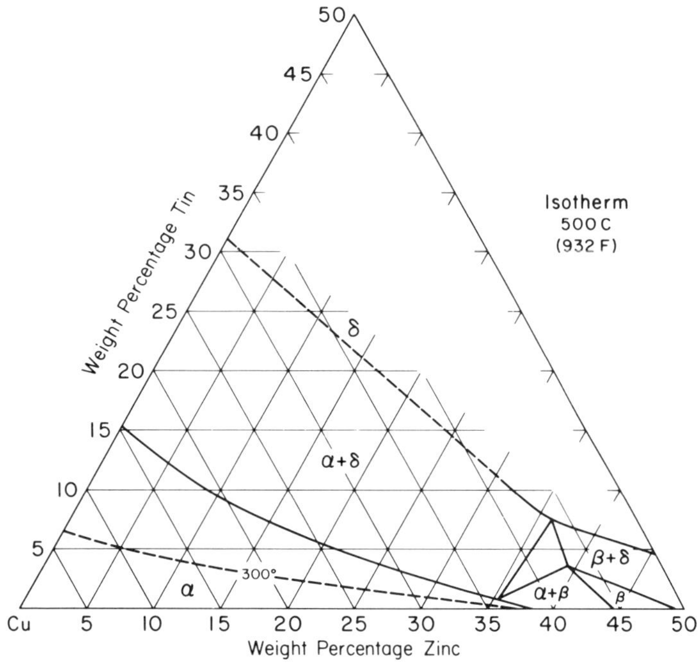

Fig. 8-33. Isotherm at 500 °C for the Cu-rich corner of the Cu-Zn-Sn phase diagram. The α solubility boundary at 300 °C is also shown. (From same source as Fig. 8-2)

Table 8-7. Mechanical Properties of Cu-Zn and Cu-Zn-Sn Solid Solution Alloys, Illustrating the Strengthening Effect of Tin (Grain Size 0.015 mm)

Alloy	Yield strength, $lb_f/in.^2$	Tensile strength, $lb_f/in.^2$	Elongation, %	Hardness, Rockwell F
Cu-10Zn	15,000	40,000	44	65
Cu-9.5 Zn-0.5Sn	18,000	45,000	43	70
Cu-30Zn	30,000	57,000	48	86
Cu-29 Zn-1Sn	33,000	58,000	40	87

Adapted from same source as Table 8-5.

8-5. Copper-Base Leaded Alloys

It was pointed out in Section 8-1 that the amount of lead must be kept low to be able to hot work copper (and also the brasses), because if the quite low solubility of lead in copper is exceeded, a lead-rich liquid film wets the grain boundaries, reducing the strength. However, if the lead content exceeds approximately 0.5%, at 25 °C the lead is essentially all located along the grain boundaries as separated particles, not spread along the grain boundaries. A typical microstructure is shown in Fig. 8-34. The leaded alloys usually do not have good hot working characteristics, but they can be cold worked and safely annealed. The important advantage of adding lead is that the machinability greatly improves. Since many objects manufactured from copper-base alloys are machined, this is an important consideration. The effect is illustrated in Fig. 8-35. Thus, there are several commercial copper-base alloys that are leaded for machinability.

It is important to realize that the lead is very finely distributed. According to the Cu-Pb phase diagram (Fig. 8-8), upon solidification of an alloy containing from 0.5 to 3% Pb, just prior to the completion of freezing, the structure will consist mainly of α crystals with a small amount of liquid of monoeutectic composition (36% Pb). This liquid will freeze by forming simultaneously more α and a lead-rich liquid. Upon further cooling, this liquid rejects more copper, which forms on the already existing α crystals, until at 326 °C, the liquid, almost pure lead now, undergoes an eutectic reaction, forming almost all pure lead. Thus in the as-cast alloy, these lead particles are trapped between the primary α dendrites. However, upon hot working, or cold working and annealing, the original α crystals lose their identity, and the lead particles are found in the configuration shown in Fig. 8-34. To give a fine and uniform distribution of the lead particles, the solidification process must be controlled to give a fine α grain size,

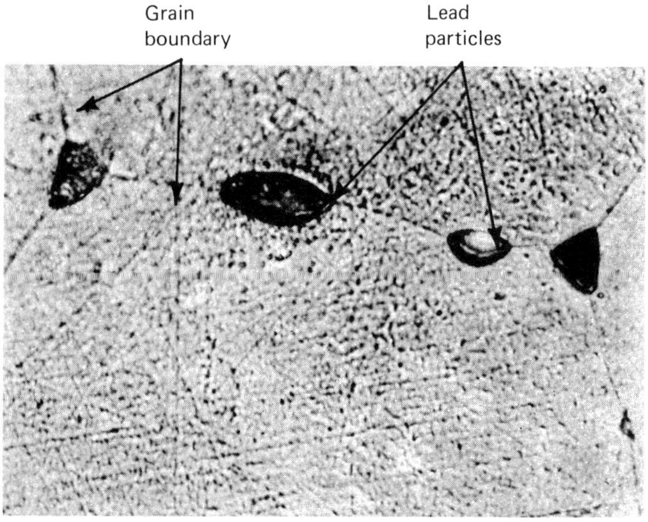

Fig. 8-34. Microstructure of a Cu-3Pb alloy, showing the configuration of the lead particles typical of leaded Cu-base alloys. (Adapted from Structure and Properties of Engineering Materials, by R. M. Brick, A. W. Pense and R. B. Gordon. Copyright © 1977, McGraw-Hill, New York. Used with the permission of McGraw-Hill Book Company)

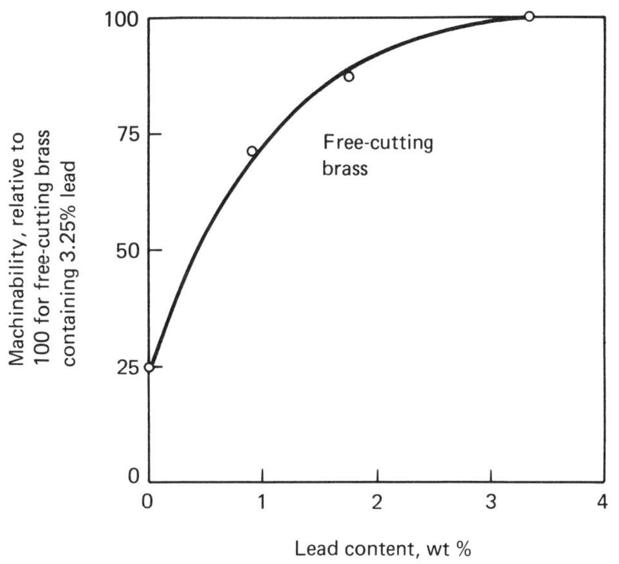

Fig. 8-35. Effect of Pb content on the machinability of yellow brass (Cu-33Zn). All the lead is present as undissolved particles. (Adapted from same source as Fig. 8-23)

and freezing must be sufficiently rapid so as not to allow the lead-rich liquid to separate under gravity but instead be trapped by the α dendrites.

To appreciate the fineness of the distribution of the lead particles, note that for particles similar in size to those shown in Fig. 8-34, an alloy containing about 3% Pb will have a particle density of about 10^{10} particles per cubic inch of material. If a 1/16-in.-diameter drill removes about 0.004 in. of this material in one revolution, the volume of material removed by the drill will contain about 150,000 lead particles (see Fig. 8-36). Thus, the cutting edge of the drill encounters a large number of lead particles during the one revolution. The finely divided lead makes the alloy relatively brittle and weak on a microscopic scale, as the lead particles are weak, and the cutting tool more easily fractures the matrix material separating the weak lead particles. In addition, the lead acts as a lubricant, so that less energy is required for machining.

8-6. Copper-Aluminum Alloys

In a general sense, the Cu-Al phase diagram is similar to that for Cu-Zn. There is a rather high solubility of Al in Cu, and several intermetallic compounds occur for higher aluminum contents, some of which have the same crystal structure and stoichiometry as ones in the Cu-Zn system. With a high atom-size difference combined with a relatively high maximum solu-

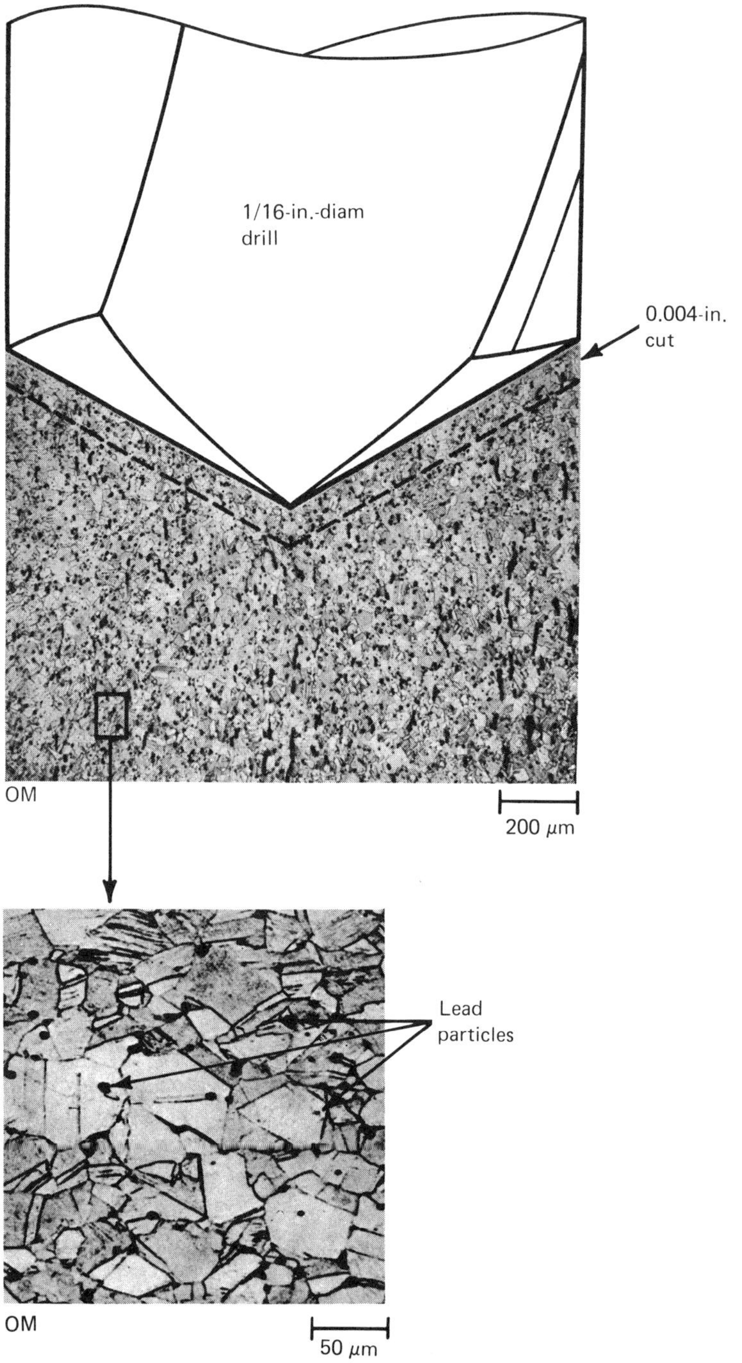

Fig. 8-36. Schematic illustration of the amount of material removed when a 1/16-in.-diameter drill penetrates 0.004 in. The number of lead particles in the volume of material removed is about 150,000 for an alloy containing 3% Pb, with the particle size shown in the higher-magnification photograph. (Upper illustration adapted from same source as Fig. 8-35; lower micrograph adapted from same source as Fig. 8-7)

bility (Table 8-3), it would be expected that commercial Cu-Al alloys which are solid solutions would be available because of their strength. Figure 8-37 shows that Al is indeed a potent strengthener, and there are two common commercial Cu-Al alloys (5 and 8% Al) which are solid solutions, their heat treatments involving only the conventional homogenization of the as-cast structure and annealing of the cold worked structure.

Figure 8-38 shows the Cu-rich portion of the Cu-Al phase diagram. The phases established at 500 °C are essentially those at lower temperatures, unless extremely long annealing times are employed. Note that alloys with aluminum contents above about 8% are subject to the formation of β at high temperature, which upon slow cooling undergoes an eutectoid reaction to α and γ_2. Therefore, we will now examine briefly this eutectoid reaction, then return to the specific commercial alloys whose compositions allow α, β and γ_2 to be present.

The β phase is body-centered cubic, like the β phase in the Cu-Zn system (see Fig. 8-10), and the γ_2 phase is similar to that of γ. The eutectoid composition is at 11.8% Al, and upon cooling an alloy of this composition slowly (e.g., 50 °C/h) after homogenization in the β region (e.g., 1 h at 800 °C), the α and γ_2 phases form in an alternate-layer configuration. This is similar in appearance to the eutectoid structure in Fe-C alloys, where the morphology is called pearlite, and we will use this term here to describe this eutectoid structure. Figure 8-39 shows a microstructure typical of the eutectoid alloy after slowly cooling from β.

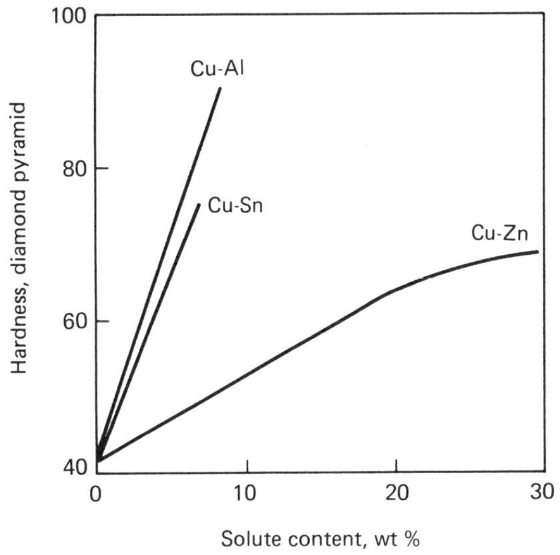

Fig. 8-37. Comparison of the effect of Al, Sn and Zn as solid solution strengtheners of Cu. (Adapted from R.M. Brick, D.L. Martin and R.P. Angier, Trans. ASM, Vol 31, p 675, 1943)

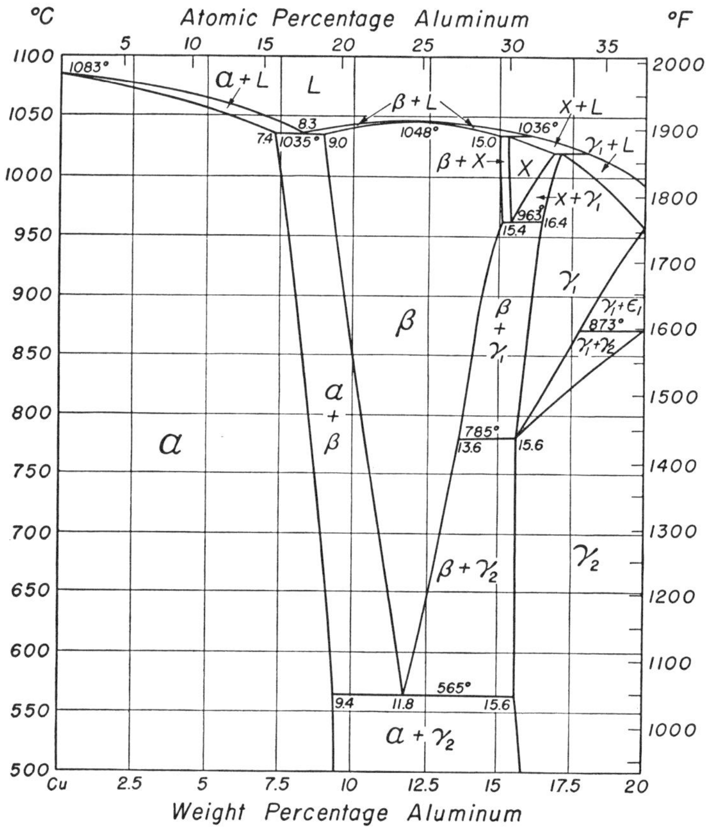

Fig. 8-38. Cu-rich end of the Cu-Al phase diagram. (From Metals Handbook, 1948 Ed., American Society for Metals, Metals Park, Ohio, 1948)

Rapid cooling from β might be expected to suppress the eutectoid reaction and to leave retained β at 20 °C. This structure then would decompose very slowly and for practical purposes be stable at this temperature. Then the properties of β could be utilized for the development of alloys. However, as in steels (Fe-C alloys), the β undergoes a transformation to another phase (β') even upon the most rapid cooling practical. This is a nonequilibrium phase, not shown on the phase diagram. It has a hexagonal crystal structure and is ordered. It forms by a martensite reaction (as in steels), with the β decomposing into long needles. Upon rapid cooling, when a temperature of about 380 °C (the martensite start temperature M_s) is reached, these needles begin to appear with great speed in the β; as cooling toward 20 °C continues, the remaining β decomposes into these fine needles. The final structure is illustrated in Fig. 8-40.

Now in steels, martensite is the hardest structure. For example, in an eutectoid alloy, Fe-0.8C, the slowly cooled steel will have a hardness (typically) of about 20 Rockwell C, whereas the martensite structure will be 65 Rockwell C. Hence the use of quenching for hardening in steels. However, in the Cu-Al eutectoid alloy here, the martensite will be approximately 70 Rockwell B, and the pearlite formed after cooling slowly will be approxi-

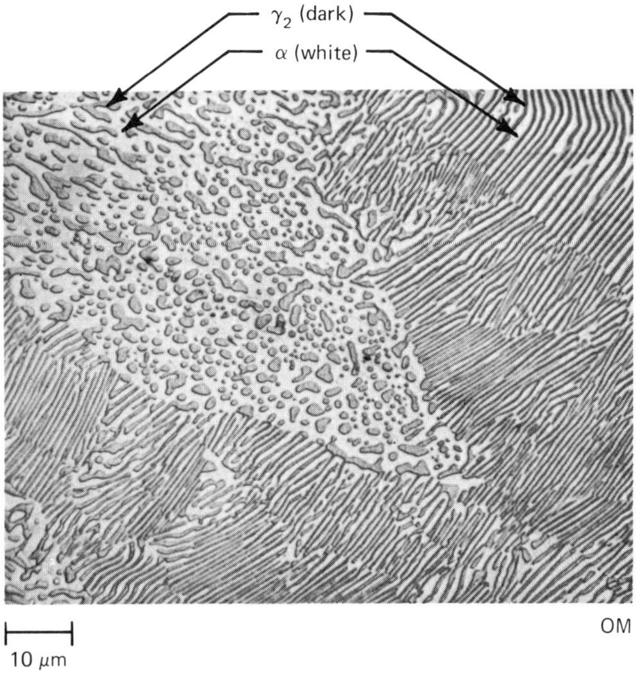

Fig. 8-39. Microstructure of a Cu-11.8Al alloy, homogenized at 800 °C for 2 h, then cooled slowly in the furnace. The structure is pearlite of alternate plates of α and γ_2. The α is the white phase, and the γ_2 is the dark.

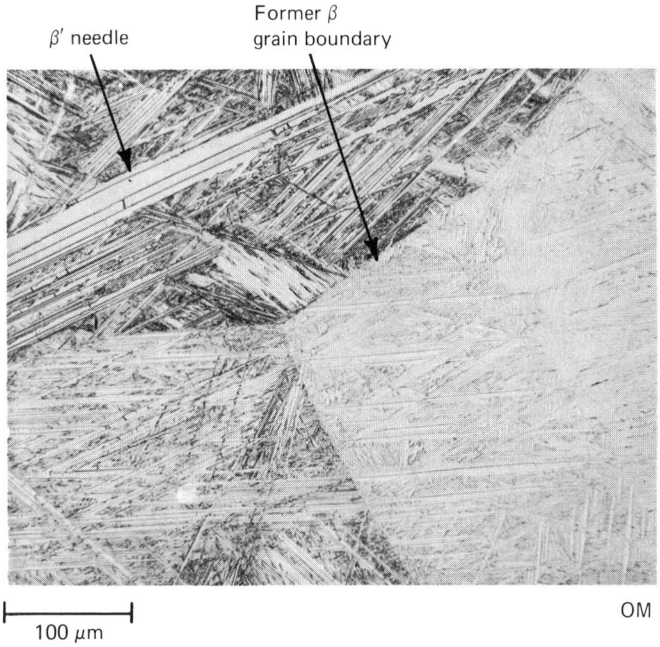

Fig. 8-40. Microstructure of a Cu-11.8Al alloy, homogenized for 2 h at 800 °C, then cooled rapidly (water quenched). The structure is thin needles of martensite, β'.

imately 80 Rockwell B. Thus, the slowly cooled product is slightly harder than that formed upon rapid cooling. This hardness change reflects the influence of the very hard γ_2 on the hardness of the structure.

Also, in steels the hardness of the martensite is reduced by heating the alloy below the eutectoid temperature to convert the martensite to the equilibrium phases, a heat treatment called tempering. However, in the Cu-Al alloy here, heating the β' martensite below the eutectoid temperature (565 °C, Fig. 8-38) will increase the hardness. If the martensite has a hardness of approximately 70 Rockwell B, then heating around 500 °C for 30 min will increase the hardness to about 90.

Table 8-8 lists the hardness and structure for different heat treatments for the eutectoid alloy. For comparison, also shown are the hardness and structure for a 10% Al alloy for similar heat treatments. We now examine this 10% Al alloy in some detail.

Figure 8-41 shows the effect of the quenching temperature on the hardness of a Cu-10Al alloy. Heating above about 850 °C results in a structure that is all β, and hence upon quenching an all-martensite structure is formed, with a Brinell hardness of about 250. The microstructure would be similar to that in Fig. 8-40. Holding below about 850 °C results in increasing amounts of β existing with the α, and hence the quenched alloy has an increasing amount of the soft α present, decreasing the Brinell hardness to about 120. A microstructure typical to that for the alloy quenched from the α-β region is shown in Fig. 8-42.

If, instead of quenching from the β region, the Cu-10Al alloy is cooled slowly (e.g., furnace cooled), a large amount (about 50% by a mass balance calculation) of primary α will form, and the rest of the structure will be the eutectoid pearlite. Even though the pearlite is harder than the β' martensite, the large amount of the soft α makes the alloy soft for this heat treatment. As the cooling rate decreases from rapid cooling, increasing amounts of the soft α and decreasing amounts of martensite appear, and the hardness decreases (see Fig. 8-43).

Table 8-8. Effect of Heat Treatment on Hardness for a Cu-11.8Al (Eutectoid) Alloy and a Cu-10.2Al Alloy(a)

	Cu-11.8Al		Cu-10.2Al	
Heat treatment	Structure produced	Brinell hardness	Structure produced	Brinell hardness
Slowly cooled (furnace cooled)	Pearlite ($\alpha + \gamma_2$)	220	Primary α + pearlite ($\alpha + \gamma_2$)	150
Rapidly cooled (water quenched)	All-β' martensite	150	All-β' martensite	240
Rapidly cooled, then tempered 30 min(b)	$\alpha + \gamma_2$	240	$\alpha + \gamma_2$	110

(a) From same source as Fig. 8-24. (b) Cu-11.8Al tempered at 500 °C; Cu-10.2Al, at 350 °C.

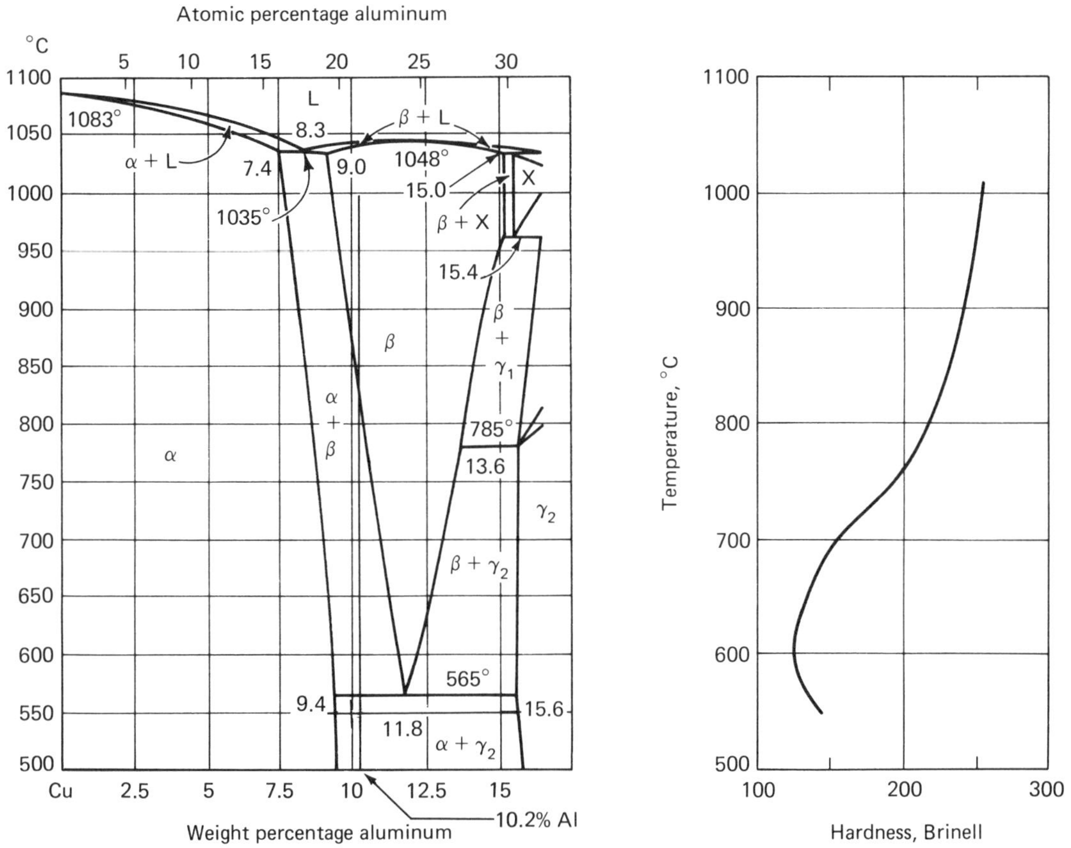

Fig. 8-41. Effect of quenching temperature on the hardness of a Cu-10.2Al alloy. The samples were initially heat treated to give a primary α and $\alpha + \gamma_2$ eutectoid structure, heated at indicated temperatures for 30 min, then quenched. (Adapted from same source as Fig. 8-24)

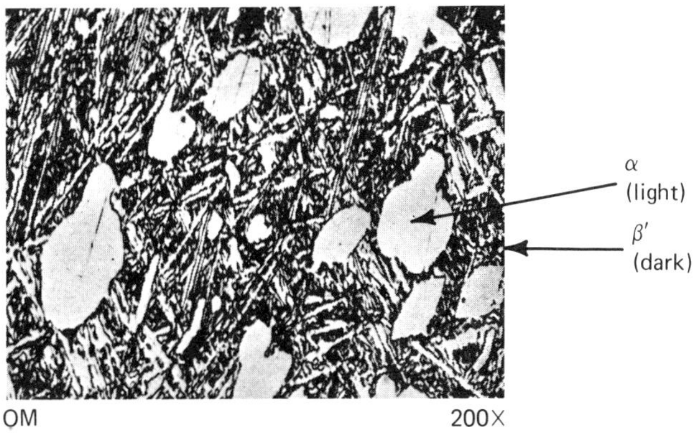

Fig. 8-42. Microstructure of a Cu-10.2Al alloy quenched from 750 °C, giving a structure of α and β'. (Adapted from same source as Fig. 8-24)

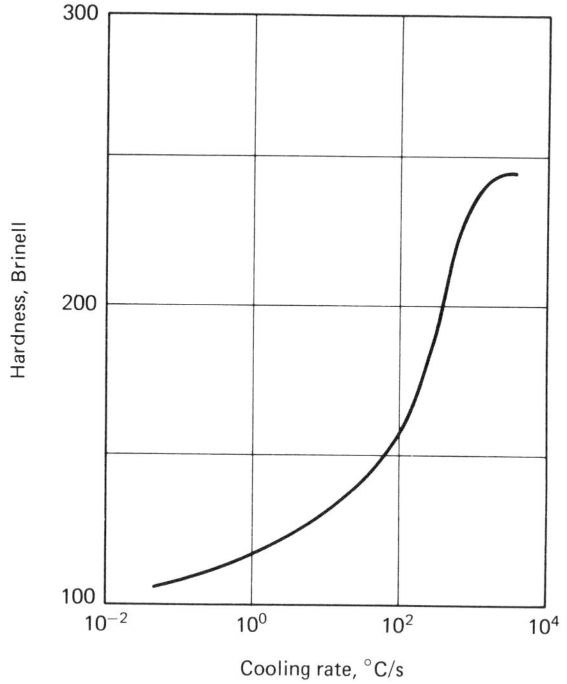

Fig. 8-43. Effect of cooling rate from 900 °C on the hardness of a Cu-10.2Al alloy. (Adapted from same source as Fig. 8-24)

Tempering of the quenched Cu-10Al alloy results in an increase in hardness if the alloy is tempered around 350 °C, but a decrease in hardness when tempered at higher temperatures (see Fig. 8-44). The increase is due to the formation of a fine dispersion of the hard γ_2 in the softer α. However, at sufficiently high tempering temperatures (i.e., above about 400 °C), the tempering time is sufficient to allow coarsening of the γ_2, and the hardness decreases.

Now we return to Table 8-8. We found that the eutectoid alloy was hardest if the structure was $\alpha + \gamma_2$, formed either by cooling slowly from β or by tempering the β' martensite. However, increasing aluminum content decreases the hardness of the β' martensite (although it is not clear why). Thus, the Cu-10Al alloy is hardest with the least amount of soft α, and the maximum amount of β'. This is in the quenched condition.

The data in the previous figures show that a Cu-Al alloy containing approximately 10% Al would have the highest hardness when water quenched from the β region, or when water quenched and tempered below approximately 400 °C. However, the toughness for these heat treated conditions must be considered. Table 8-9 shows that annealing below approximately 600 °C will embrittle the alloy. Note that the elongation at fracture indicates considerable toughness, but upon impact loading, the alloy is considerably more brittle than for the other two heat treatments listed. (Note that these data are for a 9.8% Al alloy.) Thus, a heat treating recommendation for a Cu-9Al bronze (Metals Handbook, 8th Ed., Vol 1, American

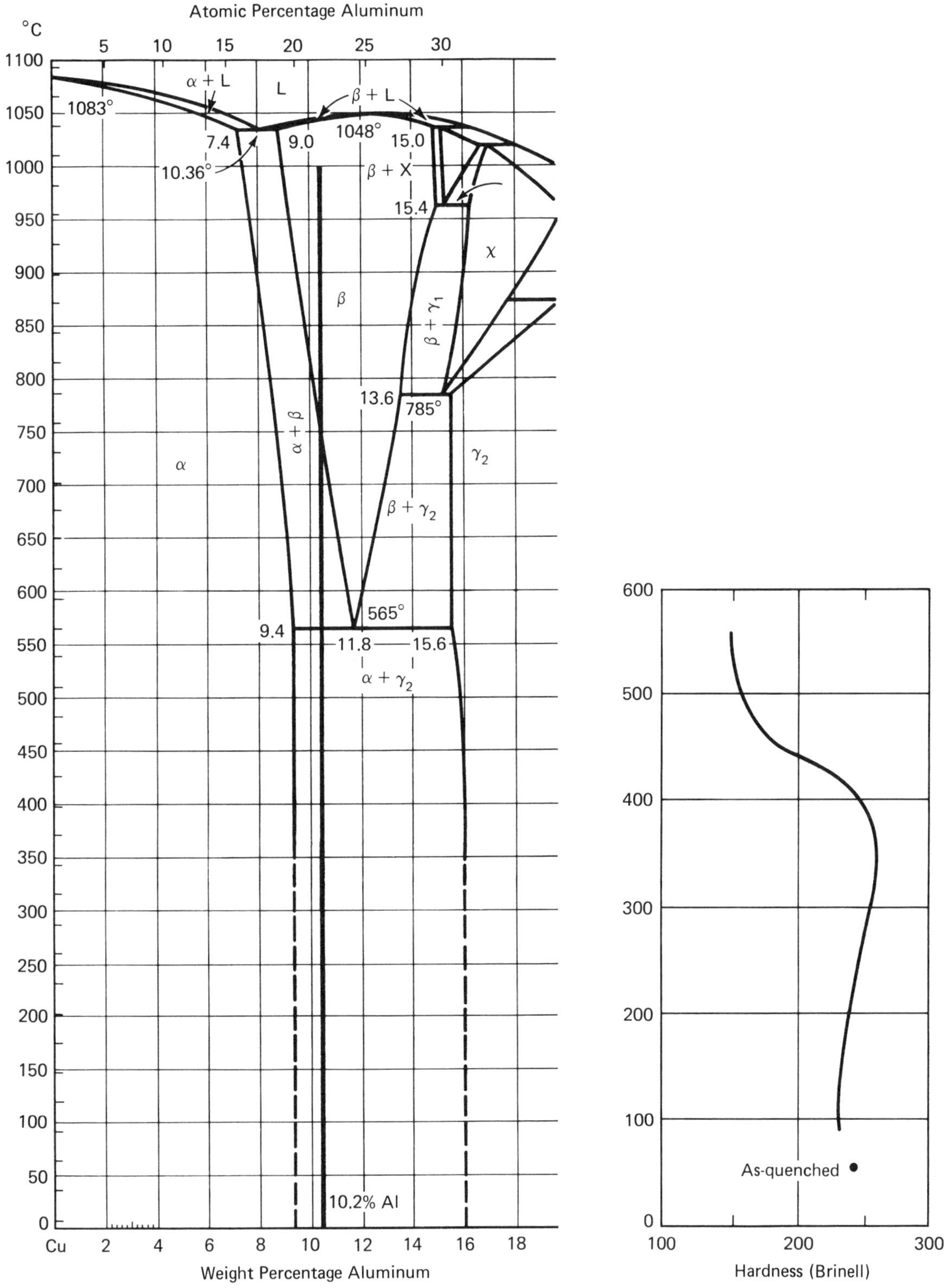

Fig. 8-44. Effect of tempering (for 30 min) on the hardness of a Cu-10.2Al alloy. The alloy was initially quenched from 900 °C. (Adapted from same source as Fig. 8-24)

Table 8-9. Effect of Heat Treatment on Some Mechanical Properties of a Cu-9.8Al Alloy

	Approximate properties		
Heat treatment	Hardness, Brinell	Elongation, %	Izod impact energy, M-Kg
Quenched from 900 °C.	155	7	4
Annealed at 650 to 700 °C. . . .	110	40	5
Annealed at less than 570 °C.	115	30	1.5

Data from same source as Fig. 8-24.

Society for Metals, Metals Park, Ohio, 1961) states that for most commercial purposes, this alloy can be annealed at 650 °C, and then quenched, and, in general, it is advised to avoid annealing below 570 °C or cooling too slowly from the annealing temperature.

It must be borne in mind that the properties of these Cu-Al alloys which involve primary α depend quite sensitively upon the primary α grain size and shape. This is demonstrated by the microstructures in Fig. 8-45. This alloy was cooled from the liquid at different rates, which affected not only

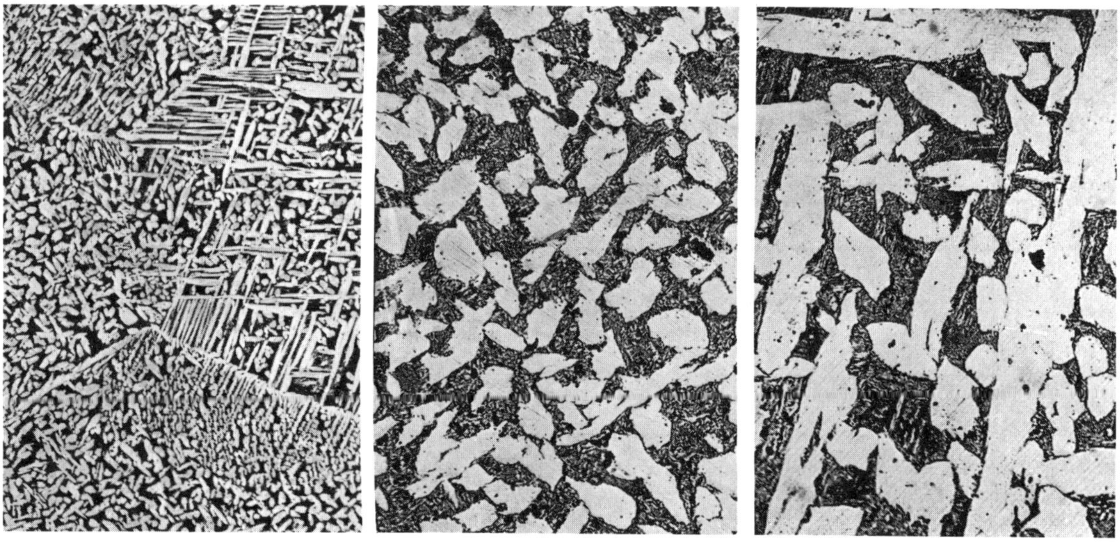

	Chill Cast ⅞-In. Bar	Ear of Chilled Test Block	Ear of Sand Cast Test Block
Tensile strength	90,000 psi.	75,000 psi.	60,000 psi.
Yield strength	40,000 psi.	37,000 psi.	43,000 psi.
Elongation	15%	12%	8%
Brinell hardness	140	140	160

Fig. 8-45. Microstructures (OM) and properties of a Cu-10.5Al-1.0Fe alloy which was cooled at different rates from the liquid. The magnification is about 100 ×. The light areas are primary α, and the dark background consists of unresolved decomposition products of β, which are either β' martensite or α and γ_2, depending on the cooling rate. (From G. K. Dreher, Metal Progress, Vol 38, p 789, 1940)

the original β grain size, but the subsequent α grain size which formed upon cooling into the $\alpha + \beta$ region. The cooling rate also affects the decomposition of the β. Note in the figure that the finer the primary α structure, the greater the tensile strength. However, the complex structure brings about some unpredicted effects on properties. The hardness is almost unaffected and the structure with the coarser primary α grain size has the least ductility. This structure was the most slowly cooled of the three sections of the casting, and hence the decomposition of β into the more brittle $\alpha + \gamma_2$ structure was favored.

8-7. Copper-Beryllium Alloys

Table 8-3 shows that there is a size difference between Be and Cu comparable to that between Cu and Al or Sn. But Fig. 8-46 shows that the solubility is quite limited, so that the addition of beryllium for solid solution strengthening is not favorable. However, alloys containing greater than 1.5% Be might be precipitation hardenable, since around 800 °C they will be single-phase α (or $\alpha + \beta$), and at lower temperature, the γ (CuBe) phase will precipitate from α. Indeed, in this alloy remarkable precipitation strengthening occurs, with hardness values above 40 Rockwell C. (To place this in perspective, this is the range normally associated with steels.)

These Cu-Be alloys show precipitation-hardening behavior similar to that observed for Cu-Al alloys. Figure 8-47 illustrates the effect of aging time and of aging temperature. For this Cu-1.9Be alloy, the optimum aging temperature is around 350 °C.

The data in Fig. 8-48 show that for maximum hardness the beryllium content should be between 1.8 and 2.0%. The aging temperature should be around 350 °C, and the aging time can be between 3 and 5 h. Note that the aging temperature, aging time and beryllium content are all rather flexible, and yet an approximate maximum hardness of about 40-45 Rockwell C can be attained.

As in many other precipitation processes (see Chapter 3), the equilibrium precipitate γ is preceded by the formation of a metastable phase. The hardening is caused by the precipitation of this phase in an extremely finely dispersed manner, as illustrated in Fig. 8-49. The precipitates are disks from 200 to 400 atoms in diameter, and approximately 50 atoms thick, and very close together. As shown in Fig. 8-47, these alloys will overage as the structure coarsens, and the equilibrium precipitate forms.

The recrystallization temperature for these alloys is in the range of 500 °C and above. If these alloys are solution annealed, then cold worked prior to aging in the range 300-385 °C, the precipitation process will occur before recrystallization (and hence softening) can occur. The effect of such a treatment on hardness is shown in Fig. 8-50. Note that the hardness in this condition is not significantly different from that of the aged material without prior cold working. However, the data in Table 8-10 show that the yield and

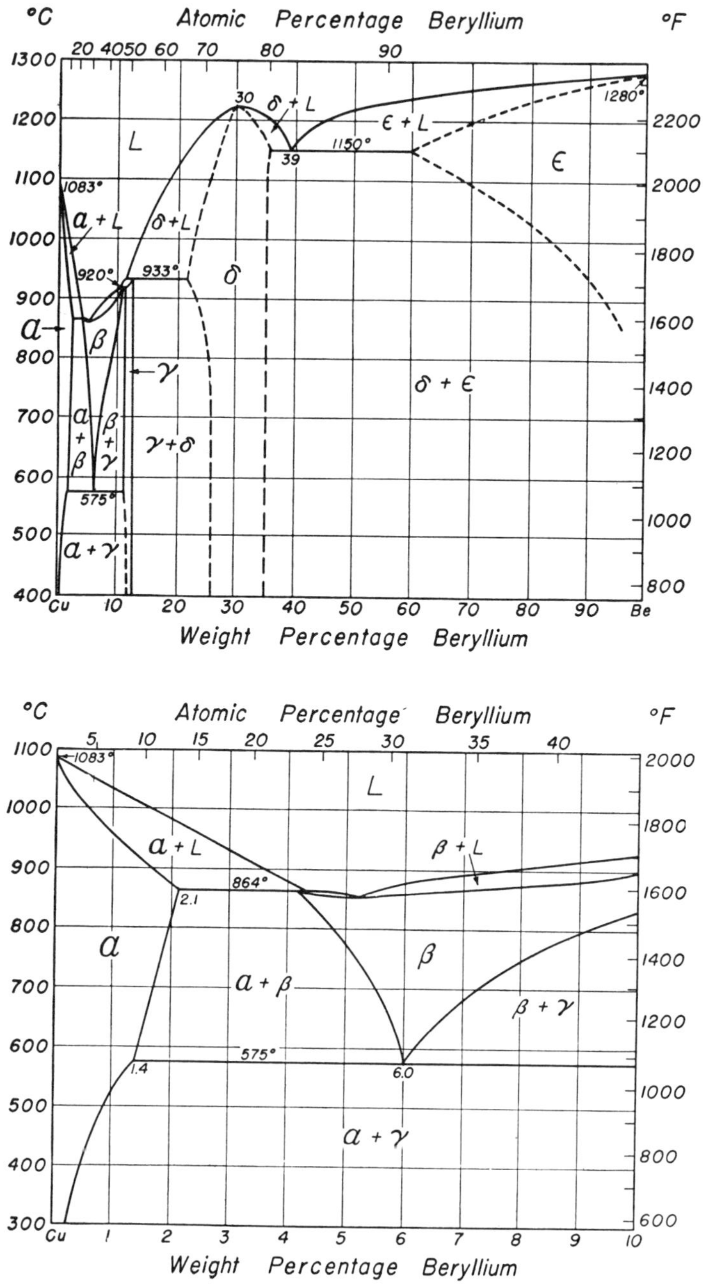

Fig. 8-46. The Cu-Be phase diagram. (From same source as Fig. 8-38)

tensile strength are greatly enhanced by prior cold working. Note that in the aged condition, with or without prior cold work, the ductility is quite low.

The commercial precipitation-hardening Cu-Be alloys have alloy additions other than beryllium. Many contain some cobalt, which forms an insoluble Be-Co compound which inhibits grain growth during solution

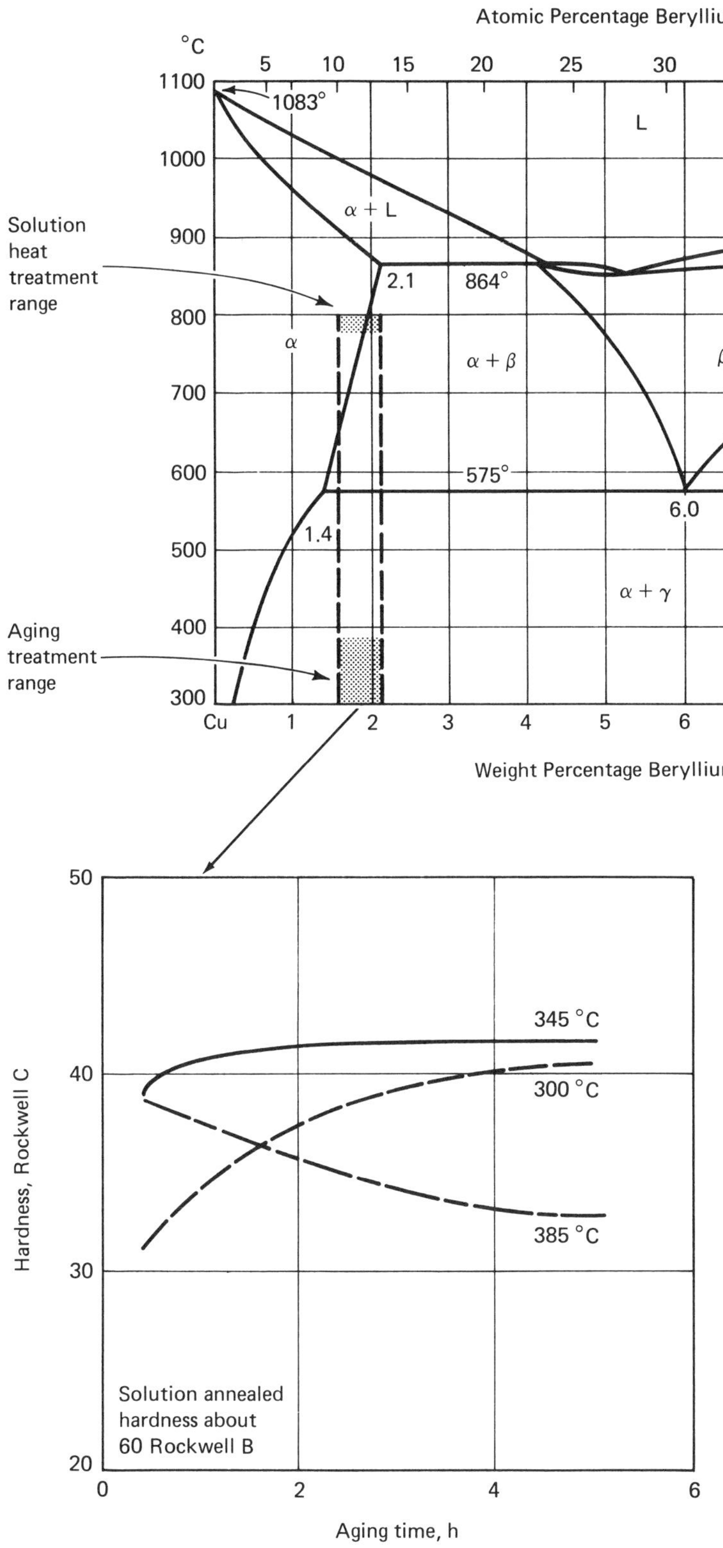

Fig. 8-47. Age-hardening response of a Cu-1.9Be alloy. It was solution heat treated for about 8 min at 800 °C, then water quenched, giving a hardness of approximately 60 Rockwell B. (Phase diagram adapted from same source as Fig. 8-38; hardness data adapted from G. R. Gohn, G. J. Herbert and J. B. Kuhn, The Mechanical Properties of Copper-Beryllium Alloys Strip, ASTM STP 367, American Society for Testing and Materials, Philadelphia, 1964)

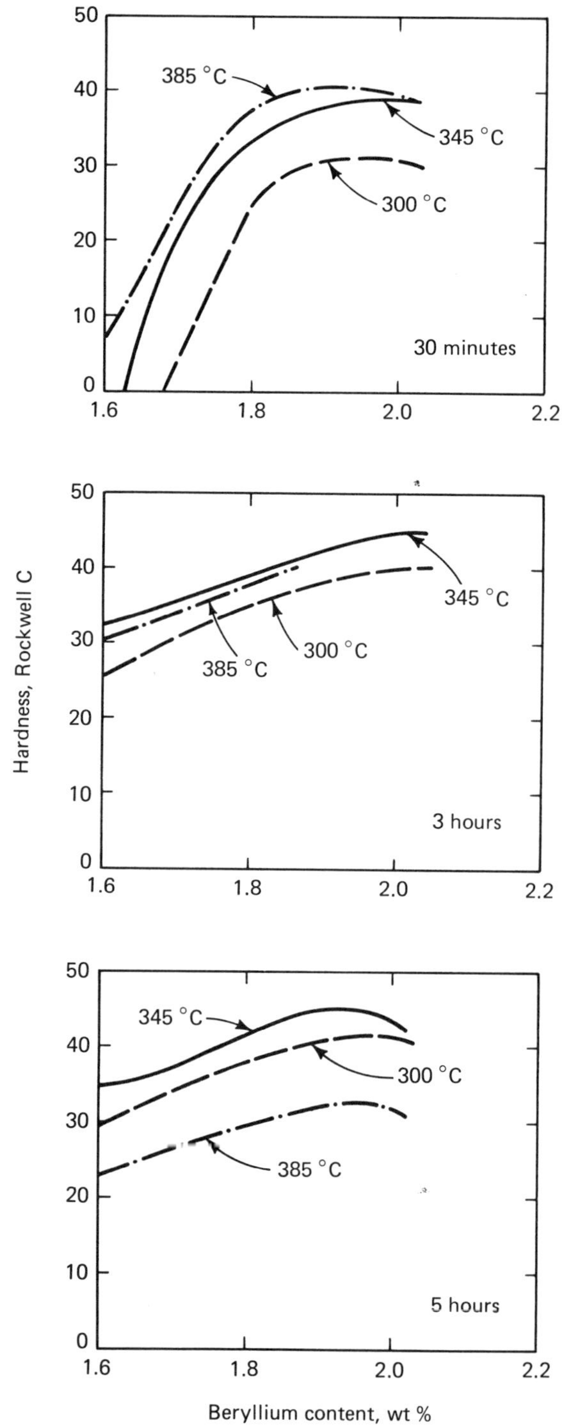

Fig. 8-48. Effect of Be content, aging temperature and aging time on the precipitation hardening of Cu-Be alloys. (Hardness data adapted from same source as data in Fig. 8-47)

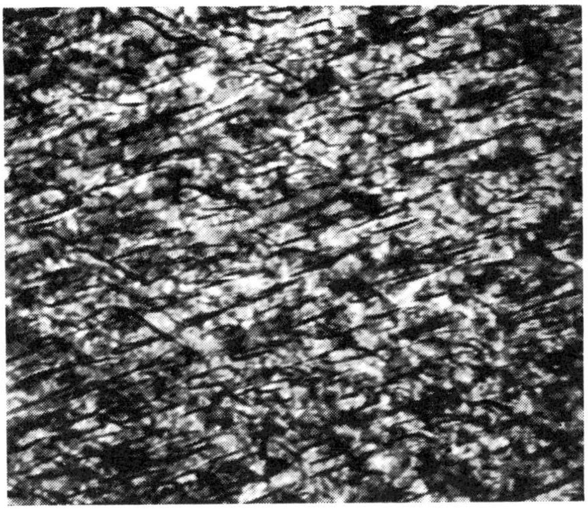

Fig. 8-49. Microstructure of a Cu-2Be alloy solution treated at 800 °C, water quenched, then aged at 350 °C for 4 h. The dark lines are the metastable precipitate. (From J. Nutting and M. A. P. Dewey, in Heat Treatment of Metals, Special Report 95, The Iron and Steel Institute, London, 1966)

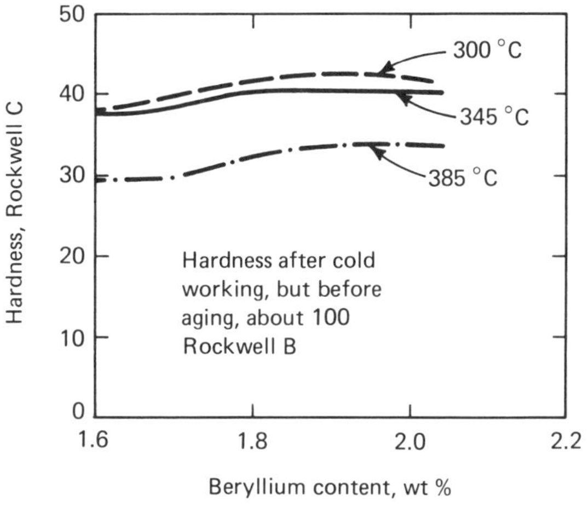

Fig. 8-50. Effect of cold working on the aging response of Cu-Be alloys. The alloys were solution annealed at 800 °C for 10 min, water quenched, then reduced in thickness about 40% by rolling, then aged for 3 h. (Hardness data adapted from same source as data in Fig. 8-47)

annealing. Also, the cast alloys have additions which act as grain refiners to control the as-cast grain size. Table 8-11 gives the recommended heat treatment for some of the commercial Cu-Be alloys. Note that the solution treatment and the aging treatment are essentially independent of the exact alloy, which would be predicted from the data in Fig. 8-49 and 8-50 (as discussed previously).

Table 8-10. Some Mechanical Properties of a Cu-1.9Be Alloy for Four Different Treatments

Treatment	Hardness	0.2% yield strength, $lb_f/in.^2$	Tensile strength, $lb_f/in.^2$	Elongation, %
Solution annealed for 8 min at 800 °C, water quenched....	61 HRB	37,000	71,000	56
Solution annealed for 8 min at 800 °C, water quenched, then cold rolled to a reduction in thickness of 38%	100 HRB	107,000	115,000	5
Solution annealed treatment, then aged for 3 h at 345 °C...........	42 HRC	168,000	187,000	4
Solution annealed treatment, cold worked 38%, then aged for 2 h at 345 °C...........	42 HRC	177,000	199,000	3

From same source as Fig. 8-47.

Table 8-11. Recommended Heat Treatments for Some Commercial Cast and Wrought Alloys

Alloy designation (or trade name)	Chemical composition, wt %	Solution treatment(a)	Aging treatment for maximum hardness
	Cast Alloys		
C82400 ...... (165C)	98Cu-1.7Be-0.3Co	800-815 °C	3 h at 345 °C
C82500 ...... (20C)	97.2Cu-2Be-0.5Co-0.25Si	790-800 °C	3 h at 345 °C
C82600 ...... (245C)	97Cu-2.4Be-0.5Co	790-800 °C	3 h at 345 °C
C82800 ...... (275C)	96.6Cu-2.6Be-0.5Co-0.3Si	790-800 °C	3 h at 345 °C
	Wrought Alloys		
C17000 ...... (165)	98Cu-1.7Be-0.3Co	775-800 °C	1 to 3 h at 315 to 345 °C, depending upon amount of cold working prior to aging
C17200 ...... (25)	Cu-1.9Ce-0.3Co	760-790 °C	1 to 3 h at 315 to 345 °C, depending upon amount of cold working prior to aging

(a) Time depends upon thickness of part. Allow 1 h for each 2.5 cm of thickness.
From Metals Handbook, 9th Ed., Vol 2, American Society for Metals, Metals Park, Ohio, 1979.

In heat treating, some precautions are mentioned. One is to avoid using too high a solution temperature, as partial melting may occur, and upon cooling the β phase will form, which is difficult to dissolve upon subsequent solution annealing. The other is to avoid too low a solution temperature, because β will form at the solution temperature. Its formation lowers the beryllium content of the α, and the hardening response is reduced. Also, this β is difficult to redissolve by resolution annealing. Also, if the alloy is not cooled sufficiently rapidly from the solution temperature, then some β may form during cooling. Thus, control of the water quenching must be maintained.

References

Metals, Vol II, H. Carpenter and J. M. Robertson, Oxford University Press, New York (1939).

Practical Metallurgy, G. Sachs and K. van Horn, American Society for Metals, Metals Park, Ohio (1940).

Copper and Copper Base Alloys, R. A. Wilkins and E. S. Bunn, McGraw-Hill, New York (1943).

Structure and Properties of Alloys, R. M. Brick and A. Phillips, McGraw-Hill, New York (1949).

Engineering Metals and Their Alloys, C. H. Samans, Macmillan Co., New York (1953).

Copper, ed. A. Butts, Reinhold Publishing Corp., New York (1954).

The Metallurgical Control of Quality in the Production of Copper-Base Alloy Castings, A. R. French, in Metallurgical Aspects of the Control of Quality in Non-Ferrous Castings, The Institute of Metals, London (1957).

Metallic Materials in Engineering, C. H. Samans, Macmillan Co., New York (1963).

Gunmetal Castings, F. Hudson and D. A. Hudson, Hart Publishing Co., New York (1967).

Understanding Copper Alloys, ed. J. H. Mendenhall, Olin Corporation, East Alton, Ill. (1977).

CHAPTER 9

Titanium-Base Alloys

High melting point (1820 °C), low density (4.5 gm/cm^3) and good corrosion resistance make titanium and its alloys important materials to consider for many applications. Unfortunately, the expense of preparation of the metal precludes its wide use compared with that of ferrous alloys and many nonferrous alloys, but it is still widely used in many applications. Especially important is the strength coupled with the low density. The data in Fig. 9-1 show that the strength-to-weight ratio (given there as the yield strength divided by the specific gravity) is similar for pure titanium and high strength aluminum- and magnesium-base alloys. Further, pure titanium can be strengthened appreciably by alloying and heat treatment, yet the alloys still retain low density. Table 9-1 compares typical properties for titanium and two titanium alloys with those of ultra-high strength steel, showing the titanium alloys to be remarkably strong on this basis.

Above 882 °C, titanium is body-centered cubic (referred to as β hereafter) and below this temperature it is hexagonal close-packed (c/a = 1.587) (referred to as α). The kinetics of the β-to-α phase transformation upon cooling strongly influences the properties of titanium and its alloys. Further, the effect of the added elements upon this phase change is important in determining desirable heat treatments.

The properties of titanium and titanium alloys are quite sensitive to even small amounts of interstitial elements (H, O, N and C). Because their control takes care during refining and heat treatment, their effect will be examined first. Then solid solution strengthened alloys will be described, and finally more complex alloys involving solid-state transformations associated with the β-to-α phase change will be examined. The table in Appendix 6 gives some properties and applications of common commercial titanium-base alloys.

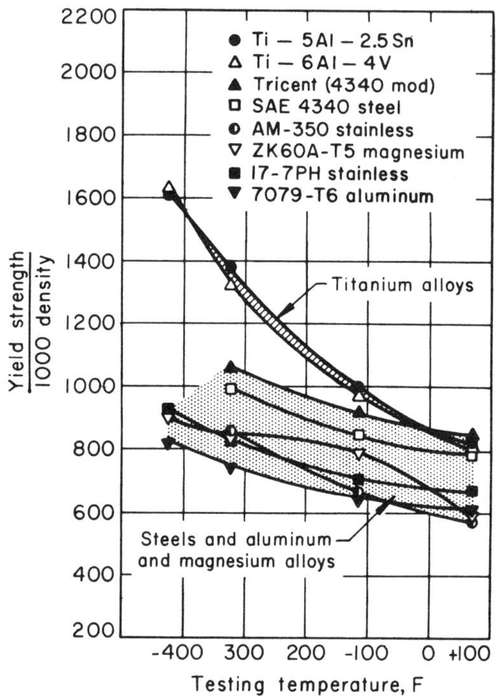

Fig. 9-1. Yield strength-to-density ratio ($lb_f/in.^2$) as a function of temperature for Ti alloys compared with that for several other alloys. (From Metals Handbook, 8th Ed., Vol 1, American Society for Metals, Metals Park, Ohio, 1961)

It is emphasized that the microstructures of titanium alloys can be quite complex, and in some cases are not well understood. In the sections which deal with commercial alloys, the microstructural description strives for a level usable in understanding the effects of heat treatment of properties, but without getting involved in some of the fine details which are still unsettled.

9-1. Commercially Pure Titanium

Like that of any single-phase alloy, the microstructure of pure titanium depends upon whether it has been cold worked and on the type of annealing. In addition, upon cooling from the β region (above 882 °C), the structure will depend upon the cooling process, as this affects the β-to-α transformation, and hence the final α grain size and shape.

Figure 9-2(a) shows the equiaxed microstructure of α titanium after annealing at 800 °C, in the α region. Here the grain size (and hence properties) can be varied only by cold working and annealing (Chapter 1). Properties typical of this microstructure are shown in the figure.

Annealing in the β region, at 1000 °C, then cooling rapidly to 25 °C (water quenching), produces the typical structure shown in Fig. 9-2(b). Even

Table 9-1. Comparison of Typical Strength-to-Weight Ratios at 20 °C for Ti and Two High Strength Ti Alloys With That for an Ultra-High Strength Steel

Metal	Specific gravity	Tensile strength, $lb_f/in.^2$	Tensile strength ÷ specific gravity
Pure Ti	4.5	58,000	13,000
Ti-6Al-4V	4.4	130,000	29,000
Ti-4Al-3Mo-1V	4.5	200,000	45,000
Ultra-high strength steel (4340)	7.9	287,000	36,000

Data from Metals Handbook, 8th Ed., Vol 1, American Society for Metals, Metals Park, Ohio, 1961.

the rapid cooling will not suppress the β-to-α transformation, so that the structure is entirely α. Note that the α grain boundaries are serrated and quite irregular. Properties typical of this structure are shown in the figure. This structure is stronger than the equiaxed structure developed by annealing only in the α region. For both treatments, the titanium is still quite ductile.

Cooling slowly (e.g., 20 h to 25 °C) produces the structure in Fig. 9-2(c). The structure is again completely α, but the grain boundaries are less irregular than those produced upon cooling rapidly. This structure is somewhat weaker than that produced upon cooling rapidly, but is still stronger than the equiaxed structure (Fig. 9-2a).

It might be expected that rapid cooling, such as produced by water quenching (e.g., cooling rates of 1000 °C/sec), would suppress the transformation of β to α, and hence the structure at 25 °C would be retained (but unstable) β. However, in pure titanium quite rapid cooling (e.g., 10,000 °C/sec) has been unable to suppress the formation of α. In the β-to-α transformation there is no change in chemical composition, of course, but there is a change in crystal structure. For this to occur easily (as reflected in the inability to suppress it) there must be a diffusionless martensitic transformation occurring, in which the titanium atoms in the body-centered cubic β shift only slightly to convert to the hexagonal close-packed α structure. That this can occur is illustrated in Fig. 9-3. The hexagonal-close packed cell obtained from the body-centered cubic structure is shown in Fig. 9-3(b). To obtain the correct dimensions of the α cell requires a slight adjustment (Fig. 9-3c) of the dimensions.

9-2. Interstitial Solid Solutions

Hydrogen, nitrogen, carbon and oxygen dissolve interstitially in titanium, and have a potent effect on mechanical properties. The phase diagrams in Fig. 9-4 show that H, N and O have a relatively high solubility in titanium, whereas C, being a strong carbide former, has a low solubility.

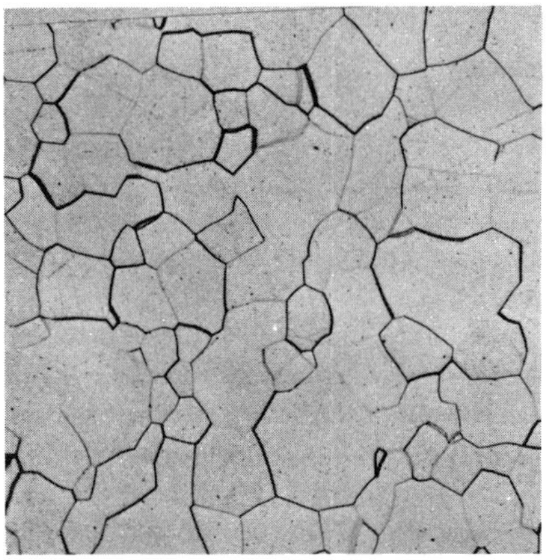

100× OM

Annealed 1 h at 800 °C, water quenched
0.2% yield strength, 18,000 psi;
tensile strength, 36,000 psi;
elongation, 80%

(a)

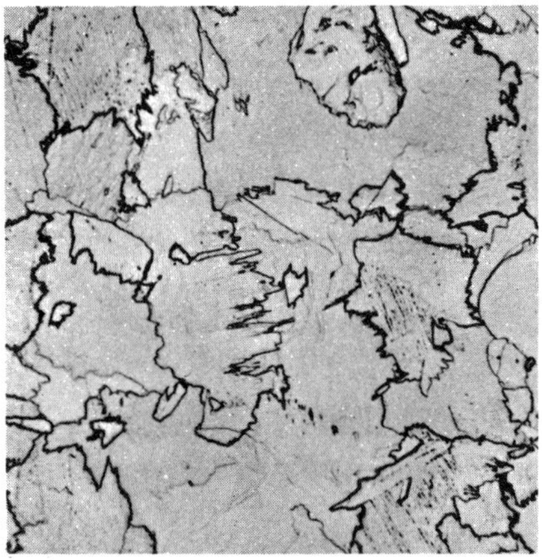

100× OM

Annealed 1 h at 1000 °C, water quenched
0.2% yield strength, 33,000 psi;
tensile strength, 42,000 psi;
elongation, 60%

(b)

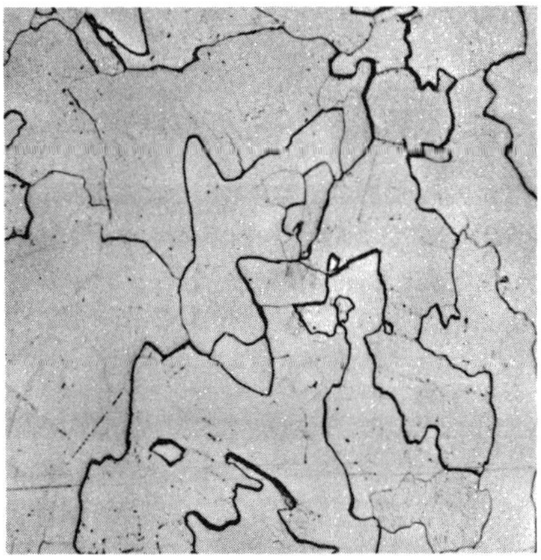

100× OM

Annealed 1 h at 1000 °C, furnace cooled
0.2% yield strength, 24,000 psi;
tensile strength, 38,000 psi;
elongation, 60%

(c)

Fig. 9-2. Microstructure and mechanical properties (at 25 °C) of pure titanium typical of those developed upon annealing in the α region, and upon annealing in the β region and cooling to the α region. (Microstructures adapted from Metallography of Titanium Alloys, TML Report No. 103, May 29, 1958, Battelle Memorial Institute, Columbus, Ohio; mechanical properties adapted from R.I. Jaffee, F.C. Holden and H.R. Ogden, Trans. AIME, Vol 6, p 1282, 1954)

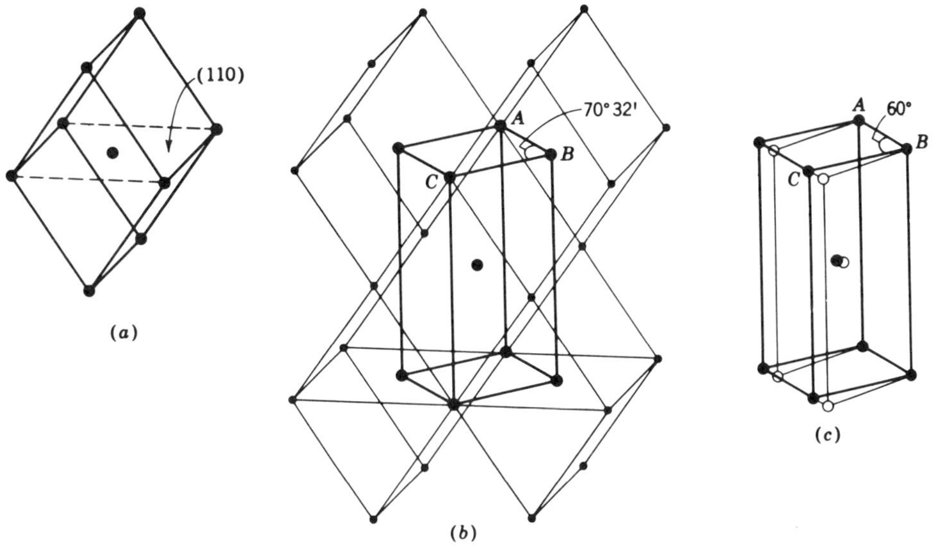

Fig. 9-3. Illustration of a possible crystallographic relation for a diffusionless β-to-α transformation in titanium. (From Structure and Properties of Engineering Materials, 4th Ed., by R. M. Brick, A. W. Pense and R. B. Gordon. Copyright © 1977 by McGraw-Hill, New York. Used with the permission of McGraw-Hill Book Company)

Hydrogen, being a rather small element, has a high diffusivity in titanium, which, with the solubility, allows a significant amount to be absorbed at quite low temperatures. Thus hydrogen absorption can occur upon heating in gases containing water vapor (by the decomposition of the water vapor into oxygen and hydrogen) or hydrocarbons, and upon pickling in certain acids. The absorbed hydrogen can be removed by annealing in sufficiently high vacuum (e.g., 10^{-5} atm).

Hydrogen in solution has little effect on the mechanical properties. Upon precipitation of the hydride, however, the ductility suffers, as shown in Fig. 9-5. Note from the phase diagram in Fig. 9-4 that the solubility of hydrogen in α-Ti at 300 °C is about 8 at.% (about 0.15 wt % or about 1000 ppm by weight). Twenty ppm corresponds to about 0.1 at.%, and the data in Fig. 9-6 show that this low a concentration has little effect on the impact strength, independent of the heat treatment, and this is in line with the data in Fig. 9-5. However, note in Fig. 9-6 that 0.5% hydrogen (about 100 ppm) causes measurable embrittlement. Slow cooling from the α region (e.g., 400 °C) to 25 °C allows sufficient hydride to precipitate to reduce the impact energy. Rapid cooling (water quenching) from 400 °C suppresses the precipitation, retaining the high impact energy. However, aging at room temperature even for a few days allows sufficient hydride to precipitate to lower the impact energy. Thus, even though the hydride precipitation may be controlled by heat treatment, aging at 25 °C will result in sufficient precipitate formation and coarsening to embrittle the alloy. The only practical approach to control of the hydrogen problem is to maintain its concentration low. Thus, commercial pure titanium will usually have a maximum

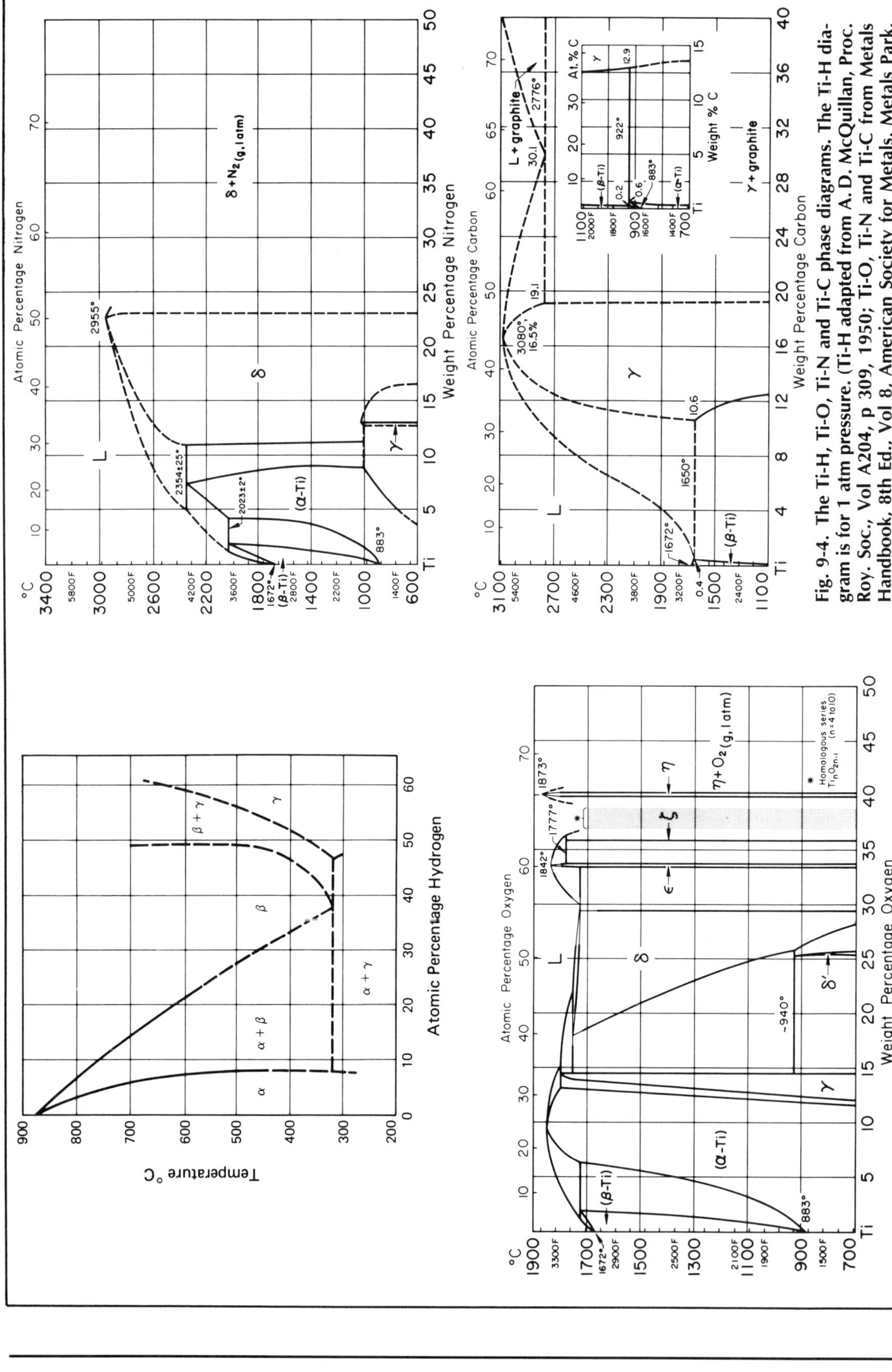

Fig. 9-4. The Ti-H, Ti-O, Ti-N and Ti-C phase diagrams. The Ti-H diagram is for 1 atm pressure. (Ti-H adapted from A. D. McQuillan, Proc. Roy. Soc., Vol A204, p 309, 1950; Ti-O, Ti-N and Ti-C from Metals Handbook, 8th Ed., Vol 8, American Society for Metals, Metals Park, Ohio, 1973)

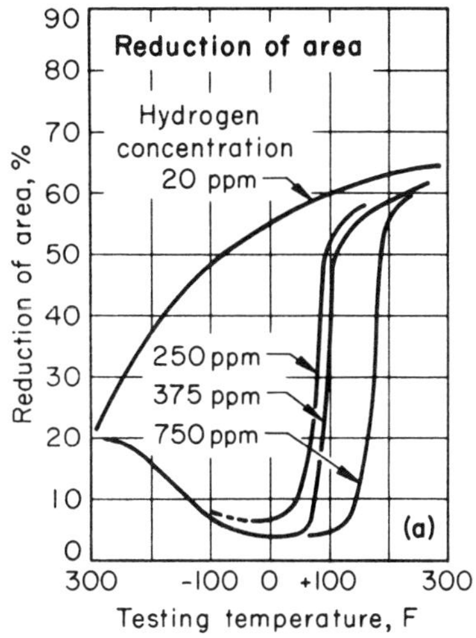

Fig. 9-5. Effect of hydrogen in α-titanium on the reduction in area in a tension test, showing the embrittling effect of hydrogen. (From same source as Fig. 9-1)

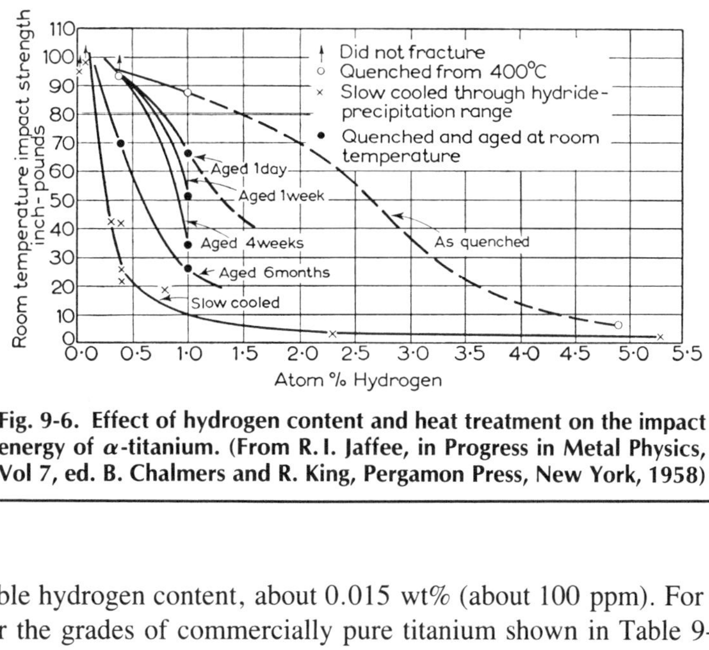

Fig. 9-6. Effect of hydrogen content and heat treatment on the impact energy of α-titanium. (From R. I. Jaffee, in Progress in Metal Physics, Vol 7, ed. B. Chalmers and R. King, Pergamon Press, New York, 1958)

allowable hydrogen content, about 0.015 wt% (about 100 ppm). For example, for the grades of commercially pure titanium shown in Table 9-2 it is about 0.01%.

Oxygen and nitrogen have a potent effect of strength, as shown in Fig. 9-7. As the amount of O and N increases, the toughness decreases, until the material eventually becomes quite brittle. From the data in Table 9-2 note that the allowed O content is higher than the allowed N content, and this is consistent with the relative effects shown in Fig. 9-7. It also can be seen from Fig. 9-7 why the grades of titanium with the higher allowable O and N contents have the higher strengths (Table 9-2). Note, though, that even the grade with the highest interstitial content has good ductility. Examination of the Ti-O and the Ti-N phase diagrams in Fig. 9-4 shows that embrittlement occurs (Fig. 9-7) at a concentration considerably below the solubility limit.

Table 9-2. ASTM Maximum Interstitial Content and Minimum Mechanical Properties of Four Grades of Commercially Pure Alpha Titanium for Bars and Billets (ASTM Specification B348-78)

	Grade 1	Grade 2	Grade 3	Grade 4
Maximum interstitial content allowed, wt %:				
Nitrogen	0.03	0.03	0.05	0.05
Carbon	0.10	0.10	0.10	0.10
Oxygen	0.18	0.25	0.35	0.40
Hydrogen	0.0125(a) 0.0100(b)	0.0125(a) 0.0100(b)	0.0125(a) 0.0100(b)	0.0125(a) 0.0100(b)
Minimum tensile strength, psi	35,000	50,000	65,000	80,000
Minimum yield strength, psi	25,000	40,000	55,000	70,000
Elongation, %	24	20	18	15
Reduction in area, %	30	30	30	25

From ASTM Specifications, Vol 8, American Society for Testing and Materials, Philadelphia, 1979.
(a) Bars only. (b) Billets only.

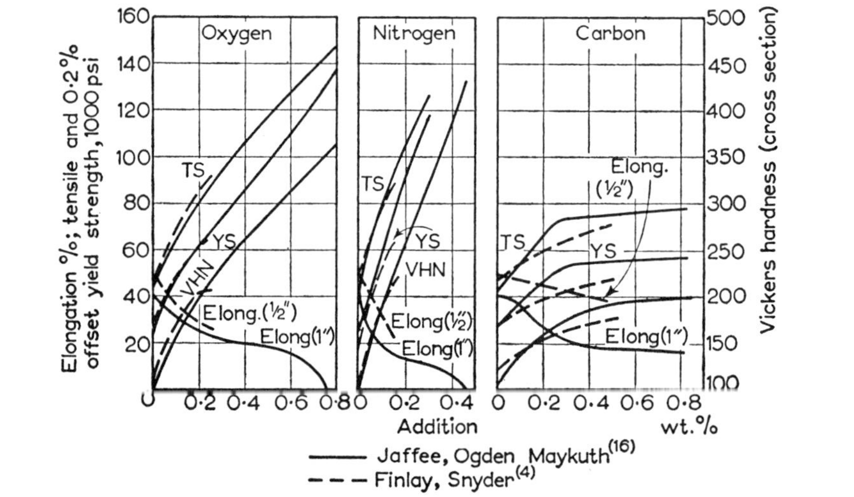

Fig. 9-7. Effect of oxygen, nitrogen and carbon content on the tensile mechanical properties and hardness of α-titanium. The data of Jaffee et al. are for samples annealed at 850 °C, and those of Finlay and Snyder are for samples annealed at 700 °C; thus both sets of data refer to alloys well annealed in the α region. (From same source as Fig. 9-6)

In titanium which contains O and N the α formed from β has a much more distinctive Widmanstätten structure (see Fig. 9-36) than in titanium essentially free of these elements (Fig. 9-8). However, this effect may have little direct bearing on the mechanical properties.

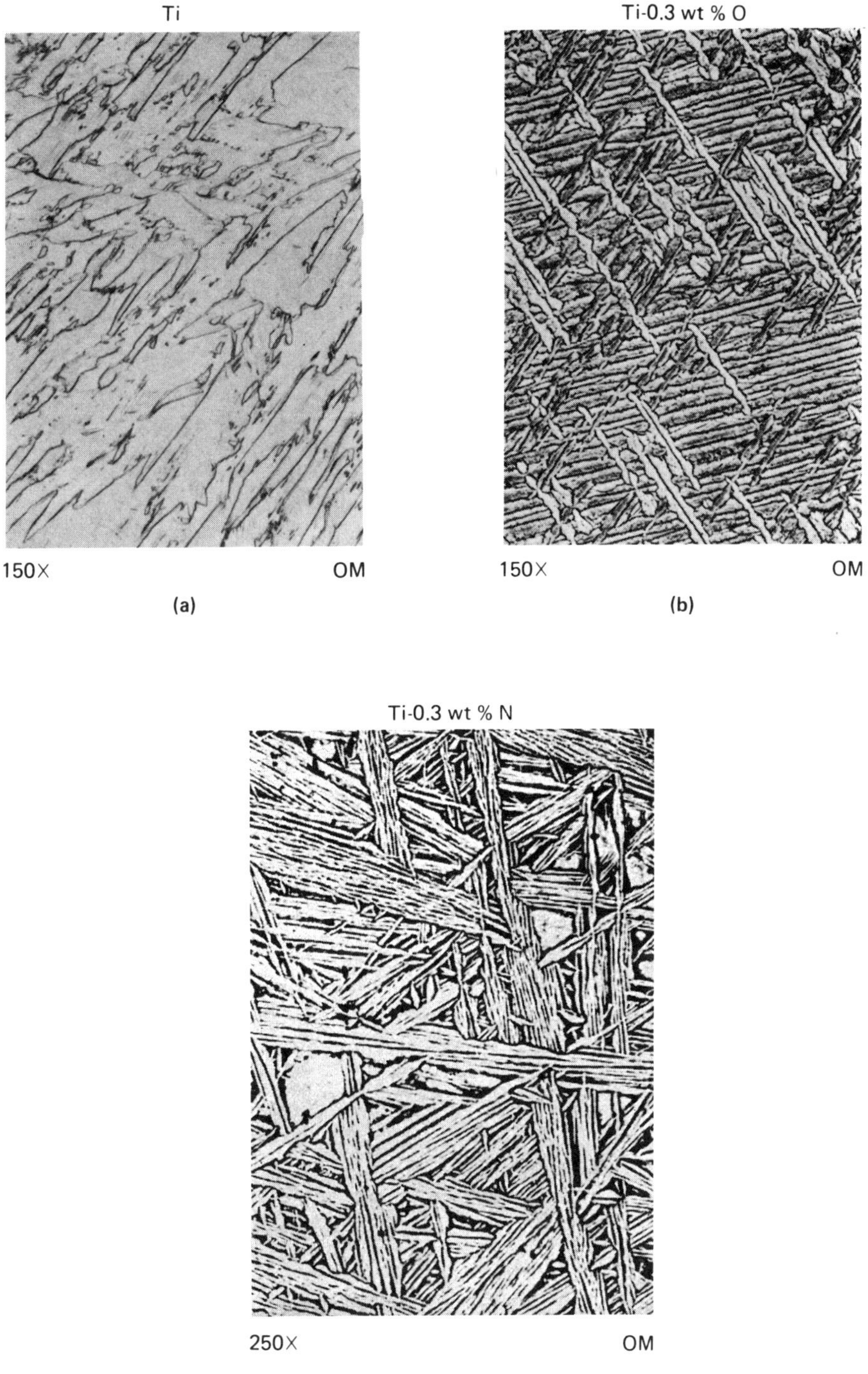

Fig. 9-8. Microstructures of (a) relatively pure titanium, (b) a Ti-0.3 wt% O and (c) a Ti-0.3 wt% N alloy, all obtained after annealing in the β region and then cooling to 25 °C. (a) and (b) adapted from E.S. Bumps, H.D. Kessler and M. Hansen, Trans. ASM, Vol. 45, p 1015, 1953; (c) adapted from R.I. Jaffee and I.E. Campbell, Trans. AIME, Vol 185, p 646, 1949.

The addition of carbon strengthens titanium greatly, up to about 0.3% C, and reduces ductility somewhat (Fig. 9-7). These data were obtained for samples annealed at 850 °C. The Ti-C phase diagram in Fig. 9-4 shows that at this temperature the solubility limit is about 0.4% C. Thus little hardening is achieved by the presence of the carbide TiC upon exceeding the solubility limit.

The concentration of the interstitial elements has been combined empirically into an oxygen equivalent, and the hardness correlated to this quantity. This is shown in Fig. 9-9. Also, Brown (see R. I. Jaffee, in *Progress in Metal Physics,* Vol 7, ed. B. Chalmers and R. King, Pergamon, New York, 1958) has obtained for the Brinell hardness the relation:

$$\text{BHN} = 196\sqrt{\%\text{N}} + 158\sqrt{\%\text{O}} + 45\sqrt{\%\text{C}} + 20\sqrt{\%\text{Fe}} + 57.$$

The previous information gives some of the facts about the mechanical properties of pure titanium, and the effects of interstitial content on them. Now we briefly examine the reasons for the behavior of these alloys. The high ductility of pure titanium is due to the ability of this metal to twin (see Chapter 1) on several planes. Thus, upon plastic deformation twinning occurs relatively readily, the reorientation allowing activation of new slip systems. However, unlike hexagonal close-packed elements like zinc, titanium also has slip planes other than the basal plane (see Chapter 1). The low c/a ratio causes planes other than the basal plane to have relatively large planar separations, favorable for slip.

The mechanism of the embrittlement due to increasing amounts of O and N, which leads to low ductility and the need to limit their concentration, is not clear. They lower the ductility by reducing the number of twin and slip

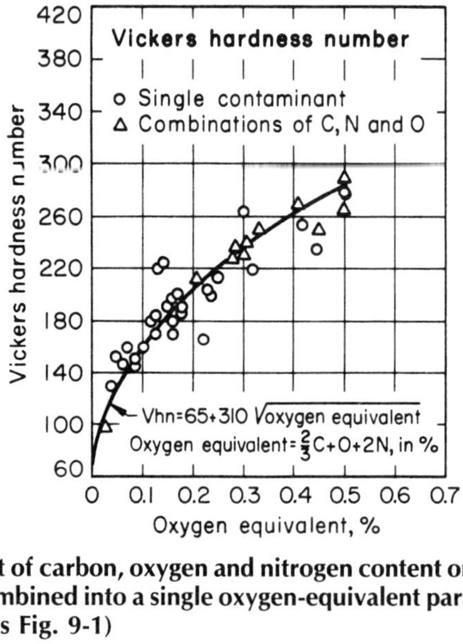

Fig. 9-9. Effect of carbon, oxygen and nitrogen content on the hardness of titanium combined into a single oxygen-equivalent parameter. (From same source as Fig. 9-1)

systems, but how this inhibition occurs is still not understood. Since titanium is hexagonal close-packed (e.g., non-cubic) a complicated dislocation structure might be expected to be present, and to have a complicated interaction with interstitial elements. In any case, the exact mechanisms are not yet clearly established.

9-3. Alpha Alloys

In discussing the physical metallurgy of titanium, it is common to separate the alloys into three categories, referring to the common phases present. The categories are α, $\alpha + \beta$, and β alloys. In discussing the alloys common to each category, we will examine the effects of elements on the stability of α and β, as revealed, to some extent, by the phase diagrams and by the kinetics of the β-to-α transformation.

The α alloys have α as their common phase at low temperature (i.e., below 800 °C), and thus the properties in general cannot be altered by heat treatment. The only strengthening mechanisms are cold work, cold work and anneal to control the α grain size, and solute additions for solid solution strengthening. To predict possible influential solute strengtheners we examine (as in other chapters) the relative atom size of possible solutes compared with titanium, as tabulated in Table 9-3. Looking at the substitutional solutes, we see that those that appear most promising are Al, Si, V, Zr, Nb and Sn, as they show high or medium solubility and at least moderate size differences, and all may be economical to use.

Figure 9-10 shows the phase diagrams for these solutes. Note that only Al, Sn and Zr stabilize the α phase, so for solid solution strengthening only

Table 9-3. Approximate Atom Diameter Size Difference and Approximate Solubility in α-Ti of Several Solutes

Solute	Approximate atom diameter size difference, % $(d_{Ti} - d_M)/d_{Ti}$, %	Approximate solubility in α-Ti, wt %	Solute	Approximate atom diameter size difference, % $(d_{Ti} - d_M)/d_{Ti}$, %	Approximate solubility in α-Ti, wt %
Hydrogen*	68	0.2	Cobalt	14	1
Carbon*	49	0.5	Nickel	15	1
Nitrogen*	51	4	Copper	12	1
Oxygen*	59	12	Gallium	16	10
Magnesium	−9	0.1	Zirconium	−8	100
Aluminum*	2	25	Niobium	2	4
Silicon	19	2	Tin*	−3	22
Vanadium*	10	4	Molybdenum*	7	1
Chromium	14	<1	Tantalum	2	1
Manganese	23	1	Tungsten	6	1
Iron	15	1			

Note. Solutes marked with an asterisk (*) are common additions to Ti.

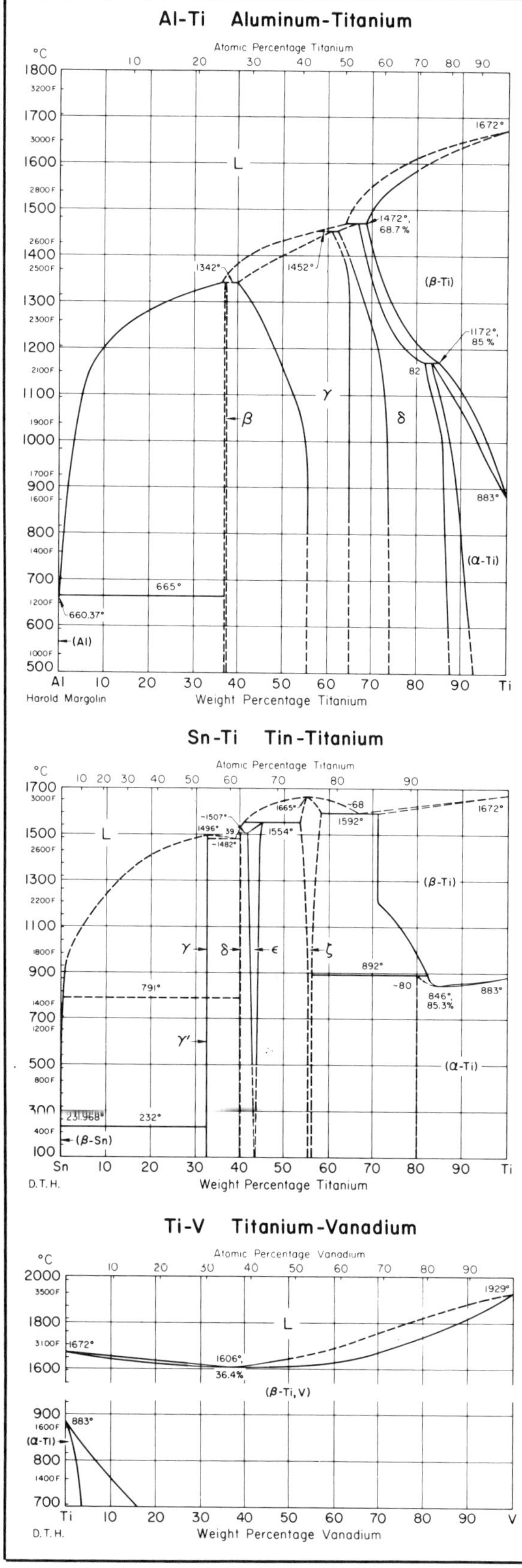

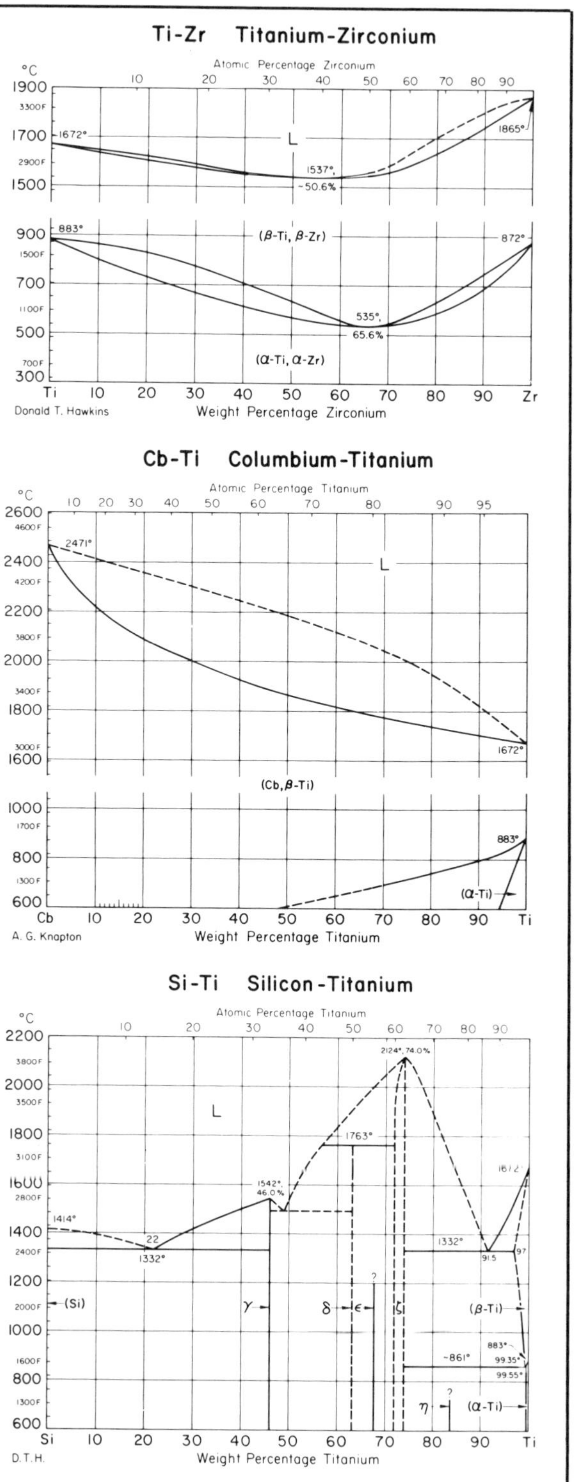

Fig. 9-10. The Ti-Al, Ti-Sn, Ti-Zr, Ti-V, Ti-Nb and Ti-Si phase diagrams. (From Metals Handbook, 8th Ed., Vol 8, American Society for Metals, Metals Park, Ohio, 1973)

these three will be considered. The effect of Al and of Sn on the mechanical properties of Ti is shown in Fig. 9-11. The strengthening is significant for both elements, being an increase in strength of approximately 8,000 psi for each percent of Al and 4,000 for each percent of Sn. However, for Zr, the effect is only about 500 psi for each percent of Zr. Thus, Zr is not used for strengthening α titanium alloys.

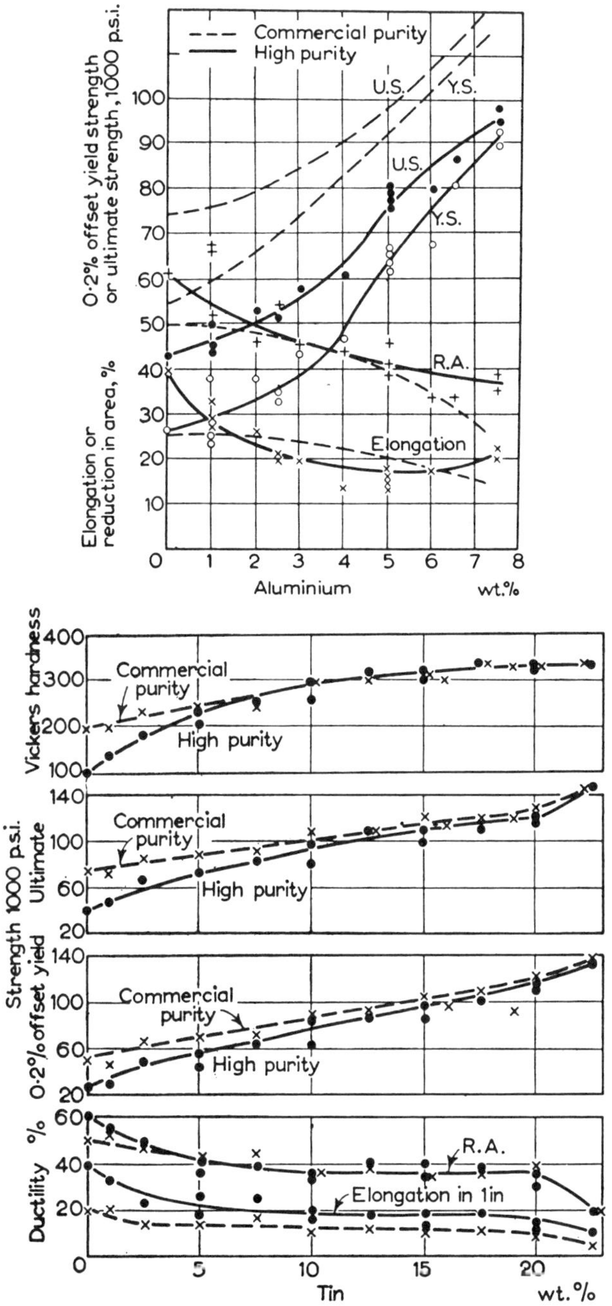

Fig. 9-11. Mechanical properties of Ti-Al and Ti-Sn annealed α solid solution alloys. (From same source as Fig. 9-6)

Note in Fig. 9-11 that the effectiveness of Sn as a strengthener begins to level off, whereas for Al it does not. However, the practical limit to the Al content is about 7%, as above this value the alloy is difficult to hot work and at low temperature embrittlement may occur. This latter effect is illustrated in Fig. 9-12.

The origin of the embrittlement in Ti-Al alloys is not in the precipitation of δ, but in the formation of Ti_3Al. Figures 9-13 shows the titanium-rich region of the Ti-Al phase diagram, as given in Fig. 9-10. Also shown is the diagram applicable for long aging times, and for slow cooling from the α region.

An important consideration in using titanium alloys is stress-assisted corrosion and stress corrosion cracking. In many alloys, especially when heat treated to high strength levels, a corrosive environment can lower the strength of the material and make it susceptible to crack formation and propagation. Stress corrosion cracking (SCC) is determined by elastically loading a specimen to a constant stress below the yield strength. Upon placing the specimen in the corrosive environment (e.g., salt water, moist air, etc.), if the material is susceptible to SCC in the environment, it will eventually show crack formation. The effect of the environment can also be revealed by determining the tensile mechanical properties in air (a standard test) and then in the corrosive environment.

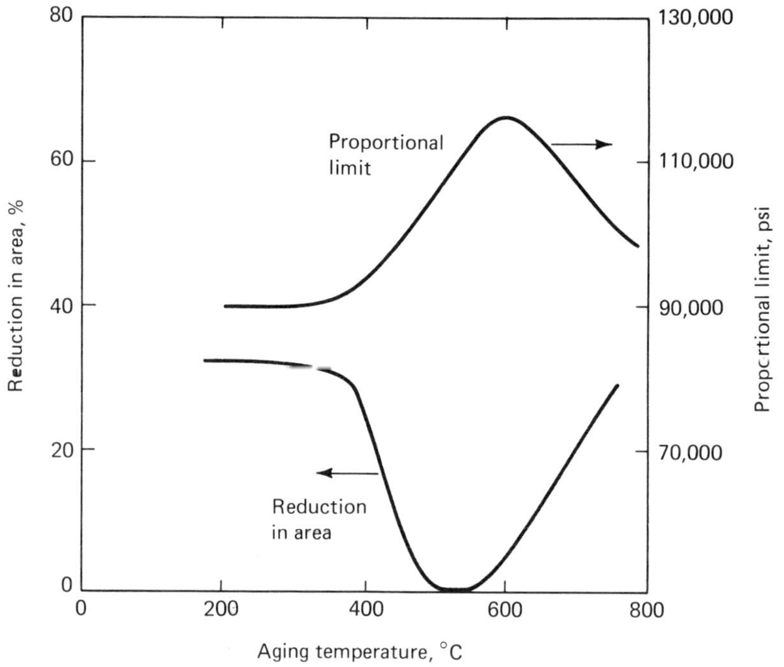

Fig. 9-12. Effect of aging on the strength (proportional limit) and ductility (reduction in area) of a Ti-8Al alloy. The aging times increased with reduced aging temperatures, being 1 h at 800 °C and 120 h at 200 °C. (Adapted from F. A. Crossley and W. F. Carew, Trans. AIME, Vol 209, p 43, 1957)

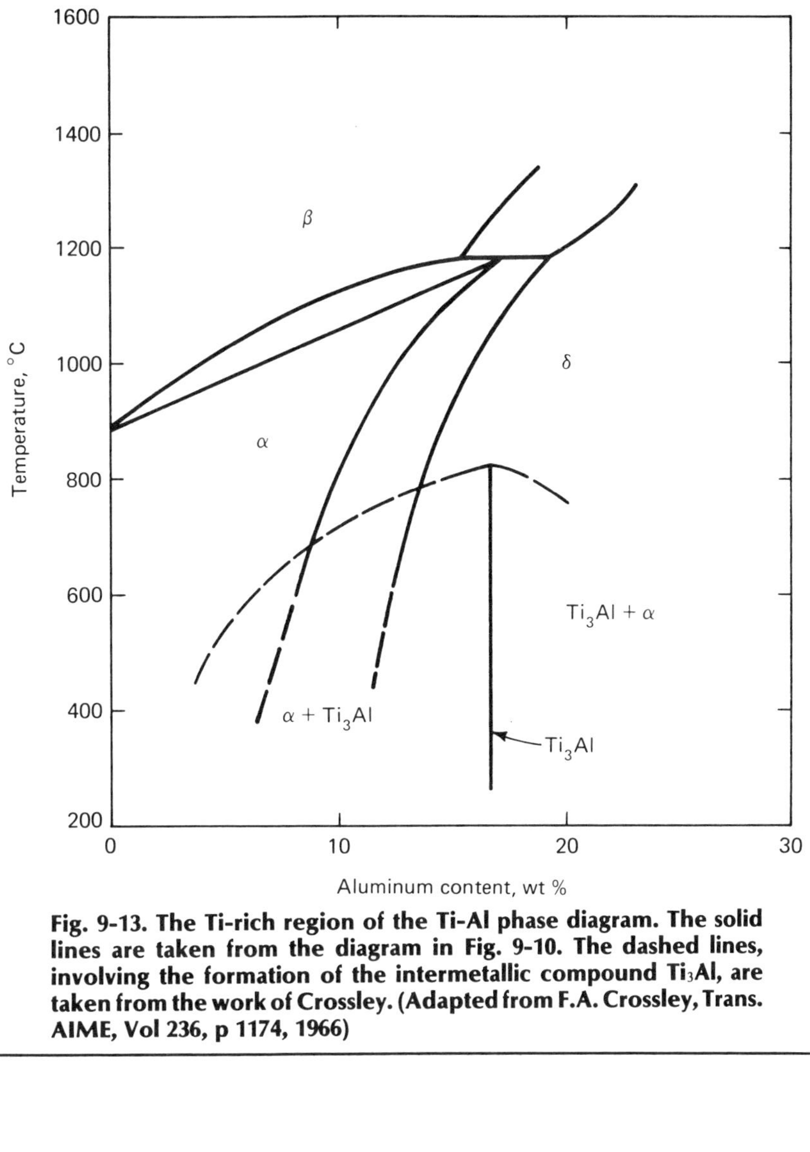

Fig. 9-13. The Ti-rich region of the Ti-Al phase diagram. The solid lines are taken from the diagram in Fig. 9-10. The dashed lines, involving the formation of the intermetallic compound Ti_3Al, are taken from the work of Crossley. (Adapted from F.A. Crossley, Trans. AIME, Vol 236, p 1174, 1966)

The effect of aluminum content and heat treatment on the sensitivity of Ti-Al alloys to stress corrosion is shown in Fig. 9-14. The alloys were annealed at 1200 °C, in the β region, then water quenched, forming α martensite. If in this condition the alloys are reheated to the α region (900 °C), then water quenched, the fracture stress (measured at 25 °C) is the same whether tested in air or seawater if the Al content is below 7%. For higher Al contents, it is believed that even rapid cooling does not suppress completely the formation of some Ti_3Al, and hence the fracture stress is lowered somewhat in air, and especially in seawater. However, the stress corrosion is increased markedly by aging for 2 h at 595 °C following the heat treatment just described. This is shown in Fig. 9-14. This temperature will put the alloys in the two-phase $\alpha + Ti_3Al$ region for Al contents above about 5%.

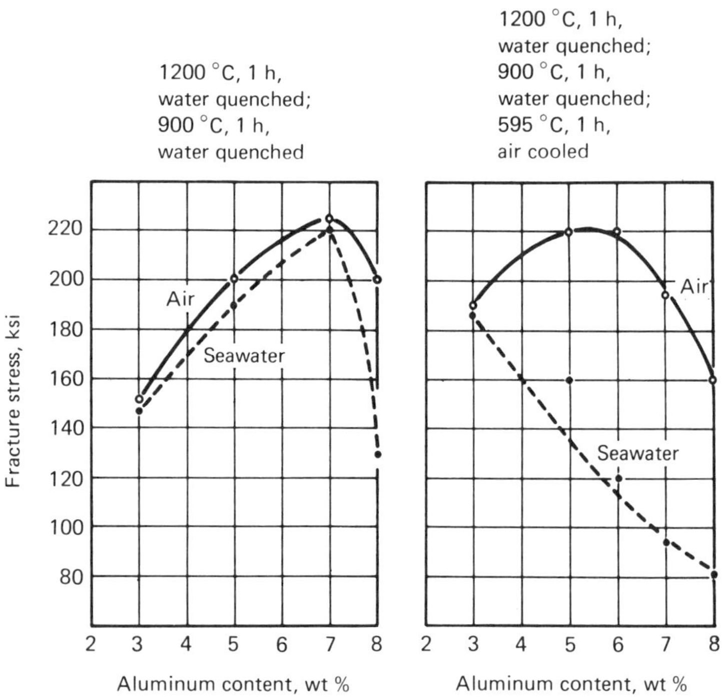

Fig. 9-14. Fracture stress in air and in seawater as a function of aluminum content in Ti-Al alloys. The susceptibility to the influence of corrosive environment for the aged alloys is clearly shown. (Adapted from I. R. Lane and J. L. Cavallaro, in Applications Related Phenomena in Titanium Alloys, STP 432, American Society for Testing and Materials, Philadelphia, 1968)

Another important consideration in using high-strength alloys is their notch sensitivity. This is a measure of the ability of the material to plastically deform locally at regions of high stress, relative to the average stress, instead of forming a crack and fracturing. The notch sensitivity is frequently measured by conducting a tensile test on a pre-notched specimen, and comparing the tensile strength (or yield strength or fracture stress) to that obtained using an unnotched specimen. Figure 9-15 shows schematically the stress-strain curves for both conditions. Here the stress is engineering stress, obtained by dividing the load by the original area. In the notched specimen, this is the area at the root of the notch. Note that for the notched specimen the yield and tensile strength are higher, and the elongation at fracture is less.

The behavior in Fig. 9-15 can be explained by the following qualitative description. At a distance removed from the notch, the axial load is evenly distributed across the cross section. Thus the stress is uniform. At the notch level, the notch area supports the load, and the free surface outside the notch supports no axial load. If the axial stress contours are determined (from elastic stress analysis), the result shown in Fig. 9-16 is obtained, where it is seen that the axial stress rises as the root of the notch is approached from the center. The stress concentration is usually defined as $K_t = \sigma_{max}/\sigma_{nom}$, where σ_{nom} is the average axial stress and σ_{max} is the axial stress right at the root

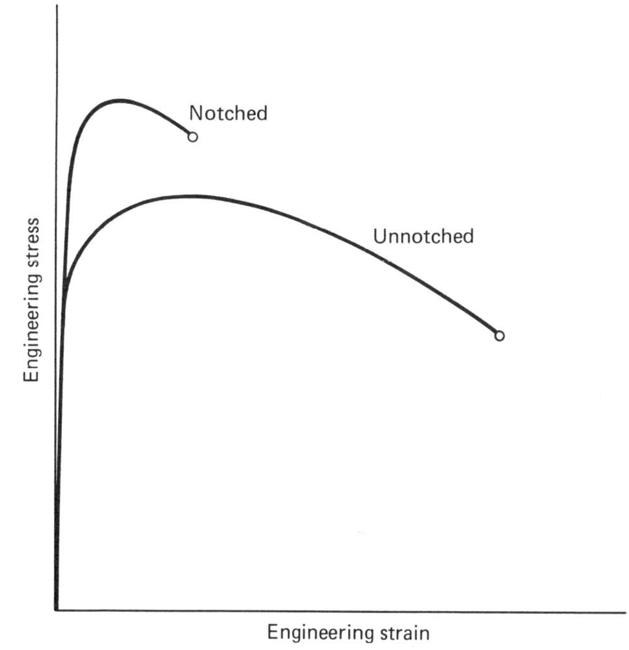

Fig. 9-15. Schematic stress-strain curves for an unnotched and a notched tensile specimen. In the notched specimen, the yield and tensile strengths are higher and the elongation at fracture is less.

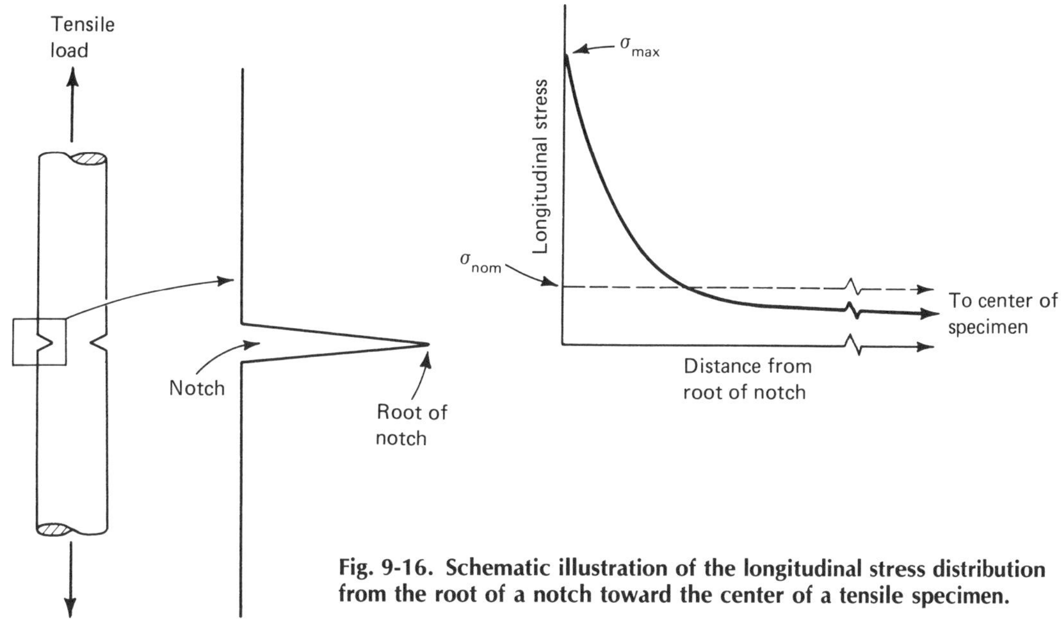

Fig. 9-16. Schematic illustration of the longitudinal stress distribution from the root of a notch toward the center of a tensile specimen.

of the notch (see Fig. 9-16). The sharper the notch (smaller root radius), the higher K_t will be.

As the axial load is increased, the axial stress at the root of the notch eventually reaches the value at which yielding would occur if the notch were not present. However, the region outside the notch supports no load at the free surface, and thus should not elongate. Because this region is intimately connected to the region of the root of the notch, it constrains the latter from elongating, and yielding is not initiated at this axial stress level. As the load increases further, eventually a level is attained where yielding occurs in spite of the constraint, but this gives higher yield strength than for an unnotched specimen (see Fig. 9-15).

The importance of comparing tensile properties of notched and unnotched specimens lies in the observation that if the strengths (yield or tensile) are similar for the two tests, then the material behaves similarly for smooth surfaces and for surfaces which have discontinuities (e.g., holes) or are rough.

The notch sensitivity of Ti-Al alloys is illustrated in Fig. 9-17. Even for relatively low Al contents, the notch fracture stress is lower than that for smooth specimens. Especially in the aged condition this sensitivity is pronounced. Again, the precipitation of Ti_3Al is the culprit. When the Al content is above about 10%, even rapid cooling from the α region cannot suppress the precipitation of some Ti_3Al, and hence some embrittlement occurs. To increase the strength and yet retain reasonable toughness a ternary alloy may be used, and the only common α alloys contain about 5% Al and

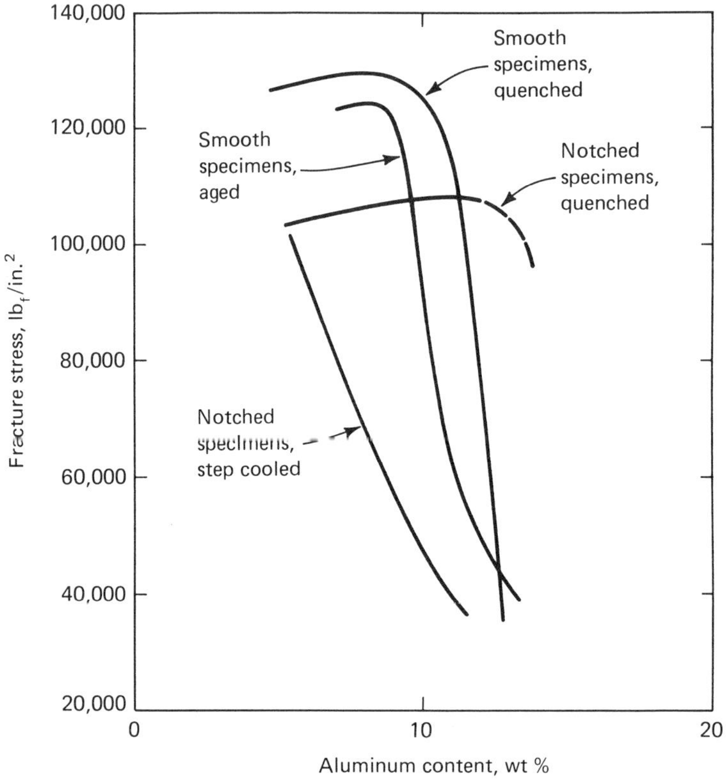

Fig. 9-17. Effect of aluminum content and of the formation of Ti_3Al (induced by aging or step cooling) on the notch sensitivity of Ti-Al alloys. (Adapted from M. J. Blackburn and J. C. Williams, Trans. ASM, Vol 62, p 398, 1969)

2.5% Sn. Figure 9-18 shows some mechanical properties of the Ti-Al-Sn ternary alloys, and Table 9-4 gives some properties for Ti-Al, Ti-Sn and Ti-Al-Sn alloys with a total solute content in each case of about 7.5%. The effectiveness of increasing the strength by limiting the Al content to 5% and adding Sn is apparent.

The interstitial elements C, O and N affect the α alloys in a way similar to the way they affect pure titanium: increasing strength and decreasing ductility. This is the basis of the difference in the properties of high-purity alloys and commercially pure alloys shown in Fig. 9-11. The effect is shown more clearly for Ti-Sn alloys in Fig. 9-19. Oxygen is particularly important, as it may be picked up during annealing in air. Note in the table in Appen-

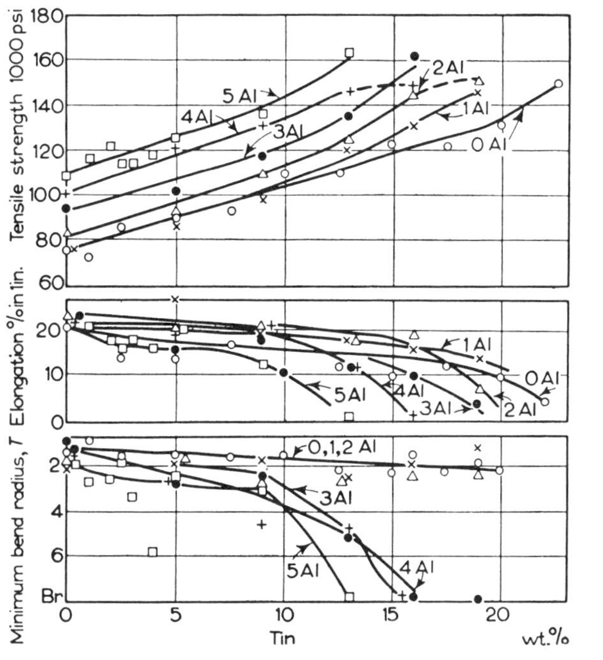

Fig. 9-18. Mechanical properties of annealed α Ti-Al-Sn alloys. (Adapted from same source as Fig. 9-6)

Table 9-4. Approximate Mechanical Properties of Commercial Ti-Al, Ti-Sn and Ti-Al-Sn Annealed Alpha Alloys

Alloy	Yield strength, psi	Tensile strength, psi	Elongation, %
Ti-5Al	90,000	98,000	20
Ti-7.5Al	115,000	120,000	20
Ti-2.5Sn	60,000	80,000	15
Ti-7.5Sn	80,000	90,000	15
Ti-5Al-2.5Sn	—	120,000	15

Values obtained from Fig. 9-11 and 9-18.

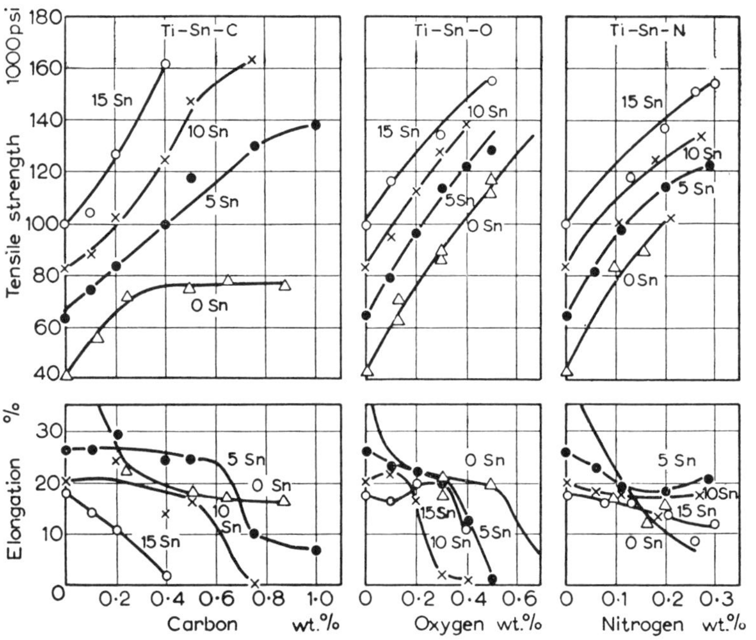

Fig. 9-19. Effect of carbon, oxygen and nitrogen content on the mechanical properties of Ti-Sn alloys. (From same source as Fig. 9-6)

dix 6, that the only α alloy listed is Ti-5Al-2.5Sn, given for two oxygen levels. The higher oxygen content has a markedly higher strength, but lower ductility. Thus in applications requiring increased ductility, the low-oxygen alloys may be used.

9-4. Alpha-Plus-Beta Alloys

The previous section dealt only with alloys which have the α structure at low temperature, and these were alloys where the β-to-α transformation could not be suppressed even upon rapid cooling. We now begin to examine the effect of alloying additions which stabilize the β phase, and the structure and properties developed from this. The phase stability can be predicted, to some extent, by the crystal structure of the alloying element. For example, body-centered cubic elements would be expected to stabilize β, and this is shown for Nb and V in Fig. 9-10. The common β-stabilizing elements are divided into two types: β isomorphous and β eutectoid. β-isomorphous means that the solubility of the element in titanium in the β form is unlimited, giving a phase diagram such as that for Ti-V (Fig. 9-10) and Ti-Mo (Fig. 9-20). β-eutectoid means that an eutectoid reaction occurs which involves the β phase, such as the Ti-Cu system shown in Fig. 9-21. A term frequently used in discussing the heat treatment of titanium alloys is β transus. This is the

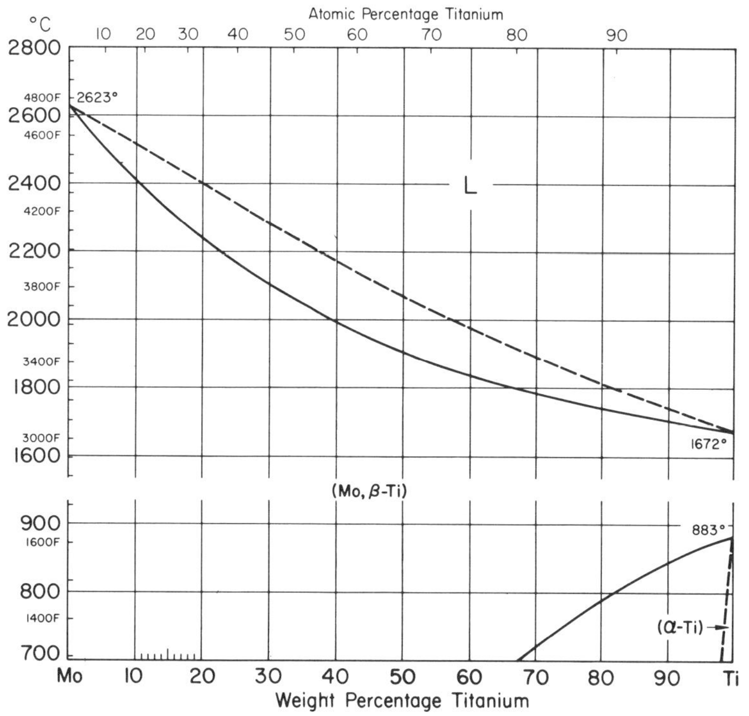

Fig. 9-20. The Ti-Mo phase diagram. (From same source as Fig. 9-10)

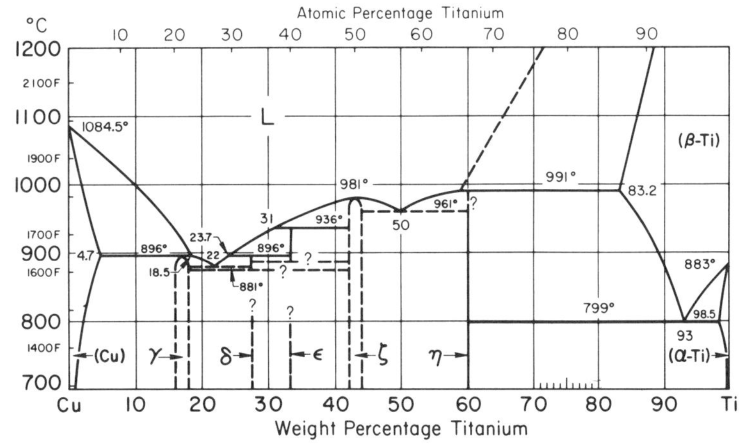

Fig. 9-21. The Ti-Cu phase diagram. (From same source as Fig. 9-10)

phase boundary between β and $\alpha + \beta$. That is, it is the temperature above which only the β phase is present at equilibrium.

To put the behavior of $\alpha + \beta$ alloys into perspective, consider the schematic diagram in Fig. 9-22. Upon slow cooling these alloys from the β region, for low alloy content the α phase will form. But with sufficient alloy content, the two-phase structure $\alpha + \beta$, with the α phase rich in titanium and the β phase rich in the alloying element, will be present. The amount of β increases with increasing alloy content, and it is found that the strength of such alloys increases. This treatment is sometimes called annealing.

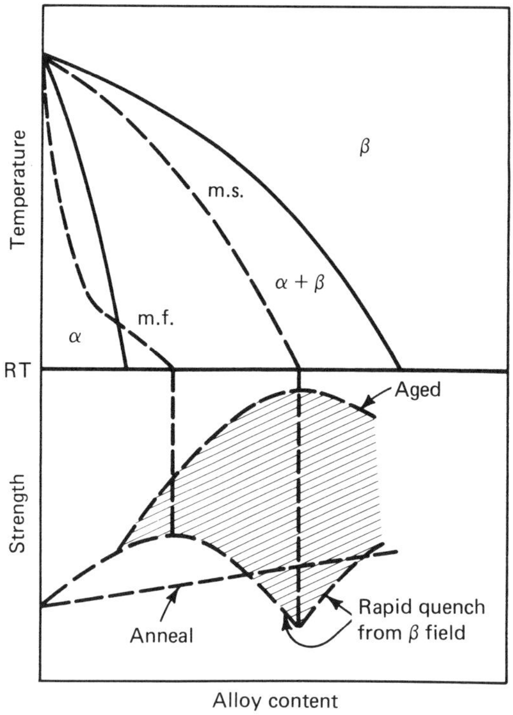

Fig. 9-22. Schematic diagram of the heat treatment of titanium alloys of a β-isomorphous system. (Adapted from R. E. Goosey, in Heat Treatment of Engineering Components, Iron and Steel Institute, London, 1969)

Now, if pure titanium is cooled rapidly from β, the β-to-α transformation occurs by a martensite reaction, beginning (M_s) and finishing (M_f) at essentially the same temperature. However, the addition of β-stabilizing elements lowers the M_s and M_f, and separates them (Fig. 9-22). As long as the alloying addition is such that M_f is above 25 °C, then the strength (at 25 °C) is higher than that of the annealed structure, because of the finer α formed and the dislocation structure induced by the volume change upon transforming. However, once the alloy content is high enough for M_f to be below 25 °C, then at 25 °C some β is retained, and the strength begins to decrease as the amount of retained β increases (Fig. 9-22). Eventually, though, a solute content is attained such that the solid solution strengthening is sufficient to cause the strength in the as-quenched condition to increase (Fig. 9-22). If the alloy content is sufficient for M_s to be below 25 °C, then the structure is entirely retained β (but having the composition of the alloy, not the equilibrium composition at 25 °C). The β has less intrinsic strength than an $\alpha + \beta$ mixture, as shown in Fig. 9-22 by the curve for the quenching treatment being below that for the annealing treatment. For alloy contents sufficiently high to place M_s below 25 °C, the strength begins to increase because of increased solid solution strengthening of the β.

These quenched structures which contain retained β are not stable, as neither retained β nor the α have the equilibrium alloy content, and thus upon reheating the structure should change. Manipulation of the aging temperature and time should allow control of the aged structure, and hence the properties. Such treatments lead to significant strengthening, as shown by the shaded region in Fig. 9-22.

9-5. The Commercial Alloy Ti-8Al-1Mo-1V

As mentioned previously, a framework in which to understand titanium alloys is to separate them into α, $\alpha + \beta$, and β alloys. Such a scheme indicates that the behavior of these titanium alloys will be quite complex, and indeed it is. It would be difficult to describe the behavior of all of these types of commercial alloys, and in some cases their behavior is not well understood. Instead, in this and the following two sections, specific commercial alloys which are illustrative of each of the $\alpha + \beta$ and the β alloys will be examined in some detail.

We will discuss the alloy Ti-8Al-1Mo-1V. The body-centered cubic Mo and V both stabilize the β phase (see Fig. 9-10 and Fig. 9-20). As an approximation the Mo and V can be considered to have similar effects, and hence we should be able to use the Ti-Al-Mo ternary phase diagram for guidance in heat treatments. However, this diagram is not well determined, so instead we will use the schematic isopleth shown in Fig. 9-23. For an alloy containing approximately 1% Mo and 1% V, the structure is β above about 1050 °C (the β transus). Upon slow cooling to 25 °C, the α phase, rich in Ti, separates from the β, the β increasing in Mo and V. Below about 500 °C, the ordered hexagonal close-packed phase α_2 (Ti_3Al) is stable, along with α and β. However, the rate of formation of α_2 is slow: it will not form at all on slow cooling but only on prolonged aging.

Now let us examine the structure and properties after cooling from the annealing temperature to 25 °C. The annealing temperature determines the amount of Mo + V in both the α and the β, and the amount of each phase present (Fig. 9-24a). The structure at 25 °C, after quenching, depends upon what happens to the β phase (Fig. 9-24b).

Figure 9-25 lists the structures present at 25 °C after cooling rapidly (air or water quenching) from the annealing temperature. Above about 1050 °C, only the β phase is present at the annealing temperature, and upon cooling it transforms completely by a martensite reaction to a hexagonal structure, designated α''. Upon annealing in the two-phase $\alpha + \beta$ region, then quenching, the α present at the annealing temperature will be present at 25 °C after cooling and is designated primary α, and the amount present increases with decreasing annealing temperature (Fig. 9-24a).

The transformation product of β depends upon the annealing temperature because this determines the amount of the β-stabilizing elements

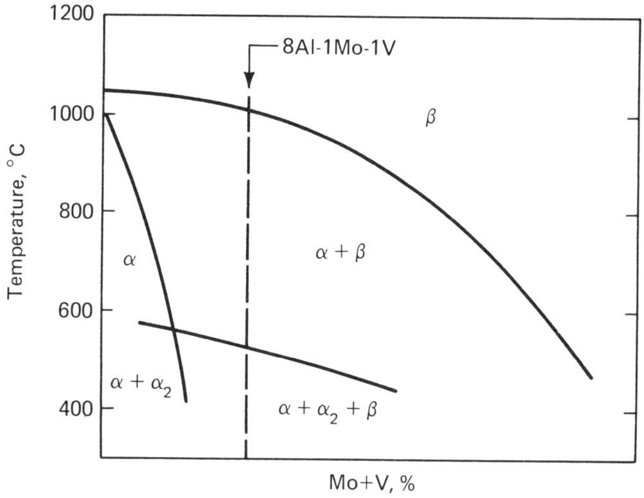

Fig. 9-23. Schematic isopleth at 8% Al. (Adapted from M. J. Blackburn, Trans. ASM, Vol 59, p 876, 1966)

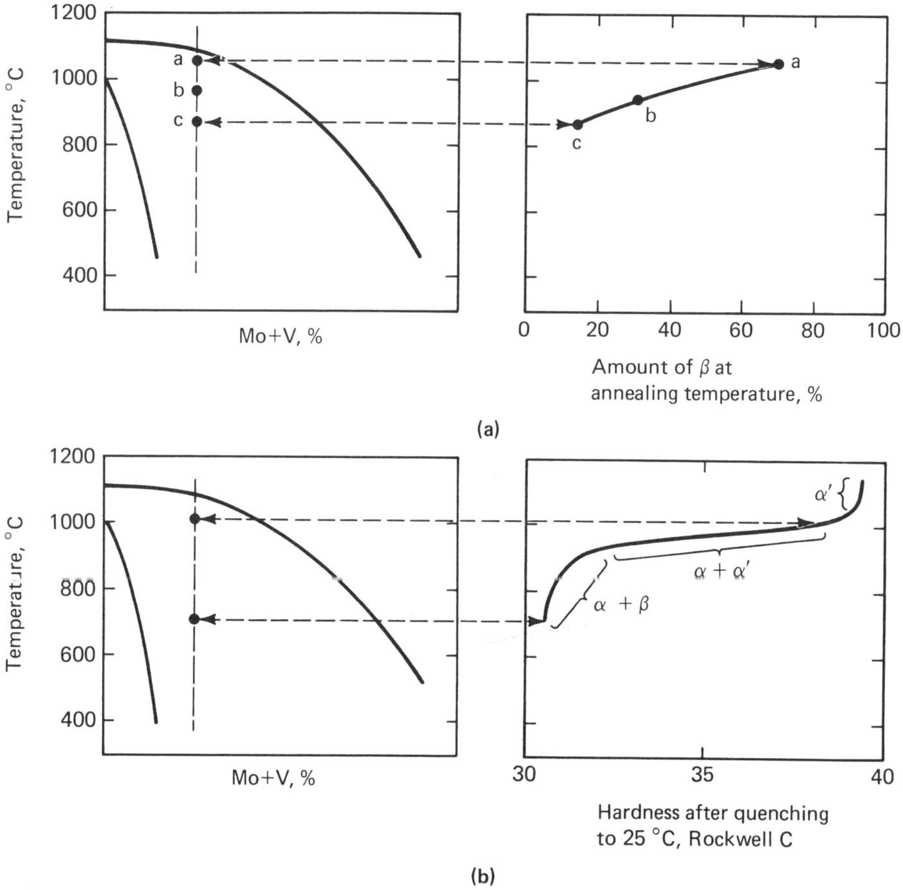

Fig. 9-24. Relationship of the phases present at the annealing temperature, those at 25 °C after quenching, and the hardness at 25 °C. Microstructures corresponding to a, b and c are shown in Fig. 9-26. (Adapted from P. J. Fopiano and C. F. Hickey, Jr., in Titanium Science and Technology, Vol 3, ed. R. I. Jaffee and H. M. Burte, Plenum, New York, 1973)

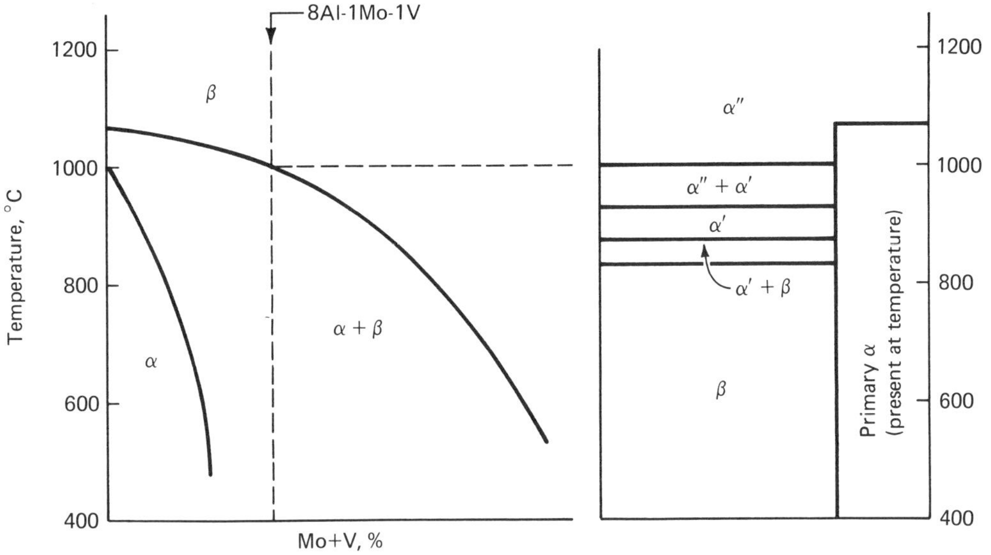

Fig. 9-25. Illustration of the phases present at annealing temperature, and at 25 °C after quenching from the annealing temperature. (Adapted from M. J. Blackburn, Trans. ASM, Vol 59, p 877, 1966, and J. C. Williams and M. J. Blackburn, Trans. ASM, Vol 60, p 373, 1967)

Mo and V present in the β. At annealing temperature just below about 1050 °C, just into the $\alpha + \beta$ region, another type of martensite appears with α''; it is designated α' and is essentially face-centered cubic.

When the annealing temperature is below about 900 °C, the Mo and V contents of the β are sufficient to put the martensite finish temperature for α' below 25 °C, and thus upon quenching to 25 °C some β is retained. Then the structure is α', retained β, and primary α. When the annealing temperature is below about 850 °C, the Mo and V contents are sufficient to put the martensite start below 25 °C, so that upon quenching to 25 °C, no β transforms, and the structure is retained β and primary α.

The characteristics described above are shown microstructurally in Fig. 9-26. Note that the amount of primary α increases as the annealing temperature decreases, and that at low annealing temperatures retained β is present. (In these microstructures, no distinction is made between α' and α''.)

Unlike the case with steel, the martensite in these alloys is not as strong as that developed upon aging. Thus, a structure developed by aging initially all β (at the annealing temperature) has less strength than that developed upon aging a structure which was initially $\alpha + \beta$. This is illustrated in Fig. 9-27. Microstructures typical of those formed for this range of annealing are shown in Fig. 9-28. Upon cooling rapidly from the β region (1065 °C), the β decomposes martensitically to α'. Upon aging at 580 °C, this martensite transforms into a fine structure of α and β, each having different Al, Mo and V contents, the amount being dictated by the phase diagram. In the microstructures in Fig. 9-28, this two-phase mixture is not resolved. As

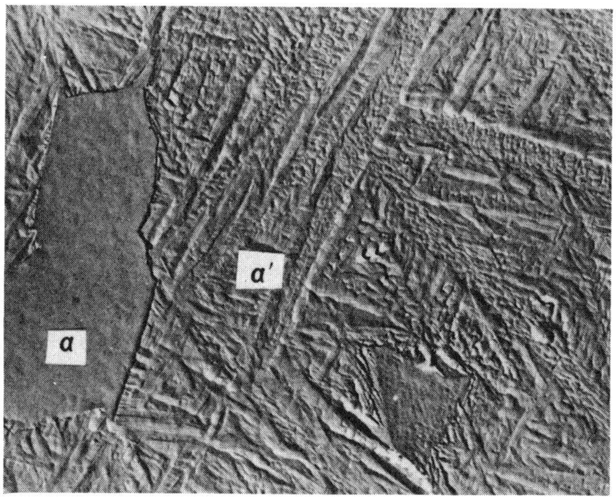

1 h at 1040 °C, brine quenched; 75% β-25% α at annealing temperature

(a)

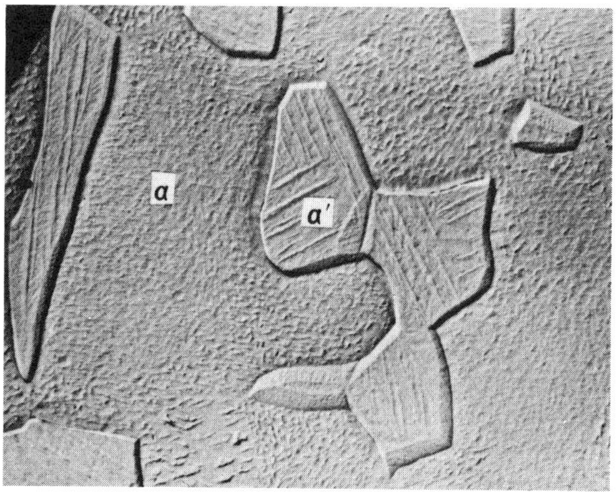

1 h at 925 °C, brine quenched; 25% β-75% α at annealing tempterature

(b)

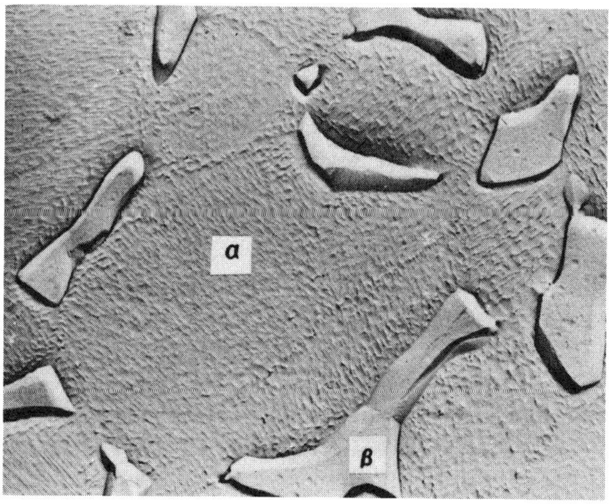

1 h at 845 °C, brine quenched; 15% β-85% α at annealing temperature

(c)

Fig. 9-26. Microstructure of Ti-8Al-1Mo-1V for different annealing temperatures. At the lower annealing temperatures, the Mo and V contents of the β are higher, and the M_s is below 25 °C, so that β is retained after quenching. The relationship between these microstructures and the phase diagram is shown in Fig. 9-24, where a, b and c are marked. Replica transmission electron micrographs; 5500×. (Adapted from same source as Fig. 9-24)

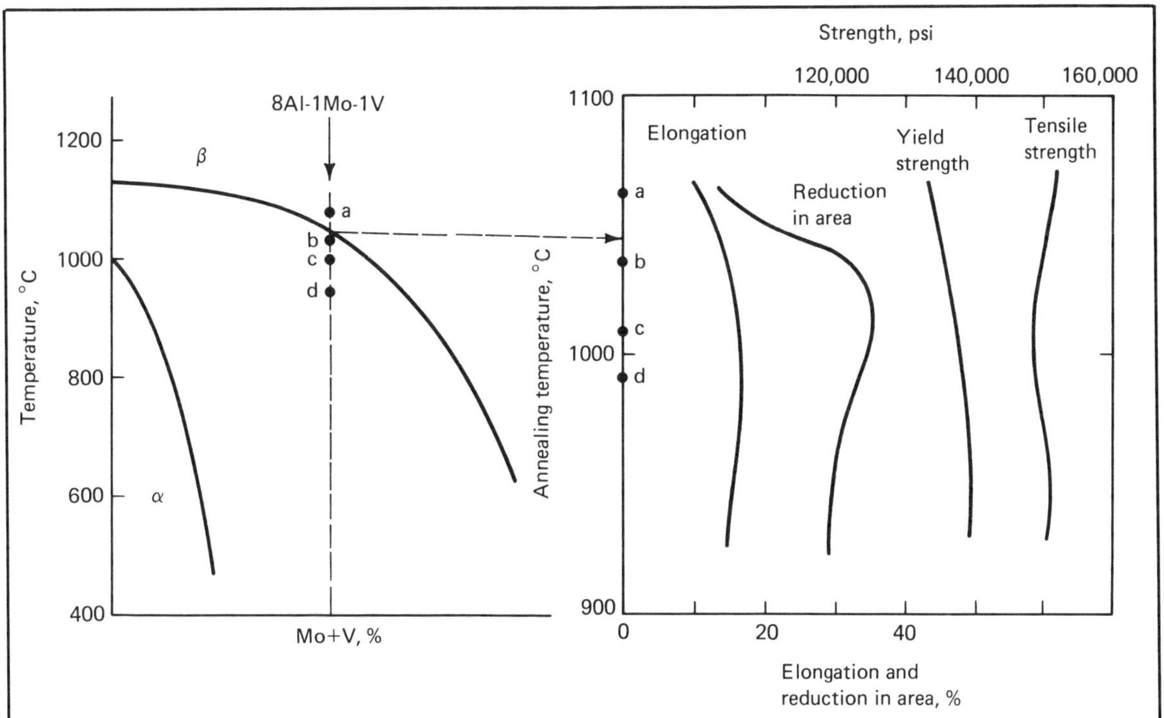

Fig. 9-27. Illustration of the effect of annealing in the α and β region on mechanical properties at 25 °C. The alloys were annealed for 1 h, oil quenched, then aged (stabilized) for 8 h at 580 °C. Note that annealing in the all-β region reduces the ductility. The letters on the vertical line in the phase diagram correspond to the microstructures in Fig. 9-28. (Adapted from D. B. Warmuth and H. D. Kellser, Metal Progress, Vol 89(3), p 191, 1966)

the annealing temperature decreases below the β transus, into the two-phase α + β region, the amount of primary α increases. This is the white phase in Fig. 9-28(a), (b) and (c). Note in Fig. 9-27 that this decreasing annealing temperature increases the yield strength slightly, and that the maximum ductility is attained for a temperature just below the β transus.

If the cooling rate from the annealing temperature is lower than that employed for the structures in Fig. 9-28 the structures in Fig. 9-29 are obtained. Air cooling from 1100 °C is not sufficiently rapid to suppress some primary α formation, which is the white Widmanstätten structure (see Fig. 9-36) in Fig. 9-29(c). However, the remaining β transforms to α'', which upon aging becomes the fine α + β structure. Annealing at lower temperature (Fig. 9-29a and b) develops prominent amounts of primary α present at the annealing temperature.

As with the binary Ti-Al alloys, the notch sensitivity and stress corrosion sensitivity of Ti-8Al-1Mo-1V must be examined. A common heat treatment for this alloy is to anneal it at 790 °C for 8 h, then furnace cool it. This heat treatment is called *mill annealing*. Figure 9-30 shows that in this condition this alloy is quite notch sensitive.

1065 °C

Former β grain boundaries; structure is fine $\alpha + \beta$ developed by the aging

(a)

1040 °C

Small amount of white primary α along the former β grain boundaries; the dark is fine $\alpha + \beta$ from aging

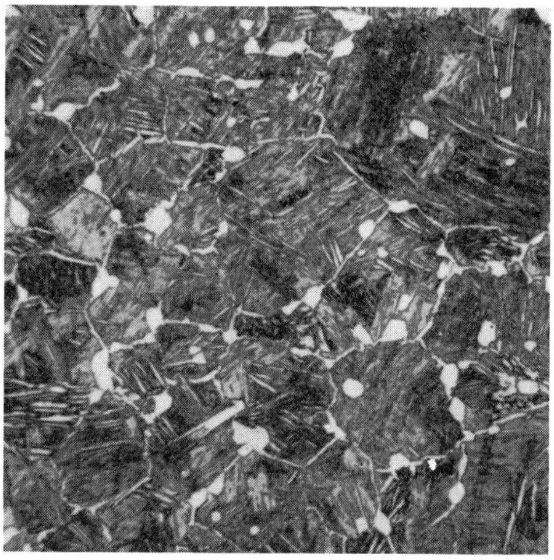

(b)

1010 °C

White is primary α; dark is fine $\alpha + \beta$ from aging

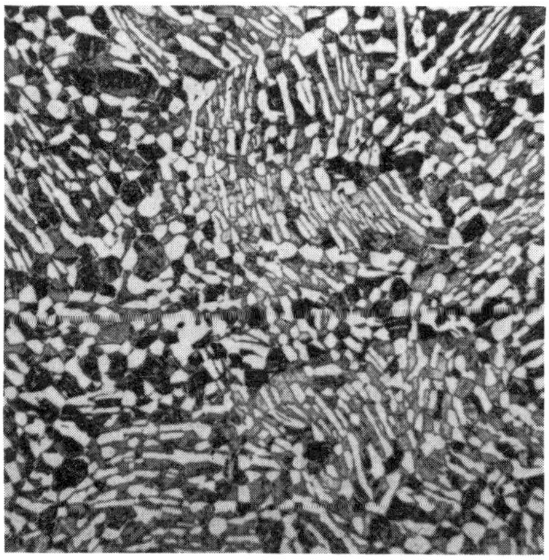

(c)

985 °C

White is primary α; dark is fine $\alpha + \beta$ from aging

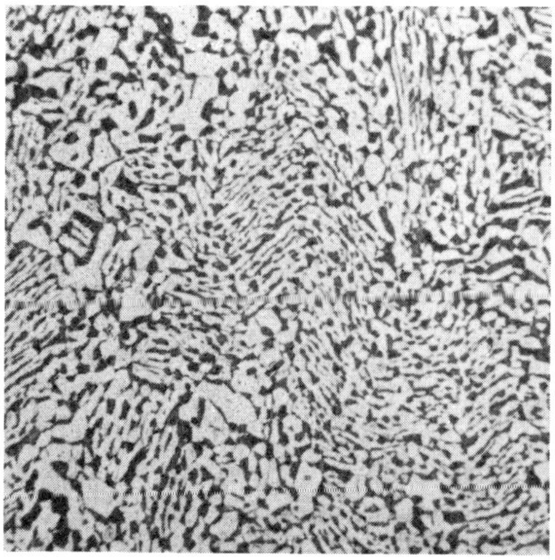

(d)

Fig. 9-28. Microstructures for the annealing temperatures shown. The alloy was annealed for 1 h at the temperatures indicated, oil quenched, then aged (stabilized) for 8 h at 580 °C. The letters locate the annealing temperature on the phase diagram in Fig. 9.27. Optical micrographs; 250×. (Adapted from same source as Fig. 9-27)

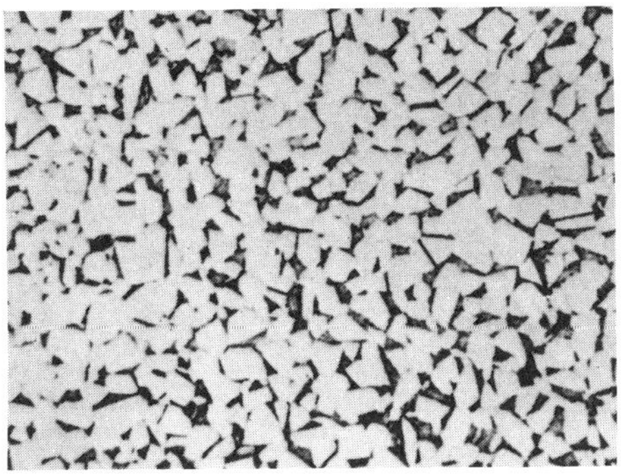

Annealed at 900 °C, air cooled.
White is primary α, dark is martensite + retained β.

(a)

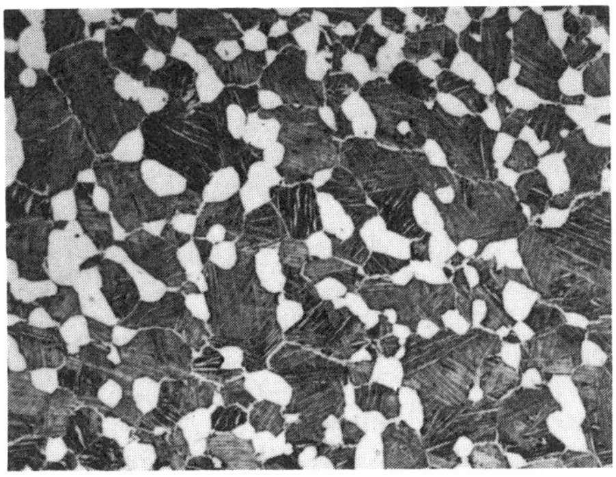

Annealed at 1000 °C, air cooled.
White is primary α, dark is martensite.

(b)

Annealed at 1100 °C, air cooled.
White is primary α formed upon cooling; dark is martensite.

(c)

Fig. 9-29. Microstructures of a Ti-8Al-1Mo-1V alloy showing the effect of annealing temperature on the microstructure. Optical micrographs; 250×. (Adapted from Metals Handbook, 8th Ed., Vol 7, American Society for Metals, Metals Park, Ohio, 1973)

A heat treatment which improves the notch sensitivity is to heat the alloy to 1010 °C for about 15 min, air cool it, then reheat it for 15 min at 750 °C, followed by air cooling. Figure 9-30 shows the marked improvement in notch sensitivity. This heat treatment is called *duplex annealing*. Table 9-5 compares some typical properties for mill annealing and duplex annealing.

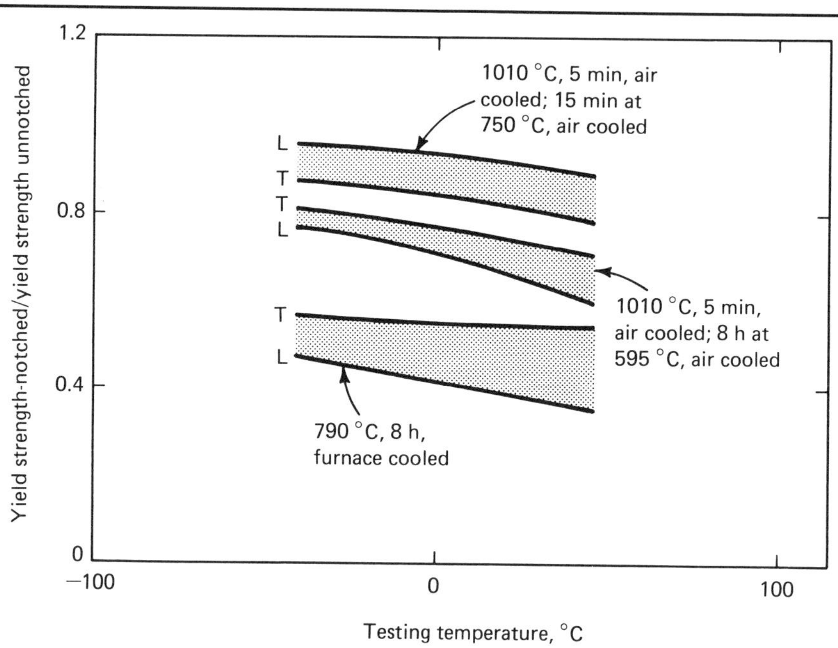

Fig. 9-30. Ratio of the sharp notch yield strength (K_t approx 17) to the smooth specimen yield strength as a function of test temperature, for mill annealing and for duplex annealing plus aging. L and T refer to longitudinal and transverse properties of the sheet material. (Adapted from R. G. Broadwell, Metals Eng. Q., Vol 3(3), p 26, 1963)

Table 9-5. Typical Mechanical Properties at 25 °C for Mill-Annealed and Duplex-Annealed Ti-8Al-1Mo-1V

Properties	Mill annealed	Duplex annealed
0.2% yield strength, psi	140,000	125,000
Tensile strength, psi	146,000	142,000
Elongation, psi	15	17
Precracked Charpy energy absorption, in.-lb/in.^{3}	1,200	1,700

Adapted from M. J. Blackburn, Trans. ASM, Vol 59, p 694, 1966.

The mill-annealed structure is stronger but less tough than the duplex-annealed structure probably because of some precipitation of Ti_3Al, as was described for the case of Ti-Al alloys (see Fig. 9-13). The effect of slow cooling is shown in Fig. 9-31. This possibility is also consistent with the notch-sensitivity data in Fig. 9-30. The slow cooling from 790 °C made the alloy quite notch sensitive. Air cooling from 1010 °C, then aging for a short time (15 min) at 750 °C, then air cooling, reduced the notch sensitivity considerably. However, annealing at 1010 °C, air cooling, then aging for 8 h at 595 °C increased the sensitivity, although not to the degree attained by very slow cooling from 790 °C.

This effect is also seen in the data in Fig. 9-32, showing the susceptibility of Ti-8Al-1Mo-1V to stress corrosion cracking. Aging above about 800 °C is above the temperature of stability of Ti_3Al, and none is present after cooling to 25 °C. Aging below about 400 °C is too low for a significant amount of precipitation of Ti_3Al to occur in 5000 min. However, aging around 650 °C for 10 min or longer will induce stress corrosion cracking, due to precipitation of the compound.

The diagram in Fig. 9-32 is really an isothermal time-temperature-transformation (TTT) diagram where the susceptibility to stress corrosion cracking indicates the beginning of the formation of Ti_3Al. This diagram (refer now to Fig. 9-33) can be used to explain the results in Fig. 9-30. Furnace cooling is sufficiently slow to allow the formation of some Ti_3Al. However, air cooling is too rapid, so that no precipitation occurs upon cooling. Note in Fig. 9-30 that aging at 750 °C for 15 min did not increase appreciably the notch sensitivity compared with aging for 8 h at 595 °C. The TTT diagram in Fig. 9-33 shows that the former treatment would allow precipitation of Ti_3Al, whereas the latter would not.

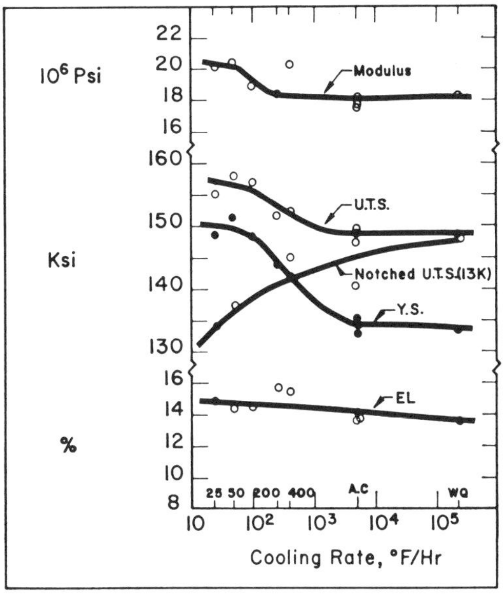

Fig. 9-31. Effect of cooling rate on mechanical properties (at 25 °C) for Ti-8Al-1Mo-1V alloy. Note that very slow cooling causes the alloy to be notch sensitive. (From S. R. Seagle and L. J. Bartlo, Metals Eng. Q., Vol 8(3), p 1, 1968)

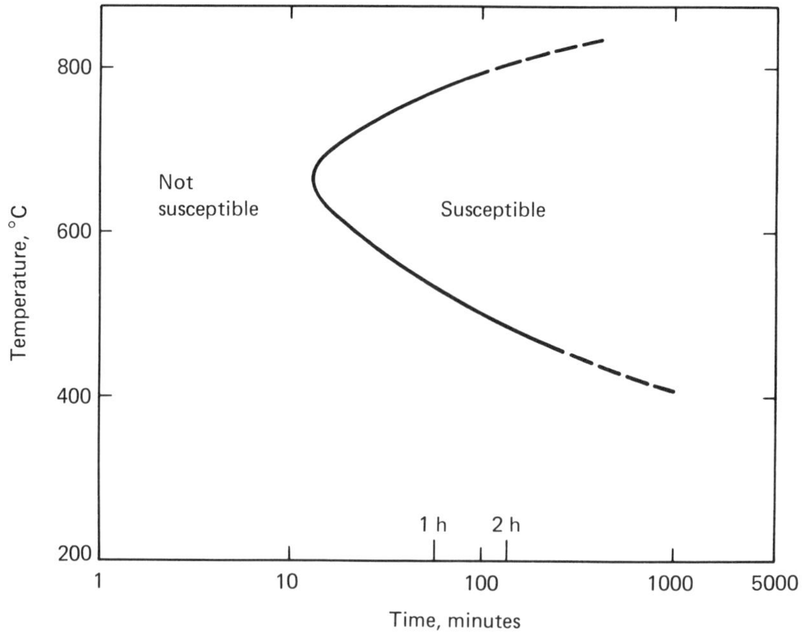

Fig. 9-32. Effect of aging on the stress corrosion incidence in Ti-8Al-1Mo-1V. The alloy was originally annealed at about 985 °C, then air cooled. (Adapted from S. R. Seagle, R. R. Seeley and G. S. Hall, in Applications Related Phenomena in Titanium Alloys, American Society for Testing and Materials, Philadelphia, 1968)

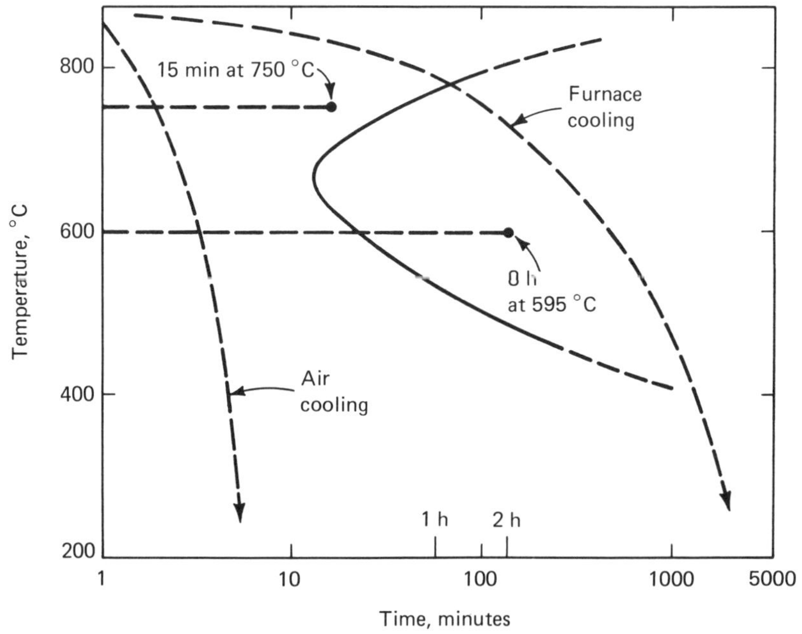

Fig. 9-33. Effect of heat treatment on the formation of Ti_3Al. The TTT diagram is taken from Fig. 9-32. The cooling curves are approximate for the type of cooling shown.

9-6. The Commercial Alloy Ti-6Al-4V

One of the most widely used titanium alloys is one containing 6% Al and 4% V. It has an excellent combination of strength and toughness and excellent corrosion resistance, and finds uses in aerospace applications, pressure vessels, aircraft-turbine and compressor blades and disks, surgical implants, etc.

The properties of this alloy are developed by relying on the refinement of the grains upon cooling from the β region, or the $\alpha + \beta$ region, and subsequent low-temperature aging to decompose martensite formed upon quenching. We begin our description of the heat treatment of this alloy by examining the Ti-Al-V ternary phase diagram. As seen in Fig. 9-10, Al stabilizes the hexagonal close-packed α, and V, being body-centered cubic, stabilizes the β phase. In ternary Ti-Al-V alloys, it is not apparent what the stabilizing effect will be with both Al and V present. However, we would expect that if both phases are present, the Al will tend to segregate in the α, and the V in the β.

Isothermal sections for the Ti-rich region of the Ti-Al-V ternary phase diagram are shown in Fig. 9-34. The composition corresponding to 6% Al–4% V is shown by the dot. At 1000 °C, only β is present; at 900 and 800 °C, this composition is in the two-phase $\alpha + \beta$ region. The chemical composition of each phase can be determined at a given temperature only if the "tie line" is known. In Fig. 9-34, the tie lines are assumed to be those given by the dashed lines. Since the 6% Al–4% V composition lies near the three-phase α-β-γ triangle, it is reasonable to assume that the tie line will be approximately parallel to the side of the triangle, as shown. Note that upon cooling from 1000 to 800 °C the Al content of the α remains about constant at 6%. The β phase, though, dissolves considerable V, the amount increasing to about 14% at 800 °C. A mass balance calculation (level-rule calculation) gives at equilibrium approximately 57% α and 43% β at 900 °C, and about 83% α and 17% β present at 800 °C. Thus, upon cooling slowly, the amount of α increases.

When this alloy is slowly cooled from the β region, α begins to form below the β transus, which is about 980 °C. The α forms in plates, with a crystallographic relationship to the β in which it forms. The α plates form with their basal (close-packed) plane parallel to the $\{110\}$ plane of the β. This is the relationship for the formation of martensite illustrated in Fig. 9-3. Upon slow cooling, a nucleus of α forms and, because of the close atomic matching along this common plane, the α phase thickens relatively slowly perpendicular to this plane but grows faster along the plane. Thus, plates are developed. Since in a given β grain there are six sets of non-parallel $\{110\}$ planes, then a structure of α plates is formed consisting of six non-parallel sets. The microstructure is illustrated in Fig. 9-35.

The formation process is shown schematically in Fig. 9-36, using a constant-composition section (isopleth) at 6% Al from the ternary phase diagram to illustrate the formation of α upon cooling. In Fig. 9-36, the darker regions are the β phase, left between the α plates which have formed.

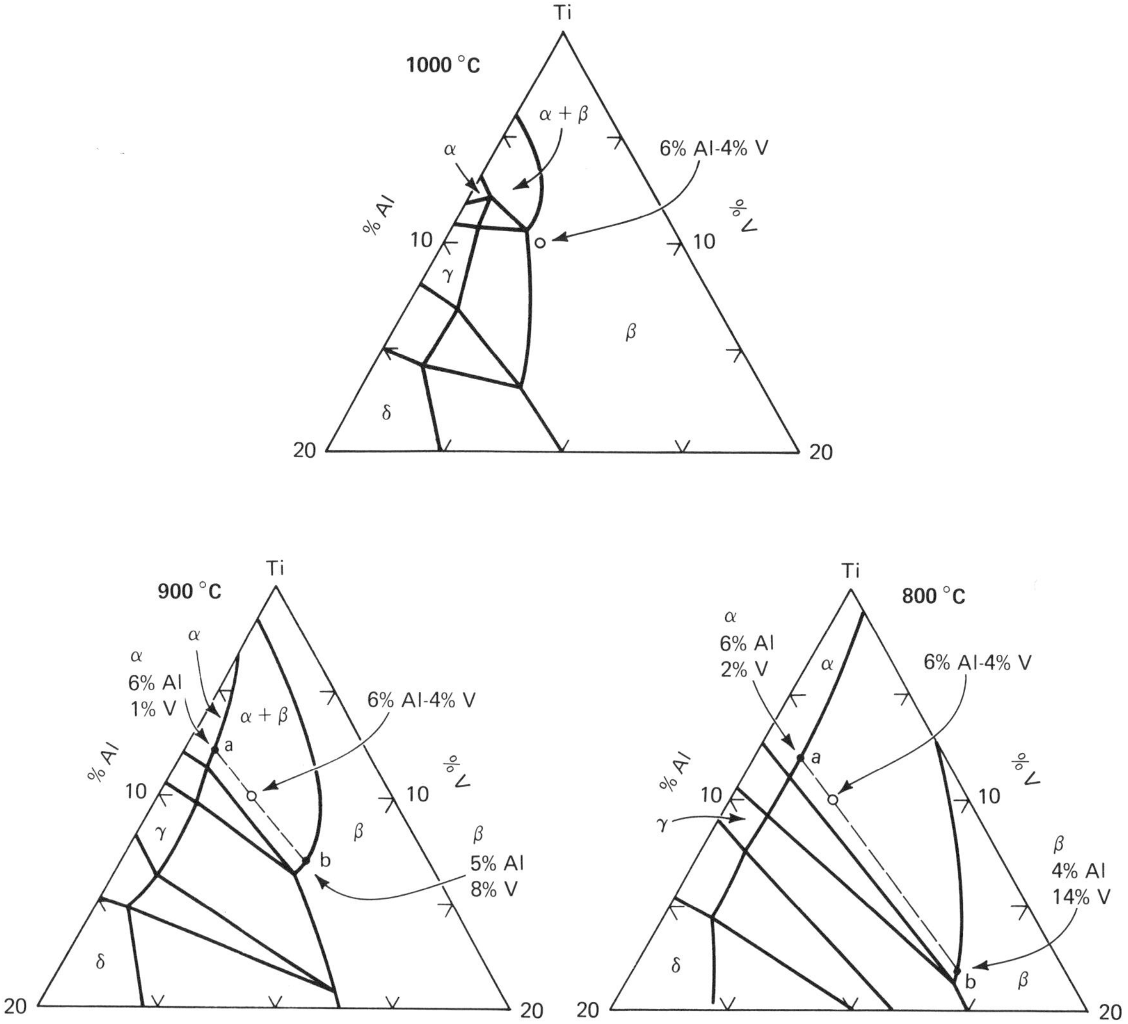

Fig. 9-34. Isothermal sections of the Ti-rich region of the Ti-Al-V phase diagram at 1000, 900 and 800 °C. (Adapted from P. A. Farrar and H. Margolin, Trans. AIME, Vol 221, p 1214, 1961)

The microstructure consists of parallel plates of α delineated by the β between them. Where α plates formed parallel to one specific {110} plane of β meet plates formed on another {110} plane, a high-angle grain boundary exists between the α crystals, and etches to reveal a line separating them.

This microstructural morphology, consisting of these sets of parallel plates which have formed with a crystallographic relationship to the phase from which they formed, is called a Widmanstätten structure, named in honor of a German scientist who observed a similar structure in Fe-rich, Fe-Ni meteorites. (See R. F. Mehl, in The Sorby Centennial Symposium on the History of Metallurgy, ed. C. S. Smith, Gordon and Breach, New York, 1965.)

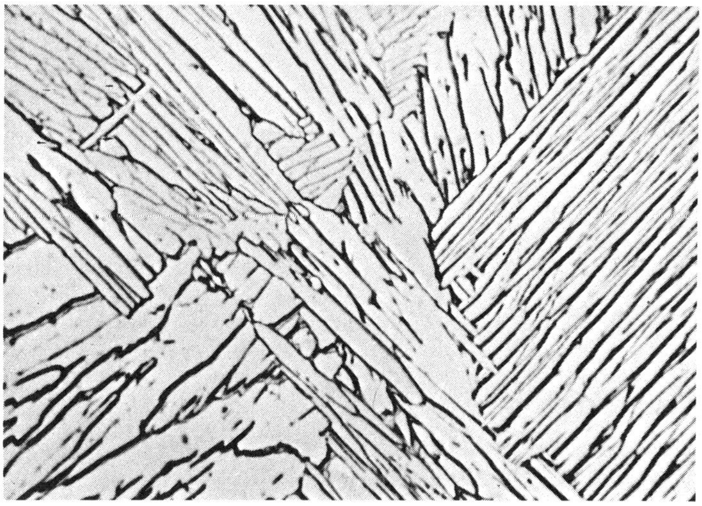

Fig. 9-35. Microstructure of a Ti-6Al-4V alloy after cooling slowly from above the β transus. The white plates are α, and the dark regions between them are β. This is a typical Widmanstätten structure. Optical micrograph; 500×. (Adapted from M. J. Blackburn, W. H. Smyrl and J. A. Feeney, in Stress-Corrosion Cracking in High Strength Steels and in Titanium and Aluminum Alloys, ed. B. F. Brown, Naval Research Laboratory, Washington, D.C., 1972)

Upon cooling rapidly, β may decompose by a martensite reaction, similar to that described for pure Ti, and for the Ti-8Al-1Mo-1V alloy in the previous section. The structure present after quenching to 25 °C depends upon the annealing temperature, and there have been inconsistencies in the reported structures. Figure 9-37 shows one report of the structures after quenching. The phase diagram there is an isopleth, a constant-composition section of the ternary diagram at 6% Al. (It is important to remember that in the two-phase region of such a section of a ternary phase diagram an isothermal line is not a tie line. That is, the intersection of a constant-temperature line with the α- and the β-phase boundaries does *not* give the chemical composition of the α and the β phases. This information must be obtained from the isothermal section and the experimentally determined tie lines.)

In Fig. 9-37, it is seen that two different types of martensite may form depending upon the quenching temperature. These are designated α', which is face-centered cubic or face-centered tetragonal (which has not been resolved), and α'', which is hexagonal. Upon quenching from above the β transus (about 980 °C), the structure is all hexagonal α'' with a small amount of β (although in some alloys the β has not been observed).

The presence of some β in the structure after quenching from above the β transus is due to the fact that the martensite finish temperature M_f is below 25 °C for this alloy. That is, vanadium, being body-centered cubic, is a β

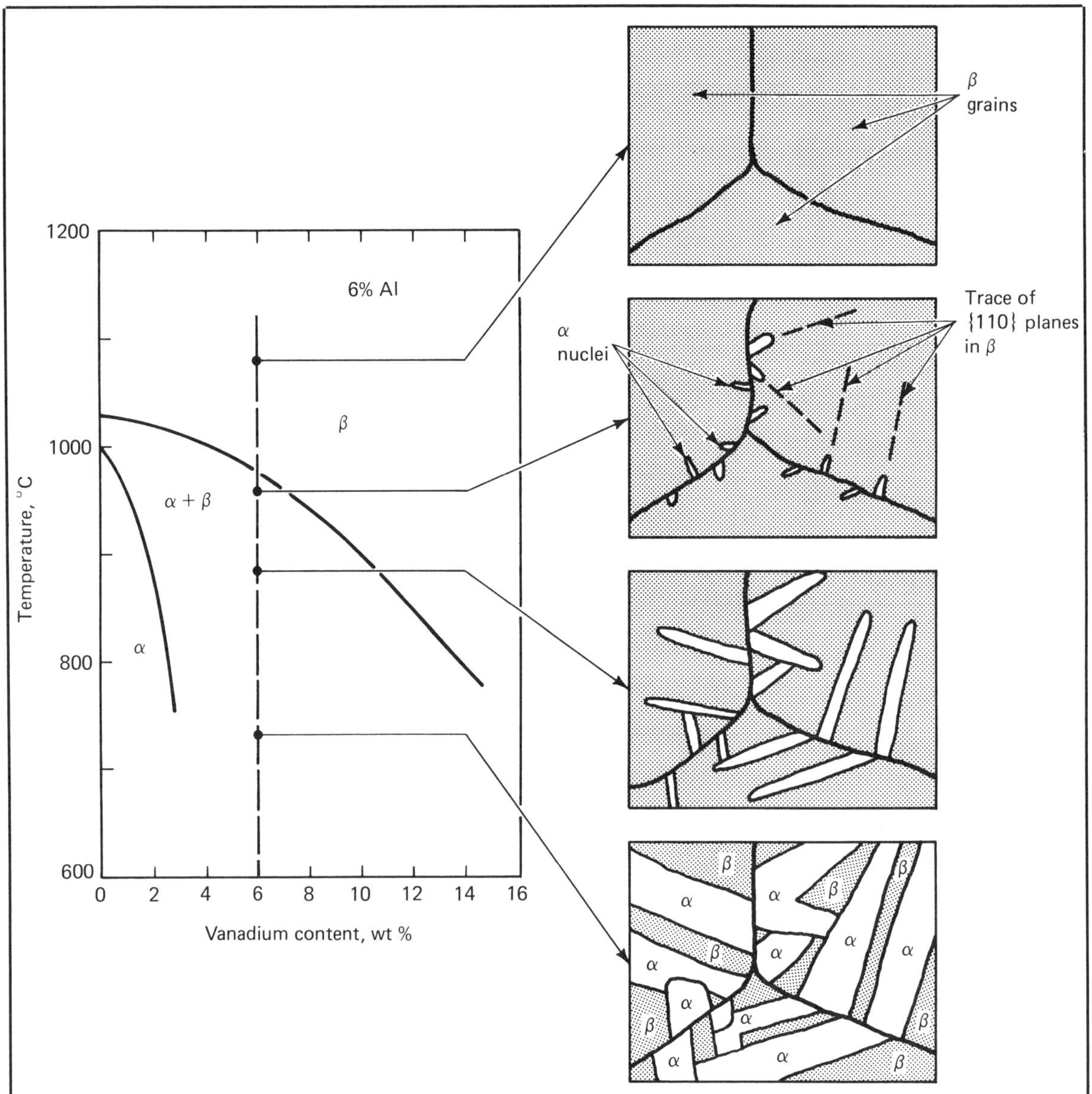

Fig. 9-36. Schematic illustration of the formation of a Widmanstätten structure in a Ti-6Al-4V alloy by cooling slowly from above the β transus. The final microstructure consists of plates of α (white) separated by the β phase (dark). An actual microstructure is shown in Fig. 9-35.

stabilizer, and the addition of 4% V to a Ti–6% Al alloy is sufficient to place the M_f below 25 °C. Thus, upon quenching to 25 °C, not all of the β is converted to α' or α''. When this alloy is annealed in the two-phase region, at the annealing temperature the α and β have different equilibrium Al and V contents. As the annealing temperature decreases, the β phase has an increasing concentration of V (Fig. 9-34) so that for this phase the M_s and M_f are progressively decreasing. At about 840 °C, the M_s is below 25 °C, so that upon quenching from lower than this annealing temperature, none of

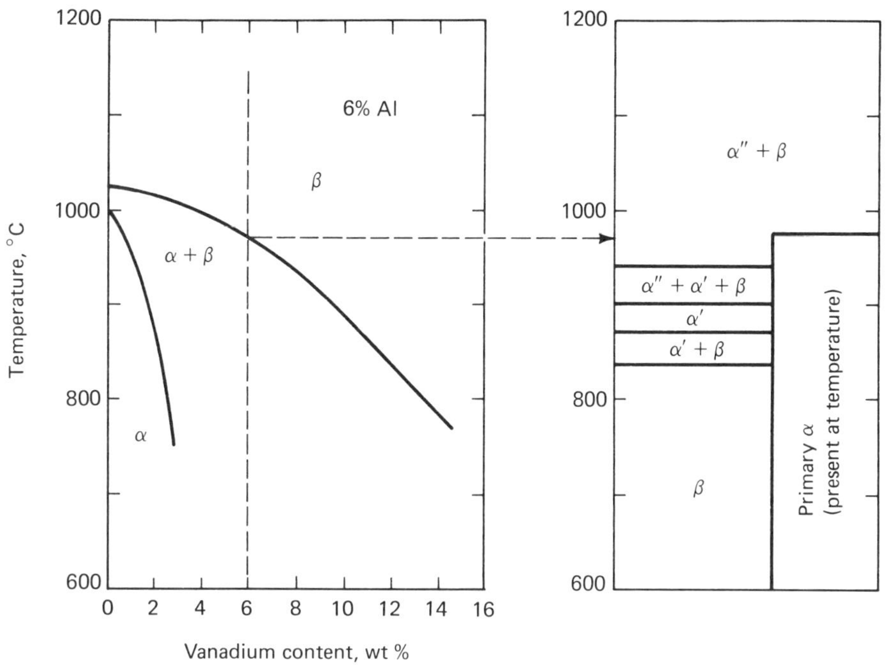

Fig. 9-37. Illustration of the structures formed in a Ti-6Al-4V alloy upon quenching from different annealing temperatures. (Phase diagram adapted from data in Fig. 9-34; structures adapted from same source as Fig. 9-25)

the β is converted to α′, and the structure consists of primary α (present at the annealing temperature) and retained β.

In Fig. 9-38(a) is shown the isopleth at 6% Al. Also shown (dashed line) is the M_s temperature as a function of V content. For β-containing 6% Al–4% V, the M_s is about 920 °C. The M_f temperature is below 25 °C, as some retained β is present (Fig. 9-37) upon quenching from above the β transus (i.e., from the β region). As the V content increases, the M_s and M_f temperatures decrease. Upon reaching about 14% V in the β, the M_s is below 25 °C. The microstructure of the Ti-6Al-4V alloy is affected through the influence of annealing temperature in the two-phase (α + β) region on the V content of the β. It is seen from the ternary isotherms (Fig. 9-34) that as the temperature decreases, the V content increases. Thus, upon annealing in the two-phase region, the V content of the β phase increases with decreasing annealing temperature. For this β the M_s decreases as the annealing temperature decreases (Fig. 9-38b). Therefore, upon quenching to 25 °C, the amount of retained β present increases as the annealed temperature decreases. At an annealing temperature of about 800 °C, the V content of the β is about 14%, and the M_s temperature is about 25 °C. Thus, upon quenching to 25 °C after annealing at 800 °C, none of the β is converted to α′. The structure at 25 °C consists of primary α (present at the annealing temperature) and retained β. It is important to remember that the amount

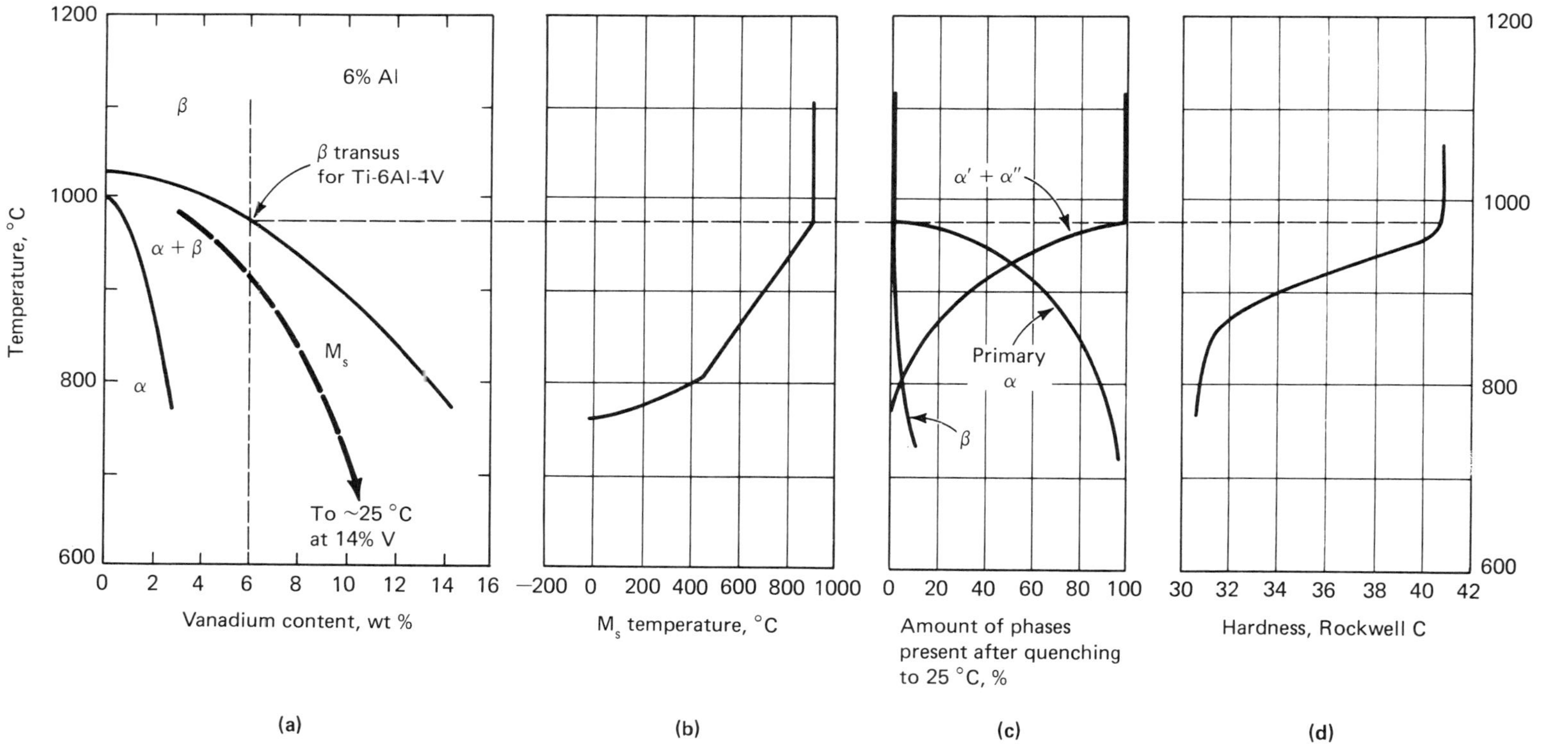

Fig. 9-38. (a) Isopleth at 6% Al for the Ti-Al-V system. The M_s temperature as a function of V content is also shown. (b) M_s temperature as a function of the annealing temperature. It decreases with decreasing annealing temperature because the V content of the β increases with decreasing annealing temperature. (c) Amount of primary α (present at the annealing temperature), retained β and α' or α'' present after quenching to 25 °C from different annealing temperatures. (d) Hardness obtained after quenching to 25 °C. (Data from P. J. Fopiano, M. B. Bever and B. L. Averbach, Trans. ASM, Vol 62, p 324, 1969)

of β present at the annealing temperature decreases with decreasing annealing temperature.

The effect the annealing temperature has on the as-quenched hardness is shown in Fig. 9-38(d). These data show that quenching from the β region (to give α'' and a small amount of retained β) gives the hardest material. However, all the mechanical properties should be examined, and Fig. 9-39 shows the effect of annealing temperature on the tensile mechanical properties for the as-quenched condition. Note that the highest tensile yield strengths are obtained by quenching from β, consistent with the hardness data in Fig. 9-38(d). However, this condition has a low elongation and reduction in area, indicating that caution should be exercised in using this alloy in this condition. A preferred treatment is to anneal just below the β transus (980 °C), about 900-950 °C, in which case the strength is lowered only slightly (if at all), and the ductility is greatly improved.

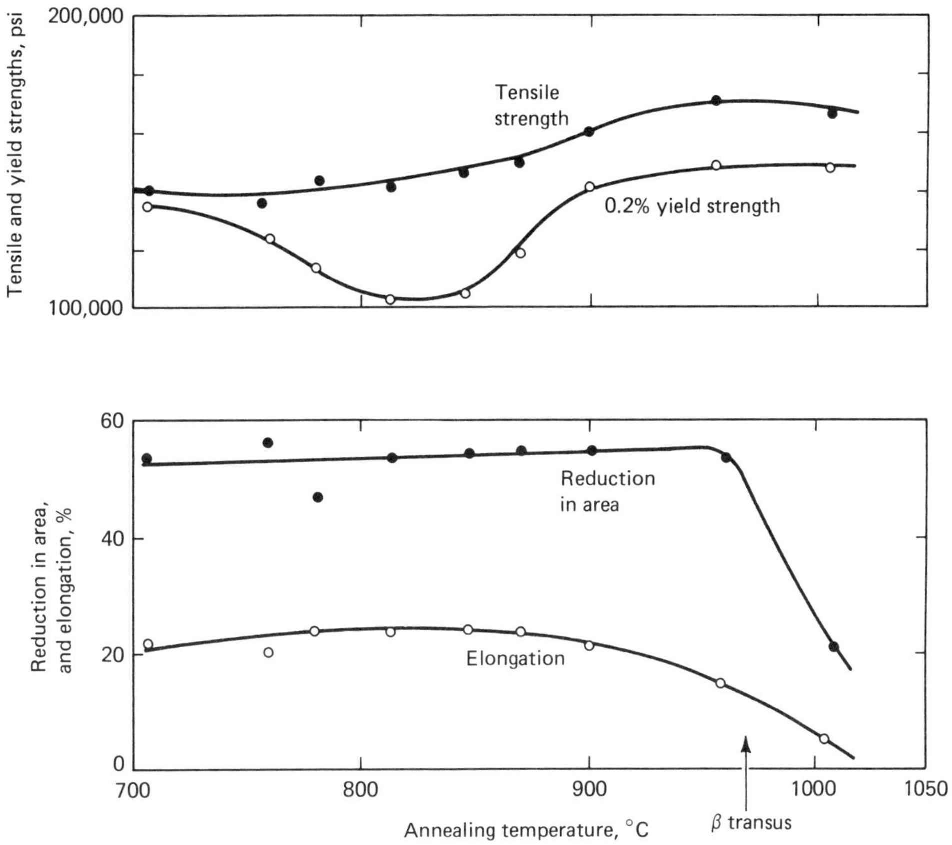

Fig. 9-39. Effect of annealing temperature on the tensile mechanical properties of Ti-6Al-4V alloy. The alloy was annealed 1 h, then water quenched. (Adapted from R. G. Sherman and H. D. Kessler, Trans. ASM, Vol 48, p 657, 1956)

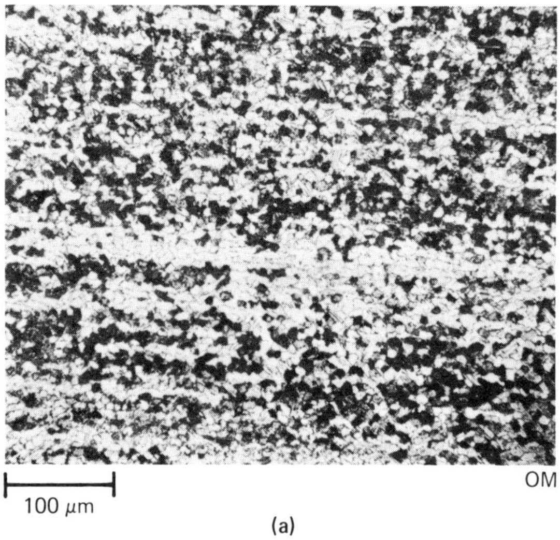

(a)

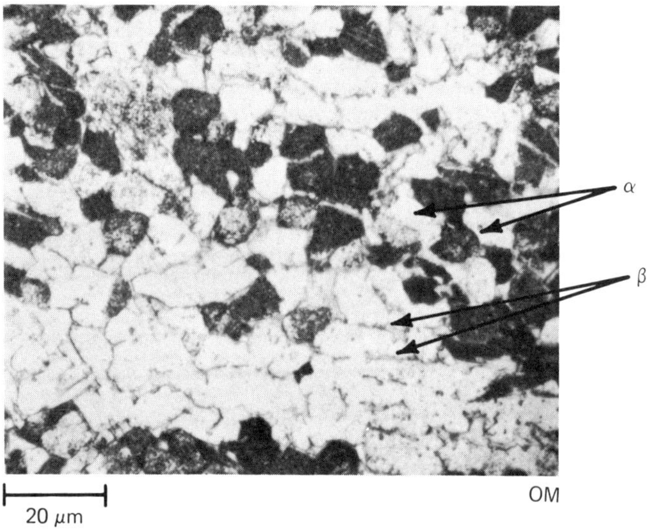

(b)

Fig. 9-40. Microstructures of a Ti-6Al-4V alloy annealed 1 h at 955 °C, then furnace cooled.

Figure 9-40 shows the microstructure of a Ti-6Al-4V alloy which was annealed at 955 °C, then furnace cooled. The structure consists of about 10% β and 90% α. However, the optical micrographs (Fig. 9-40a and b) are misleading, because the dark grains and the light grains are both α. The contrast is caused by the dependence of the etching effect on the orientation of the crystal. That is, the etch will attack different crystallographic planes differently: after etching some will have a relatively smooth surface and hence appear light in optical microscopy, and others will be roughened, a surface topology which will scatter the impinging light out of the optical

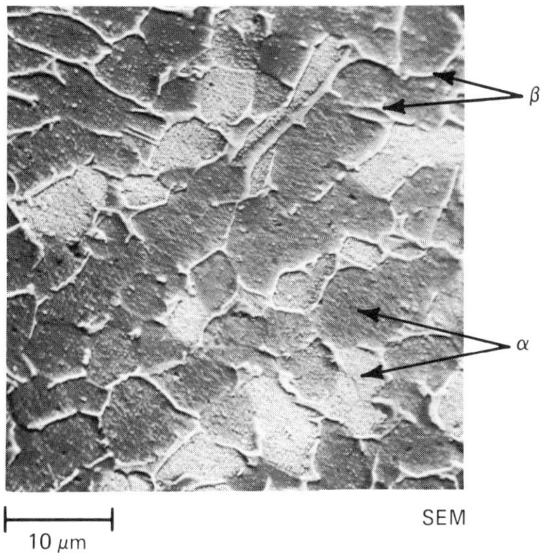

(c) **Fig. 9-40 (continued).**

system, and therefore these grains will appear dark. Higher-magnification scanning electron micrographs (Fig. 9-40c) show this surface roughening clearly. Also, here the β (relatively unetched) between the α grains is clearly seen. Looking back now at Fig. 9-40(b), the β can be identified.

Upon quenching from 955 °C, the primary α will be unaffected by the cooling process. The β, though, will tend to transform to $\alpha' + \alpha''$ martensite. However, the V content of the β at the annealing temperature is sufficient to place the M_f below 25 °C, so that some retained β will be present. Typical microstructures for this treatment are shown in Fig. 9-41. In the optical micrograph (Fig. 9-41a), the primary α is dark, and the martensite and retained β mixture is light. The martensitic appearance of the transformed β is shown more clearly in the higher-magnification pictures (Fig. 9-41b, c and d). For this treatment, the amount of retained β is small (Fig. 9-38c) and is not resolved in the microstructures in Fig. 9-41.

Upon water quenching from 955 °C, the amount of primary α is fixed by the annealing temperature, and is about 40% (Fig. 9-41). The amount of primary α present after furnace cooling is greater than this, as some of the β transforms to α during the slow cooling. Figure 9-40 shows about 90% primary α for furnace cooling.

The microstructure after slow cooling from 900 °C is shown in Fig. 9-42. Note again that the light and dark grains are the primary α; the contrast is due to the dependence of the effect of the etching on the orientation of the grain. This is especially apparent in Fig. 9-42(c), where a twin in the HCP α is present. On one side of the twin the etch has created a smooth surface; on the other, the surface is roughened.

Upon comparing the microstructure for slow cooling from 955 °C (Fig. 9-40) with that for 900 °C (Fig. 9-42), it is seen that the amount of α

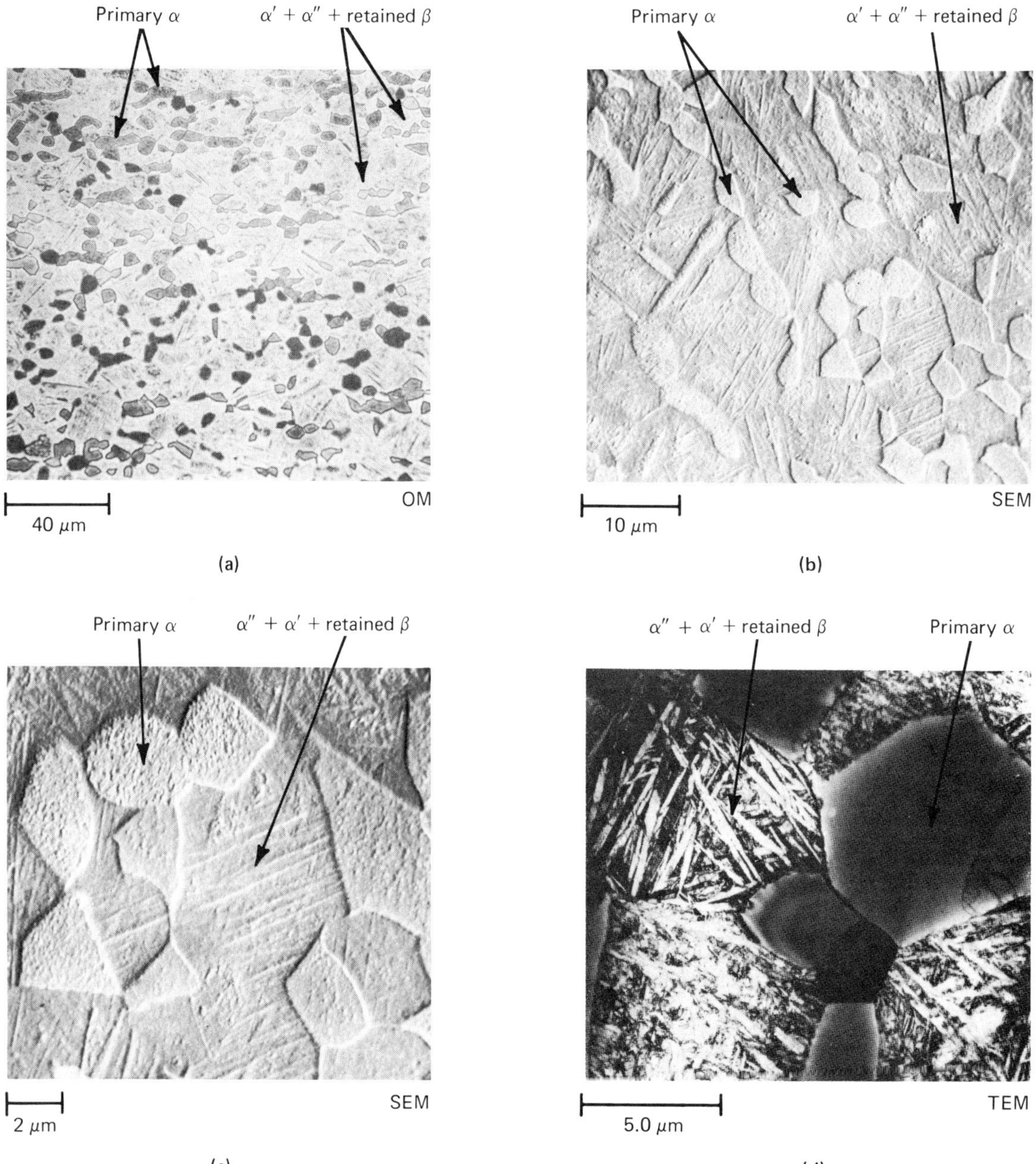

(a) (b) (c) (d)

Fig. 9-41. Microstructures of a Ti-6Al-4V alloy after annealing 1 h at 955 °C, then water quenching. (d) adapted from C. Hammond and J. Nutting, Metal Science, Vol 11, p 483, 1977.

is about 10% in both cases. This is expected, as the β transforms to α upon cooling slowly until a temperature is reached where diffusion is negligible (e.g., 300 °C), leaving about 10% β in both cases.

Quenching from either 955 or 900 °C gives similar microstructures (Fig. 9-41 and 9-43). The structures consist of primary α and $\alpha' + \alpha''$

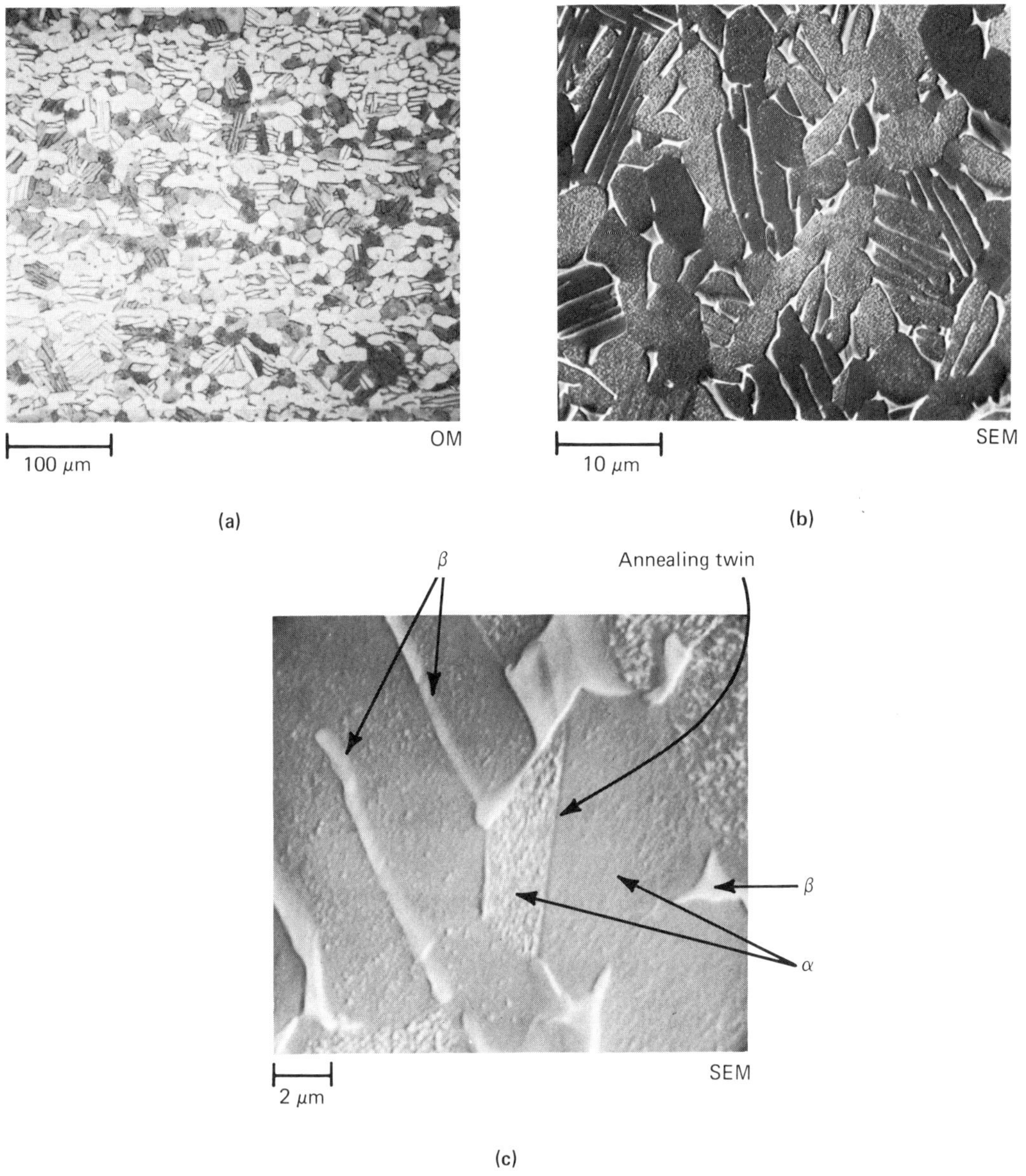

Fig. 9-42. Microstructures of a Ti-6Al-4V alloy annealed 1 h at 900 °C, then furnace cooled.

martensite dispersed in some retained β. For both annealing temperatures, the retained β is very difficult to distinguish in the martensite.

Table 9-6 lists some tensile mechanical properties typical of the Ti-6Al-4V alloy after annealing at 955 and 900 °C. The properties are approximately the same for either annealing temperature for a given cooling method, and this is consistent with the similarity in the microstructures. For both annealing temperatures and cooling methods, the ductility is high. Furnace cooling from either annealing temperature gives a weaker structure,

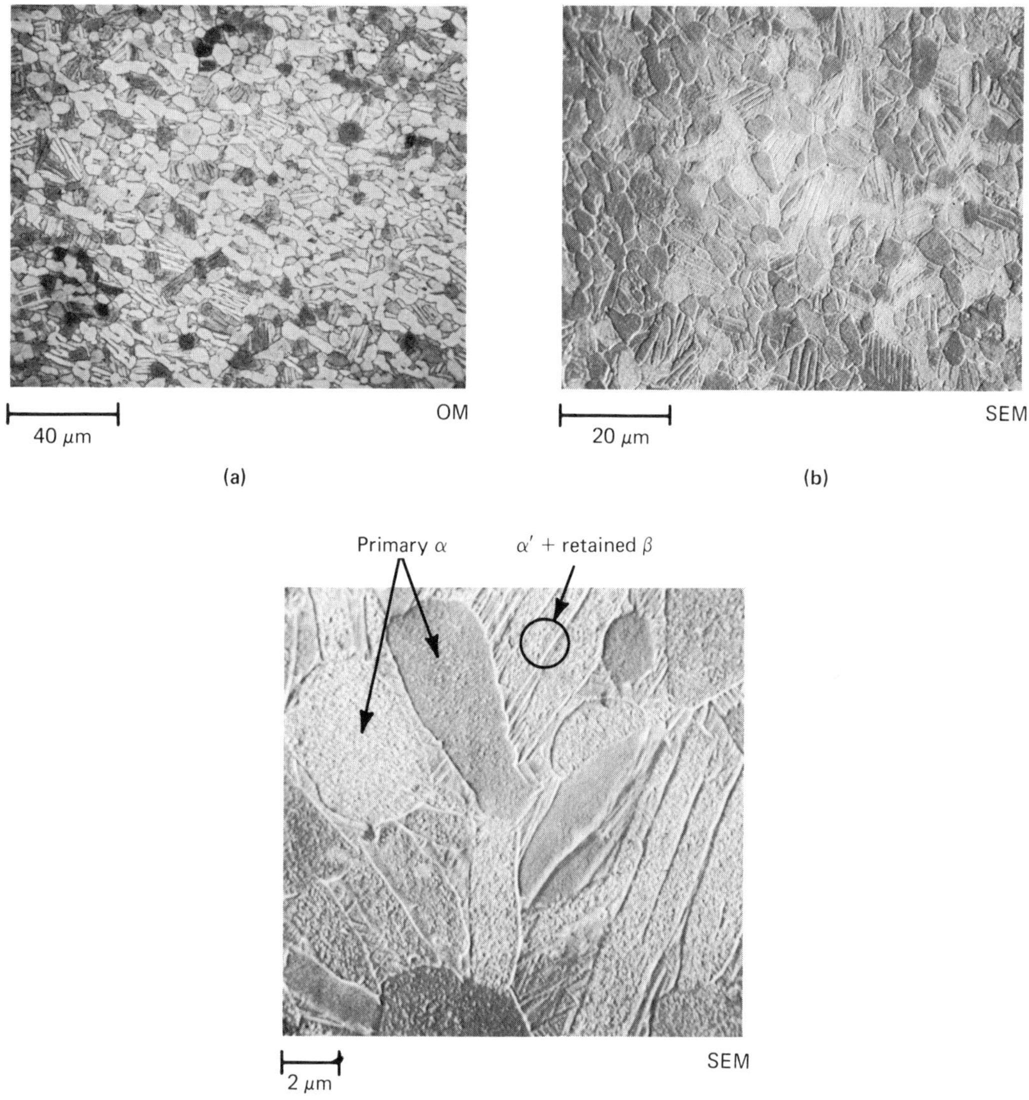

(c)

Fig. 9-43. Microstructures of a Ti-6Al-4V alloy annealed 1 h at 900 °C, then water quenched.

so this heat treatment is used prior to certain mechanical deformation and machining operations. Although quenching produces a stronger structure, it is only about 20% stronger than that produced by furnace cooling. It is reasonable to expect that heating the quenched structure (aging) in a moderate temperature range (e.g., 400-600 °C) to decompose the β, α'' and α' would produce a stronger structure, and it is this heat treatment which we examine now.

It is clear from Fig. 9-39 that the annealing temperature should be in the upper temperature range of the $\alpha + \beta$ region, to attain almost maximum

Table 9-6. Typical Tensile Mechanical Properties and Microstructure at 25 °C of a Ti-6Al-4V Alloy for Different Thermal Treatments

Thermal treatment	Yield strength, psi	Tensile strength, psi	Elongation at fracture, %	Reduction in area, %	Microstructure at 25 °C (approx. vol % phases)
955 °C, furnace cooled	121,000	136,000	19	46	90% α; 10% β
955 °C, water quenched	138,000	162,000	17	60	50% primary α; 60% α′ + α″ + retained β
900 °C, furnace cooled	124,000	140,000	17	43	90% α; 10% β
900 °C, water quenched	134,000	162,000	15	54	60% primary α; 40% α′ + α″ + retained β

Mechanical properties adapted from Metal Progress, Vol 107(3), p 72, March 1975.

strength (in the as-quenched condition) yet avoid the loss of toughness associated with quenching from β (from above the transus). The structures obtained consist of primary α (present at the annealing temperature) and martensitic α′ or α′ + α″, formed from the β, and a small amount of retained β. The retained β, α′ and α″ each have the same chemical composition. Also, none of the phases has its equilibrium composition. Thus, upon aging below the annealing temperature (e.g., around 400-700 °C), the retained β should precipitate some α to become enriched in V, and the martensitic α′ and α″ should precipitate some β and convert to the HCP α. Since the amount of retained β is small (Fig. 9-38), any strengthening during aging will be derived from the precipitation of β in the α′ and α″, and their conversion to α.

Figure 9-44 shows the effect of annealing temperature and of aging time at 540 °C on the hardness of a Ti-6Al-4V alloy. Note that quenching from above the β transus gives the highest as-quenched hardness, but remember that the toughness is low (Fig. 9-39). For a solution treatment (anneal) at 950 °C, the hardness after aging about 2 h is at the maximum. For an annealing temperature of 900 °C, aging about 10 min increases the hardness about 15%.

The strengthening shown in Fig. 9-44 is due mainly to the precipitation of a fine dispersion of β in the martensitic α′ and/or α″. As this precipitate nucleates and grows, it absorbs V from the surrounding martensite, which converts to α. Figure 9-45 shows a microstructure typical of the β precipitate. Note that the β particles are about 200 Å in size, a very fine dispersion. The β nucleates at dislocations which form because of the volume change when the martensite forms from the β, and on sub-boundaries and twin interfaces.

The data in Table 9-7 show the effect of aging 4 h at 540 °C on the tensile mechanical properties of this alloy. Note that aging increases the yield

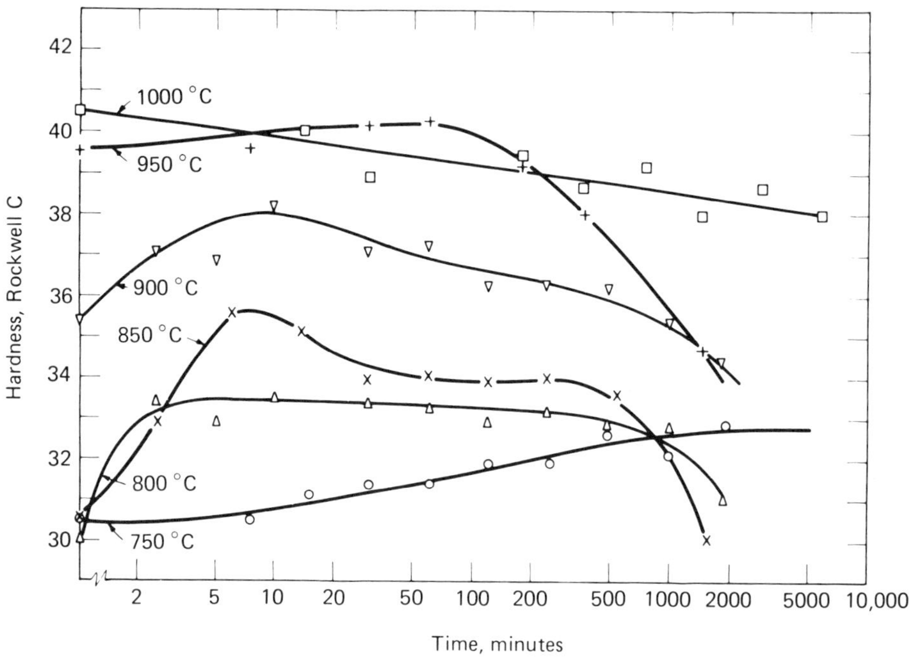

Fig. 9-44. Effect of aging time (at 540 °C) and of annealing temperature on the hardness of a Ti-6Al-4V alloy following water quenching from the annealing temperature. (Adapted from same source, p 324, as Fig. 9-38)

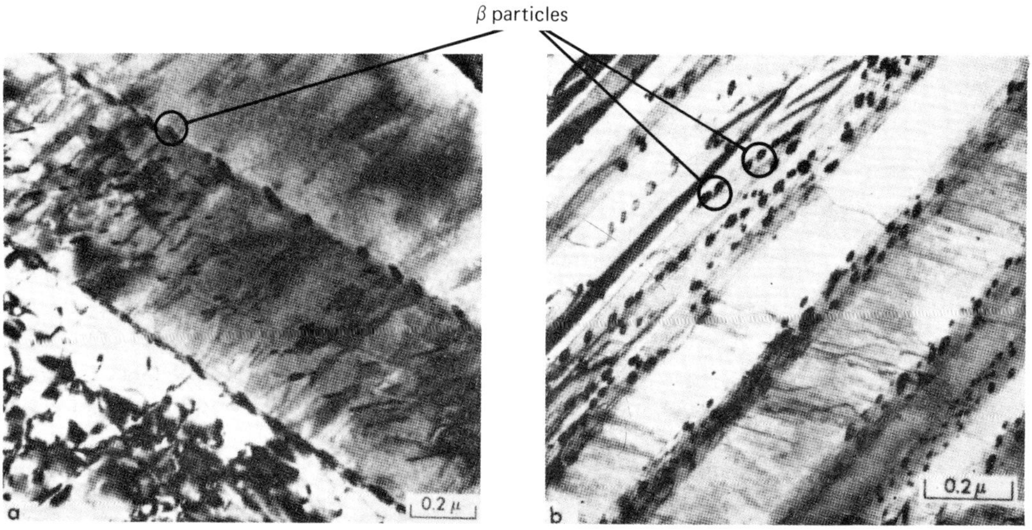

Fig. 9-45. Transmission electron micrograph of a Ti-6Al-4V alloy aged 48 h at 500 °C. The structure is a fine dispersion of β in α formed from α'. The contrast is due to dislocations and differences in subgrain orientation. (Adapted from J. C. Williams and J. J. Blackburn, Trans. ASM, Vol 60, p 381, 1967)

Table 9-7. Effect of Aging on the Tensile Mechanical Properties at 25 °C of a Ti-6Al-4V Alloy

Thermal treatment	Yield strength, psi	Tensile strength, psi	Elongation at fracture, %	Reduction in area, %
955 °C, water quenched	138,000	162,000	17	62
955 °C, water quenched; aged 4 h at 540 °C	155,000	172,000	17	56
900 °C, water quenched	134,000	162,000	15	54
950 °C, water quenched; aged 4 h at 540 °C	147,000	162,000	15	48

Adapted from same source as Table 9-6.

strength by about 10%, and that the elongation and reduction in area remain about the same.

For the commercial Ti-6Al-4V alloy, there are some commonly used heat treatments. For each of these, the following descriptions are typical of the temperatures and times used. The actual practice varies with alloy producer and user.

To place the alloy in a soft, relatively machinable condition, the alloy is heated to about 730 °C in the lower range of the $\alpha + \beta$ region, held for 4 h, then furnace cooled to 25 °C. This treatment, called mill annealing, produces a microstructure of globular crystals of β in an α matrix. A typical microstructure is shown in Fig. 9-46.

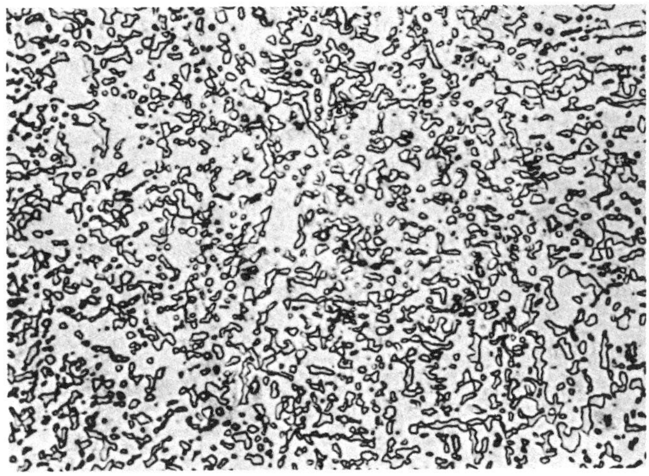

Fig. 9-46. Microstructure of a Ti-6Al-4V alloy which has been mill annealed (4 h at 750 °C, furnace cooled). The structure is globular particles of β in a matrix of α. Optical micrograph; approx 500×. (Adapted from M. J. Blackburn, J. A. Feeney and T. R. Beck, in Advances in Corrosion Science and Technology, Vol 3, ed. M. G. Fontana and R. W. Staehle, p 100, Plenum, New York, 1973)

Another annealing treatment is duplex annealing. Several variants of this treatment are used. Typically, the alloy is heated to 955 °C for 10 min, then air cooled. It is then heated to 675 °C for 4 h and air cooled to 25 °C.

With the aging treatment called solution treating and aging, typically the alloy is heated at 955 °C for 10 min, water quenched, then aged for 4 h at a temperature between 540 and 675 °C, followed by air cooling to 25 °C.

Typical tensile mechanical properties for the three treatments are compared in Table 9-8. The strongest alloy is the one that has been solution treated and aged. The mill-annealed condition is stronger than the duplex-annealed condition, but the difference is slight.

9-7. The Commercial Alloy Ti-13V-11Cr-3Al

The effect of the body-centered cubic elements in stabilizing the β phase raises the possibility of the development of an eutectoid alloy which will form a martensite structure upon rapid cooling from β and will be strong and hard, as is developed in steels. Indeed, some elements will form an eutectoid region with titanium with a martensite structure formed upon rapid cooling. The M_s and M_f temperatures are decreased as the alloy content increases, similar to the influence of carbon on iron. Sufficient alloy content will allow retained β to be present at 25 °C after quenching. (See Fig. 9-47.)

However, the martensite is the α' structure and is not particularly strong. Thus, the expectation that this heat treatment can be used to strengthen titanium alloys is not realized. However, when heated below the eutectoid temperature, the martensite will decompose into a structure of α and other phases, depending upon the exact alloy, and this should provide more strengthening than the martensitic structure. But significant strengthening is not obtained, so this approach is not used.

These alloys, however, can be used because of the relative formability of the body-centered cubic structure compared to the hexagonal-close packed structure (see Chapter 1.) Sufficient alloy content will allow retention of a completely β structure after quenching from the β region to 25 °C (see Fig. 9-47). In this condition the alloy can be fabricated by plastic defor-

Table 9-8. Tensile Mechanical Properties at 25 °C for a Ti-6Al-4V Alloy in the Mill-Annealed, Duplex-Annealed, and Solution-Treated and Aged Conditions

Condition	Yield strength, psi	Tensile strength, psi	Elongation at fracture, %
Mill annealed	137,000	155,000	10
Duplex annealed	133,000	140,000	18
Solution treated and aged	160,000	167,000	13

Adapted from same source as Fig. 9-46.

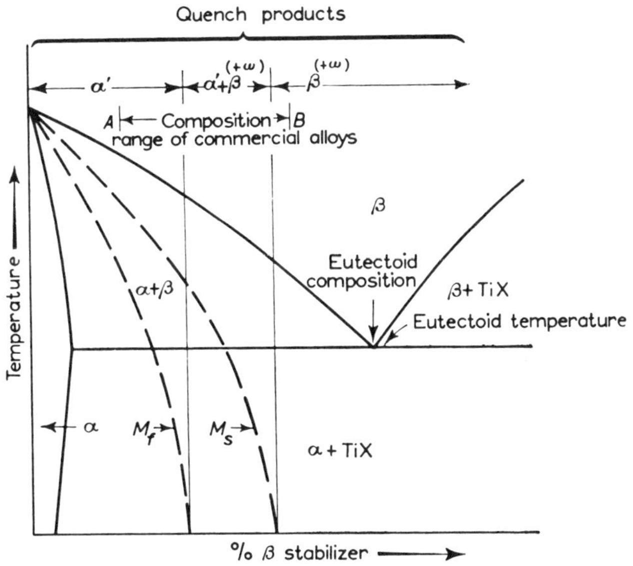

Fig. 9-47. Schematic phase diagram for a typical β-stabilized eutectoid titanium alloy. (Adapted from same source as Fig. 9-6)

mation. Then the component can be reheated below the eutectoid temperature to decompose the retained β to a multiphase structure of α and other phases which depend upon the exact alloy, providing considerable strengthening over that of the retained β. It is this approach which is the basis of a few commercial alloys, and in this section the physical metallurgy of one of these, Ti-13V-11Cr-3Al, is examined.

The addition of Cr should maintain the desirable corrosion and oxidation resistance characteristic of titanium alloys. Table 9-9 shows the effect of several elements on the eutectoid temperature and composition and the

Table 9-9. Effect of Alloying Elements in Several Titanium Binary Alloys on Eutectoid Temperature and Composition, and Content Needed to Retain β

Alloying element	Eutectoid temperature, °C	Eutectoid composition, wt %	Alloy content needed to retain β after quenching to 25 °C, wt %
Manganese	550	20	6.5
Iron	600	15	4
Chromium	675	15	8
Cobalt	585	9	7
Nickel	770	7	8
Copper	790	7	13
Silicon	860	0.9	—

Adapted from same source as Fig. 9-6.

required alloy content to lower the M_f to 25 °C. Note that Cr is relatively effective in retaining β.

Figure 9-48(a) shows the Ti-Cr phase diagram, with the eutectoid composition at about 15% Cr. The problem with using a binary Ti-Cr alloy is that a structure of complete β may not be obtained by quenching to 25 °C, which means that the M_s remains above 25 °C even for high Cr contents. Thus, what is sought is another addition which will further stabilize the β and lower the M_s to below 25 °C. The element used here is V. The Ti-V phase diagram (Fig. 9-48b) shows that V greatly stabilizes the β. Also, Cr and V are completely soluble in each other (Fig. 9-48c).

Isothermal sections of the Ti-V-Cr phase diagram are shown in Fig. 9-49. For the Ti-13V-11Cr alloy, the structure at equilibrium is β until between 650 and 500 °C the three-phase region, $\beta + \alpha + TiCr_2$, is entered. Quenching from the β region will retain a completely β structure; reheating the β to strengthen it will depend upon the formation of α and $TiCr_2$. Figure 9-50 is an isopleth showing these phase relations for a 12% Cr alloy.

Before examining the effect of aging the β structure in the ternary Ti-V-Cr alloys, we will examine the hardening of the Ti-V and Ti-Cr binary alloys. If a Ti-12.5V alloy is heated to about 900-1000 °C, then cooled rapidly to 25 °C (e.g., water quenched), the structure consists mainly of retained β, small amounts of α and a metastable phase designated ω, the latter two formed during quenching.

Upon reheating this structure, an appreciable increase in hardness occurs, the magnitude depending on the aging temperature and the time. The curves in Fig. 9-51 show a typical hardening response (see Chapter 3) associated with precipitation, with the overaging occurring generally after 50 min for this alloy.

The temperature and time dependence of the transformation of β into a structure of α and β (each with their equilibrium V contents) can be depicted by a TTT diagram, as is done for steels. For the Ti-12.5V alloy being considered here, the diagram is shown in Fig. 9-52. Note that the M_s temperature for the formation of α' by a martensite reaction is at about 290 °C, but the M_f is below 25 °C. Thus retained β is present upon cooling rapidly from β. Figure 9-52 does not show the amount of α' formed (only the M_s), but at 25 °C it is about 5% or less. The formation of α from β is rapid, being completed at 550 °C in about 1 min. In the range 200-400 °C, aging promotes the formation of more ω. Note that the rapid hardening shown in Fig. 9-51 is reflected in the rapid decomposition of β shown in the TTT diagram.

The hardening which occurs in the Ti-Cr alloys is typified by the data in Fig. 9-53. Upon quenching a 7.5% Cr alloy from the β region, the β is completely retained. Isothermally transforming to form α and Ti_2Cr increases the hardness for some transformation temperatures and times. Transforming at 450 °C for 1 h will increase the hardness above that of the β structure (as-quenched value) by about 40%.

Figure 9-54 shows the TTT diagram for this 7.5% Cr alloy. Note that some α forms in the β prior to the formation of the intermetallic compound

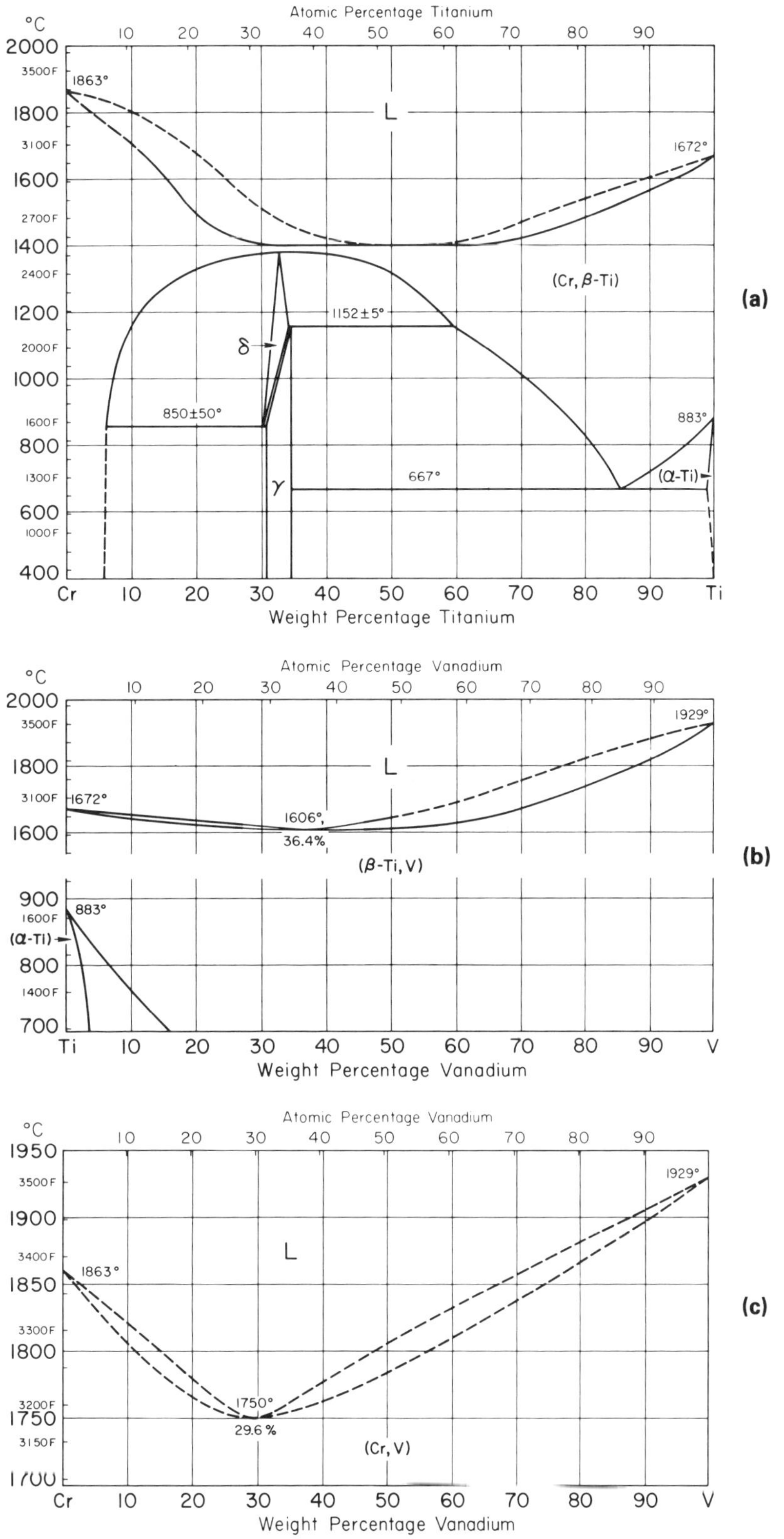

Fig. 9-48. The Ti-Cr, Ti-V and Cr-V phase diagrams. (From same source as Fig. 9-10)

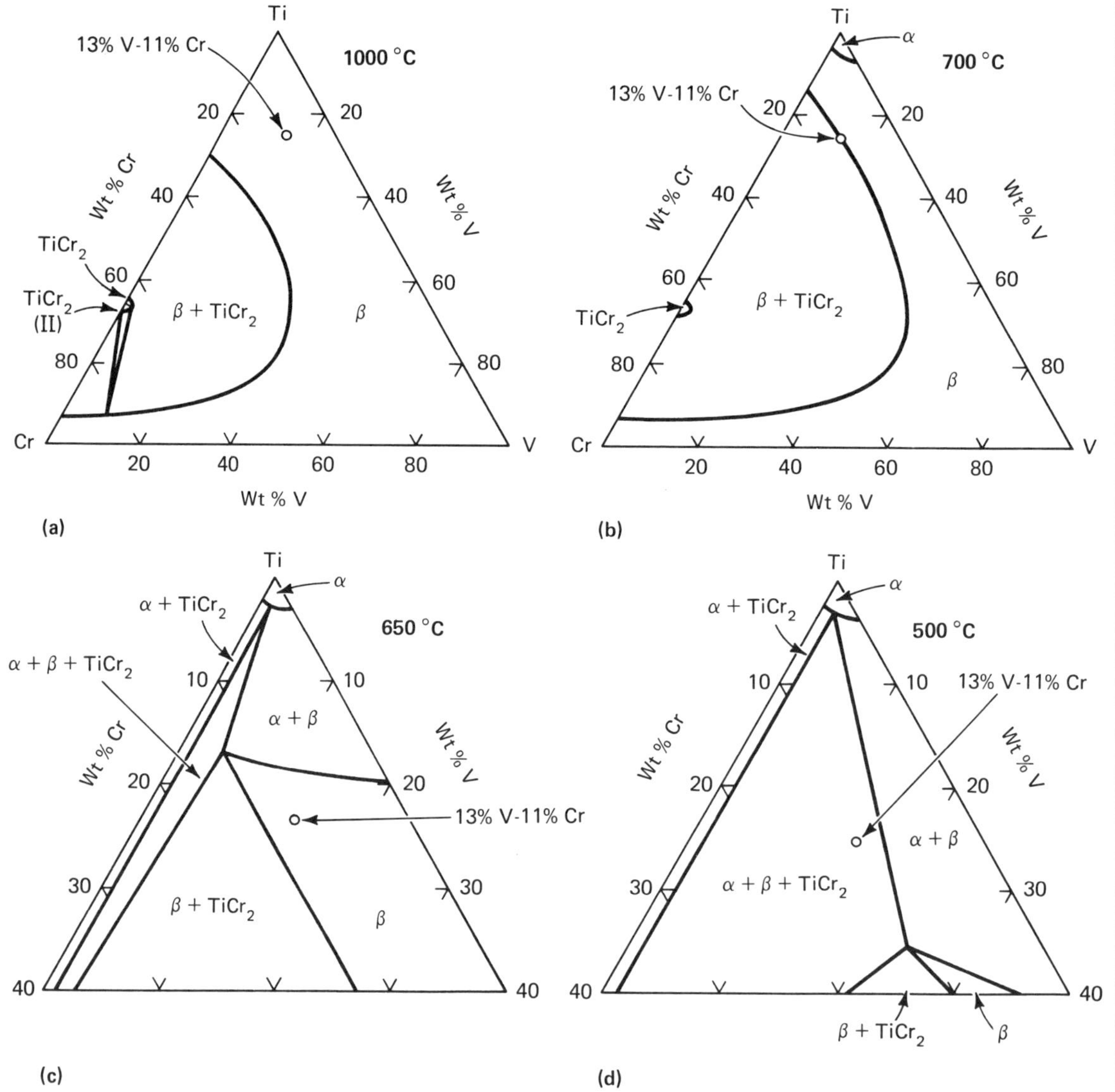

Fig. 9-49. Isothermal sections of the Ti-Cr-V phase diagram. The alloy Ti-13V-11Cr is marked by the dot. (Adapted from P.A. Farrar and H. Margolin, Trans. ASM, Vol 60, p 57, 1967, and from same source as Fig. 9-10)

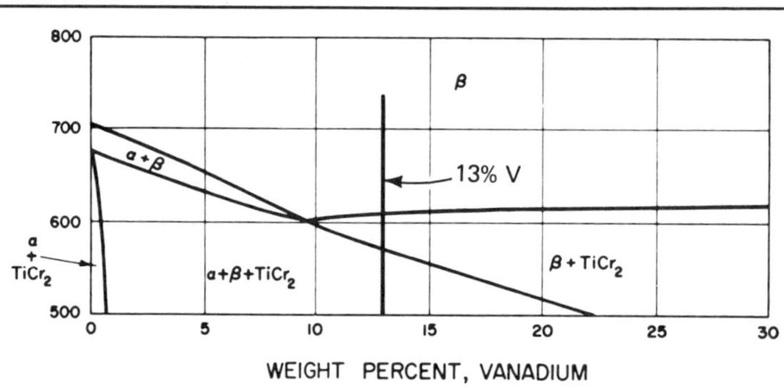

Fig. 9-50. Isopleth at 12% Cr from the Ti-Cr-V phase diagram. (From same source, p 62, as Fig. 9-49)

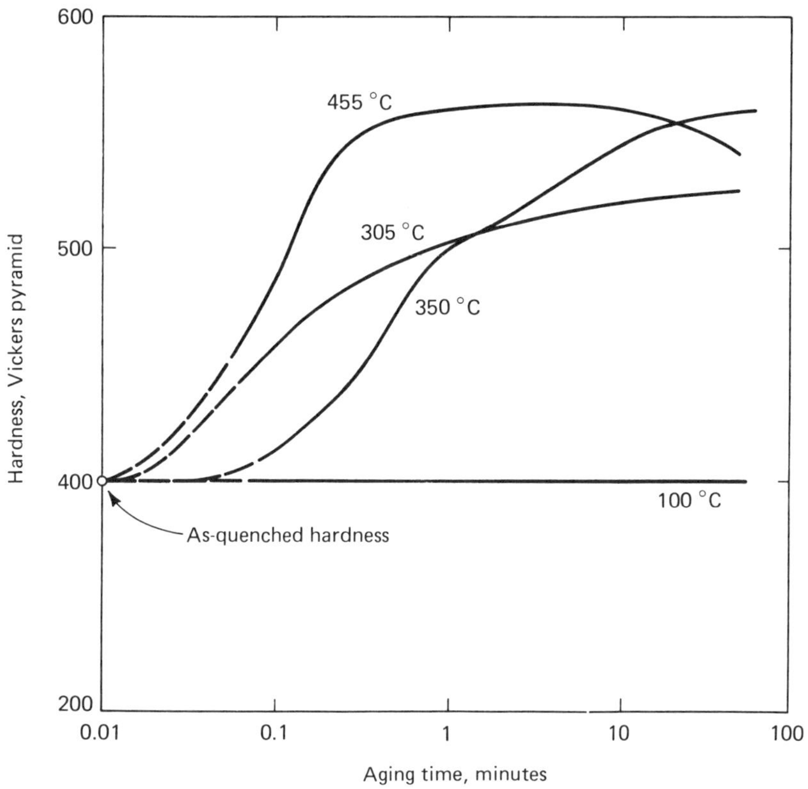

Fig. 9-51. Effect of aging temperature and time on the hardness of a Ti-12.5V alloy. The alloy was homogenized for 20 min at 975 °C, then quenched to 0 °C prior to aging. (Adapted from F. R. Brotzen, E. L. Harmon and A. E. Troiano, Trans. AIME, Vol 203, p 413, 1955)

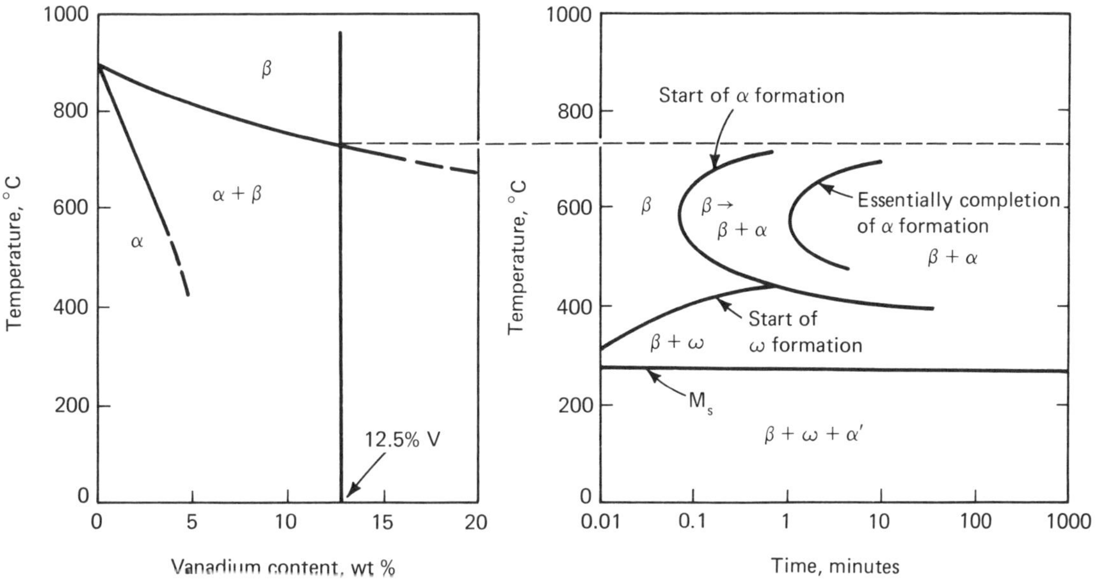

Fig. 9-52. Approximate TTT diagram for the decomposition of *β* in a Ti-12.5V alloy. The alloy was solution treated for 20 min at 975 °C, quenched to 0 °C, then reheated to transform the *β*. (Adapted from same source as Fig. 9-51)

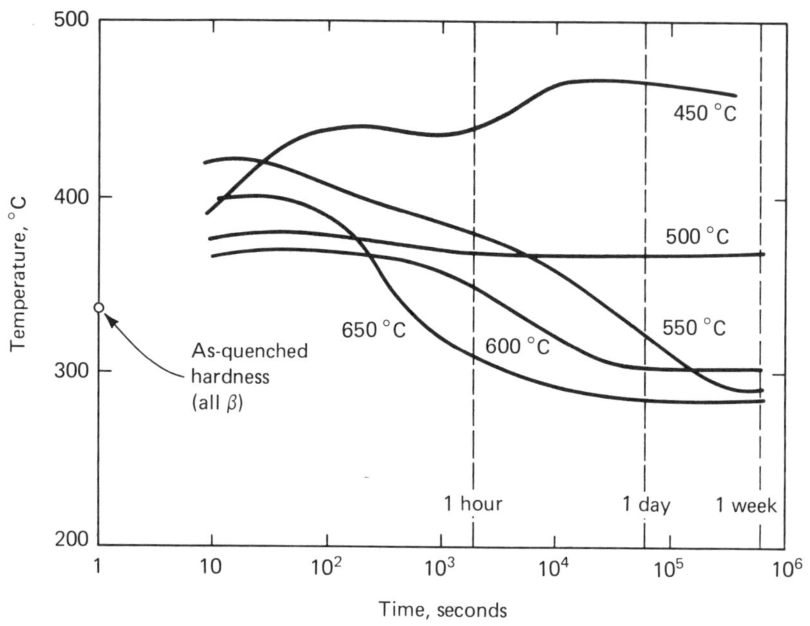

Fig. 9-53. Effect of aging temperature and time on the hardness of a Ti-7.5Cr alloy. The alloy was solution treated at 950 °C for 30 min, then quenched to the transformation temperature. (Adapted from P. D. Frost, W. M. Parris, L. L. Hirsch, J. R. Doig and C. M. Schwartz, Trans. ASM, Vol 46, p 231, 1954)

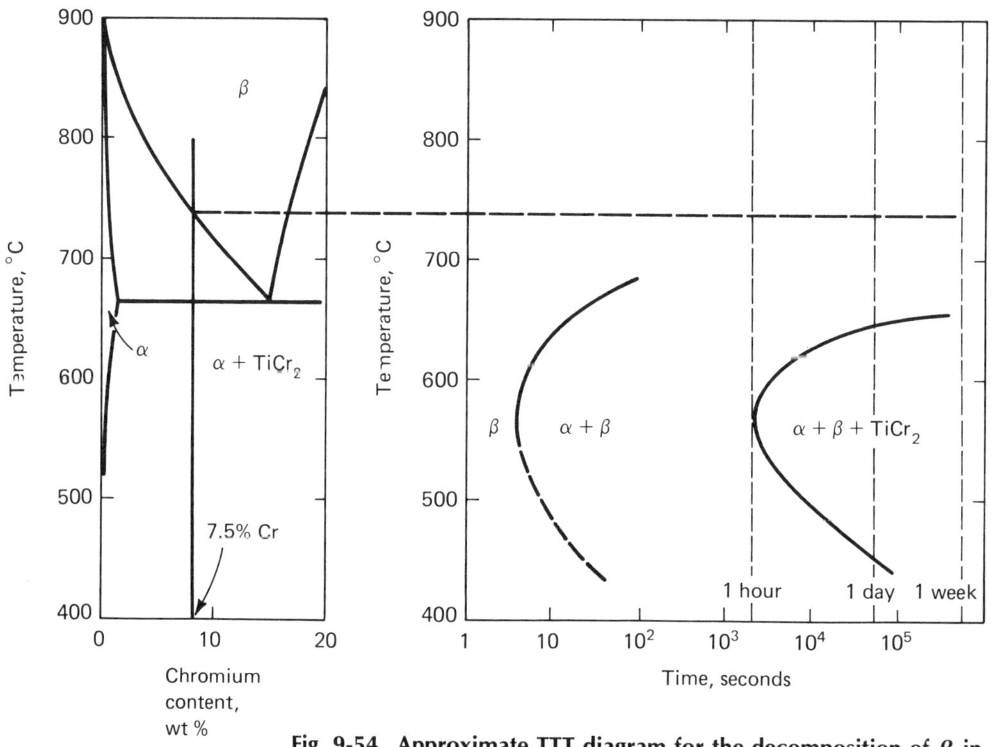

Fig. 9-54. Approximate TTT diagram for the decomposition of β in a Ti-7.5Cr alloy. The alloy was solution treated for 30 min at 950 °C, then quenched to the transformation temperature. (Adapted from same source as Fig. 9-53)

Ti_2Cr. This diagram shows that considerable hardening is associated with the formation of α.

The data just presented for Ti-V and Ti-Cr alloys have shown that the β can be retained upon quenching from the β region. In the Ti-V alloys hardening occurs upon aging because of the formation of α in the β and the appearance of the intermediate ω phase. In the Ti-Cr alloys, hardening is associated with the formation of α in the β, and also subsequently $TiCr_2$. Thus, in the ternary alloy Ti-Cr-V it is expected that β can be retained upon quenching from the β region to 25 °C, and that upon aging, hardening will occur associated with the formation of α and $TiCr_2$, and perhaps the intermediate ω phase. Figure 9-55 is the TTT diagram for a Ti-13V-11Cr-4Al alloy, and it is seen that the transformation products are α and $TiCr_2$. However, at the lower temperature regions, the β phase separates into two body-centered cubic β phases, β_1 and β_2, instead of the ω phase. These two phases are metastable and disappear upon prolonged aging.

The hardening of this alloy upon aging is shown in Fig. 9-56. The alloy was solution treated for 2 h in the β region, then air cooled to 25 °C, giving a structure of all β. The TTT diagram in Fig. 9-55 shows that the shortest time for the beginning of the decomposition of β is about one hour. Thus, air cooling from β—taking, for example, about 10 min to cool to 25 °C—is sufficiently rapid to prevent any β decomposition. Thus, in this

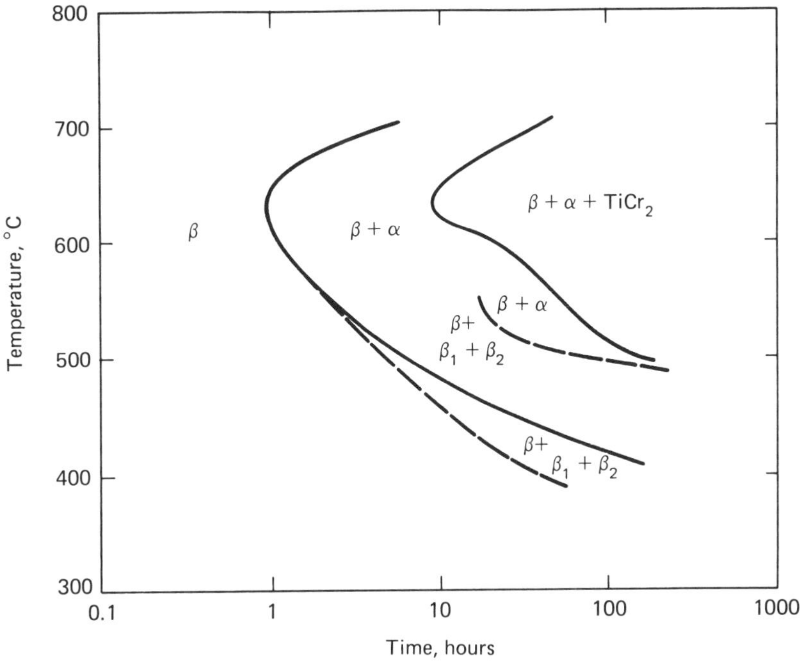

Fig. 9-55. Approximate TTT diagram for a Ti-13V-11Cr-4Al alloy. The alloy was initially solution treated in the β region for 2 h at 760 °C, air cooled to 25 °C, then aged. (Adapted from R. A. Rawe, J. M. Dupouy and M. B. Bever, Trans. AIME, Vol 218, p 821, 1960, and G. H. Narayanan and T. F. Archbold, Met. Trans., Vol 1, p 2281, 1970)

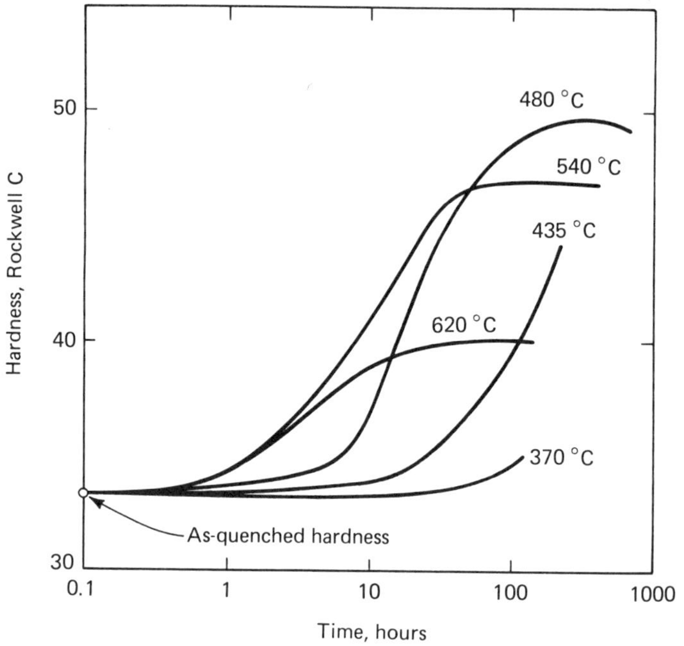

Fig. 9-56. Effect of aging temperature and time on the hardness of a Ti-13V-11Cr-4Al alloy. The alloy was solution treated in the β region for 2 h at 760 °C and air cooled to 25 °C. (Adapted from same source as Fig. 9-55)

alloy, the decomposition of β is considerably slower than in either the Ti-V alloy (Fig. 9-52) or in the Ti-Cr alloy (Fig. 9-54).

Figure 9-57 shows the effect of aging time at 480 °C on the hardness of several Ti-Cr-V and Ti-Cr-V-Al alloys. Note that several hours are required to develop appreciable hardening, consistent with the TTT diagram for the Ti-13V-11Cr-4Al alloy shown in Fig. 9-55. In the as-quenched condition, the alloys containing higher amounts of Cr, V and Al have higher as-quenched (all β) hardnesses because of the solid solution strengthening caused by these solutes. The Ti-12V-18Cr alloy has the highest hardness, in both the as-quenched and the aged conditions, but it is the most expensive because of its high Cr content. Note that the Ti-12V-13Cr alloy shows no hardening upon aging up to 64 h, whereas the addition of 3% Al to a Ti-13V-11Cr alloy allows hardening to occur during within 60 h. The addition of Al to the ternary alloy probably accelerates the precipitation of α (and retards the formation of β_1 and β_2). The addition of Al also decreases the density of the alloy, compensating for the increased density due to the V and Cr additions.

In considering the effect of aging on the mechanical properties, it is important to examine the tensile properties. Table 9-10 lists some tensile mechanical properties (at 25 °C) for the solution treated condition (all β) and for an aged condition. Note that all the alloys in the aged condition show considerably reduced ductility. For the aging treatments used, the Ti-13V-11Cr-3Al alloy has the highest yield and tensile strength.

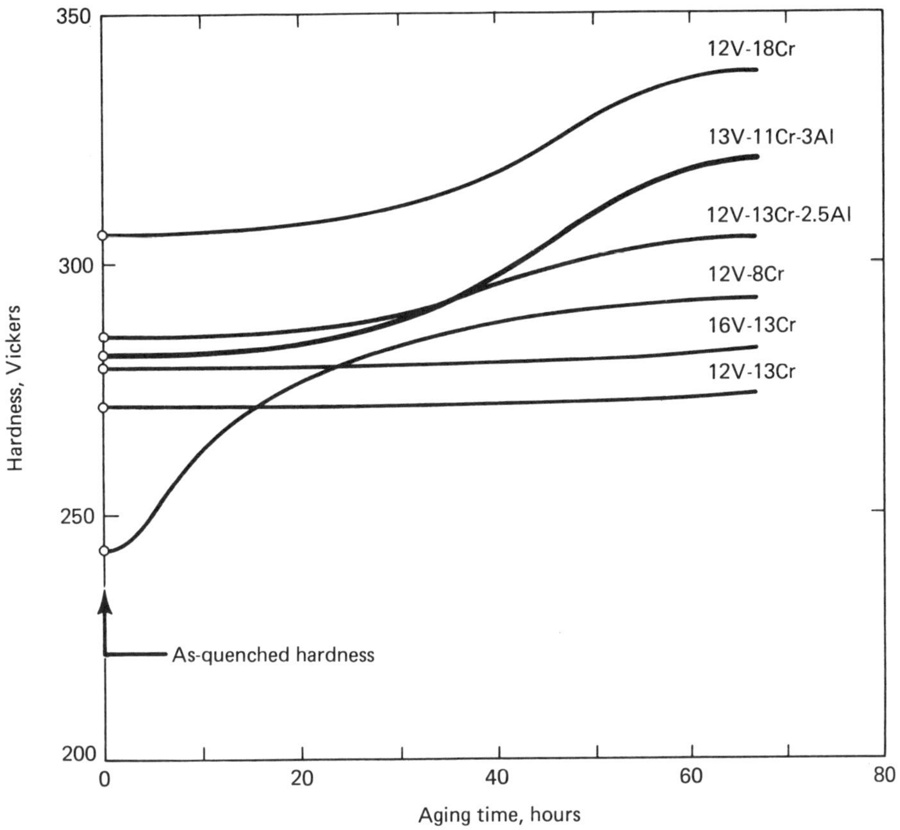

Fig. 9-57. Effect of aging time at 480 °C on the hardness of several Ti-V-Cr and Ti-V-Cr-Al alloys. The alloys were solution treated at 705 °C, in the β region, and quenched to 25 °C to retain the β structure prior to aging. (Adapted from A. G. Imgram, D. N. Williams and H. R. Ogden, Trans. AIME, Vol 227, p 742, 1963)

The recommended commercial heat treatment is to solution heat treat in the β region from about 760 to 815 °C for 0.2 to 1 h, then air cool or quench (depending upon the size of the part) to retain the β structure. Then aging is accomplished around 480 °C, the time usually between 2 and 100 h, depending upon the properties desired. The use of aging temperatures around 480 °C is based on data such as in Fig. 9-58, which show that this is the optimum range to use for maximum strength for aging times up to 100 h. Below this temperature, the rate of formation of α is too low (Fig. 9-55) to give appreciable hardening. Above this temperature, the structure is too coarse to attain maximum strengthening.

References

Titanium, A. D. McQuillan and M. K. McQuillan, Academic Press, New York (1956).

Table 9-10. Tensile Mechanical Properties at 25 °C for Several Ti-V-Cr and Ti-V-Cr-Al Alloys

Alloy	Heat treatment	Yield strength, psi	Tensile strength, psi	Elongation at fracture, %
Ti-12V-8Cr	β quenched	106,000	114,000	26
Ti-12V-8Cr	β quenched and aged for 100 h at 455 °C	147,000	154,000	6
Ti-12V-13Cr	β quenched	116,000	125,000	36
Ti-12V-13Cr	β quenched and aged for 100 h at 455 °C	115,000	125,000	22
Ti-12V-18Cr	β quenched	124,000	130,000	28
Ti-12V-18Cr	β quenched and aged for 100 h at 455 °C	118,000	125,000	28
Ti-16V-13Cr	β quenched	119,000	128,000	24
Ti-16V-13Cr	β quenched and aged for 100 h at 455 °C	117,000	126,000	26
Ti-12V-13Cr-2.5Al	β quenched	126,000	127,000	40
Ti-12V-13Cr-2.5Al	β quenched and aged for 100 h at 480 °C	140,000	155,000	10
Ti-13V-11Cr-3Al	β quenched	118,000	122,000	38
Ti-13V-11Cr-3Al	β quenched and aged for 100 h at 480 °C	159,000	172,000	4

Adapted from same source as Fig. 9-57.

The Physical Metallurgy of Titanium Alloys, R. I. Jaffee, Progress in Metal Physics, Vol 7 (1958).

Titanium Metallurgy, H. Margolin and J. P. Nielsen, in Modern Materials, Vol 2, ed. H. H. Hausner, Academic Press, New York (1960).

Phase Transformations in Titanium and Its Alloys, M. K. McQuillan, Metallurgical Reviews, Vol 8(29), p 41 (1963).

The Formation of Omega Phase in Titanium and Zirconium Alloys: A Review, B. S. Hickman, Journal of Materials Science, Vol 4, p 554 (1969).

The Science, Technology and Application of Titanium, ed. R. I. Jaffee and N. E. Promisel, Pergamon Press, New York (1970).

Titanium Alloys, M. J. Blackburn, W. H. Smyrl and J. A. Feeney, in Stress-Corrosion Cracking in High Strength Steels and in Titanium and Aluminum Alloys, ed. B. F. Brown, Naval Research Laboratory, Washington (1972).

Beta Titanium Alloys, R. A. Wood, MCIC Report 71-11, Metals and Ceramics Information Center, Battelle Columbus Laboratory, Columbus, Ohio (1972).

Titanium Science and Technology, Vol 1-4, ed. R. I. Jaffee and H. M. Burte, Plenum Press, New York (1973).

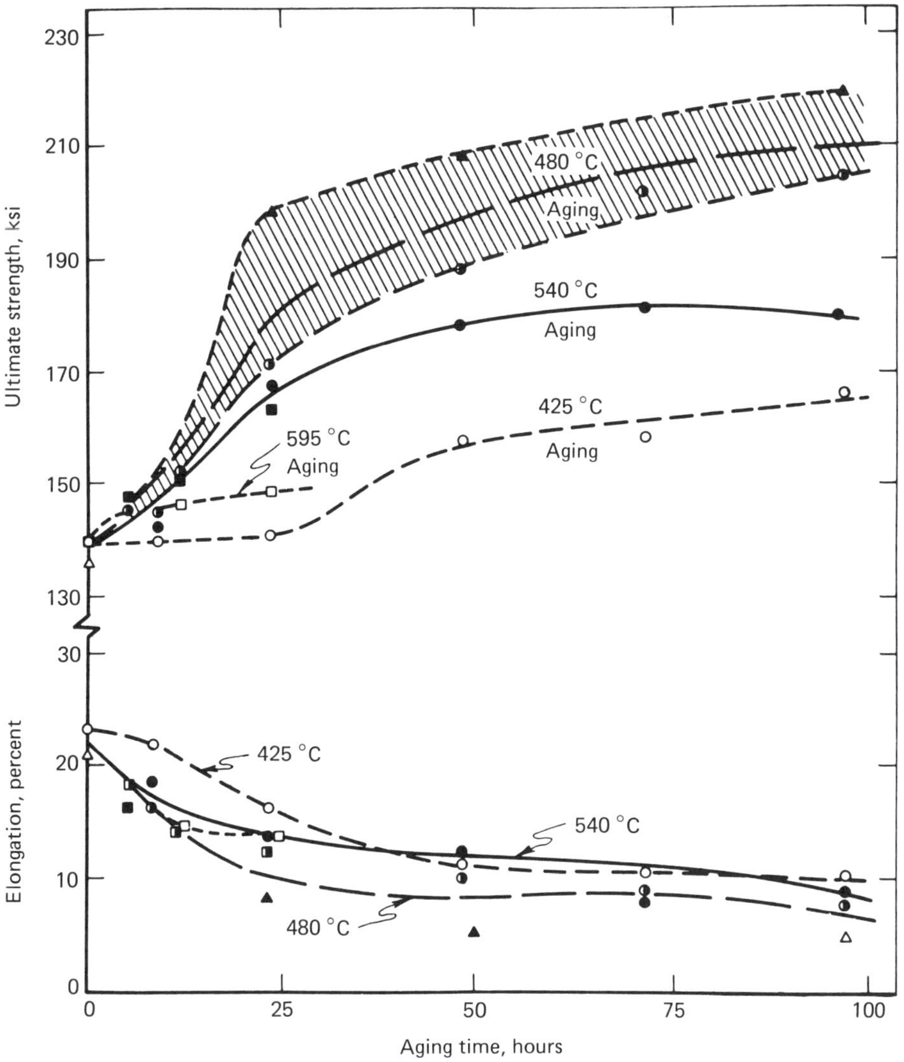

Fig. 9-58. Effect of aging time on tensile mechanical properties of Ti-13V-11Cr-3Al alloy. The data are from different suppliers. (Adapted from R. A. Wood, Beta Titanium Alloys, MCIC Report 71-11, Metals and Ceramics Information Center, Battelle Columbus Laboratory, Columbus, Ohio, 1972)

Titanium Alloys for Engineering Structures, P. H. Morton, Trans. Roy. Soc. (London), Vol 282, p 401 (1976).

The Physical Metallurgy of Superalloys and Titanium Alloys, C. Hammond and J. Nutting, Metal Science, Vol 11, p 474 (1977).

Structure and Properties of Engineering Materials, 4th Ed., R. M. Brick, A. W. Pense and R. B. Gordon, McGraw-Hill, New York (1977).

Structure and Properties of Engineering Alloys, W. F. Smith, McGraw-Hill, New York (1981).

Titanium '80, Vol I-IV, ed. H. Kimura and O. Izumi, The Metallurgical Society of AIME, Warrendale, Pa. (1980).

APPENDIX 1: General References

Engineering Metals and Their Alloys, C. H. Samans, Macmillan Co., New York (1953).

Metals Handbook, 8th Ed., Vol 1, Properties and Selection of Metals, American Society for Metals, Metals Park, Ohio (1961).

Structure and Properties of Alloys, 3d Ed., R. M. Brick, R. B. Gordon and A. Phillips, McGraw-Hill, New York (1965).

Metals Handbook, 8th Ed., Vol 7, Atlas of Microstructures of Industrial Alloys, American Society for Metals, Metals Park, Ohio (1972).

Metals Handbook, 8th Ed., Vol 8, Metallography, Structures and Phase Diagrams, American Society for Metals, Metals Park, Ohio (1973).

Physical Metallurgy Principles, 2d Ed., R. E. Reed-Hill, Van Nostrand, New York (1973).

Metals Handbook, 9th Ed., Vol 2, Properties and Selection: Nonferrous Alloys and Pure Metals, American Society for Metals, Metals Park, Ohio (1979).

Metals Handbook, 9th Ed., Vol 5, Heat Treating, American Society for Metals, Metals Park, Ohio (1982).

Structure and Properties of Engineering Alloys, William F. Smith, McGraw-Hill, New York (1981).

APPENDIX 2: English/Metric Conversion Factors

1 N/m^2	= 1 pascal (Pa)
	= 10^{-3} M · Pa
	= 10^{-6} N/mm^2
	= 1.4505×10^{-4} $lbf/in.^2$ (psi)
	= 1.4505×10^{-7} ksi
	= 2.0886×10^{-2} lbf/ft^2
	= 7.2520×10^{-8} $tons/in.^2$
	= 1.0198×10^{-7} kgf/mm^2
1 $lbf/in.^2$	= 6.89417×10^3 N/m^2
	= 7.03068×10^{-4} kgf/mm^2
1 kgf/mm^2	= 1.4223363×10^3 $lbf/in.^2$
1 $ton/in.^2$	= 1.378929×10^7 N/m^2
1 Joule	= 1.019716×10^{-1} kgf/m
	= 7.3756×10^{-1} ft · lbf
	= 2.7777×10^{-4} watt · hour
	= 1 watt · sec
	= 9.48451×10^{-4} Btu
1 ft · lbf	= 1.3558 Joule
1 MPa · $m^{1/2}$	= 1.09885 ksi · $in.^{1/2}$
	= 3.1623×10^{-2} MN · $m^{-3/2}$
1 ksi · $in.^{1/2}$	= 9.10038×10^{-1} MPa · $m^{1/2}$

1 km	= 10^3 m
	= 10^5 cm
	= 10^6 mm
	= 10^9 μm
	= 10^{12} angstroms (A)
	= 3.93700788×10^4 in.
	= 3.2808399×10^3 ft
1 m	= 10^2 cm
	= 10^3 mm
	= 10^6 μm
	= 10^{10} A
1 cm	= 10 mm
	= 10^4 μm
	= 10^8 A
1 μm	= 10^{-6} m
	= 10^{-4} cm
	= 10^{-3} mm
	= 10^4 A
°C	= $\frac{5}{9}$(°F − 32)
K	= $\frac{5}{9}$(°F − 32) + 273.15
°F	= $\frac{9}{5}$(°C) + 32
K	= °C + 273.15
°R	= °F + 460

From Heat Treatment of Ferrous Alloys, Charlie R. Brooks, Hemisphere Publishing Corp./McGraw-Hill Book Co., New York, 1979.

APPENDIX 3: Chemical Compositions and Typical Applications of Several Nickel-Base Superalloys

From Metal Progress Databook 1979, American Society for Metals

Alloy(a)	Nominal Composition, %										
	C	Mn	Si	Cr	Ni	Co	Mo	W	Cb	Ti	Al
AF2-IDA	0.35	0.1(b)	0.1(b)	12	Bal	10	3.0	6.0	—	3.0	4.6
Alloy 713C(c)	0.12	—	—	12.5	Bal	—	4.2	—	2.0	0.8	6.1
Alloy 713LC(c)	0.05	—	—	12.0	Bal	—	4.5	—	2.0	0.6	5.9
Astroloy(d)	0.06	—	—	15.0	Bal	15	5.25	—	—	3.5	4.4
B-1900(c)	0.1	0.2(b)	0.25(b)	8.0	Bal	10	6.0	0.1(b)	0.1(b)	1.0	6.0
DS-Nickel (sheet)	—	—	—	—	Bal	—	—	—	—	—	—
B-1910	0.10	—	—	10.0	Bal	10.0	3.0	—	—	1.0	6.0
Ford 406(c)	—	—	—	6.0	Bal	10	1.0	8.5	2.0	2.0	4.5
GMR-235(c)	0.15	0.25(b)	0.60(b)	15.5	Bal	—	5.25	—	—	2.0	3.0
GMR-235D(c)	0.15	0.10(b)	0.30(b)	15.5	Bal	—	5.0	—	—	2.5	3.5
Hastelloy alloy C-276	0.02(b)	1.0(b)	0.08(b)	15.5	Bal	2.5(b)	16.0	3.8	—	—	—
Hastelloy alloy C-4	0.015(b)	1.0(b)	0.08(b)	16.3	Bal	2.0(b)	15.3	—	—	0.5	—
Hastelloy alloy G	0.05(b)	1.5	1.0(b)	22.3	Bal	2.5(b)	6.5	1.0(b)	—	—	—
Hastelloy alloy N	0.06	0.8(b)	1.0(b)	7.0	Bal	0.2(b)	16.5	0.5(b)	—	—	0.05(b)
Hastelloy alloy S	0.02(b)	0.50	0.40	15.5	Bal	2.0(b)	14.5	—	—	—	0.2
Hastelloy alloy W	0.12(b)	1.0(b)	1.0(b)	5.0	Bal	2.5(b)	24	—	—	—	—
Hastelloy alloy X	0.10	1.0(b)	1.0(b)	21.8	Bal	1.5	9.0	0.6	—	—	—
Hastelloy alloy B-2	0.02(b)	1.0(b)	0.10(b)	1.0(b)	Bal	1.0(b)	28	—	—	—	—
Illium G(c)	0.20	—	—	22	Bal	—	6	—	—	—	—
Illium 98(c)	0.05	—	—	28	Bal	—	8	—	—	—	—
Illium B(c)	0.05	—	3.5	28	Bal	—	8	—	—	—	—
Illium W(c)	0.08	—	—	16	Bal	—	17	4	—	—	—
IN-100(c)	0.18	—	—	10.0	Bal	15.0	3.0	—	—	4.7	5.5
IN-102	0.06	0.75(b)	0.4(b)	15.0	Bal	—	2.9	3.0	2.9	0.5	0.5
IN-162(c)	0.12	0.10(b)	0.20(b)	10	Bal	—	4.0	2.0	1.0	1.0	6.5
IN 643(c)	0.50	—	—	25.0	Bal	12.0	0.5	9.0	2.0	0.25	—
IN-731(c)	0.18	0.2(b)	0.2(b)	9.5	Bal	10	2.5	—	—	4.65	5.5
IN-738(c)	0.17	0.2(b)	0.3(b)	16	Bal	8.5	1.75	2.6	0.9	3.4	3.4
IN-792(c)	0.21	—	—	12.7	Bal	9.0	2.0	3.9	—	4.2	3.2
Inconel MA 754	0.05	—	—	20.0	78.5	—	—	—	—	0.5	0.3
Inconel 600	0.08	0.5	0.2	15.5	76.0	—	—	—	—	—	—
Inconel 601	0.05	0.5	0.2	23	60.5	—	—	—	—	—	1.4
Inconel 617	0.1	0.5	0.5	22.0	52.0	12.5	9.0	—	—	0.3	1.2
Inconel 625	0.05	0.2	0.2	21.5	61.0	—	9.0	—	3.6	0.2	0.2
Inconel 671	0.05		—	48.0	53.5	—	—	—	—	0.4	—
Inconel 690	0.03	—	—	30.0	60.0	—	—	—	—	—	—
Inconel 706	0.03	0.2	0.2	16	41.5	—	—	—	2.9	1.8	0.2
Inconel 718	0.04	0.2	0.2	18.5	52.5	—	3.0	—	5.1	0.9	0.5
Inconel X-750	0.04	0.5	0.2	15.5	73.0	—	—	—	1.0	2.5	0.7
Inconel 751	0.05	0.5	0.2	15.5	72.5	—	—	—	1.0	2.3	1.2
M-252, J1500	0.15	0.5(b)	0.5(b)	19.0	Bal	10	10	—	—	2.6	1.0
MAR-M200(c)	0.15	—	—	9.0	Bal	10	—	12.5	1.8	2.0	5.0
MAR-M211(c)	0.15	—	—	9.0	Bal	10	2.5	5.5	2.7	2.0	5.0
MAR-M246(c)	0.15	0.10	0.05	9.0	Bal	10	2.5	10.0	—	1.5	5.5
MAR-M421(c)	0.15	0.20(b)	0.20(b)	15.5	Bal	10	1.75	3.5	1.75	1.75	4.25
MAR-M432(c)	0.15	—	—	15.5	Bal	20	—	3.0	2.0	4.3	2.8
MM-002(c)	0.15	—	—	9.0	Bal	10.0	10.0	—	—	1.5	5.5
MM-004(c)	0.05	—	—	12.0	Bal	—	4.5	—	2.0	0.6	5.9
MM-007(c)	0.10	—	—	8.0	Bal	10.0	6.0	0.10(b)	0.10(b)	1.0	6.0
MM-008(c)	0.07	—	—	14.6	Bal	15.2	4.4	—	—	3.35	4.3
Nicrotung(c)	0.10	—	—	12	Bal	10	—	8	—	4	4
NX 188(c) (f)	0.04	—	—	—	Bal	—	18	—	—	—	8
RA-333	0.05	1.5	1.25	25	Bal	3.0	3.0	3.0	—	—	—
Rene 41	0.09	—	—	19	Bal	11	10	—	—	3.1	1.5
Rene 77(d)	0.15(b)	—	—	15	Bal	18.5	5.2	—	—	3.5	4.25
Rene 80	0.17	—	—	14	Bal	9.5	4.0	4.0	—	5.0	3.0
Rene 85	0.27	—	—	9.3	Bal	15	3.25	5.35	—	3.25	5.25
Rene 95	0.15	—	—	14	Bal	8.0	3.5	3.5	3.5	2.5	3.5

B	Zr	Fe	Other	Characteristics and Typical Applications
0.015	0.10	0.5(b)	1.5 Ta	Turbine discs.
0.012	0.10	—	—	Jet engine blades, parts.
0.01	0.10	—	—	Jet engine blades, parts.
0.03	—	—	—	Forgings for high temperatures.
0.015	0.08	0.35(b)	0.015 S(b), 4.3 Ta	Jet engine blades.
—	—	—	2.1 ThO_2	Jet engine parts.
0.015	0.10	—	7.0 Ta	Modified B-1900.
—	—	—	6.0 Ta	Integrally cast turbine wheels.
0.06	—	10	—	Jet engine parts.
0.05	—	4.5	—	Jet engine parts.
—	—	5.5	0.35 V(b)	Resists severe corrosive conditions.
—	—	3.0(b)	—	High-temperature stability and corrosion resistance.
—	—	19.8	2.0 Cu, 2.1 Cb + Ta	Resists pitting and stress-corrosion cracking and both acid and alkaline environments.
0.01(b)	—	5.0(b)	0.5 Al + Ti(b), 0.35 Cu(b)	Resistant to hot, molten fluoride salts.
0.009	—	1.0	0.02 La	Gas turbine parts, welding wire for high-temperature components; high stability, low thermal expansion alloy.
—	—	5.5	0.6 V(b)	Welding wire for high-temperature components.
—	—	18.5	—	Jet engine sheet parts; good oxidation resistance.
—	—	2.0(b)	—	Resists hot hydrochloric acid and hydrogen chloride gas.
—	—	5	6 Cu	Resists hot sulfuric, hydrofluoric, nitric, and phosphoric acids.
—	—	—	5 Cu	Same as Illium G, but to higher temperature.
—	—	—	5 Cu	Hardenable form of Illium 98, resists wear and galling.
—	—	6	—	Corrosion resistance and high-temperature strength.
0.014	0.06	—	1.0 V	Jet engine blades and wheels.
0.005	0.03	7.0	0.02 Mg	Superheater and jet engine parts.
0.020	0.10	0.50(b)	2.0 Ta	Gas turbine blades, vanes.
—	0.25	3.0	—	Tubes in chemical reforming units.
0.015	0.06	0.5(b)	0.95 V	Modification of IN-100.
0.01	0.10	0.5(b)	1.75 Ta (Fe, Mn, S, Si, low as possible)	Good hot corrosion resistance.
0.02	0.10	—	3.9 Ta	Good hot corrosion resistance, blades.
—	—	—	0.6 Y_2O_3	Mechanically alloyed from powder, turbine vanes.
—	—	8.0	0.2 Cu	Heat treating and chemical processing equipment.
—	—	14.1	0.2 Cu	Furnace, heat treat fixtures.
—	—	1.5	0.2 Cu	Gas turbine aircraft engine parts.
—	—	2.5	—	Aircraft engines and structures, chemical processing.
—	—	—	—	For extremely corrosive environments.
—	—	9.5	—	Resists nitric and hydrofluoric acids.
—	—	40.0	0.2 Cu	Similar to Inconel 718, more machinable.
—	—	18.5	0.2 Cu	Jet engine and rocket parts.
—	—	7.0	0.2 Cu	Gas turbine parts, bolts.
—	—	7.0	0.2 Cu	Diesel engine valves.
0.005	—	—	—	Gas turbine blades, parts, sheets.
0.015	0.05	—	—	Jet engine blades.
0.015	0.05	—	—	Integrally cast turbine wheels, blades.
0.015	0.05	0.15	1.5 Ta, 0.10 Cu(b)	Jet engine blades; better ductility than MAR-M200.
0.015	0.05	1.0(b)	—	Superior hot corrosion resistance and stability.
0.015	0.05	—	2.0 Ta	Integrally cast turbine wheels.
0.015	0.05	1.0	2.5 Ta, 1.5 Hf	Hafnium-containing alloy; improved ductility for turbine parts.
0.010	0.10	—	1.3 Hf	Hafnium-modified Alloy 713LC; improved ductility for turbine parts.
0.015	0.075	—	4.25 Ta, 1.3 Hf	Hafnium-modified B-1900; improved ductility for turbine parts.
0.015	0.03	—	1.3 Hf	Hafnium-containing alloy; improved ductility for turbine parts.
0.050	0.05	—	—	High-strength parts.
—	—			Jet engine vanes.
—	—	18	—	Turbine parts, radiant tubes.
0.010(b)	—	—	—	Jet engine blades, parts.
0.05(b)	—	1.0(b)	—	Jet engine parts.
0.015	0.03	—	—	Turbine blade alloy.
0.015	0.03	—	—	Compressor disc alloy.
0.01	0.05	—	—	Turbine or compressor disc alloy.

(Continued on the next page)

APPENDIX 3: Chemical Compositions and Typical Applications of Several Nickel-Base Superalloys *(continued)*

From Metal Progress Databook 1979, American Society for Metals

Alloy(a)	Nominal Composition, %										
	C	Mn	Si	Cr	Ni	Co	Mo	W	Cb	Ti	Al
SEL	0.08	0.3(b)	0.5(b)	15	Bal	26	4.5	—	—	2.4	4.4
SEL-15	0.07	0.3(b)	0.5(b)	11	Bal	14.5	6.5	1.5	0.5	2.5	5.4
TAZ-8(c)	0.125	—	—	6.0	Bal	—	4.0	4.0	—	—	6.0
TAZ-8A(c)	0.125	—	—	6.0	Bal	—	4.0	4.0	2.5	—	6.0
TAZ-8B(c) (f)	0.125	—	—	6.0	Bal	5.0	4.0	4.0	1.5	—	6.0
TRW 1800(c)	0.09	—	—	13.0	Bal	—	—	9.0	1.5	0.6	6.0
TRW 1900(c)	0.11	—	—	10.3	Bal	10.0	—	9.0	1.5	1.0	6.3
TRW VI A(c)	0.13	—	—	6	Bal	7.5	2.0	5.8	0.5	1.0	5.4
TRW VI A(h)	0.13	—	—	6	Bal	7.5	2.0	5.8	0.5	1.0	5.4
Udimet 500	0.08	—	—	19	Bal	18.0	4	—	—	3.0	3.0
Udimet 520	0.05	—	—	19	Bal	12.0	6	1.0	—	3.0	2.0
Udimet 630	0.04	0.2(b)	0.2(b)	17	Bal	1.0(b)	3.1	3.0	6.0	1.1	0.6
Udimet 700(d)	0.07	—	—	15	Bal	18.5	5.0	—	—	3.5	4.4
Udimet 710	0.07	0.1(b)	0.2(b)	18	Bal	15	3.0	1.5	—	5.0	2.5
Udimet 720	0.035	—	—	18.0	Bal	15.0	3.0	1.25	—	5.0	2.50
Unitemp C-300	0.12	—	—	8.7	Bal	9.0	2.0	7.6	—	0.7	3.4
Unitemp 1753	0.24	0.05	0.10	16.25	Bal	7.2	1.6	8.4	—	3.2	1.9
Waspaloy A(g)	0.07	0.5(b)	0.5(b)	19.5	Bal	13.5	4.3	—	—	3.0	1.4
Waspaloy B(g)	0.07	0.75(b)	0.75(b)	19.5	Bal	13.5	4.3	—	—	3.0	1.4
WAZ 16(c)	0.2	—	—	—	Bal	—	2	16	2	—	7
WAZ-20(c) (f)	0.15	—	—	—	Bal	—	—	18.5	—	—	6.2

(a) Some superalloys such as Inconel 718, Rene 41, Udimet 500, and others are made by more than one manufacturer. The proprietary designation for such alloys has been used in this compilation; (b) Maximum; (c) Cast alloy; (d) Compositions of Astroloy, Rene 77, and Udimet 700 are very similar. Certain elements are controlled to prevent sigma phase formation. (e) Estimated from limited data using Larson-Miller time-temperature parameter; (f) Directionally solidified; (g) Waspaloy A has a higher solution temperature and longer time at stabilization than Waspaloy B. (h) Prealloyed powder product prepared by strength enhancement technique.

B	Zr	Fe	Other	Characteristics and Typical Applications
0.015	—	1.0(b)	—	High-temperature parts.
0.015	—	0.5(b)	—	High-strength parts.
—	1.0	—	8.0 Ta, 2.5 V	Non-vacuum melted blade alloy.
0.004	1.0	—	8.0 Ta	Good thermal fatigue and oxidation resistance.
0.004	1.0	—	8.0 Ta	Improved ductility blade alloy.
0.07	0.07	—	—	Gas turbine blades.
0.03	0.10	—	—	Gas turbine blades.
0.02	0.13	—	9.0 Ta, 0.5 Re, 0.43 Hf	Turbine blade alloy.
0.02	0.13	—	9.0 Ta, 0.5 Re, 0.43 Hf	Ultrahigh-strength disc alloy.
0.005	—	0.5(b)	—	Gas turbine parts, sheets, bolts.
0.005	—	—	—	Properties equal to Udimet 500; improved workability.
0.005	—	17.5	—	Applications to 1,000 F.
0.025	—	0.5(b)	—	Jet engine parts.
0.02	—	0.5(b)	—	Sulfidation-resistant disc alloy.
0.033	0.030	—	—	Sulfidation and impact-resistant gas turbine alloy.
0.02	0.1	—	10 Ta, 1 Hf, 0.5 V	Experimental, high strength 1,200 F alloy.
0.008	0.06	9.5	—	Gas turbine parts, blades, discs, rings, spacers.
0.006	0.09	2.0(b)	0.03 S(b), 0.10 Cu(b)	Jet engine blades.
0.006	0.07	2.0(b)	0.02 S(b), 0.10 Cu(b)	Jet engine discs.
—	0.5	—	—	High temperature stator vane alloy.
—	1.5	—	—	High temperature stator vane alloy.

Source: Ward P. Simmons, Battelle-Columbus; additional data supplied by Elgiloy Co.; Huntington Alloys Inc.; Stellite Div., Cabot Corp.; Sherritt Gordon Mines Ltd.; Special Metals Corp.; Stainless Foundry & Engineering Inc.; NASA-Lewis Research Center. (Revised April 1979.)

APPENDIX 4: Chemical Compositions and Typical Applications of Common Cobalt-Base Superalloys

Adapted from Metal Progress Databook 1980, American Society for Metals

Alloy(a)	Nominal Composition, %								
	C	Mn	Si	Cr	Ni	Co	Mo	W	Cb
AiResist 13(b)	0.45	0.5(c)	—	21	1.0(c)	Bal	—	11	2.0
AiResist 213	0.18	—	—	19	—	Bal	—	4.7	—
AiResist 215(b)	0.35	—	—	19	—	Bal	—	4.5	—
Elgiloy	0.15	2.0	—	20.0	15.0	40.0	7.0	—	—
FSX-414(b)	0.25	1.0(c)	1.0(c)	29.5	10.5	Bal	—	7.0	—
FSX-418(b)	0.25	1.0(c)	1.0(c)	29.5	10.5	Bal	—	7.0	—
FSX-430(b)	0.40	—	—	29.5	10.0	Bal	—	7.5	—
X-40(b)	0.50	0.50	0.50	25	10	Bal	—	7.5	—
Haynes alloy 150	0.08	0.65	0.75	28	3.0(b)	Bal	1.5(b)	—	—
Haynes alloy 188 (sheet)	0.10	1.25(b)	0.3	22	22	Bal	—	14	—
Illium X(b)	0.75	—	—	29	—	55	—	14	—
MAR-M302(b)	0.85	0.10	0.20	21.5	—	Bal	—	10.0	—
MAR-M322(b)	1.00	0.10	0.10	21.5	—	Bal	—	9.0	—
MAR-M509(b)	0.60	0.10(c)	0.10(c)	21.5	10	Bal	—	7.0	—
MAR-M918	0.05	0.2(c)	0.2(c)	20	20	Bal	—	—	—
MP35N	—	—	—	20.0	35.0	35.0	10.0	—	—
NASA Co-W-Re(b)	0.40	—	—	3	—	Bal	—	25	—
S-816	0.38	1.20	0.40	20	20	Bal	4.0	4.0	4.0
V-36	0.27	1.00	0.40	25	20	Bal	4.0	2.0	2.0
WF-11, L605, Haynes alloy 25	0.10	1.50	0.50	20	10	Bal	—	15	—
WF-31	0.15	1.42	0.42	20	10	Bal	2.6	10.7	—
WI-52(b)	0.45	0.50(c)	0.50(c)	21	1.0(c)	Bal	—	11	2.0
X-45(b)	0.25	1.0(c)	—	25.5	10.5	Bal	—	7.0	—

(a) Some superalloys are made by more than one manufacturer. The proprietary designation for such alloys has been used in this compilation; (b) Cast alloy; (c) Maximum composition; (d) At 1,740 F; (e) Estimated from limited data using Larson-Miller time-temperature parameters.

Ti	Al	B	Zr	Fe	Other	Characteristics and Typical Applications
—	3.5	—	—	2.5(c)	0.1 Y	High-temperature parts.
—	3.5	—	0.15	—	6.5 Ta, 0.1 Y	Sheets, tubing; resistant to hot corrosion.
—	4.3	—	0.13	—	7.5 Ta, 0.17 Y	Nozzle vanes; resistant to hot corrosion.
—	—	—	—	Bal	0.04 Be	Springs; corrosion resistance, high strength.
—	—	0.012	—	2.0(c)	—	Gas turbine vanes.
—	—	0.012	—	2.0(c)	0.15 Y	Gas turbine vanes; improved oxidation resistance.
—	—	0.027	0.9	—	0.5 Y	Gas turbine vanes; improved strength and ductility.
—	—	—	—	1.5	—	Gas turbine parts, nozzle vanes.
—	—	—	—	20.0	—	Resists thermal shock; high-temperature corrosion (air and air-SO_2)
—	—	—	—	3.0(b)	0.04 La	Better oxidation resistance than Hastelloy X.
—	—	—	—	—	—	Corrosion and abrasion resistant.
—	—	0.005	0.15	—	9.0 Ta	Jet engine blades, vanes.
0.75	—	—	2.25	—	4.5 Ta	Jet engine blades, vanes.
0.2	—	0.010(c)	0.50	1.0	3.5 Ta	Jet engine blades, vanes.
—	—	—	0.10	0.5(c)	7.5 Ta	High-temperature sheets.
—	—	—	—	—	—	Stress-corrosion resistant, high-strength fasteners.
1.0	—	—	1.0	—	2.0 Re	High-temperature space applications.
—	—	—	—	4	—	Gas turbine blades, bolts, springs.
—	—	—	—	3	—	High-temperature sheets.
—	—	—	—	3.0(b)	—	Jet engine parts, sheets.
1.0	—	—	—	—	—	High-temperature sheets.
—	—	—	—	2.0	—	Gas turbine parts, nozzle vanes.
—	—	0.010	—	2.0(c)	—	Nozzle vanes.

Source: Ward P. Simmons, Battelle-Columbus; additional data supplied by Elgiloy Co.; Huntington Alloys Inc.; Stellite Div., Cabot Corp.; Sherritt Gordon Mines Ltd.; Special Metals Corp.; Stainless Foundry & Engineering Inc.; NASA-Lewis Research Center.

APPENDIX 5: Chemical Compositions and Some Applications of Several Commercial Magnesium-Base Alloys

Adapted from Metal Progress Databook 1977, American Society for Metals

Alloy	Al	Mn, min	Composition, % Zn	Zr	Other(a)	Applications
SAND AND PERMANENT MOLD CASTINGS						
AM100A	10.0	0.10	—	—	—	Pressure-tight sand and permanent mold castings. Good combination of room-temperature properties.
AZ63A	6.0	0.15	3.0	—	—	Sand castings requiring good room-temperature strength with ductility.
AZ81A	7.6	0.13	0.7	—	—	Tough, leak-proof sand castings.
AZ91C	8.7	0.13	0.7	—	—	Sand and permanent mold castings requiring good room-temperature strength with ductility.
AZ92A	9.0	0.10	2.0	—	—	Pressure-tight sand and permanent mold castings. Good combination of room-temperature properties.
EZ33A	—	—	2.6	0.7	3.2 RE	Pressure-tight sand and permanent mold castings. For applications at 350-500 F.
HK31A	—	—	—	0.7	3.2 Th	Sand castings used at 400 F and higher.
HZ32A	—	—	2.1	0.7	3.2 Th	Sand castings used at 400 F and higher.
K1A	—	—	—	0.6	—	Sand castings. Excellent damping capacity.
QE22A	—	—	—	0.7	2.2 RE, 2.5 Ag	Sand castings. Good room and elevated-temperature properties.
ZE41A	—	—	4.2	0.7	1.2 RE	Sand castings. Good strength at room temperature. Improved castability over ZK alloys.
ZE63A	—	—	5.7	0.7	2.5 Be	Sand castings. Good strength at room temperature. Improved castability over ZK alloys.
ZH62A	—	—	5.7	0.7	1.8 Th	Sand castings. Good strength at room temperature. Improved castability over ZK alloys.
ZK51A	—	—	4.6	0.7	—	Sand castings. Good strength and ductility at room temperature.
ZK61A	—	—	6.0	0.8	—	Sand castings. Good strength and ductility at room temperature.
DIE CASTINGS(b)						
AM60A	6.0	0.13	—	—	—	Automobile wheels.
AS41A	4.2	0.35	—	—	1.0 Si	Automobile engines and housings.
AZ91A and AZ91B	9.0	0.13	0.7	—	—	Die castings. Parts for cars, lawnmowers, luggage, business machines, tools.
SHEETS AND PLATES						
AZ31B	3.0	0.20	1.0	—	—	General-purpose sheet and plate alloy.
AZ61A	6.5	0.15	1.0	—	—	Reserve-type batteries.
HK31A	—	—	—	0.7	3.2 Th	Missile and aircraft skins; for other uses at 400 to 700 F.
HM21A	—	0.80	—	—	2.0 Th	Missile and aircraft uses up to 800 F.
EXTRUDED BARS, RODS, SHAPES, TUBE, AND WIRE						
AZ21A	2.0	0.15 max	1.0	—	0.2 Ca	Anode material for primary batteries.
AZ31B	3.0	0.20	1.0	—	—	General-purpose extrusions.
AZ61A	6.5	0.15	1.0	—	—	Extrusions requiring higher properties than AZ31B.
AZ80A	8.5	0.12	0.5	—	—	Highly stressed extrusions.
HM31A	—	1.20	—	—	3.0 Th	Missile and aircraft use as high as 800 F.
ZK40A	—	—	4.0	0.45(c)	—	
ZK60A	—	—	5.5	0.45	—	For highly stressed conditions requiring better ductility than AZ80A.
FORGINGS						
AZ31B	3.0	0.20	1.0	—	—	For low-stressed forgings.
AZ61A	6.5	0.15	1.0	—	—	For forgings with higher strength than AZ31B.
AZ80A	8.5	0.12	0.5	—	—	Heat-treatable forgings.
HM21A	—	0.80	—	—	2.0 Th	Aircraft and missile forgings up to 800 F.
ZK60A	—	—	5.5	0.45	—	For forgings of maximum strength in aircraft and military applications.

(a) RE = rare earths; (b) Tentative typical properties determined by four laboratories with separately cast ASTM ¼ in. diameter bars cast by NL Industries Inc. Properties may vary as more data are amassed from other producers; (c) Min.

Source: Dow Chemical Co.; Kaiser Magnesium, Div. of Kaiser Aluminum & Chemical Corp.; Magnesium Research Center, Battelle-Columbus; Magnesium Div., NL Industries Inc.

APPENDIX 6: Typical Mechanical Properties (at 25 °C) of Several Commercial Titanium-Base Alloys

Adapted from Metal Progress, Vol 106, Metal Property Data Book, 1974

Nominal Composition, %	Modulus of Elasticity, 10^6 Psi	Condition	Tensile Strength, Psi	Yield Strength, Psi	Elongation, %	Reduction in Area, %	Charpy Impact Strength, Ft-Lb	Hardness	Applications and Characteristics
Commercially Pure									
99.5 Ti	14.9	Annealed	48,000	35,000	30	55	—	Bhn 120	Airframes; chemical, desalination, and marine parts; plate-type heat exchangers; cold-spun or pressed parts; platinized anodes; high formability.
99.2 Ti	14.9	Annealed	63,000	50,000	28	50	32	Bhn 200	Airframes; aircraft engines; marine and chemical parts; heat exchangers; condenser and evaporator tubing; high formability.
99.1 Ti	15.0	Annealed	75,000	65,000	25	45	28	Bhn 225	Chemical, marine, airframe, and aircraft engine parts which require formability strength, weldability, and corrosion resistance.
99.0 Ti	15.1	Annealed	96,000	85,000	20	40	15	Bhn 265	Chemical, marine, airframe, and aircraft engine parts; surgical implants; high-speed fans; gas compressors; good formability and corrosion resistance, high strength.
0.20 Pd	14.9	Annealed	63,000	50,000	28	50	32	Bhn 200	Good corrosion resistance for chemical industry applications where media are mildly reducing or vary between oxidizing and reducing.
Alpha Alloys									
5 Al, 2.5 Sn	16.0	Annealed	125,000	117,000	16	40	19	Rc 36	Weldable alloy for forgings and sheet-metal parts such as aircraft engine compressor blades and ducting; steam turbine blades; good oxidation resistance and strength at 600 to 1,100 F; good stability at elevated temperatures.
5 Al, 2.5 Sn (low O_2)	16.0	Annealed	117,000	108,000	16	—	20	Rc 35	Special grade for high-pressure cryogenic vessels operating down to −423 F.
Near Alpha Alloys									
8 Al, 1 Mo, 1 V	18.0	Duplex annealed	145,000	138,000	15	28	24	Rc 35	Airframe and jet-engine parts requiring high strength to 850 F; good creep and toughness properties; good weldability.
11 Sn, 1 Mo, 2.25 Al, 5.0 Zr, 1 Mo, 0.2 Si	16.5		160,000	144,000	15	35	—	Rc 36	Airframes; blades, discs, wheels, spacers, and fasteners for turbine engines.
6 Al, 2 Sn, 4 Zr, 2 Mo	16.5		142,000	130,000	15	35	—	Rc 32	Parts and cases for jet-engine compressors; airframe skin components.
5 Al, 6 Sn, 2 Zr, 1 Mo, 0.2 Si	16.5	1,800 F (½ hr), AC +1,100 F (2 hr), AC	152,000	140,000	13	—	—	—	Jet engine parts; high creep strength to 1,000 F.
6 Al, 2 Cb, 1 Ta, 1 Mo	17.5	As-rolled (1 in. plate)	124,000	110,000	13	34	23 (−80 F)	Rc 30	High toughness; moderate strength; good resistance to seawater and hot-salt stress corrosion; good weldability.
6 Al, 2 Sn, 1.5 Zr, 1 Mo, 0.35 Bi, 0.1 Si	—	Beta forge + duplex anneal	147,000	137,000	11	—	—	—	Jet engine discs and blades requiring extra creep resistance and stability.
Alpha-Beta Alloys									
8 Mn	16.4	Annealed	137,000	125,000	15	32	—	—	Aircraft sheet components, structural sections, and skins; good formability, moderate strength.
3 Al, 2.5 V	15.5	Annealed	100,000	85,000	20	—	—	—	Aircraft hydraulic tubing, foil; combines strength, weldability, and formability.
6 Al, 4 V	16.5	Annealed	144,000	134,000	14	30	14	Rc 36	Rocket motor cases; blades and discs for aircraft turbines and compressors; structural forgings and fasteners; pressure vessels; gas and chemical pumps; cryogenic parts; ordnance equipment; marine components; steam-turbine blades.
		Solution + age	170,000	160,000	10	25	—	Rc 41	
6 Al, 4 V (low O_2)	16.5	Annealed	130,000	120,000	15	35	18	Rc 35	High pressure cryogenic vessels operating down to −320 F
6 Al, 6 V, 2 Sn	16.0	Annealed Solution + age	155,000 185,000	145,000 170,000	14 10	30 20	13 —	Rc 38 Rc 42	Rocket motor cases; ordnance components; structural aircraft parts and landing gears; responds well to heat treatments; good hardenability.
7 Al, 4 Mo	16.5	Solution + age	160,000	150,000	16	22	13 —	Rc 38 Rc 42	Airframes and jet engine parts for operation at up to 800 F; missile forgings; ordnance equipment.
6 Al, 2 Sn, 4 Zr, 6 Mo	16.5	Solution + age	184,000	170,000	10	23	—	—	Components for advanced jet engines.
6 Al, 2 Sn, 2 Zr, 2 Mo, 2 Cr, 0.25 Si	17.7	Solution + age	185,000	165,000	11	33	—	—	Strength, fracture toughness in heavy sections.
Beta Alloys									
1 Al, 8 V, 5 Fe	14.2	Solution + age	177,000	170,000	8	—	—	—	High-strength fasteners.
13 V, 11 Cr, 3 Al	14.7	Solution + age	185,000	175,000	8	—	8	Rc 40	High-strength fasteners, aerospace components, honeycomb panels; good formability, heat treatable.
8 Mo, 8 V, 2 Fe, 3 Al	15.5	Solution + age	190,000	180,000	8	—	—	Rc 40	High-strength, tough airframe sheet, plate, fasteners, and forged components.
3 Al, 8 V, 6 Cr, 4 Mo, 4 Zr	15.3	Solution + age	210,000	200,000	7	—	7.5	Rc 42	High-strength fasteners, torsion bars, aerospace components.
	14.0	Annealed	128,000	121,000	15	—	—	—	Parts requiring formability and corrosion resistance.
11.5 Mo, 6 Zr, 4.5 Sn	15.0	Solution + age	201,000	191,000	11	—	—	—	High-strength fasteners, high-strength aircraft sheet parts.

APPENDIX 7: Temper Designations for Aluminum Alloys

Adapted from Metal Progress Materials and Processing Databook 1981, American Society for Metals

F **As fabricated.** Applies to products of shaping processes in which no special control over thermal conditions or strain-hardening is employed.

H **Strain-hardened (wrought products only).** Applies to products which have their strength increased by strain-hardening, with or without supplementary thermal treatments to produce some reduction in strength. The H is always followed by two or more digits.

H1 **Strain-hardened only.** Applies to products which are strain-hardened to obtain the desired strength without supplementary thermal treatment. The number following this designation indicates the degree of strain-hardening.

H111 Applies to products which are strain-hardened less than the amount required for a controlled H11 temper.

H112 Applies to products which acquire some temper from shaping processes not having special control over the amount of strain-hardening or thermal treatment, but for which there are mechanical property limits.

H2 **Strain-hardened and partially annealed.** Applies to products which are strain-hardened more than the desired final amount, and then reduced in strength to the desired level by partial annealing. For alloys that age-soften at room temperature, the H2 tempers have the same minimum tensile strengths as the corresponding H3 tempers. For other alloys, the H2 tempers have the same minimum tensile strength as the corresponding H1 tempers and slightly higher elongation. The number following this designation indicates the degree of strain-hardening remaining after the product has been partially annealed.

H3 **Strain-hardened and stabilized.** Applies to products which are strain-hardened and whose mechanical properties are stabilized by a low temperature thermal treatment which results in slightly lowered tensile strength and improved ductility. This designation is applicable only to those alloys which, unless stabilized, gradually age-soften at room temperature. The number following this designation indicates the degree of strain-hardening before the stabilization treatment.

H311 Applies to products which are strain-hardened less than the amount required for a controlled H31 temper.

H321 Applies to products which are strain-hardened less than the amount required for a controlled H32 temper.

H323
H343 Applies to products which are specially fabricated to have acceptable resistance to stress corrosion.

O **Annealed.** Applies to wrought products which are annealed to obtain the lowest strength temper, and to cast products which are annealed to improve ductility and dimensional stability. The O may be followed by a digit other than zero.

T **Thermally treated to produce stable tempers other than F, O, or H.** Applies to products which are thermally treated, with or without supplementary strain-hardening, to produce stable tempers. The T is always followed by one or more digits.

T1 **Cooled from an elevated temperature shaping process and naturally aged to a substantially stable condition.** Applies to products which are not cold worked after cooling from an elevated temperature shaping process, or in which the effect of cold work in flattening or straightening may not be recognized in mechanical property limits.

T2 **Cooled from an elevated temperature shaping process, cold worked, and naturally aged to a substantially stable condition.** Applies to products which are cold worked to improve strength after cooling from an elevated temperature shaping process, or in which the effect of cold work in flattening or straightening is recognized in mechanical property limits.

T3 **Solution heat-treated, cold worked, and naturally aged to a substantially stable condition.** Applies to products which are cold worked to improve strength after solution heat-treatment, or in which the effect of cold work in flattening or straightening is recognized in mechanical property limits.

T4 **Solution heat-treated and naturally aged to a substantially stable condition.** Applies to products which are not cold worked after solution heat-treatment, or in which the effect of cold work in flattening or straightening may not be recognized in mechanical property limits. (T42 indicates material is solution heat-treated from the O or F temper to demonstrate response to heat-treatment, and naturally aged to a substantially stable condition.)

T5 **Cooled from an elevated temperature shaping process and then artificially aged.** Applies to products which are not cold worked after cooling from an elevated temperature shaping process, or in which the effect of cold work in flattening or straightening may not be recognized in mechanical property limits.

T6 **Solution heat-treated and then artificially aged.** Applies to products which are not cold worked after the solution heat-treatment, or in which the effect of cold work in flattening or straightening may not be recognized in mechanical property limits. (T62 indicates material is solution heat-treated from the O or F temper to demonstrate response to heat-treatment, and artificially aged.)

T7 **Solution heat-treated and stabilized.** Applies to products which are stabilized after solution heat-treatment to carry them beyond the point of maximum strength to provide control of some special characteristic.

T8 **Solution heat-treated, cold worked, and artificially aged.** Applies to products which are cold worked to improve strength, or in which the effect of cold work in flattening or straightening is recognized in mechanical property limits.

T9 **Solution heat-treated, artificially aged, and cold worked.** Applies to products which are cold worked to improve strength.

T10 **Cooled from an elevated temperature shaping process, cold worked, and artificially aged.** Applies to products which are cold worked to improve strength, or in which the effect of cold work in flattening or straightening is recognized in mechanical property limits.

Additional Digits for T Tempers

T__51 **Stress-relieved by stretching.** Applies to the following products when stretched the indicated amounts after solution heat-treatment or cooled from an elevated temperature shaping process.

Plate .... 1½ to 3% permanent set
Rod, bar, shapes, extruded tube .... 1 to 3% permanent set
Drawn tube .... ½ to 3% permanent set

Applies directly to plate and rolled or cold finished rod and bar, which receive no further straightening after stretching.

Applies to extruded rod, bar, shapes, tubing, and to drawn tubing when designated as follows:

T__510 Products that receive no further straightening after stretching.

T__511 Products that may receive minor straightening after stretching to comply with standard tolerances.

T__52 **Stress-relieved by compressing.** Applies to products which are stress-relieved by compressing after solution heat-treatment, or cooled from an elevated temperature shaping process to produce a permanent set of 1 to 5%.

T__54 **Stress-relieved by combined stretching and compressing.** Applies to die forgings which are stress-relieved by restriking cold in the finish die.

W **Solution heat-treated.** An unstable temper applicable only to alloys which spontaneously age at room temperature after solution heat-treatment.

Source: Aluminum Assn. (Reviewed January 1981)

APPENDIX 8: Temper Designations for Coppers and Copper Alloys

By Richard Dale Smith, Kennecott Copper Corp.
Adapted from Metals Handbook 9th Ed., Vol 2, American Society for Metals, 1979

DESIGNATION of tempers for copper and its alloys must allow for highly diversified and complex variation in product forms and processing methods. Out of this diversity, one group of temper designations can be specified on the basis of cold reduction by rolling or drawing. Such a scheme is presented in Table 1, where the nominal temper designations are related directly to the amount of reduction stated in Browne & Sharpe (B&S) gage numbers. In the B&S gage system, each gage is related to both adjacent gages by a fixed ratio. As the gage numbers increase, the thickness (or diameter) decreases by 10.95% for each step. For example, a strip of flat rolled material 0.032 in. thick or drawn wire 0.032 in. in diameter is designated B&S gage 20; cold reduction one B&S number (to 21 gage) reduces the thickness or diameter 10.95% (to 0.0285 in.), and a further reduction to 22 gage reduces the diameter or thickness another 10.95% (to 0.0254 in.). Because the reductions are based on thickness for flat material and on diameter for wire, the reduction in cross-sectional area is less for strip than for wire for a given gage interval.

Temper designations for coppers and copper alloys require a broad-based classification scheme because (*a*) some alloys have much higher work-hardening rates than the coppers and brasses for which the scheme summarized in Table 1 was developed and (*b*) a temper-designation scheme based on cold reduction cannot be applied to product forms such as rod, tube, extrusions and castings or to heat treatable copper alloys. For example, some manufacturers of high-strength alloys with high work-hardening rates have established nominal temper designations (¼ hard, ½ hard, etc.) based on strength level rather than on dimensional reduction. In an extreme case for

Table 1. Scheme of Temper Designations for Wrought Copper and Brass Based on Cold Reduction

		Rolled sheet		Drawn wire		
Nominal temper designation	Increase in B&S gage numbers	Reduction in thickness and area, %	True strain (a)	Reduction in diameter, %	Reduction in area, %	True strain (a)
1/4 hard	1	10.9	0.116	10.9	20.7	0.232
1/2 hard	2	20.7	0.232	20.7	37.1	0.463
3/4 hard	3	29.4	0.347	29.4	50.1	0.694
Hard	4	37.1	0.463	37.1	60.5	0.926
Extra hard	6	50.1	0.696	50.1	75.1	1.39
Spring	8	60.5	0.928	60.5	84.4	1.86
Extra spring	10	68.6	1.16	68.6	90.2	2.32
Special spring	12	75.1	1.39	75.1	93.8	2.78
Super spring	14	80.3	1.62	80.3	96.1	3.25

(a) True strain equals ln A_0/A, where A_0 is initial cross-sectional area and A is final area.

rod, "hard" temper can mean anything from "annealed to a fine grain size" to a cold reduction of 10, 15, 20, 35 or even 40%. "Half hard" and "quarter hard" tempers are also inconsistent with the scheme of temper designations based on dimensional reduction.

Similar problems exist for annealed tempers. With simple single-phase alloys, annealed tempers are generally specified on the basis of grain size of recrystallized material. This system cannot be applied to grain-size-stabilized alloys because there may be considerable variation in properties for material annealed at various temperatures up to several hundred degrees Fahrenheit above the recrystallization temperature, even though the grain size remains quite stable. For these alloys, the terms "light annealed," "soft annealed" and "annealed to temper" are generally used instead of grain-size designations.

In order to clarify this complex situation, the American Society for Testing and Materials has published Standard Recommended Practice ASTM B601, "Temper Designation for Copper and Copper Alloys—Wrought and Cast," which is intended to be unambiguous and applicable to all product forms currently in widespread use. An alpha-numeric code has been assigned to each of the standard descriptive temper designations (see Table 2). Wherever possible in this Handbook, data for different tempers have been listed using these codes for identification of material condition.

Table 2. ASTM B601 Temper Designation Codes for Copper and Copper Alloys

Temper designation	Temper name or material condition
Cold Worked Tempers	
H00	1/8 hard
H01	1/4 hard
H02	1/2 hard
H03	3/4 hard
H04	Hard
H06	Extra hard
H08	Spring
H10	Extra spring
H12	Special spring
H13	Ultra spring
H14	Super spring
H50	Extruded and drawn
H52	Pierced and drawn
H55	Light drawn; light cold rolled
H58	Drawn general purpose
H60	Cold heading; forming
H63	Rivet
H64	Screw
H66	Bolt
H70	Bending
H80	Hard drawn
H85	Medium hard-drawn electrical wire
H86	Hard-drawn electrical wire
Cold Worked and Stress-relieved Tempers	
HR01	H01 and stress relieved
HR02	H02 and stress relieved
HR04	H04 and stress relieved

Table 2. ASTM B601 Temper Designation Codes for Copper and Copper Alloys (continued)

Temper designation	Temper name or material condition
HR06	H06 and stress relieved
HR08	H08 and stress relieved
HR10	H10 and stress relieved
HR50	Drawn and stress relieved
Cold Worked and Order-strengthened Tempers	
HT04	H04 and order heat treated
HT06	H06 and order heat treated
HT08	H08 and order heat treated
As- manufactured Tempers	
M01	As sand cast
M02	As centrifugal cast
M03	As plaster cast
M04	As pressure die cast
M05	As permanent mold cast
M06	As investment cast
M07	As continuous cast
M10	As hot forged and air cooled
M11	As hot forged and quenched
M20	As hot rolled
M30	As hot extruded
M40	As hot pierced
M45	As hot pierced and rerolled
Annealed Tempers(a)	
O10	Cast and annealed(b)
O20	Hot forged and annealed
O25	Hot rolled and annealed
O30	Hot extruded and annealed
O40	Hot pierced and annealed
O50	Light annealed
O60	Soft annealed
O61	Annealed
O65	Drawing annealed
O68	Deep drawing annealed
O70	Dead soft annealed
O80	Annealed to temper — 1/8 hard
O81	Annealed to temper — 1/4 hard
O82	Annealed to temper — 1/2 hard
Annealed Tempers(c)	
OS005	Average grain size 0.005 mm
OS010	Average grain size 0.010 mm
OS015	Average grain size 0.015 mm
OS025	Average grain size 0.025 mm
OS035	Average grain size 0.035 mm
OS050	Average grain size 0.050 mm
OS070	Average grain size 0.070 mm
OS100	Average grain size 0.100 mm
OS120	Average grain size 0.120 mm
OS150	Average grain size 0.150 mm
OS200	Average grain size 0.200 mm
Solution-treated Temper	
TB00	Solution heat treated
Solution-treated and Cold Worked Tempers	
TD00	TB00 cold worked to 1/8 hard
TD01	TB00 cold worked to 1/4 hard

Table 2. ASTM B601 Temper Designation Codes for Copper and Copper Alloys (continued)

Temper designation	Temper name or material condition
TD02	TB00 cold worked to 1/2 hard
TD03	TB00 cold worked to 3/4 hard
TD04	TB00 cold worked to full hard
Precipitation-hardened Temper	
TF00	TB00 and precipitation hardened
Cold Worked and Precipitation-hardened Tempers	
TH01	TD01 and precipitation hardened
TH02	TD02 and precipitation hardened
TH03	TD03 and precipitation hardened
TH04	TD04 and precipitation hardened
Precipitation-hardened and Cold Worked Tempers	
TL00	TF00 cold worked to 1/8 hard
TL01	TF00 cold worked to 1/4 hard
TL02	TF00 cold worked to 1/2 hard
TL04	TF00 cold worked to full hard
TL08	TF00 cold worked to spring
TL10	TF00 cold worked to extra spring
TR01	TL01 and stress relieved
TR02	TL02 and stress relieved
TR04	TL04 and stress relieved
Mill-hardened Tempers	
TM00	AM
TM01	1/4 HM
TM02	1/2 HM
TM04	HM
TM06	XHM
TM08	XHMS
Quench-hardened Tempers	
TQ00	Quench hardened
TQ50	Quench hardened and temper annealed
TQ75	Interrupted quench hardened
Tempers of Welded Tubing(d)	
WH00	Welded and drawn to 1/8 hard
WH01	Welded and drawn to 1/4 hard
WM00	As welded from H00 strip
WM01	As welded from H01 strip
WM02	As welded from H02 strip
WM03	As welded from H03 strip
WM04	As welded from H04 strip
WM06	As welded from H06 strip
WM08	As welded from H08 strip
WM10	As welded from H10 strip
WM15	WM50 and stress relieved
WM20	WM00 and stress relieved
WM21	WM01 and stress relieved
WM22	WM02 and stress relieved
WM50	As welded from O60 strip
WO50	Welded and light annealed
WR00	WM00, drawn and stress relieved
WR01	WM01; drawn and stress relieved

(a) To produce specified mechanical properties. (b) Homogenization anneal. (c) To produce prescribed average grain size. (d) Tempers of fully finished tubing that has been drawn or annealed to produce specified mechanical properties or that has been annealed to produce a prescribed average grain size are commonly identified by the appropriate H, O or OS temper designation.

Index

NOTE: The symbol (F) or (T) following a page number in this index indicates that information on the subject is presented in a Figure or Table.